American Institute of Chemical Engineers

10th Process Plant Safety Symposium

Topical Conference at the
2008 AIChE Spring National Meeting

April 6-10, 2008
New Orleans, Louisiana, USA

Printed from e-media with permission by:

Curran Associates, Inc.
57 Morehouse Lane
Red Hook, NY 12571
www.proceedings.com

ISBN: 978-1-60560-209-7

Some format issues inherent in the e-media version may also appear in this print version.

ISBN: 978-1-60560-209-7

American Institute of Chemical Engineers
Proceedings
Three Park Avenue
New York, NY 10016-5991
Phone: 212-591-8100

www.aiche.org

American Institute of Chemical Engineers

10th Process Plant Safety Symposium
2008

TABLE OF CONTENTS

PROCESS PLANT SAFETY SYMPOSIUM

Welcome to the 10th Process Plant Safety Symposium! This is the 16th year since the first Process Plant Safety Symposium (PPSS) was held in 1992. The PPSS was originally organized by the South Texas section of AIChE, and today continues to focus on process safety and risk issues and solutions for immediate application at process plants and throughout industry. PPSS is now an annual symposium and is part of the AIChE 4th Annual Global Congress on Process Safety. PPSS sessions and focus areas this year include:

- **Global Congress Keynote Session** (Joint Session with LPS and CCPS) - Mr. Edwin Foulke, Assistant Secretary of Labor, OSHA
- **Applications of Safety Culture** provides new insights into safety culture and operational discipline, and how they can affect process safety.
- **Plant Process Safety Management Systems** goes beyond compliance with OSHA 1910.119. Topics include addressing multiple management system requirements, employing Six Sigma methodology to PSM, improving PSM and using metrics, and implementing the DHS Chemical Facility Antiterrorism Standards.
- **Maintaining Instrument and Mechanical Integrity** is an always important topic. This year's session will focus on topics such as how to manage unforeseen delays in Mechanical Integrity inspection schedules, recent improvements in Risk-Based inspection planning technology, and an analysis of relief valve proof testing data.
- **Conduct of Operations for Process Safety** has not been an area of focus by OSHA in the past. An interesting range of topics in this year's session include improving process safety by addressing the human element, how to evaluate operating discipline in PSM audits, and how to manage PSM knowledge across sister plants
- **Hazard Identification and Risk Assessment and Tools** is a session where a variety of risk techniques are presented – from lessons learned in LOPA and evaluating protection layers that are not independent to integrating process infrastructure hazards analysis with facility siting and using process safety concepts to manage security concerns.
- **Case Histories and Lessons Learned** (Joint Session with LPS and CCPS) is always a favored session reviewing accidents that have occurred and the lessons that we can learn from them. The presentations will include dust explosions, a tank explosion and a runaway reaction with resulting vapor cloud explosion.
- **Poster Session on Risk and Process Design**. This session will be held in conjunction with the Global Congress on Process Safety reception. Nearly 20 different poster presentations on many diverse topics will be on display. Come enjoy the reception and the posters, and talk with the authors face-to-face.

There are many opportunities to extend your knowledge, with process safety and risk applications you can take back with you to the plant from over 50 technical papers presented. Copies of these papers are generally contained in this publication (and CD) for your reference.

We thank all who have contributed their time and effort to make this a successful symposium, including the session chairs and co-chairs, the organizing committee, and presenters and authors. We also thank you for joining us in New Orleans.

Enjoy the symposium and all it has to offer you with the latest innovations in process safety.

Jack Chosnek
PPSS Chair

John Champion
PPSS Co-Chair

10th Process Plant Safety Symposium

Part of the 4th Global Congress on Process Safety

AIChE 2008 Spring National Meeting – New Orleans, LA - April 6-10, 2008

Symposium Chair:
Jack Chosnek
Principal, Knowledge*One*
P.O. Box 580465
Houston, TX 77258
Ph: (281) 538-0220
Fx: (281) 538-0220
jc@knowledge1.net

Symposium Co-Chair:
John W. Champion
Senior Risk Analyst, Rohm and Haas
6519 State Hwy 225
Deer Park, TX 77536
Ph: (281) 228-8265
Fx: (281) 228-8675
jchampion@rohmhaas.com

Hazard Identification and Risk Assessment Tools

Chair: Philip M. Myers
President
Advantage Risk Solutions, Inc.
Ph: (740) 965-6304
Fx: (740) 965-9159
pmyers@arisksolution.com

Co-Chair: Sanjeev Saraf
Senior Engineer
Exponent, Failure Analysis Associates
Ph: (832) 325-5716
Fx: (832) 325-5700
saraf@exponent.com

Conduct of Operations for Process Safety

Chair: James R. Thompson
Senior Consultant
ABSG Consulting Inc.
Ph: (281) 673-2853
Fx: (281) 673-2789
jthompson@absconsulting.com

Co-Chair: Douglas J. Ferguson, P.E.
Consultant_PHA & Risk Management
DuPont Engineering
Ph: (281) 586-6818
Fx:
Douglas.J.Ferguson@usa.dupont.com

Plant Process Safety Management Systems

Chair: Bill Vogtmann
Senior Consultant
SIS-Tech Solutions, LP
Ph: (281) 922-8324 Ext 22
Fx: (281) 922-4326
bvogtmann@sis-tech.com

Co-Chair: Donald Abrahamson
Global Process Safety Manager
Celanese International Corp
Ph: (972) 443-4942
Fx:
donald.abrahamson@celanese.com

Applications of Safety Culture

Chair: Dr. Colin (Chip) S. Howat
Director, Kurata Thermodynamics Lab
University of Kansas
Ph: (785) 218-3718
Fx: (785) 864-4967
cshowat@ku.edu

Co-Chair: Dr. Sam Mannan
Director, Mary Kay O'Connor Process Safety Center
Texas A&M University
Ph: (979) 862-3985
Fx: (979) 845-6446
mannan@tamu.edu

Maintaining Instrument and Mechanical Integrity

Chair: Katherine E. Pearson
Senior Risk Analyst
Rohm and Haas Company
Ph: (281) 228-8236
Fx: (281) 228-8675
katherinepearson@rohmhaas.com

Co-Chair: Steven D. Emerson
President
Emerson Technical Analysis, LLC
Ph: (361) 880-8145
Fx:
www.EmersonAnalysis.com

Case Histories and Lessons Learned

Chair: John F. Murphy
Process Safety Services
2304 Kenya Lane
Punta Gorda, FL 33983

Co-Chair Lisa Long
OSHA, Office of General
Industry Enforcement, DOL
200 Constitution Ave., NW

Co-Chair Albert I. Ness
Rohm and Haas
3100 State Road
Croydon, PA 19021

Co-Chair Antoinette Wenzel
Sunoco
100 Green Street
Marcus Hook, PA 19061

PPSS Committee Members at Large:

Terri A. Bob HES Lead Chevron International E&P 1500 Louisiana, Rm #3086A Houston, TX 77002 Ph: (832) 854-3635 Fx: (832) 854-4179 tebo@chevron.com	Donald J. Connolley Global Process Safety Advisor – Refining Technology BP Products North America 2401 5th Ave. South, NOB-3D Texas City, TX 77590 Ph: (409) 949-3139 Fx: (409) 942-4091 don.connolley@bp.com
Kenneth H. Harrington Senior Process Safety Advisor – EHS 10001 Six Pines Dr., Suite 4025 The Woodlands, TX 77380 Ph: (832) 813-4968 Fx: (832) 813-4680 harrken@cpchem.com	Lisa Morrison Process Safety BP International Group Safety and Operations Auditor Ph: (724) 759-1107 Fx: (724) 625-8075 lisa.morrison@bp.com
Michael Livingston HSE Group Lead and Chief Engineer WS Atkins, Oil and Gas Division 12121 Wickchester Lane, Suite 550 Houston, Texas 77079-1230 Ph: (713)463-6180 Fx: (713)463-6557 Michael.Livingston@AtkinsAmericas.com	Dr. Stephanie Payne Associate Professor Department of Psychology Texas A&M University 283 Psychology Building College Station, TX 77843 Ph: (979) 845-2090 Fx: (979) 845-4727 scp@psyc.tamu.edu
Dr. William J. Rogers Laboratory Director/TEES Research Specialist Mary Kay O'Connor Process Safety Center Chemical Engineering Department Texas A&M University College Station, TX 77843 Ph: (979) 845-3330 Fx: (979) 846-6446 wjrogers@tamu.edu	Dr. Vic Edwards Process Safety Director Aker Kvaerner 3600 Briarpark Drive Houston, TX 77042 Ph: (713) 270-2817 Fx: (713) 270-3195 vic.edwards@akerkvaerner.com

RELEVANCE OF SAFETY, HEALTH AND ENVIRONMENTAL (SHE) FACTORS IN SITING OF CHEMICAL INDUSTRIES

Dr. R. K. Elangovan
E mail : elanrk@yahoo.co.in ; elanrk @ hotmail.com

ABSTRACT

The Siting of chemical industries takes into account of various factors like any other industry, ranging from availability of raw material, land and water to proximity to market and transportation. Even though Safety, Health and Environmental (SHE) factors are very much relevant to Siting of any industry, they attain more significance in siting of chemical industries due to their relevance in operation and survival of the industry.

The consideration of SHE factors in Siting of chemical industries ensures a distance between the factory site and built up area, thereby ensures separation distance between hazards and risks in them from the community apart from extending the benefit of mitigating the effects of any release. The releases from a chemical industry are generally due to the breach of safety, thereby resulting in impact on the health and environment. The releases may be intended like effluents and discharges and unintended like accidental releases, and they may be intermittent or continuous. Therefore, the releases having impact on health and environment are part of any chemical industry. The effects of any release decay rapidly with distance. The separation distance attained by the proper siting tends to create more opportunity for protection there by containing and controlling the effect of releases.

This paper discusses on the relevance and importance of SHE factors in Siting as well as describes a methodology to incorporate SHE factors in site selection of chemical industries.

KEY WORDS

Siting, Accident, Releases, Safety, Health and Environment.

INTRODUCTION

The Siting of chemical industries takes into account of various factors like any other industry, ranging from availability of raw material, land and water to proximity to market and transportation. Even though Safety, Health and Environmental (SHE) factors are very much relevant to Siting of any industry, they attain more significance in siting of chemical industries due to their relevance in operation and survival of the industry. The SHE factors are equally important like business factors in deciding the site selection of the chemical industries. This paper discusses on the relevance of SHE factors and their associated factors in deciding on site selection for chemical industries.

CONCEPTUAL FRAME WORK

RISK

Risk is the likelihood of a specified event occurring within a specified period or in specified circumstances. The different types of risks from chemical factories are as follows:

1) Individual Risk
2) Societal Risk
3) Property Risk

Out of the above, the societal risk attains high significance in plant Siting, since it is the relationship between frequency and the number of people suffering from a specific level of harm in a given population from the realization of specified hazards. The concept of societal risk reflects the likelihood of accidents involving multiple casualties and long- term detriment. Any risk within an area surrounding a dangerous process, as in chemical industries, depends upon its frequency and consequence not only to the individual but also to the number of individuals likely to be involved [6,8] .

COMMUNITY HEALTH

Community Health concerns preventing illness and managing the illness of the community and promoting health. Immunization campaigns, controlling infectious diseases, ensuring safe drinking water, food hygiene, public Hygiene, free from illness or diseases out of industrial activity, ensuring availability of adequate antidotes for the chemicals used in the industry, health centers, first - aid facilities, emergency services like ambulances etc., are some of the activities under community health.

Health Impact Assessment (HIA)

The Health Impact Assessment (HIA) is defined as a combination of procedures or methods by which a policy, program or project may be judged as to the effects it may have on the health of a population. The HIA seeks to predict the health impact of any industrial activity before implementation and it aims at facilitating the avoidance or reduction of negative effects on community health. The HIA should consider both positive and negative effects on human health of any industrial activity and the likely human health effects should be identified to decide on corrective actions and finally aid in the plant siting,

Environmental Impact Assessment (EIA)

The Environmental inventory is a complete description of the environmental as it exists in an area where a particular proposed action, like setting up a chemical factory, is being considered The Environmental Impact Assessment is an attempt to evaluate the consequences of a proposed action on each of the description in the environmental inventors.

The essential steps in an environmental impact assessment are:

- Prediction of the anticipated change in an environmental descriptor
- Determination of the magnitude or scale of the particular change.
- Application of an importance or significance factor to the change [3] .

LEGAL PROVISIONS

Safety

Various Indian Statutes on Safety are as follows:

- The Factories Act 1948
- The Chemical Accidents (Emergency Planning, Preparedness and Response) Rules, 1996
- Manufacture, Storage and Import of Hazardous Chemical Rules, 1989

Health

Various Indian Statutes on Health are as follows:

- The Factories Act 1948
- The Public Liability Insurance Act, 1991
- The Disaster Management Act, 2005

Environment

Various Indian Statutes on Environment are as follows [7]:

- The Environment (Protection) Act 1986
- The Environmental (Protection) Rules, 1986
- The Water (Prevention And Control of Pollution) Act, 1974
- The Air (Prevention & Control of Pollution) Act., 1981
- Environmental impact Assessment of Development Project, Notification Hazardous Wastes (Management and Handling) Rules, 1989
- Manufacture, Storage and Import of Hazardous Chemical Rules, 1989
- Manufacture, Use, Import, Export and Storage of Hazardous Micro-Organisms Scheme of Labelling of Environment Friendly Products (ECO-MARKS)
- Bio-Medical Waste (Management and Handling) Rules, 1998
- Re-Cycle Plastics Manufacture and Usage Rules, 1999
- Coastal Regulation Zone, Rules

- Environment (Siting for Industrial Projects) Rules, 1999
- Noise Pollution (Regulation and Control) Rules, 2000
- Municipal Solid Wastes (Management & Handling) Rules, 2000
- Ozone Depleting Substances (Regulation) Rules, 2000
- Manufacture, Storage and Import of Hazardous Chemical (Amendment) Rules 2000
- The Prevention and Control of Pollution (Uniform Consent Procedure) Rules, 1999.
- The Batteries (Management and Handling) Rules, 2001
- The Disaster Management Act 2005
- National Environment Policy 2006

INTERNATIONAL AGREEMENTS ON ENVIRONMENTAL ISSUES

India is signatory to a number of multilateral environment agreements (MEA) and conventions. An overview of some of the major MEAs and India's obligations under these are presented below:

a) Montreal Protocol on substances that deplete the Ozone Layer (to the Vienna Convention for the Protection of the Ozone Layer), 1987

The Montreal Protocol to the Vienna Convention on Substances that deplete the Ozone Layer came into force in 1989. The protocol set targets for reducing the consumption and production of a range of ozone depleting substances (ODS). India accepted to the Montreal Protocol along with its London Amendment in September 1992. The Ministry of Environment and Forests (MoEF), Government of India, has established an Ozone Cell and a steering committee on the Montreal Protocol to facilitate implementation of the India Country Program, for phasing out ODS production by 2010. The Gazette of India on 19 July 2000 notified rules for regulation of ODS phase- out called the Ozone Depleting Substances (Regulation and Control) Rules, 2000. They were notified under the Environment (Protection) Act, 1986.

b) Basel Convention on Transboundary Movement of Hazardous Wastes, 1989

Basel Convention, which entered into force in 1992, has three key objectives.

- To reduce transboundary movements of hazardous wastes.
- To minimize the creation of such wastes; and
- To prohibit their shipment to countries lacking the capacity to dispose hazardous wastes in an environmentally sound manner.

India ratified the Basel Convention in 1992, shortly after it came into force. The Indian Hazardous Wastes Management Rules Act 1989 encompasses some of the Basel provisions related to the notification of import and export of hazardous waste, illegal traffic, and liability.

c) UN Framework Convention on Climate Change (UNFCCC), 1992

The Primary goals of the UNFCCC were to stabilize greenhouse gas emissions at levels that would prevent dangerous anthropogenic interference with the global climate. The convention embraced the principle of common but differentiated responsibilities which has guided the adoption of a regulatory structure. India signed the agreement in June 1992, which was ratified in November 1993. India has initiated the preparation of its First National

Communication (base year 1994), the potential vulnerability to climate change, adaptation measures and other steps being taken in the country to address climate change.

d) UN Convention on Désertification, 1994

Delegates of the 1992 UN Conference on Environment and Development (UNCED) recommended establishment of an intergovernmental negotiating committee for the elaboration of an international convention to combat desertification in countries experiencing serious drought and / or desertification. The UN General Assembly established such a committee in 1992 that later helped formulation of Convention on Desertification in 1994.

NATURAL FACTORS IN PLANT SITING

The following natural factors need to be considered in plant siting not only because of the hazards associated with them but also their impact on SHE factors from the consequences of such hazards.

a) Earthquake

Earthquake can generate secondary effects such as fire, explosion and toxic release as a result of combined actions of the earthquake wave and environmental conditions. Additional safety factors for considering base shear force which tends to topple building or structures of the chemical plants should be taken into account at the design stage itself.

b) Hurricane / Tornadoes

Hurricane and Tornadoes can induce hazardous chemical releases in chemical industries by high winds.

c) Flooding

Flooding can destabilize the equipments, storage systems, piping, and facilities handling hazardous chemicals there by creating hazardous situations. The chemicals that are highly soluble in water can dissolve in the flood water creating health and environmental hazards.

d) Lightning

The released flammable and explosive chemicals in the gaseous state in the form of vapour cloud and liquid state in the form of spills can find ignition source due to lightning leading to a hazardous situation.

e) Tsunami

A tsunami (pronounced tsoo – nah – mee) is a wave train, or series of waves, generated in a body of water by an impulsive disturbance that vertically displaces the water column. Earthquakes, landslides, volcanic eruptions, explosions, and even the impact of cosmic bodies, such as meteorites, can generate tsunami. Tsunami can savagely attack coastlines, causing devastating property damage and loss of life.

METHOD

The step – by – step procedure in plant siting is given in Figure No. 1. The step- by- step procedure should include details on Safety, Health and Environment to make the site selection process realistic and meaningful [1]. The various general factors relevant to the siting of industries are shown in Table No. 1.

SAFETY FACTORS

The Safety Factors in chemical plant siting are provided in Figure No. 2.

Process Hazard Identification [5]

Various process hazards identification methods used in chemical industries to identify the potential hazards with nature of results are tabled in Table No.2.

Types of Chemical Plant Accidents

The Chemical plant accidents are broadly classified as fires, explosions and toxic releases. The different types of chemical accidents along with their probability and potential for economic loss are provided in Table No. 3. Among the accidents fire accidents is the most common in industries followed by explosion and toxic release. The most damaging type of explosion is an unconfined vapour cloud explosion when a large cloud of volatile and flammable vapour is released and dispersed followed by ignition and explosion. Toxic releases typically results in little damage to capital equipment but personal injures and clean up liabilities are significant[4,8]

The Accident Process

The Accidents in industries follow a three step process and they are:

- Initiation - The event that starts the accident.
- Propagation - The event or events that maintain or expand the accident.
- Termination -The event or events that stop the accident or diminish it in size.

Safety Engineering involves eliminating the initiating events and replacing the propagation steps by termination events[8]

Scenario Identification

The Scenario identification is resorted without regard to importance of initiating event. The unidentified scenario cannot be analyzed and hence it is very much important to identify all probable and worst case scenarios to ensure that the study is complete in all respects.

Accident Consequence

The primary components in consequence analysis are source and dispersion models, fire and explosion models, effect model and mitigating factors. Source and dispersion models provide quantitative information on source rates and dispersion of vapour clouds to some concentration levels. Fire and explosion models convert the information on the cloud for flammable releases into hazard potentials such as thermal radiation and explosion over pressures.

Effect models convert these incident specific results into effects on people and structures. Additional refinement is provided by mitigation factors such as sheltering or evacuation, which tend to reduce the magnitude of potential effects in real accident [10].

HEALTH FACTORS

The Chemical industries can create health hazards to the community by the accidents and chemical releases. The fire can create thermal radiation hazard resulting in `thermal burn injuries and an explosion can create over pressure there by creating hazards by overpressure resulting in injuries and even death. The major health hazard can be created by the exposure to the toxic chemicals utilized and manufactured in the factory. The methodology for incorporating health factors in plant siting is given in Figure No. 3. Health Impact Assessment incorporating Risk Assessment, Exposure Assessment and a quantitative and / or quantitative evaluation of the degree of intake of toxic chemical likely to occur should be carried out to decide on plant Siting. The availability of medical facilities, antidotes, emergency vehicles and ambulances are also some of the health factors relevant to Plant Siting [5].

ENVIRONMENTAL FACTORS

The Methodology in incorporating environmental factors in plant siting for chemical industries is provided in Figure No.4. The Environmental impacts of releases due to accidents as well as process emissions like flue gas, fumes and effluents should be assessed to take actions, which require systematic, reproducible and interdisciplinary approaches. The potential impacts on the physical, biological, cultural and socio-economic aspects of the environment should be thoroughly considered in the EIA study for effective decision making [2, 9]

FINDINGS

The findings of the study are:

- The Safety, Health and Environmental (SHE) Factors are very much relevant in deciding chemical plant siting and equally important like other factors.
- Even though the study considered SHE factors separately, these factors are dependent on each other. Any accident in a chemical industry would affect health and environmental factors and hence a holistic approach must be provided.
- The natural factors need not only affect safety in chemical industry alone. They have the potential to create accidents thus playing a role in complicating health and environmental factors.
- The SHE factors should be considered as a whole instead of considering them individually, for effective decision making on Plant Siting.
- The legal compliances, statutes, codes and standards on SHE factors relevant to Plant Siting should be considered in detail for effective decision making.

CONCLUSION

The Safety, Health and Environmental (SHE) factors are very important like business industries. The consideration of SHE factors in plant Siting ensures the smooth function of the chemical industry and helps in emergency preparedness and legal compliances. The incorporation of SHE factors at the initial stage of a siting study ensures least disturbance / or impact to the society, thereby enhancing effectiveness of all control systems in the chemical industries. The SHE factors need to be integrated with other factors for effective plant Siting.

REFERENCES

1 Apple M, *Plant Layout and Material Handling*, Third Edition, John Wiley & Sons, 1977

2 Bhatia S C , *Managing Industrial Pollution*, Macmillan India Ltd, New Delhi, India, 2003

3 Canter W, *Environmental Impact Assessment*, McGraw – Hill Book Company, New York, 1977

4 Goetsch L, *Occupational Safety and Health*, Prentice – Hall, Englewood Cliffs, New Jersey, Second Edition, 1996

5 Gould J, Glossop M and Loannides A, Review *of Hazard Identification Techniques,* Sheffield: Health and Safety Laboratory, Health and Safety Executive, 2000

6 Illuing .P, *Toxicity and Risk, Context Principles and Practice*, London, Taylor & Francis, 2001

7 Mohanty S K, *Environmental & Pollution Law Manual*, Universal Law Publishing Co. Pvt. Ltd, Third Edition, New Delhi, 2002.

8 Skelton B, *Process safety Analysis An Introduction,* Warwickshire: Institution of Chemical Engineers, 1997

9 Sincero P and Sincero A, *Environmental Engineering; A Design Approach*, Prentice – Hall of India Limited, New Delhi, India, 1999

10 *Guidelines for Chemical Process Quantitative Risk analysis*, Center for Chemical Process Safety, American Institute of Chemical Engineers, Newyork 1989

A. Market

1. Location
2. Population Trends
3. Income trends
4. Consumer characteristics
5. Retail sales trends
6. Industrial Markets
7. Cometition

B. Raw Material

1. Type
2. Location of sources
3. Characteristics
4. Price
5. Terms of sale
6. Availability
7. Storage facilities

C. Labour

1. Cost
2. Attitudes
3. Union situation
4. Supply
5. Location
6. Personnel policies in area
7. Legislation
8. Recruiting
9. Community pattern
10. Relocation

D. Transportation

1. Modes of transportation
2. Methods of operation
3. Handling methods
4. Facilities
5. Costs
6. Freight rates

E. Utilities

1. Power
2. Fuel
3. Water
4. Waste
5. Communication
6. Others

F. Climate

1. Maximum and minimum temperatures
2. Direction of winds
3. Weather changes
4. Humidity
5. Evaluation
6. Climate effects
7. Special weather hazards

G. Federal activities in area

H. Representation in congress

I. Financing

1. Financial aid available
2. Requirements
3. Credit factors
4. Factors effecting loan terms
5. Special inducements – taxes, loans, building

J. Stage government and taxes

1. Structure
2. Financial assistance
3. Government attitudes
4. State regulation and legislation
5. State taxes
6. Future tax Prospects

K. Local government and taxes

1. Structure of government
2. Financial conditions
3. Civil attitudes
4. Local taxes

L. Community Facilities

M. Community appearance

N. The size of community

O. Population trend

P. Community planning and Zoning

1. Planning commission
2. Industrial zoning
3. Building codes
4. Traffic and parking
5. Streets

Q. Others

1. Cost of living
2. Insurance
3. Construction costs
4. Location relative to competition
5. Size and character of other local industries
6. Buying habits

R. Individuals sites

1. Requirements
2. Size and shape of site
3. Geologic considerations
4. Cost of development
5. Location within the community
6. Availability to transportation
7. Topography
8. Price of land
9. Availability of building
10. Personal factors
11. Intangible considerations

Table No: 1.Selected Factors for Consideration in Plant Siting [1]

Ref	Hazard Identification Techniques	Nature of Results
1.	Concept Hazard Analysis	Qualitative
2.	Concept Safety Review	Qualitative
3.	Preliminary Hazards Analysis	Qualitative
4.	Pre-HAZOP (Hazard And Operability Study)	Qualitative
5.	HAZOP	Qualitative
6.	"What if?" Analysis	Qualitative
7.	Check List	Qualitative
8.	Standards / Codes of practice	Qualitative
9.	FMEA (Failure Mode and Effect Analysis)	Qualitative
10.	FMECA (Failure Modes, Effects, and Criticality Analysis)	Qualitative
11.	Critical Examination of System Safety	Qualitative
12.	Cause-Consequences Analysis	Qualitative and Quantitative
13.	Fault Tree Analysis	Qualitative and Quantitative
14.	Structural Reliability Analysis	Qualitative
15.	Reliability Block diagram	Qualitative
16.	Human reliability Analysis	Qualitative
17.	Predictive Human Error Analysis	Semi-qualitative
18.	Vulnerability Assessment	Qualitative
19.	Matrices	Qualitative
20.	GOFA (Goal Oriented Failure Analysis)	Semi-Qualitative
22.	MOSAR (Method Organized Systematic Analysis of Risk)	Semi-Qualitative
22.	Inherent Hazard Analysis	Qualitative

Table No.2 Hazard Identification Methods Applicable to Chemical Industries

Type of Accident	Probability of Occurrence	Potential for fatalities	Potential for Economic Loss
Fire	High	Low	Intermediate
Explosion	Intermediate	Intermediate	High
Toxic Release	Low	High	Low

Table No. 3 Types of Chemical Accidents

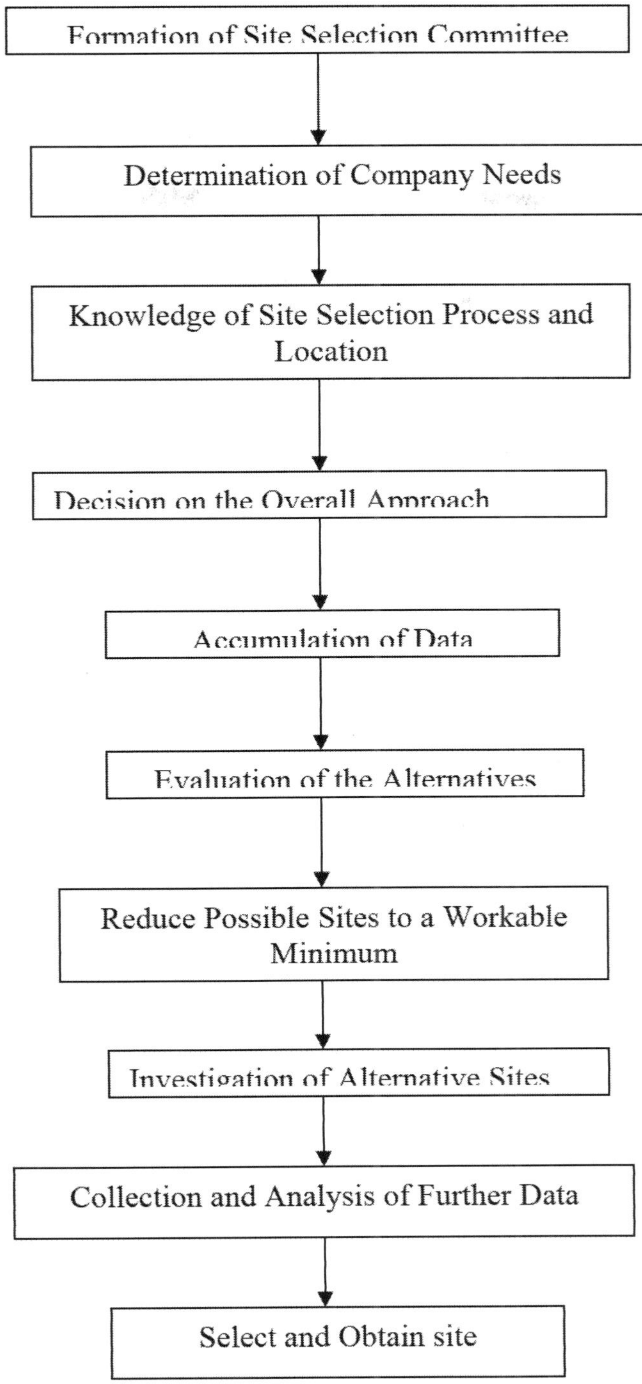

Figure No .1 Step by Step Procedures in Plant Siting.

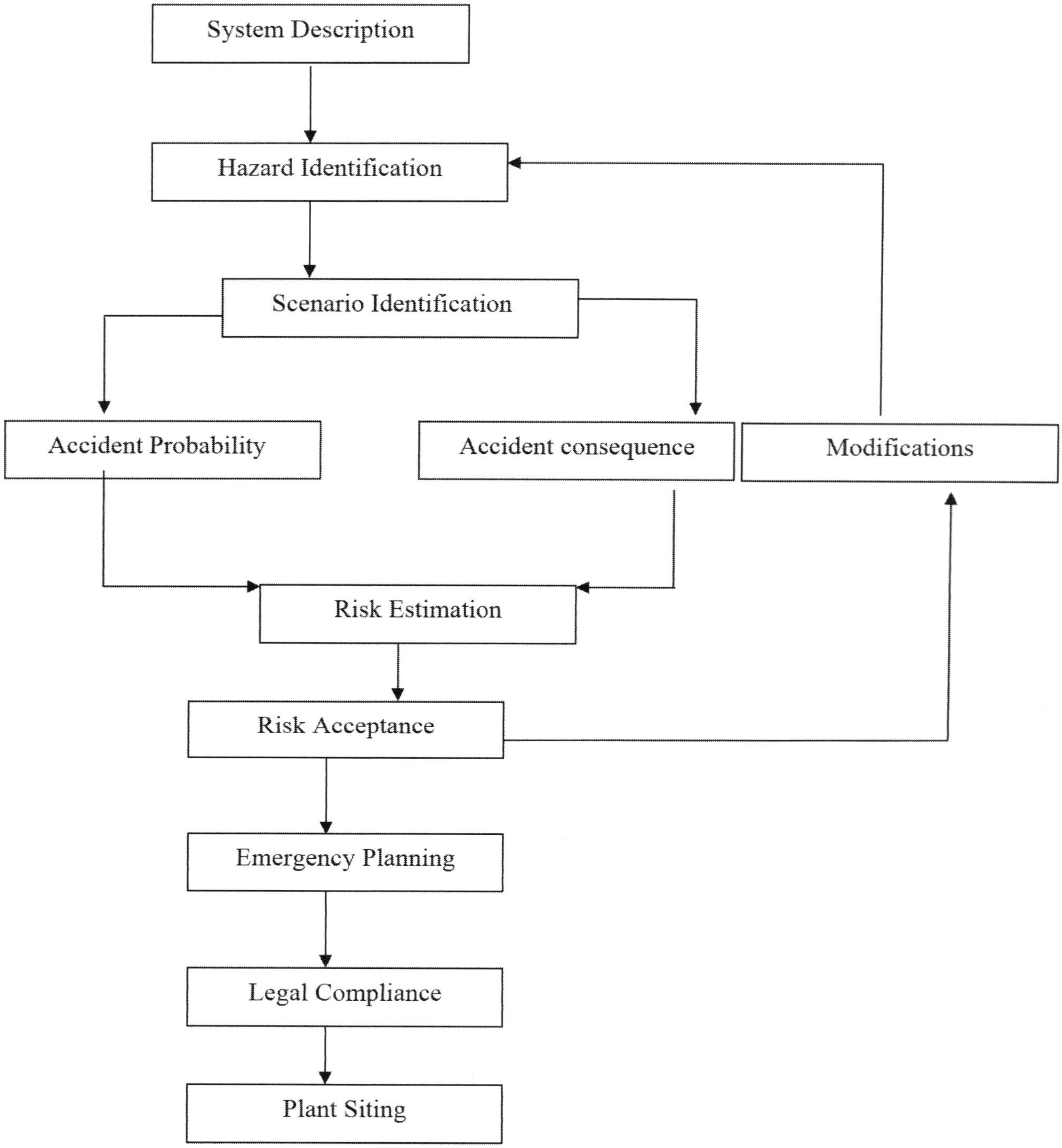

Figure No. 2 Logic of safety Factors in Plant Siting

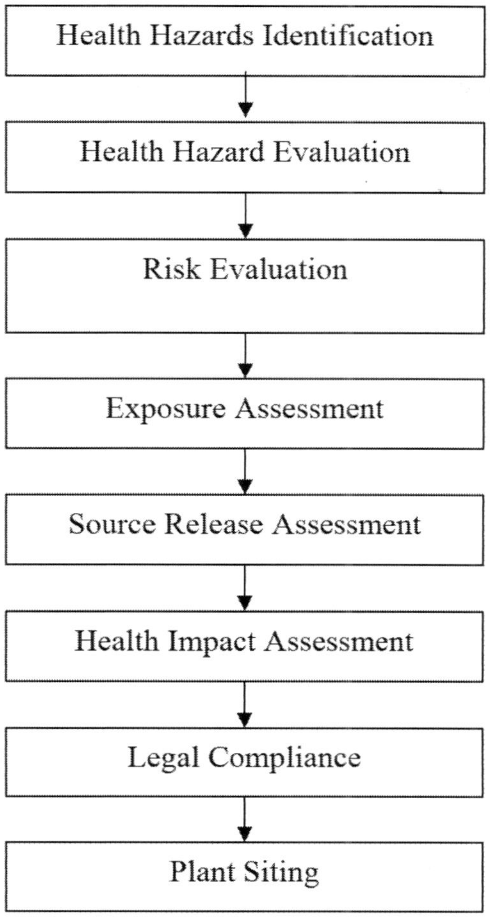

Figure No 3 Health Factors in Plant Siting

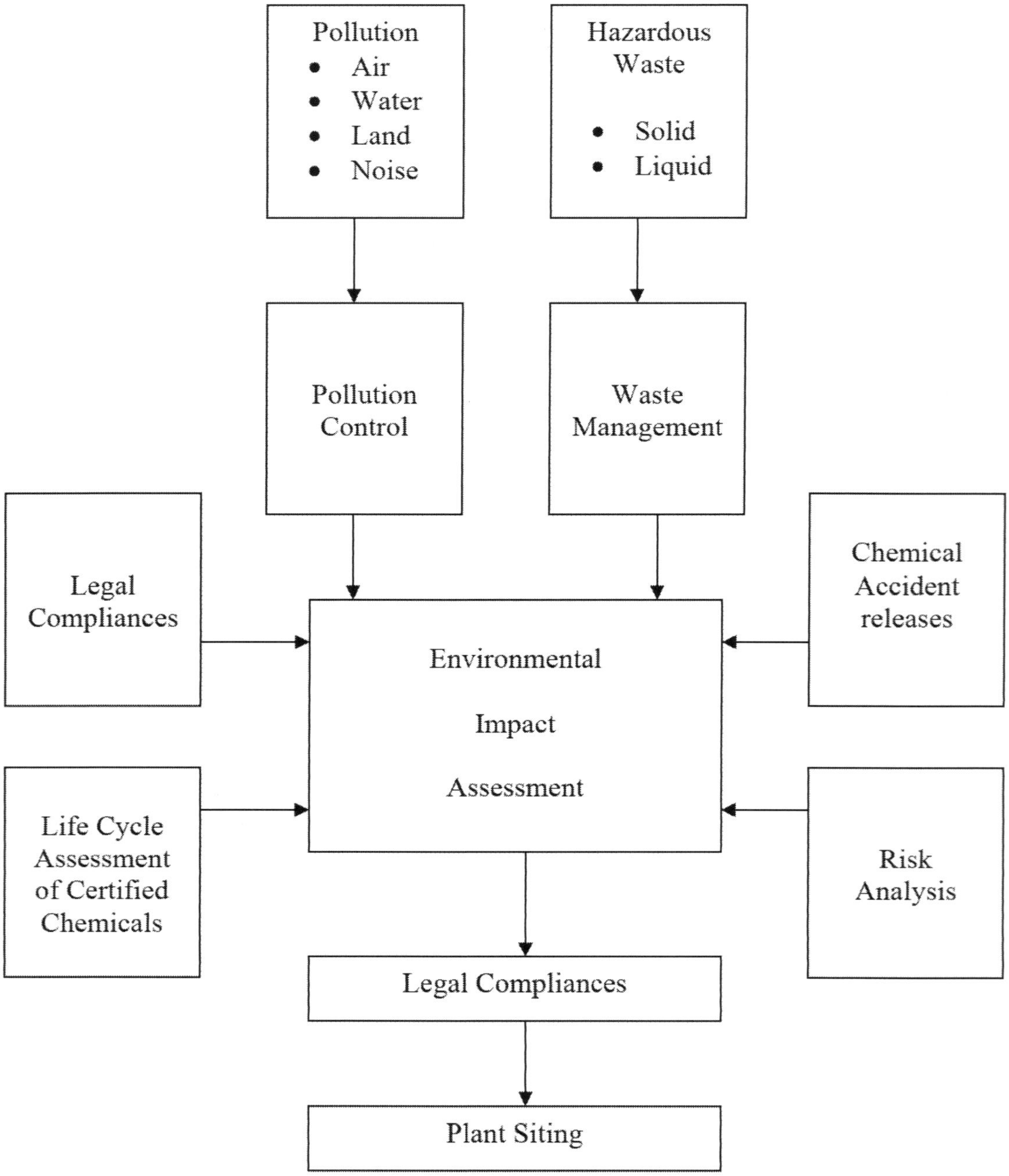

Figure No. 4. Environmental Factors in Plant Siting.

A modeling approach to assess the effectiveness of BLEVE prevention measures

Gabriele LANDUCCI[1], Menso MOLAG[2], Johan REINDERS[2], Alessandro TUGNOLI[3], Valerio COZZANI[3]*

(1) CONPRICI - Dipartimento di Ingegneria Chimica,
Chimica Industriale e Scienza dei Materiali
Università degli Studi di Pisa
Via Diotisalvi n.2, 56126 Pisa (Italy)

(2) Nederlandse Organisatie voor toegepast-natuurwetenschappelijk onderzoek TNO
Laan van Westenenk 501 P.O. Box 342 7300 AH Apeldoorn (The Netherlands)
Princetonlaan 6 NL-3584 P.O. Box 80015 NL-3508 TA Utrecht

(3) CONPRICI - Dipartimento di Ingegneria Chimica,
Mineraria e delle Tecnologie Ambientali
Alma Mater Studiorum - Università di Bologna
Viale Risorgimento n.2, 40136 Bologna (Italy)

(*) Author to whom correspondence should be addressed.
tel. (+39)-050-511215; fax (+39)-050-511266
e-mail: gabriele.landucci@ing.unipi.it

Submitted for
LOSS PREVENTION SYMPOSIUM

ABSTRACT

Distribution and storage of liquefied pressurised gases is a critical safety issue, often resulting in a very high individual and societal risk, at least in densely populated zones. Several risk assessments pointed out that this is mostly due to the possible occurrence of a Boiling Liquid Expanding Vapour Explosions (BLEVE) of the tank that may be exposed to an intense fire as a consequence of release scenarios. In the present work, a modelling approach is presented for the calculation of fired BLEVE probability. Simplified models were obtained for the estimation of the vessel time to failure with respect to the radiation intensity on the vessel shell. The time to failure was compared to a reference time required for effective mitigation actions and thus related to the escalation probability. The analysis of the effect of protective measures, such thermal insulating coating and pressure relief devices, on the time to failure allowed the identification of the necessary requirements for effective BLEVE prevention.

Keywords: domino effect, escalation probability, vessel time to failure, passive fire protection.

1 Introduction

Domino effect is responsible of severe accidents that took place in the chemical and process industry [1-3]. Several studies pointed out that the more critical step in the quantitative assessment of domino hazards is the availability of reliable models to estimate the possibility of escalation due to the effects of the primary accident [4-7]. In the case of fires caused by the accidental releases of flammable substances, it is well known that secondary events with catastrophic consequences may result as a consequence of flame impingement on equipment and pipes. Moreover, in the case of storages where pressurised liquefied gases (such LPG, chlorine, ammonia, etc.) are present, the escalation may result in a BLEVE, that may have severe consequences due to the extremely high damage distances.

An important benefit to the safety management of possible escalation hazards would come form the availability of an approach for the calculation of the possibility and probability of vessel rupture following external fires, based on simplified correlations able to take into account specific protection factors.

The present study was focused on the development of a simplified methodology for the calculation of the damage probability of process vessels, aimed to the quantitative assessment of domino effect triggered by fire scenarios. The methodology is based on simple analytical functions relating the time to failure of vessels to the radiation intensity. These were validated by an integrated approach, based on:

- available experimental data;
- a commercial finite elements code for complete thermal and mechanical simulations of the behaviour of pressurised vessels exposed to fires;
- a simplified model for vessel failure based on thermal nodes.

Specific correction factors were introduced in order to take into account the effect of protection materials. Damage probability was estimated by a probabilistic function. An approach derived from the Layer of Protection Analysis (LOPA) was used to estimate the probability of effective mitigation on the basis of the calculated time to failure and of site-specific factors. The presence and the delay time for the activation of protection systems was also considered. A fundamental issue in the development of the correlations resulted the presence of thermal protection layers on storage or process vessels. In particular, the possible use of innovative materials for passive protection systems resulted in a high impact on the time to failure. The effect of the protection systems on the time/temperature profile of vessel wall was thus implemented in the simplified threshold correlations for the assessment of time to failure.

2 The approach to the estimation of escalation probability

The use of simplified models, based on a limited number of parameters, is required to limit the computational time required for QRA procedures used in escalation assessment, as they involve the analysis of a high number of accidental scenarios [6,8]. Thus, equipment vulnerability models are usually based on probit functions that relate a "dose" of physical effects to the escalation or damage probability of an equipment item. In the case of radiation as an escalation vector, vessel failure is caused by wall temperature raise and/or by internal pressure increase due to vessel heat-up. This is a rather slow process (time to failure of the vessel is usually of the order of minutes or higher), thus a time lapse exists between the primary event and the secondary events caused by escalation.

Quite clearly, the time to failure (ttf) is a fundamental parameter in the analysis of the domino accidents of equipment exposed to fire: a higher time to failure results in a lower credibility of escalation, since a higher time lapse is available for effective mitigation actions by safety systems or emergency teams (e.g. wall cooling by water deluges). The ttf may thus be compared to a characteristic time for an effective mitigation (tte). The analysis of the effects of protection systems (dumping aimed to depressurization, water curtains by automatic systems, and additional water protection by emergency teams) may be performed by LOPA techniques, and specific damage probability functions may be obtained for the plant under examination. A site-specific or a general evaluation of the tte and its comparison to the ttf may be easily turned out in a probabilistic assessment of the success of mitigation actions. This approach was used to develop specific models relating the radiation intensity on a target vessel to the escalation probability.

3 Estimation of vessel time to failure

Fire may affect a process or storage vessel by one or more of the following modes: i) distant source radiation; ii) full or partial fire engulfment; and iii) jet fire flame impingement. Modelling vessel time to failure in these three situations is extremely difficult given the high complexity of the flame geometries. The wall temperature behaviour in a vessel exposed to an external fire would require a detailed three-dimensional analysis of the thermal flow over the vessels shell, of the thermal gradients in the fluid contained in the vessel and of the effects due to the mixing of the content by natural convection. However, a model based on this approach would require a huge time to run, not justified in a QRA framework due to the uncertainties that usually affect the characterization of the fire scenario. In this specific context, a simplified model able to yield a conservative estimation of the time to failure by a straightforward approach would be more useful.

In the present study, a lumped model was developed for the assessment of vessel time to failure. The model was based on a lumped approach for the calculation of the time and temperature profile, but was improved by an extended validation work, based on an experimental data set for the vessel time to failure, expanded by the use of results obtained from a specific finite elements model implemented for this purpose.

The development of the lumped model was based on thermal nodes modelling. Depending on the fire scenario, this approach attempts to divide the equipment in different zones (or nodes), each of which can be described by a simple set of parameters. The parameters represent physical quantities (e.g. temperature, pressure, thermal conductivity, etc.) averaged over each node. Boundary conditions together with global conservation laws lead to a system of equations which determines the parameters of interest and, in particular, the temperature at each node. This allows the calculation of temperature-time profiles as a function of the radiation mode and intensity on the vessel.

The estimation of these parameters allows the evaluation of the mechanical stress generated in each zone of the vessels and to compare it with the admissible tensile strength of the vessels material (that is strictly dependent on the material temperature). The failure conditions are strictly dependent on the structural design: geometry, material, boundary condition. For the equipment of interest (horizontal cylindrical vessels, vertical cylindrical vessels, etc.) the failure due to an external heat flow is mostly caused by the wall-thinning due to hoop stress and high temperature material degradation. A specific failure criterion was thus established to assess the time to failure of the vessel. This was derived by a direct comparison between parameters representative of the stress field over the vessels shell and of parameters representative of the tensile strength of the material. Thus, the failure criterion requires the detailed analysis of the stress field over the vessel shell and on the boundary structures that may be calculated only by more detailed modeling approaches. This means that with a lumped model only simplified failure criteria may be adopted, based on data fitting of available times to failure for similar vessels and for similar fire scenarios. Further details on the model developed are reported elsewhere [9].

An extended validation of the lumped model was thus necessary to ensure the reliability of the failure criteria implemented. A first set of validation runs was carried out using experimental data available in the literature. Although a significant number of case-studies resulted available, the number of available experiments was not sufficient to carry out an extended validation over the entire field of vessel geometries and of radiation modes and radiation intensities of interest. Thus, a finite elements model (FEM) was developed and validated on the basis of the experimental data. The FEM was used to generate a second data set used to extend the validation of the simplified lumped model.

The FEM allowed the detailed simulation of the thermal and mechanical conditions on vessel shells under fire radiation. Adopting proper failure criteria it was possible to use the knowledge of the temperature and of the stress conditions in each point of the structure to estimate more precisely the ttf. The first step in the simulations was the detailed calculation of the temperatures on the vessel shell as a function of time and radiation mode. A detailed definition of the radiation conditions on vessel shell was possible: Figure 1-(a) shows an example of detailed temperature simulations for a horizontal cylindrical LPG tank engulfed in flames.

The second step of the modeling was the calculation of the stress field as a function of the local temperatures and of the other loads present on the equipment shell (Figure 1-(b)). The calculation of the detailed temperature and stress maps allowed the application of the correct failure conditions and thus a quite accurate calculation of the equipment time to failure, also accounting for the decrease in the allowable stress due to the high temperatures of the vessel walls. The comparison of the results of the simulations carried out with the first set of experimental validation data showed that the predicted ttfs were always conservative and showed a relative error lower than 10%. Thus, the results suggested that the finite elements code could be reasonably used to extend the validation data set available.

The validation of the lumped model carried out using both the experimental and finite elements data for the vessel time to failure pointed out that the lumped model always yields credible and conservative values for the ttf of the vessels. An average relative error lower than 15% is present between the ttf calculated by the lumped model and those obtained by the FEM simulations (Figure 2).

Therefore, the lumped model developed in the present approach may be used with sufficient confidence in the assessment of the ttf of the equipment exposed to fire in the framework of domino quantitative risk assessment, also considering the uncertainties in the characterization of the fire scenario. The ttfs calculated by this simplified approach have a sufficient precision to be compared with the times required for an effective mitigation. The lumped model developed was thus the starting point for the calculation of the escalation probability due to fire and of the threshold values for escalation. As a matter of fact, even if the RADMOD model is characterised by relatively low computational time, its use may require a relevant effort in the analysis of extended areas, also considering that the model, although simplified, requires defining and inputting several parameters of each vessel and radiation mode considered.

The availability of further simplified tools to carry out at least a preliminary assessment of the time to failure is thus useful, in particular to identify the credible domino targets in a complex lay-out. To face this problem, a specific approach was used to define simple analytical functions for the evaluation of the time to failure of the equipment exposed to fire. The more important categories of pressurised equipment involved in domino accidents were identified, and their geometrical characteristics were defined (shape, range of sizes, etc.). Vessel sizes were obtained from typical design data used by engineering companies in the oil and gas sector. The volumes and diameters were derived from vessels typically used for LPG, chlorine, and ammonia pressurized storages, road tankers and tank wagons (Table 1). Cylindrical vessels with horizontal axis and design pressures of 1.5, 2.0 and 2.5 MPa were considered. The design data were verified with respect to section VIII of the ASME codes, and the relief valves were considered to provide the vent area required by API RP 520 standards.

An extended matrix of case studies was thus defined, considering the more credible primary events, the radiation modes and a credible range of radiation values. The lumped model was used to analyze each case study, estimating the time to failure of the selected equipment exposed to a specific fire scenario. A fitting procedure was implemented to obtain analytical functions that directly relate the time to failure of the equipment (ttf) to the value of the radiating heat flow (I):

$$\log_{10}(ttf) = a \log_{10}(I) + b \qquad (1)$$

where the a and b coefficients depend on the vessel type and geometry and on the radiation mode. The a and b coefficients may thus be calculated from the fitting of the results of numerical simulations to eq.(1).

First results obtained from the approach are envelope correlations obtained for different vessel categories that evidence the minimum credible value of the ttf as a function of radiation intensity. Figure 3 shows the envelope correlations obtained for vessels fully engulfed in fire.

4 Escalation thresholds and probability

The modeling approach carried out in the present study allowed an extended revision of the escalation thresholds for fire scenarios. The envelop correlations shown in Figure 3 define the minimum credible time to failure for vessels exposed to fire. Thus, if the ttf is compared to a maximum time required for effective mitigation (tte), the threshold values of stationary radiation required for credible escalation may be identified. Table 2 reports the threshold values estimated by this approach.

The Table evidences that the conditions leading to escalation due to fires are highly dependent on the target vessel and on the primary fire scenario. The more the equipment is resistant to the fire, the less credible is the escalation, since more time is available for effective mitigation actions by active safety systems or by the operation of emergency teams. Thus, the probability of damage and escalation may be related to the time to failure of the equipment, comparing the ttf to the tte. These could be estimated implementing a specific analysis of the internal or the external emergency plan of the site. In a simplified approach, a specific probit function may be built to relate the time to failure to the probability of escalation, taking into account the tte:

$$Pr = c - d \ln(ttf) \tag{2}$$

In the absence of site specific data, tte_1 was assumed of 5 min and tte_2 was assumed of 20 min. The following probit function was thus used in the present study to estimate escalation probability:

$$Pr = 9.25 - 1.857 \ln(ttf) \tag{3}$$

where ttf is expressed in minutes.

5 Mitigation by passive protection systems

The correlations and the data reported above are conservative and do not take into account the protection systems of the storage vessels. Even if active systems have an important role in emergency management, active protections, although compulsory, have a limited reliability due to delayed activation, possible damage in the primary event, possible failures of critical components. Therefore, passive protection of pressurized storage tanks is normally adopted to mitigate the effects of external fires. It is thus important to take into account the role of passive protections in the determination of vessel time to failure.

A preliminary analysis of the influence of the thermal insulation layer, performed assuming full engulfment conditions of pressurised vessels, evidenced that the time to failure of insulated vessels (ttf_{pv}) could be obtained simply adding the time to failure of the insulation layer (ttf_p) and the time to failure of the uninsulated vessel (ttf_v) without introducing a relevant error:

$$ttf_{pv} = ttf_p + ttf_v \tag{4}$$

Thus, the effect of a protection layer on the vessel may be accounted with sufficient precision simply adding a further term to the ttf estimated for the unprotected vessel. In Figure 4 an example of simulation is reported for LPG tanks of different volumes (5, 20, 50 and 100 m^3). The ttfs without the coating (Figure 4-(a)) result much lower in the case of the implementation of a protective layer. In particular, in Figure 4-(b), a 50 mm thick glass wool coating was supposed to be installed on the vessel. Hence, the increment in ttf might allow the possible effective intervention of the emergency measures, to mitigate the fire scenario, avoiding the BLEVE.

6 Conclusions

A simplified approach was developed for the management of the hazard due to external fires involving pressurized storage vessels and the possibility of escalation to BLEVE events. The approach was based on the development of tools for the straightforward assessment of vessel time to failure due to fire radiation in different radiation modes. The estimation of the vessel time to failure allowed the evaluation of the time available for effective mitigation actions and the timing of emergency response in order to prevent the BLEVE. The approach was extended to include the effect of passive protection. The results evidenced that specific and reliable values are needed to characterize the performance of the passive protection systems, since the effective prevention of fired BLEVEs critically depends on the available time to failure.

References

[1] F.P. Lees, *Loss Prevention in the Process Industries*, II Ed., Butterworth-Heinemann, Oxford, 1996.

[2] CCPS, *Guidelines for chemical process quantitative risk analysis*, II Ed., AIChE, New York, 2000.

[3] F.I. Khan and S.A. Abbasi, The world's worst industrial accident of the 1990s what happened and what might have been: A quantitative study, *Proc Safety Prog* 18:3 (2004) 135 - 145.

[4] C. Delvosalle, *A methodology for the identification and evaluation of domino effects*, Rep. CRC/MT/003, Belgian Ministry of Employment and Labour, Bruxelles, 1998.

[5] J. Gledhill and I. Lines, *Development of methods to assess the significance of domino effects from major hazard sites*, CR Report 183, Health and Safety Executive, London, 1998.

[6] F.I. Khan and S.A. Abbasi, Models for domino effect analysis in chemical process industries, *Proc Safety Prog* 17:2 (2004), 107 – 123.

[7] V. Cozzani and S. Zanelli, "An Approach to the Assessment of Domino Accidents Hazard in Quantitative Area Risk Analysis", Proc. *10th Int. Symp. on Loss Prevention and Safety Promotion in the Process Industries*, Elsevier, Amsterdam (2001), pp.1263.

[8] V. Cozzani, G. Gubinelli and E. Salzano, Escalation Thresholds in the Assessment of Domino Accidental Events, *J of Hazardous Materials* 129:1 (2006), 1-21.

[9] G. Gubinelli, *Domino effect in the process industries: quantitative methodologies for the evaluation of consequences*, Ph.D. thesis in Chemical Engineering, University of Pisa, Pisa, 2005.

[10] M.A. Persaud, C.J.Butler, T.A. Roberts, L.C. Shirvill and S. Wright, "Heat-up and failure of Liquefied Petroleum Gas storage vessel exposed to a jet-fire", Proc. *10th Int. Symp. on Loss Prevention in the Process Industries*, Stockholm (2001), pp.1069-1106.

[11] A. M. Birk, D. Poirier and C. Davison, On the response of 500 gal propane tanks to a 25% engulfing fire, *J. Hazardous Materials* 19 (2006), 527-541.

[12] B. Droste and W. Schoen, Full scale fire tests with unprotected and thermal insulated LPG storage tanks, *J of Hazardous Materials* 20 (1988), 41-53.

Nomenclature

a $[s \cdot m^{2} \cdot W^{-1}]$	first fitting coefficient for the radiation/time to failure correlation
b [s]	second fitting coefficient for the radiation/time to failure correlation
c [-]	first probit coefficient
d [-]	second probit coefficient

FEM [-]	finite element model
I [$W \cdot m^{-2}$]	radiation intensity
LOPA [-]	layer of protection analysis
Pr [-]	value of the probit function
ttf [s]	time to failure
ttf_p [s]	time to failure of the insulating coating layer
ttf_v [s]	time to failure of the uninsulated vessel
ttf_{pv} [s]	time to failure of the insulated vessel

List of figure captions

List of table captions

FIGURES

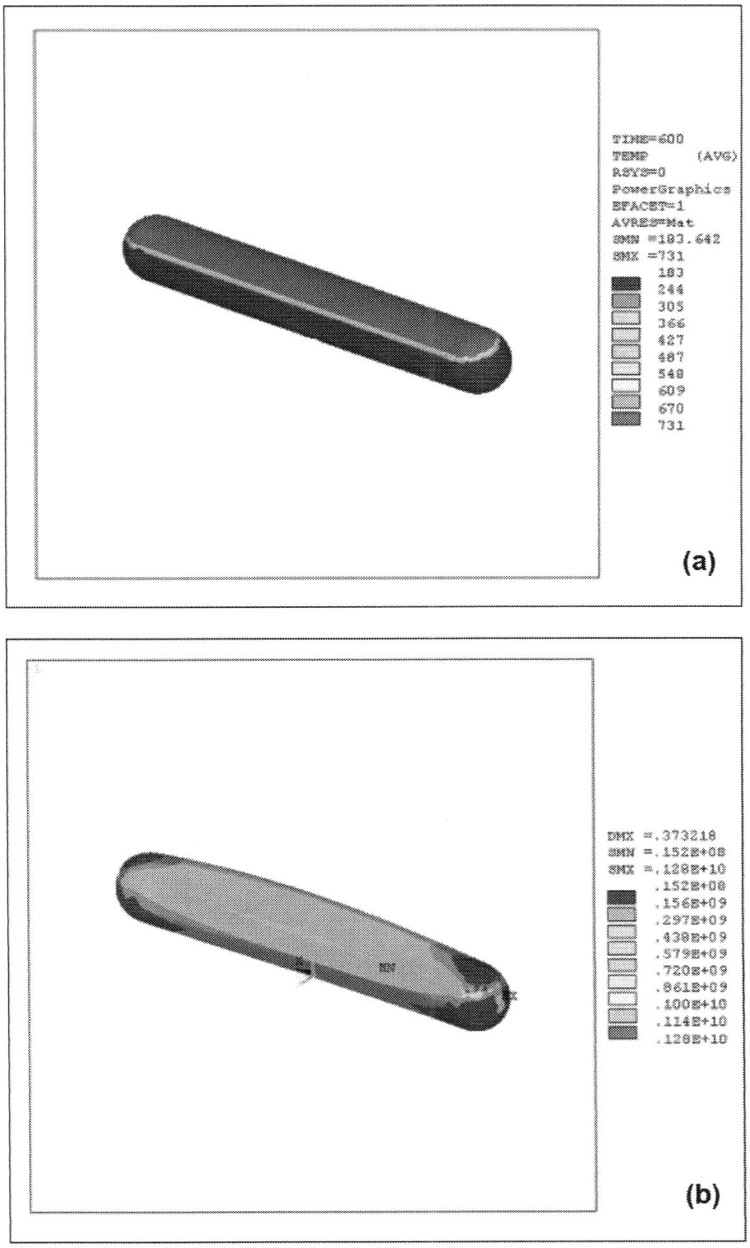

Figure 1: (a) Detailed simulation of the shell temperature (°C) calculated for a 100 m^3 LPG tank under radiation conditions due to flame engulfment (110 kW/m^2, 600 s); (b) detailed simulation of the correspondent shell stresses (Pa).

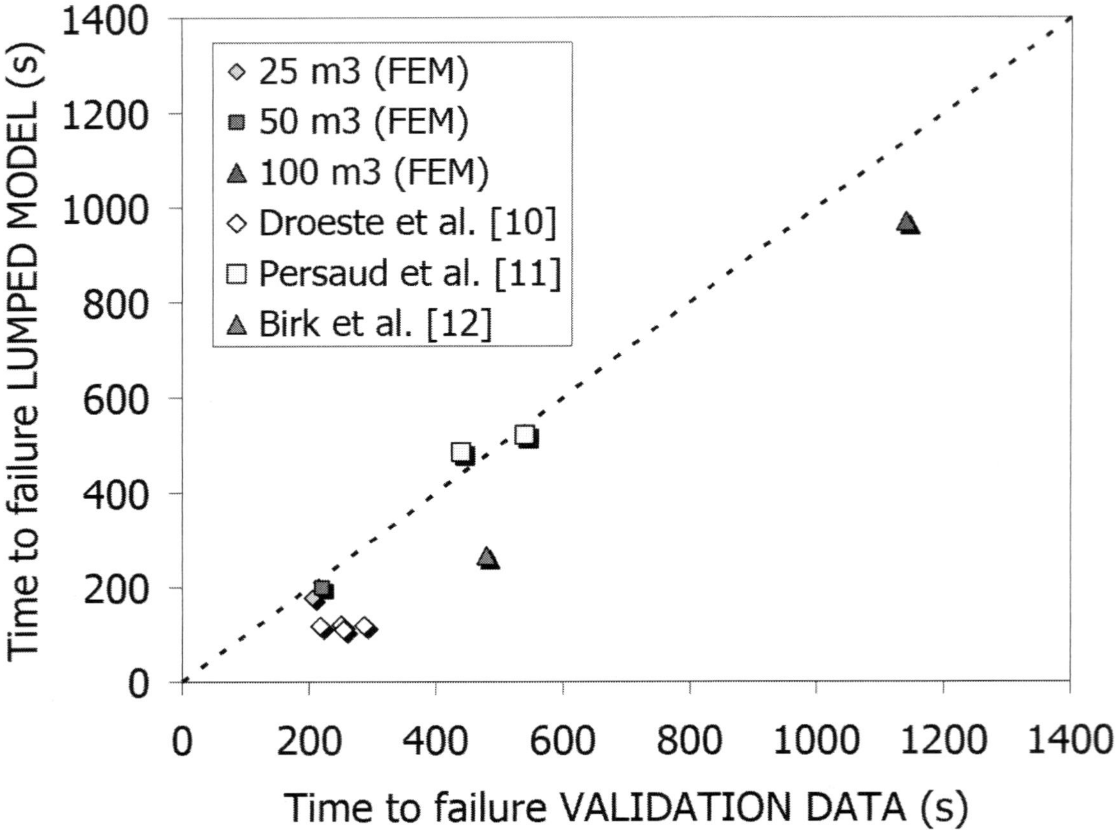

Figure 2: Examples of comparison between ttfs obtained with the lumped model runs and the validation data set. Validation data are both obtained from the FEM model results (labelled with FEM and referred to 25, 50 and 100 m^3 tank volume) and from experimental data [10-12].

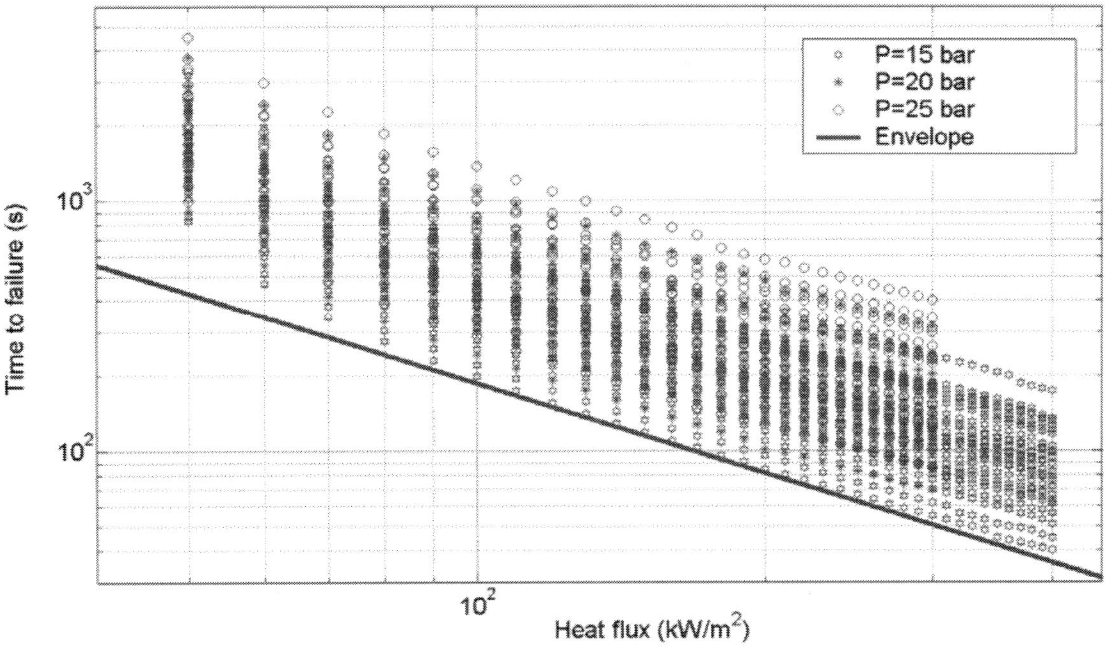

Figure 3: Envelop correlation of times to failure obtained for horizontal cylindrical pressurized vessels (1.5 to 2.5 MPa design pressure) under full engulfment conditions.

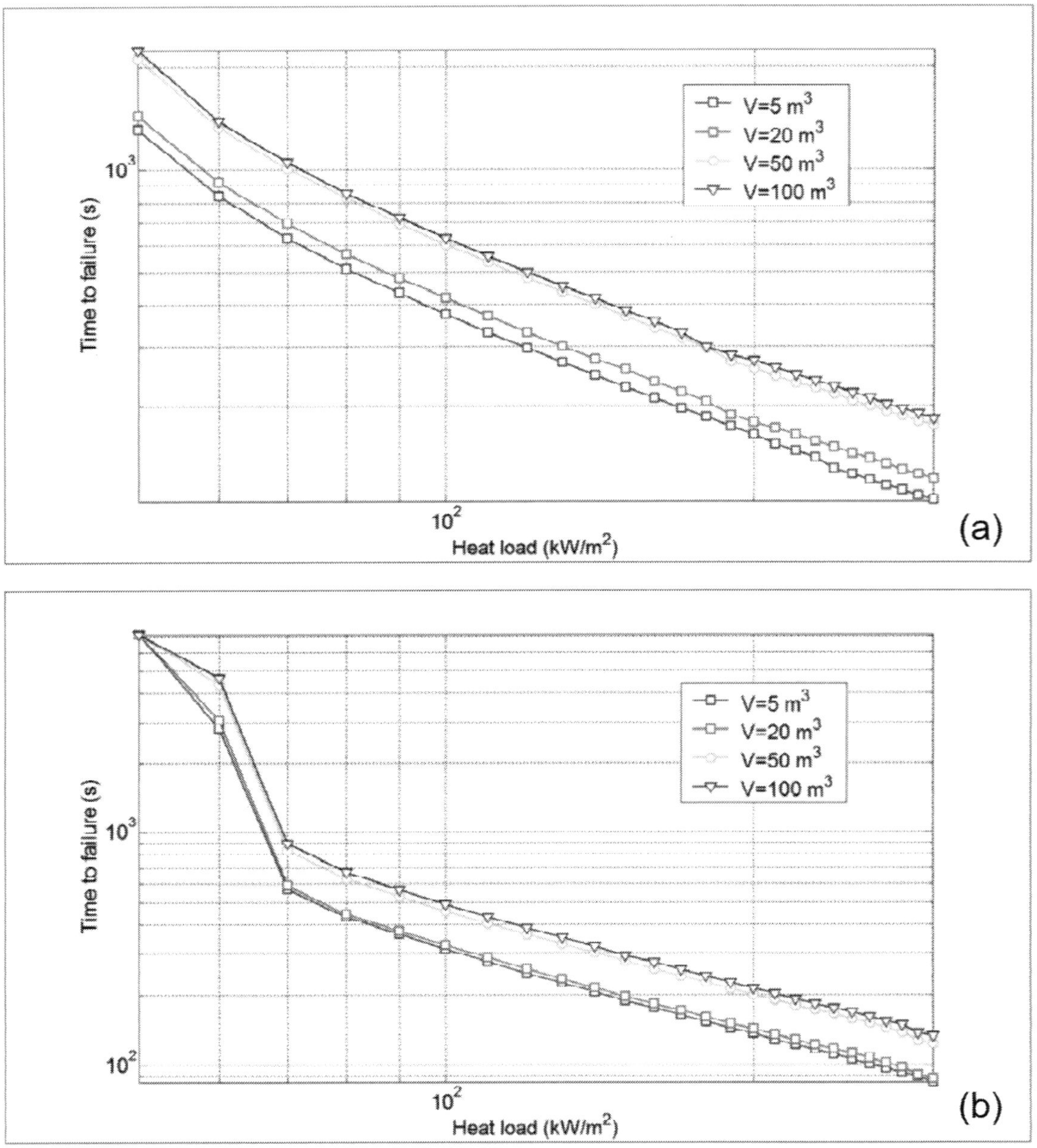

Figure 4 Comparison between the ttf of pressurized vessels (vessel volume: 5, 20, 50 and 100 m^3) without thermal shielding (a) and with a 50 mm thick glass wool coating (b).

TABLES

Table 1: Design data of horizontal pressurized cylindrical vessels

Volume (m^3)	Length (m)	Diameter (m)	Shell thickness (mm)	Area pressure relief valve (m^2)
5	3	1.6	11	0.073
20	7.2	1.9	12	0.164
50	10	2.7	17	0.292
100	18	2.8	18	0.456

Table 2: Estimated threshold values for escalation caused by fire radiation

Time to failure (s)	Radiation Threshold, (kW/m^2)
600	60
900	50
1800	40

A Semi-quantitative Risk Assessment Methodology to Prioritize Recommendations

Kumar R. Bhimavarapu[1] **and William W. Doerr**
Risk, Reliability and Failure Prevention, Research, FM Global
Norwood, Massachusetts 02062, USA

A 10th PPSS Anniversary Paper

ABSTRACT

Process safety engineers promote loss prevention by providing recommendations when deficiencies are observed. Deficiencies are undesirable conditions that aggravate the likelihood and/or the severity of loss events. Deviations in establishing and implementing the design intents are generally the causes of deficiencies. Any improvement needs resources, and since the resources are always limited, process safety engineers have to establish priorities for loss prevention recommendations. One appropriate way to prioritize the recommendations is to evaluate and compare the cost of risk reduction associated with the improvements. While qualitative risk analysis techniques may not be able to provide the necessary resolution to differentiate the improvements, quantitative techniques need much expertise, time, and funds. A feasible approach would be a semi-quantitative methodology with the simplifications necessary to make it easy to use, but at the same time providing the needed resolution for prioritization. This paper presents such a methodology to prioritize loss prevention recommendations. The methodology uses the concept of engineered systems, which are simply hardware (equipment or facility), procedures, or a combination put in place to serve specific functional intents. The basis of the methodology is founded in the premise that deficiencies are always associated with engineered systems and deficiencies increase their failure likelihoods. The high level steps of the methodology include: building scenario paths associated with the deficiencies, evaluating the likelihood of loss events with and without deficiencies, and finally, estimating the associated cost of risk reduction to facilitate prioritization of recommendations. A few examples are provided with the details of the methodology and the incorporated simplifications.

Keywords: Engineered Systems, Deficiencies, Scenario Events, Scenario Path, Risk Evaluation, Prioritization

1. INTRODUCTION

Process safety engineers provide recommendations to address deficiencies or to effect improvements. Obviously, process safety engineers routinely ask themselves the question, "How does this deficiency increase the risk?" or, "On which recommendation

---

1 . *To whom the correspondence should be sent: Kumar.Bhimavarapu@FMGlobal.com; Phone: (781) 255-4919*

should we spend our limited resources?" This paper presents a simplified single-path scenario-based (See Figure 1) semi-quantitative risk evaluation methodology to improve the prioritization of recommendations. Once the deficiency and the details of the facility are known, the methodology aids to establish the scenario path that leads to the specific loss event of concern, assign simplified single digit order of magnitude likelihoods for the scenario events, and estimate the likelihoods of the loss scenarios. Further, using the available loss expectancy values, risk reduction cost can be calculated for prioritization of the recommendations.

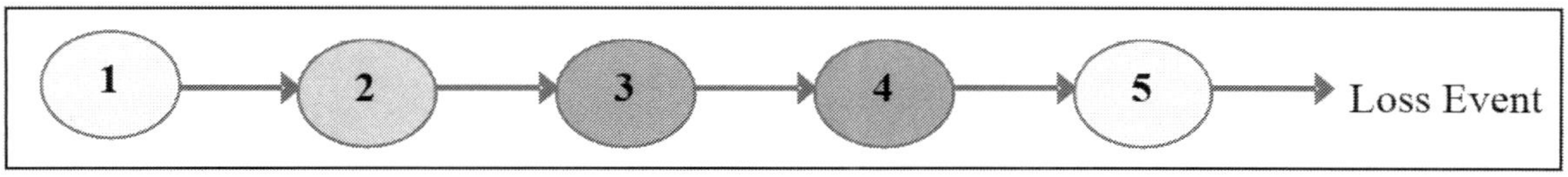

Figure 1: Scenario Path, a Series of Scenario Events

Scenario events are the events on a scenario path that leads to a loss event. Scenario events are essentially successes and failures of engineered systems, which are the building blocks of the process industry. Engineered systems can be defined as the systems that are engineered (and in place) to operate process plants productively and safely. Thus, all equipment and facilities (related to hardware) as well as procedures, and their combinations constitute engineered systems. Identification of the engineered systems and scenario events, and evaluation of the impact of a deficiency on the failure likelihood of the engineered system are important steps of the methodology presented in this paper.

Section 2 reviews the basic information used to establish the methodology and covers in brief the concepts, failures and components of engineered systems, impact of deficiencies on failures, etc. Section 3 outlines the principles and the basic steps of the methodology. Section 4 provides the details of the adopted simplified procedures to estimate the impact of deficiencies on the failure likelihood of engineered systems, which is an important part of the methodology. Section 5 provides the details of the methodology starting from the deficiency (or recommendation) to prioritization of the recommendations. Finally, conclusions are found in Section 6.

2. ENGINEERED SYSTEMS AND DEFICIENCIES

The term "engineered system" is defined intentionally to be encompassing. The objective is to facilitate a consistent and uniform approach to address a variety of risk assessment requirements. Examples of engineered systems are process control systems, pressure relief systems, safety instrumented systems, process vessels, fire suppression systems, dikes, emergency responses, fire fighting systems, hot work permit systems, and operating procedures. In other words, any system that an engineer may come across in a process plant, whether it is hardware or procedure based, is an engineered system. While some address process functions and some address safety functions, in some cases, it may be difficult to clearly distinguish between the two functions. Irrespective of the nature of the function, the common feature of all these systems is that they can fail to fulfill their functional intents.

2.1 Failures of Engineered Systems

Engineered systems are designed/established to meet one or more functional intents to the extent possible. They need maintenance or upkeep for a productive and loss-free operation. From a loss analysis perspective, an engineered system is considered to have failed if it fails to fulfill its functional intent. Generally, losses occur only when engineered systems fail. A good understanding of engineered systems and their failures is essential to understand the loss scenario mechanisms and risk.

2.2 Failure Modes and Scenario Events

Functional failure mode is the umbrella failure mode of interest to a process safety engineer. Engineered systems can fail in a variety of specific failure modes, and each mode can have different severities. For example, a compressor can fail to compress the process medium to meet the functional intent of reaching the specified pressure. In addition, a compressor can have a mechanical integrity failure, and can release material either from a small hole or from a major rupture. Further, failures may occur during active operation, during testing, or during a response from a standby condition. While most of the times process safety engineers will be interested in failure events, sometimes, success events are also needed to build scenarios, particularly to evaluate mitigated outcomes.

Each type of failure or success of an engineered system construes a scenario event. Thus, each engineered system may be associated with more than one type of scenario event, and the scenario event of interest may change with the scenario. When the functional intents are clear and the failure modes are understood, scenario events can be identified easily.

2.3 Failures Due to Deficiencies

Engineered systems fail when demands above the design intent are made, or due to deficiencies. In the process industry, standards, guidelines, and good engineering practices are generally used to establish the design intent. Unfortunately, economic and technology limitations generally impose a gap between the functional and the design intents. Bhimavarapu [1] provides additional details on functional and design intents. Deficiencies, which are our interest in this paper, are undesirable conditions that aggravate the failure likelihoods of engineered systems. Because of deficiencies, engineered systems fail to fulfill their design intents, and in turn fail to fulfill their functional intents. For example, a corroded sprinkler protection system may not deliver the water to fulfill its design intent when there is a demand; as a result, it may fail to fulfill its functional intent as well.

Deficiencies are always associated with engineered systems. Since engineered systems include hardware and procedures, deficiencies are always associated with hardware and/or procedures. In the case of hardware, deficiencies could be associated with design, fabrication and installation, operation, inspection, testing, and maintenance as applicable. In a similar fashion, in the case of procedures, deficiencies could be associated with establishing, implementation, and upkeep. The failure modes change as a function of deficiencies.

2.4 Severity of Deficiencies

It is not necessary that deficiencies be revealed. Deficiencies may remain unrevealed if they are not severe enough. When the deficiency is not severe enough, the engineered system's functioning may not be affected to any perceivable extent, and it may still fulfill its design intent and in turn the functional intent as well to an acceptable level. At the same time, failures can also occur because of unrevealed (unnoticed and unnoticeable) deficiencies. In fact, many loss events occur because of unrevealed deficiencies. The extent of identification of the deficiencies depends on the training engineers receive, the time and effort they can spare, and also nature of the deficiency. Bhimavarapu [1] provides details and additional insight into this topic.

2.5 Subordinate Engineered Systems

Engineered systems are generally comprised of many components, which we designate for convenience as subordinate engineered systems. Subordinate systems[2] are also engineered, and they function in a similar fashion. Success or failure of any engineered system depends on the success or failure of its subordinate systems. For example, a sprinkler protection system consists of a number of subordinate systems such as firewater storage, firewater pumping system, firewater headers, control valves, and sprinklers. Further, a firewater pumping system can have subordinate systems such as main pumps, auxiliary pumps, associated instrumentation, etc. If there is a deficiency associated with the maintenance of the motor of the pump, sprinkler protection may fail.

3. BASICS OF THE METHODOLOGY

In order to evaluate the aggravation of risk due to a deficiency, one has to identify the associated loss events, their likelihoods, and the impacts. To facilitate such analysis, scenario path(s) have to be built for the loss event(s) resulting from or contributed by the deficient condition. As a function of the need and the available resources, the scenario-based analysis could be quantitative or qualitative in approach. Analyses may involve single or multiple scenario paths to evaluate loss event/s and associated risk. Event tree analyses use multiple scenario paths to evaluate the multiple outcomes of an undesirable event. Semi-quantitative techniques can be used to derive many benefits of the quantitative techniques, and to minimize the resources necessary for a rigorous quantification of risk. Single path scenarios (See Figure 1) are amenable to semi-quantitative evaluations and are suitable to analyze the specific loss event arising from a deficiency.

The failure likelihood of an engineered system increases with the severity of a deficiency. Since failure of an engineered system is a scenario event, a deficiency increases the likelihood of the scenario event on the scenario path. When the likelihood of a scenario event is higher due to a deficient condition, the likelihood of the associated scenario or the loss event will also be higher. Consequently, the risk[3] increases. The needed high

---

[2] *Throughout this document, whenever possible, the word 'engineered' is omitted for brevity.*

[3] *Risk is a function of the likelihood and the impact (or, consequence) associated with the loss event.*

level steps to evaluate the aggravation of the risk due to a deficient condition, and to prioritize the recommendations, are as follows:

1. From the deficiency or the recommendation, establish the engineered system (i.e., the deficient engineered system) whose failure is of concern.
2. Establish the scenario event (i.e., the deficient scenario event) associated with the deficient engineered system.
3. Establish the scenario path (i.e., the series of scenario events) that leads to the specific loss event of concern.
4. Evaluate the likelihood of the scenario path (or the loss event) before and after the removal of deficiency.
5. Estimate the risk reduction cost for prioritization of recommendation.

In order to develop risk based prioritization of various recommendations, the likelihood evaluations can be simplified considerably. The methodology depends on how the impact of a deficiency on the scenario event likelihood is evaluated. The background information necessary for such evaluation and the adopted simplified procedures are presented in Section 4. Then, the five steps mentioned above are discussed in detail in Section 5. Examples are provided to explain the adopted terms and concepts.

4. EVALUATION OF THE IMPACT OF A DEFICIENCY ON SCENARIO EVENT LIKELIHOOD

Scenario events shown on scenario paths are generally based on macro engineered systems, and for easy reference we call them primary engineered systems. Most of the times primary systems comprise subordinate systems at different hierarchical levels (see Section 2.5). When a deficiency is associated with one of the subordinate systems, we have to assess the impact of the subordinate system on the failure likelihood of the primary system shown on the scenario path. Instead of using time consuming methods such as fault tree analyses, we established simplified rules to quantitatively evaluate the impact of the deficient subordinate system on the failure likelihood of the primary system. In other words, in order to keep the analysis simple and limited to a single scenario path, we develop scenario events based on primary systems only and assign them likelihood values. The three factors used for the assignment of likelihoods are: i) the type of engineered system, ii) the nature of the deficiency (associated with the primary system itself or with one of the subordinate systems), and iii) the criticality of the deficient subordinate system when applicable. The details of these three factors and the adopted simplifications are given in the following sections.

4.1 Types of Engineered Systems

The impact of a deficiency varies with the nature of the engineered system. The impact will be less severe when the affected functional capabilities are recoverable through some means. Procedural systems are relatively more resilient than hardware and the failure associated with the deficiency may be avoidable. Operators may rise to the occasion and fulfill the functional intent even when there are deficiencies in the established procedures.

It may not be the case with a hardware system. For example, the impact of a design deficiency on the failure likelihood of a relief system will be more definitive as compared to the impact of a procedural deficiency on a hot work permit system. Thus, on a relative basis, for procedure-based systems, the impact of the deficiency is considered to be less severe.

With this background, engineered systems are categorized under three types to allocate failure likelihoods. They are hardware-based, procedure-based, and mixed. Examples of hardware-based systems are relief systems, compressors, and automatic sprinkler protection systems. Examples of procedure-based systems are hot work permit systems (ignition controls), operator actions in normal operation, and operator responses in emergencies. Mixed systems comprise both hardware and procedural components. For example, a fire fighting system needs both facilities/hardware and procedures. Mixed systems succeed only when both components function successfully. Failure of any one of the components would fail the system.

4.2 Criticality of the Subordinate Engineered Systems

When a deficiency is associated with a subordinate system, the likelihoods to be assigned to the deficient scenario event (failure of primary system) vary depending on the criticality of the subordinate system. Subordinate systems are divided simply under two categories: 'critical' and 'non-critical'. A critical system is considered as one whose failure will result in the failure of the primary system with certainty. A non-critical subordinate system is considered as one whose failure may result in the failure of primary system, but not with certainty. Tables 1 and 2 show a few examples of critical and non-critical subordinate systems respectively. For simplicity, the hierarchical level of the subordinate systems is not considered while evaluating their criticality.

Table 1: Examples of Critical Subordinate Engineered Systems

Subordinate Engineered Systems	Primary Engineered Systems
Non-redundant fire pump	Fire fighting
Automatic fire detection at an unattended plant	Automated deluge system
Non-redundant temperature sensor	Process control
Non-redundant relief valve	Relief system

Table 2: Examples of Non-critical Subordinate Engineered Systems

Subordinate Engineered Systems	Primary Engineered Systems
Spare fire water pump	Fire fighting
Automatic fire detection at an attended plant	Fire suppression system

4.3 Nature of the Deficiencies

Deficiencies are categorized under three types: i) lack or non-availability (of the system), ii) design deficiency, and iii) maintenance deficiency, which are self explanatory. See a few examples in Table 3.

Table 3: Examples of Recommendations, Engineered Systems, and Nature of Deficiencies

Recommendation	Engineered System and the Nature of Deficiency
Fire water pump is to be upgraded	Fire water pump; deficient design
Capacity of relief discharge header is to be verified	Relief discharge header; deficient design
Electrical equipment is to be upgraded for explosion proof equipment	Electrical area classification; deficient design
The maintenance of sprinklers is to be improved	Sprinkler heads; deficient maintenance
Sprinklers are to be provided	Sprinklers; lack of system
Relief valve design has to comply with DIERS	Relief valve; deficient design

4.4. Likelihood Data

In order to estimate the likelihood of the loss event in terms of frequency, the likelihood values to be assigned to the scenario events are essentially of two types: Frequencies (# of failures in specific time period) and conditional probabilities. Conditional probabilities are used for occurrence of events given that some precursor event has happened, and we use PFDs (probabilities of failure on demand) for this purpose.

For most of the engineered systems, generic ranges of failure likelihoods are available in the literature *[2, 3, 4, 5, 6, 7, and 8]*. In view of the limited objective of the methodology, we have simplified the allotment of likelihood values to the scenario events. One could consider that the high failure likelihood values that are available in the literature are associated with systems that have deficiencies. Thus in principle, for any engineered system with a known deficiency, a high likelihood value can be assigned and, once the deficiency is removed, the failure likelihood can be reverted back to an average likelihood value. Accordingly, for all the scenario events, we considered essentially two likelihood values: an 'average' and a 'high' likelihood value. Where PFDs are not readily available, the same are estimated from frequency data using simple equations *[7 and 8]*. Further, the resolution of the likelihood values is adjusted to a single digit order of magnitude (e.g., 2×10^{-4}).

4.5 ALLOTMENT OF LIKELIHOOD VALUES

Only one event on the scenario path is assigned a frequency value, and the PFD values are used for all the others. Generally, the first event on the scenario path (called an initiating event) is the one that is assigned a frequency value. The engineered systems that operate on demand are assigned PFD values.

4.5.1. Allocation of Likelihoods for Non-deficient Scenario Events

As long as the scenario event is not associated with a known deficiency, an average failure likelihood value is assigned. The likelihood could be a frequency or PFD depending on the sequential location of the scenario event on the scenario path.

4.5.2 Allocation of Likelihoods to Deficient Scenario Events

The following simplified guidelines are established to assign the likelihoods for the deficient scenario events, i.e., the failures of deficient primary systems.

I) Procedural Systems

The deficiency could be associated with a primary system itself or one of its subordinate systems. However, for procedural systems, in view of their resilient nature, it is not considered necessary to go into the details of the nature of deficiency or the criticality of the subordinate system. When there is a deficiency, simply an applicable 'high' likelihood ('high' frequency or 'high' PFD) value is allotted.

II) Hardware-based Systems

The likelihood values for hardware-based primary systems are allotted considering that the design deficiencies affect the failure likelihood more than the maintenance deficiency, which is procedure oriented. Further, a differentiation is made based on the engineered system (primary or subordinate) with which the deficiency is associated, and the criticality of the subordinate system when applicable.

When the deficiency is maintenance related or when the deficiency is associated with a non-critical subordinate system, a high likelihood ('high' frequency or 'high' PFD) value is allotted.

A design deficiency associated with a primary system itself or with a critical subordinate system is considered to be more serious. In such cases, for systems that function on demand and need conditional failure probabilities, a PFD value of 1.0 is allotted. However, for systems that operate on a continuous basis and need failure frequency values, simply the 'high' frequency values are allotted[4].

---

[4] *It may be noted that the frequency values are generally needed for initiating events which are associated with failures of systems that are in continuous operation. A conscious decision was taken not to use a value greater than the established 'high' frequency value even when the deficiency is serious. In addition to the need to limit the number of likelihood values to keep the methodology simple, the reasons for such a decision include the higher revealability of the deficiencies associated with continuous operating systems as compared to those systems that operate on demand.*

II) Mixed Systems

For mixed primary systems, if the deficiency is associated with the procedural component, simply 'high' failure likelihood (frequency or PFD) is allotted. If the deficiency is associated with the hardware, then the likelihood allotment guidelines that are established for hardware-based systems are used (see Section 4.5.2. II).

4.5.3. Other Considerations

Guidelines developed for other recurrent situations are as follows:

- When an engineered system is absent (and a recommendation is given to provide the same) the associated scenario event will be shown on the scenario path with a PFD of 1.0.
- A scenario event could be a success event on demand. In view of the low failure likelihoods of engineered systems in the process industry, the conditional probability of a success event is always considered to be 1.0.

The likelihood allocations described in sections 4.5.1 to 4.5.3 cover most situations. However, one could come across situations where additional decisions may have to be made. Since the defined objective of the methodology is only relative prioritization of recommendations, as long as the decisions are made consistently, the variation in the prioritization may not be significant.

5. DETAILED STEPS OF THE METHODOLOGY

Described below are the details of the five steps of the methodology to prioritize recommendations starting from a deficiency or a recommendation.

5.1. Identification of Deficient Engineered Systems

The first step of the methodology is to establish the engineered system that is associated with a deficiency. Through a simple but objective examination, an engineer can identify how a deficiency would result in, or contribute to, the failure of a specific engineered system. See a few examples provided in Tables 4 and 5 for the identification of the deficient engineered system from deficiencies and recommendations respectively.

5.2 Establishing the Deficient Scenario Event

Failures of the primary systems (or the macro systems) are only shown on the scenario path as deficient scenario events. Hence, when the deficiency is associated with a subordinate system, it is necessary to identify the associated primary system. See a few examples provided in Table 6. In some cases, the deficiency can be associated with the primary system itself. See a few examples provided in Table 7.

Table 4: Examples of Deficiencies and Deficient Engineered Systems

Deficiency	Failure of the Deficient Engineered System
Inadequate number of sprinklers	Inadequate number of sprinklers can result in failure of sprinkler protection
Non-compliance with DIERS[5] methodology for relief- system design	Non-compliance with DIERS methodology can result in failure of relief system
Improper inspection of storage tank	Improper inspection can lead to mechanical integrity failure of the storage tank and in turn release of flammable materials
Poor release controls	'Release control' is a broad engineered system and failure can occur due to failures of subordinate systems such as process equipment, operator actions, or operator responses.

Table 5: Examples of Recommendations and Deficient Engineered Systems

Recommendation	Deficient Engineered System
Upgrade the fire water pump	Fire water pump
Verify the capacity of the relief discharge header	Relief discharge header
All the electrical equipment is to be upgraded to meet the electrical area classification	Electrical area classification (or, electrical ignition control)
Improve the maintenance of sprinkler heads	Sprinklers
Provide additional sprinklers	Sprinklers
Provide annual inspection for the relief system	Relief system
Provide sprinkler protection	Sprinkler protection system

Table 6: Examples of Deficient Subordinate and Primary Engineered Systems

Deficient (Subordinate) Engineered System	Deficient Primary Engineered System
Relief discharge header	Relief system
Fire water pump	Fire fighting
Sprinklers	Sprinkler protection system
Relief valve	Relief system

[5] *DIERS: Design Institute for Emergency Relief Systems.*

Table 7: Deficiencies associated with Primary Engineered Systems

Deficiency	Primary Engineered System
Deficient design basis of relief system	Relief system
Lack of sprinkler protection	Sprinkler protection

Once the primary system is identified, by associating it with the failure mode of concern, the potential deficient scenario event can be identified. See a few examples in Table 8. *Bhimavarapu [1]* and *Bhimavarapu and Doerr [9]* provide additional details on engineered systems and scenario events.

Table 8: Identification of Potential Scenario Events from the Primary Engineered Systems

Primary Engineered System	Failure Mode	Scenario Event	Remarks
Sprinkler protection	Functional failure	Failure of sprinkler protection	Due to the failure, the sprinkler protection system will not be able to extinguish/control the fire.
Relief system	Functional failure	Failure of relief system	Due to the failure, the relief system will not be able to relieve the pressure, which may lead to the failure of a process vessel.
Storage tank	Mechanical integrity Failure	Release from the tank	Mechanical integrity failure is one type of functional failure. The main outcome of interest is loss of containment. Non-availability of the tank could also be a concern.
Release control	Functional failure	A generic release	Release control is an important aspect in process industry and is achieved through minimizing the failures of many engineered systems (See Table 4 for details). Hence the failure mode is considered broadly as a functional failure.

5.3 Identification of the Scenario Path that Leads to the Loss Event.

Once the scenario event associated with a deficient primary system is identified, the development of the scenario path generally starts with the identification of the event that places the demand on the deficient primary system. Table 9 presents examples of the events that place the demand on the first two deficient primary systems shown in Table 8.

The remaining scenario events on the scenario path can be identified by considering the loss event of concern, the nature of the process, and other engineered systems in place.

Table 9: Identification of Demand Events

Deficient Primary Engineered System	Event that Places the Demand on the Deficient Primary Engineered System	Remarks
Sprinkler protection	Fire *	Every potential fire that has to be mitigated by the sprinkler protection (as per the functional intent) should be considered.
Relief system	Over-pressurization*	Every potential over-pressurization event that places the demand on the relief system (as per the functional intent) should be considered.
* Note: The events fire and over-pressurization are the outcomes of the failures of a few or more engineered systems. If the likelihoods of fire or over-pressurization are known, then there should be no need to include failures of all those individual engineered systems on the scenario path.		

The engineered systems in place may change with the location, and accordingly the scenario events may change. See the illustrative scenario paths shown in Figures 2 and 3 for the example cases shown in Table 9.

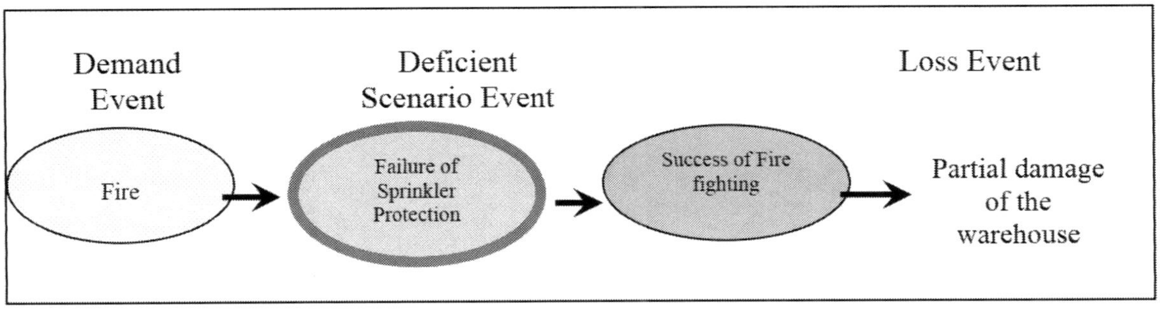

Figure 2: From Fire to Loss Event, Scenario Path

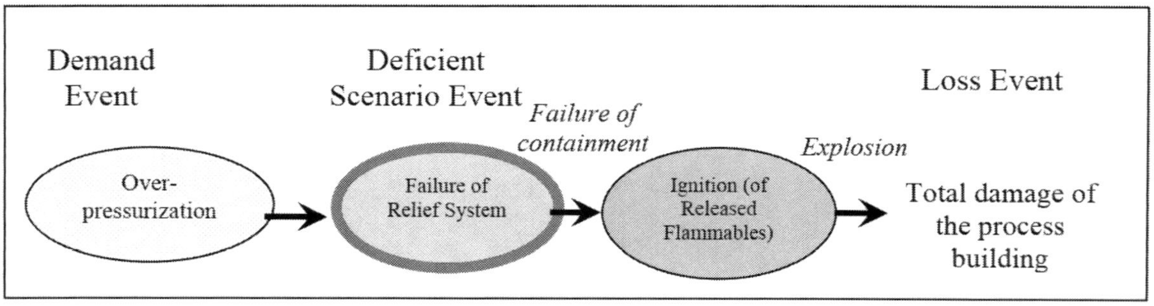

Figure 3: From Over-pressurization to Loss Event, Scenario Path

The scenario path of interest shown in Figure 2 comprises success of fire fighting after the failure of sprinkler protection. One may build the scenario for the failure of fire fighting, in which case the loss event would be different. In the case of the second scenario associated with the failure of the relief system (see Figure 3), the failure of containment is considered as certain outcome. The same is the case with the outcome of explosion in case of an ignition. Outcomes are considered as certain events with a conditional probability of 1.0; hence, they are not shown as scenario events. No further mitigation event is considered on this path.

5.4 Evaluation of Scenario likelihoods with and without Deficiencies

The example depicted in Figure 4 explains the simplified methodology to estimate the impact of deficiencies on scenario likelihoods.

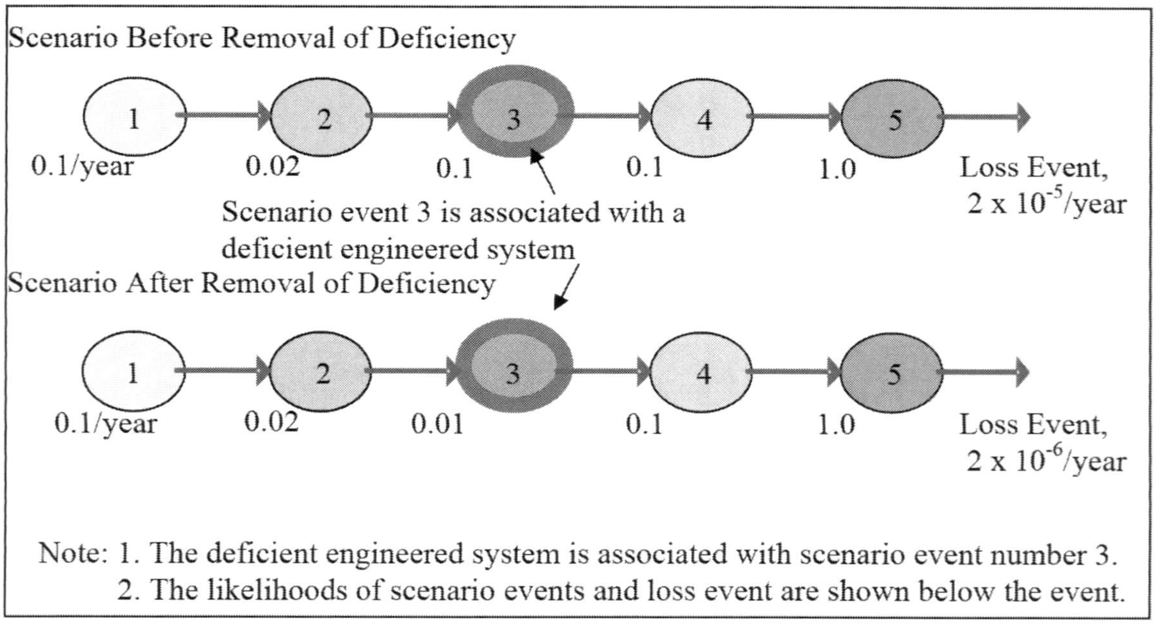

Figure 4: Likelihoods with and without Deficiency

Shown in the figure is a scenario path before and after removal of a deficiency. There are five scenario events on the scenario path and each scenario event could be a success or failure of an engineered system. The engineered system associated with scenario event 3 is considered to have a known deficiency. Other scenario events are considered to be independent of the deficient scenario event.

As can be seen, the scenario path and the loss event (with specific impact) do not change after the removal of the deficiency. The 'average' likelihood values (frequencies or PFDs) assigned to the non-deficient scenario events (as a function of the engineered systems they are associated with) remain the same. The only change that occurs is in the likelihood of the scenario event associated with a deficiency and accordingly the likelihood of the loss event. Due to the presence of a (known) deficiency, the likelihood of scenario event 3 will be greater than what it would be without that known deficiency. When the deficiency is removed, the likelihood of the scenario event is expected to be reduced to a lower value, in specific, to an average failure likelihood value[6].

In terms of quantified likelihoods (i.e., in terms of single digit order of magnitude), with the removal of the deficiency, the failure likelihood of scenario event 3 is reduced from 0.1 to 0.01. Accordingly the loss event failure likelihood reduces from 2×10^{-5}/year to 2×10^{-6}/year.[7]

5.5 Risk-based Prioritization of Recommendations

Risk reduction cost is one appropriate measure to prioritize recommendations. The likelihood values estimated for a loss event before and after the implementation of recommendations can be used to estimate the risk values before and after the implementation of the recommendation. The loss expectancy value can be chosen by the user as appropriate, and it does not change with the implementation of the recommendation. It could be any quantifiable value of interest to the user such as property damage, business interruption, loss of market share, injuries, illnesses, fatalities etc.

The adopted formulae for the estimation of risk reduction cost are shown below in terms of financial loss ($). The cost of the implementation of the recommendation is annualized to provide a dimensionless risk reduction cost.

> Risk, $/year, (before OR after the implementation of recommendation)
>
> = {loss expectancy, $/event} x {likelihood of the scenario, # events/year, before OR after the implementation of the recommendation}

---

[6] *Just by the removal of the deficiency, the failure likelihood will not become zero because engineered system can still fail due to some other deficiency, or even due to an excessive demand. In fact, there could be other subordinate systems with unknown deficiencies that can lead to the failure of the engineered system.*

[7] *The objective of this paper is only to present the principles and steps of the proposed methodology. The likelihood values shown in Figure 4 are only for demonstration purpose. Organizations' culture and philosophy, and discretion can play a credible role in the selection of likelihood values on the lines suggested in this paper. However, as long as there is consistency in the selection and the assignment of the likelihood values, the variation in the prioritization may not be significant*

Risk Reduction, \$/year

= {Risk before the implementation of the recommendation, \$/year} – {Risk after implementation of the recommendation, \$/year}

Risk reduction cost

= {Annualized recommendation implementation cost, \$/year} / {Risk reduction, \$/year}

Even though the risk reduction cost is estimated using the basic principles of risk, it may not be advisable to use these numbers in an absolute fashion in view of the simplifications incorporated into the methodology.

6. CONCLUSIONS

Process safety engineers should be cognizant that hardware (equipment or facilities), procedures, and their combinations constitute engineered systems with specific functional and design intents and specific failure modes. Engineered systems can be defined appropriately to facilitate generation of clear-cut success or failure events, i.e., scenario events. Any loss event can be described as a combination of failures and successes of a few engineered systems. Quantitative risk assessment can be performed by assigning likelihoods to all such scenario events. In view of the use of fundamental principles, this approach helps to improve the efficacy and versatility, and thus utility, of scenario-based analyses.

Risk aggravation from a deficiency will be a function of many associated scenario paths with different loss events with different loss values and likelihoods. The methodology described in this paper uses a single scenario path of interest and single digit order of magnitude likelihoods in a simplified fashion. The notable feature of the proposed methodology is that the estimated likelihood values of the loss event take into account the impact of identified deficiencies (or recommendations), and provides a quantitative understanding of the potential relative variation in the risk with improvements. In general, the resolution of the likelihood values proposed for use in the methodology will be adequate to improve the prioritization over the traditional prioritization techniques that use loss expectancy (\$) values, qualitative likelihoods for the loss events, or even a combination. Without the need for rigorous quantification, the proposed methodology helps to prioritize recommendations. At the same time, the methodology allows incorporation of virtually any type of improvement to suit individual requirements. Such improvements include enhancement of the resolution of likelihood values, evaluation of the impact of deficiency using fault tree analyses, use of more than one scenario to better understand the impact of deficiency, use of one scenario to evaluate the impact of multiple deficiencies, evaluation of the impact of site specific design, procedures, and practices while assigning the likelihoods, etc.

REFERENCES

[1] Bhimavarapu, K., *Account for Engineered Systems Properly to Analyze the Loss Scenarios*, ASME International Mechanical Engineering Congress and Exposition, 2007.

[2] European Industry Reliability Data Bank EIReDA 1998, Procaccia, H., Arsenis, S., and Aufort, P., Crete University Press, Crete, Greece, 1998.

[3] OREDA, Offshore Reliability Data Handbook, 3rd edition, SINTEF Industrial Management, Det Norske Veritas Hovik, Norway, 2002.

[4] Guidelines for Process Equipment Reliability Data with Data Tables, AIChE CCPS, New York, NY 1989.

[5] Loss Prevention in the Process Industries: Hazard Identification, Assessment and Control, Second Edition, Lees, Frank P., Butterworth Heinemann, Oxford, 1996.

[6] Layer of Protection Analysis: Simplified Process Risk Assessment, AIChE, CCPS, New York, 2001.

[7] Reliability, Maintainability and Risk: Practical Methods for Engineers, Smith, David J., Fourth Edition, Butterworth-Heinemann, Boston, 1993.

[8] Hazop and Hazan: Identifying and Assessing Process Industry Hazards, Fourth edition, Kletz, T., Taylor & Francis, 1999.

[9] Bhimavarapu, K., and Doerr, W., *Do You Know Your Engineered Systems?* Spring National Meeting, American Institute of Chemical Engineers, 2006.

Fire Protection for Extremely Corrosive Industrial Duct Environments

Paul Su
FM Global
Norwood, MA 02062

William W. Doerr
FM Global

A 10th PPSS Anniversary Paper

ABSTRACT

*Due to the extremely corrosive environments inside many industrial exhaust ducts, sprinkler systems have not been successfully designed to protect against fires originating within these ducts. Damage to the exhaust ductwork system caused by fire can lead to interruption of the plant operations for an extended period of time with substantial financial losses. Research has been performed to evaluate the corrosion resistance properties of various materials and coatings such that corrosion resistant sprinklers and fire protection hardware can be developed for use in these flammable exhaust ducts and other corrosive industrial environments. Laboratory tests, including electrochemical techniques and crevice coupon immersion tests, were used to select suitable alloys and coatings for in-situ field test rack studies. These results are providing the basis for the development of fire protection systems suitable for these industrial exhaust ducts.**

1. INTRODUCTION

Many industrial facilities generate extremely corrosive exhaust fumes including high concentrations of sulfuric, hydrochloric, nitric and hydrofluoric acids, smoke and high velocity solid particles.[1,2] Various ductwork systems are designed to remove or filter these hazardous materials. In addition, the ducts can be up to 12 feet in diameter and hundreds of feet long, with interconnecting vessels.[3] These systems can be found in applications such as metallurgical smelters, steel mill pickling lines, power generation plants, semiconductor fabrication facilities, pulp and paper plants, and chemical processing facilities. Table 1 shows the typical fume/flue properties inside such exhaust ducts. Since these exhaust fumes are extremely corrosive to most metals, at present, most ducts are fabricated from chemical resistant combustible plastic

*P. Su and J. C. Harrington, US Patent Application number 36737-252820 "Corrosion Resistant Sprinklers, Nozzles, and Related Fire Protection Components and Systems", 2007.

materials such as polypropylene (PP), chlorinated polyvinyl chloride (CPVC) and fiberglass reinforced plastic (FRP).[3]

Table 1. Typical fume/flue properties inside industrial exhaust ducts.

	Typical Exhaust Fume/Flue Properties
Temperature	20 - ~ 97 °C
Acids	HCl, HF, H_2SO_4, HNO_3,...; vapor conc. from high ppm to 20% by volume; pH < 2
Gases	SO_2, SO_3, CO_2, NO_x, Cl_2, F_2, ...; Conc. from ppm to %
Particle Composition	Cu, Fe, Pb, Zn, As, Sb, Ca, Hg, Ni, ...as oxides or salts
Condensation	Vapor saturated with moisture droplets and wet/dry cycles
Velocity	0-18.3 m/s (0-41 miles/hr)

Damage to a ductwork system due to fire will usually result in significant shutdown of the plant for an extended period, which can lead to substantial financial losses for the operation[4] as shown in Figures 1 (a)[5] and (b)[6]. According to FM Global's Understanding the Hazard – *Fire in Industrial Exhaust Systems,*[4] "From 1988 to 2003, FM Global clients reported a total of 369 fire incidents involving industrial exhaust systems, totaling more than US$255 million in gross losses (indexed to 2003)".[4]

(a) **(b)**

Figure 1. Photographs of fire incidents. (a) An un-sprinklered steel pickling facility.[5] (b) An un-sprinklered plastic duct work system.[6]

To date, no sprinkler system has been successfully designed to survive for an extended time in these corrosive environments and protect against fires that originate inside the ductwork.[4] It is necessary to evaluate corrosion resistant properties of candidate materials and coatings in order to develop corrosion resistant sprinklers and fire protection hardware to protect these flammable plastic duct and exhaust systems.

2. EXPERIMENTS

2.1 Laboratory Tests

In addition to structural plastic materials, there are limited types of alloys that can survive these extremely corrosive industrial exhaust fumes. Ten alloys, including high nickel alloys and stainless steels, were selected for evaluation, based on the results of literature reviews.[4,7,8,9,10] The chemical compositions of these alloys and their unified numbering system (UNS) numbers are given in Table 2. In addition, due to the extremely aggressive nature of certain exhaust environments (such as HF/HNO_3 pickling) toward metallic and non-metallic materials, several coating systems were selected for evaluation, including fluoropolymers of polytetrafluorethylene (PTFE), ethylene-chlorotrifluoroethylene (ECTFE, or Halar®), ethylene-tetrafluoroethylene (ETFE) and perfluoralkoxy (PFA), along with one epoxy type coating. Specimens with dimensions of 2"x¾"x⅛" having a ⅜" diameter center hole were polished to a 120-grit finish, washed with de-ionized water, and ultrasonically cleaned in acetone.

A corrosive fluid was selected as the simulated testing environment for liquid immersion tests and electrochemical measurements in the laboratory. The test fluid consisted of 11.4% H_2SO_4 + 1.2% HCl + 1% $FeCl_3$ + 1% $CuCl_2$, which is used as a simulant for the flue gas desulfurization atmosphere.[11]

Table 2. Nominal chemical compositions (wt%) and UNS numbers for tested alloys.

Alloy	C276	C 22	I686	CW6M	625	316L	20Cb-3	2205	904L	825
UNS	N10276	N06022	N06686	N30107	N06625	S31603	N08020	S31803	N08904	N08825
C	0.01	0.01	0.01	0.07	0.1	0.03	0.07	0.03	0.01	0.05
Co	2.5	2.5			1					
Cr	15.5	22	21	19	21.5	16-18	20	22	20	21.5
Cu	0.5	0.5			0.5		3.3		1.5	2.3
Fe	5.5	3	5	1	5	63	Bal.	Bal.	Bal.	22
Mn	1	0.5	0.75	0.5	0.5	2	2	2	2	1.0
Mo	16	13	16	17.5	9	2-3	2.2	3	4.5	3
Ni	Bal.	Bal.	Bal.	Bal.	Bal.	10-14	34	5.5	25	Bal.
P	0.04	0.02	0.04	0.04	0.015	0.045	0.045	0.03	0.045	0.03
S	0.03	0.02	0.02	0.03	0.015	0.03	0.03	0.02	0.03	0.03
Si	0.08	0.08	0.08	0.5	0.5	1	1	1	1	0.5
Ti			0.02		0.4					0.9
V	0.35	0.35								
W	4	3	3.7							
Nb					3.5					

Metal coupons with the multiple crevice assembly described in ASTM G48-00 (Method D), consisting of Teflon segmented washers, nuts and bolts, were used to study the crevice corrosion resistance of the alloys.[12] These assembled coupons were immersed in bottles filled with test fluid and maintained at 75°C in a water bath for one week. Periodically, after 24, 48, 72 and 168 hours of immersion, coupons were removed from the bottles, disassembled, cleaned, weighed and photographed to evaluate their corresponding corrosion properties.

Electrochemical techniques including cyclic potentiodynamic polarization (CP), linear polarization (LP), electrochemical impedance spectroscopy (EIS) and long-term open circuit potential measurement were utilized to study the alloys' corrosion properties in the laboratory.[7,13,14,15] The electrochemical configuration is schematically shown in Figure 2. It consists of a Solartron® Model 1287 potentiostat interfaced to a computer using Corrware/Corview® software. The testing cell was a Princeton Applied Research® Model K0235 Flat Cell. The cell contained a platinum counter electrode, a silver-silver chloride (SSC) reference electrode, a working electrode (test specimen) and 300 ml of test solution. The polarization scan rate was 0.17 mv/s. Each specimen was immersed in the test solution for 1 hour prior to the start of the polarization scan. The test solution was open to air at room temperature.

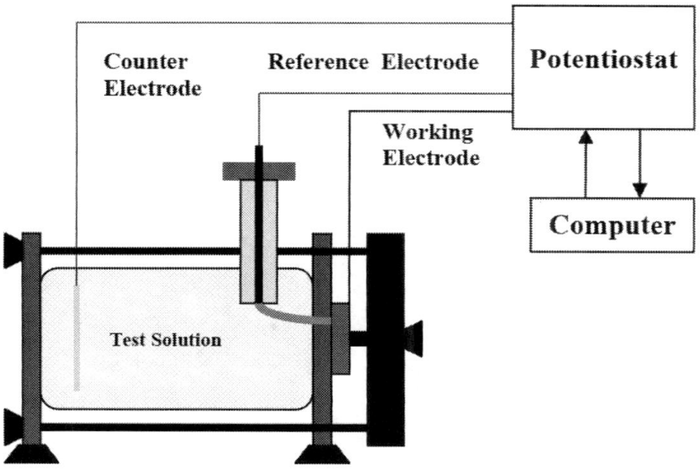

Figure 2. A schematic drawing of the electrochemical configuration.

2.2 Field Tests

Several industrial sites participated in the *in situ* field tests, including metal pickling locations with a sulfuric and HF/HNO_3 acid pickling process and the sulfuric acid plant of a mining company. Two sets of test racks were fabricated for field tests, one set was assembled with bare metal coupons and the second set with coated metal coupons.

Representative photographs of a test rack and its field installation are shown in Figures 3 (a) and (b). Test racks were installed in the exhaust ducts for periods ranging from two weeks to three months depending on the maintenance schedule of different facilities.

® Trade name.

(a)

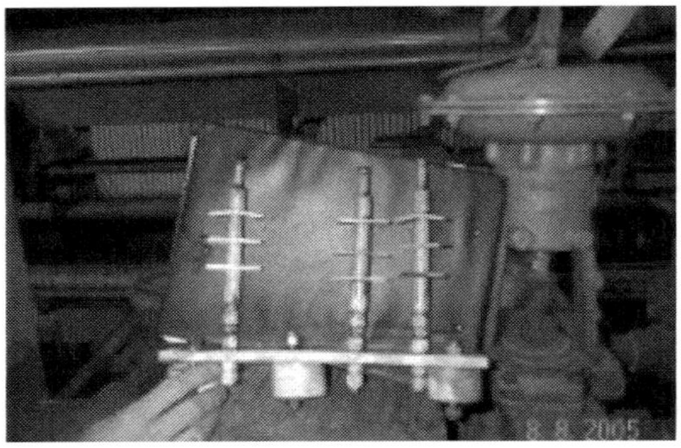

(b)

Figure 3. Representative photographs of field test rack installation. (a) Test rack in the exhaust duct of HF/HNO_3 pickling environment at the metal pickling facility. (b) Photograph of the test rack removed from the exhaust duct (a) after exposure.

3. RESULTS AND DISCUSSION

3.1 Crevice Corrosion Assembly Evaluation - Immersion Test

Figure 4 shows the crevice corrosion assembly testing results on Alloy 625 after up to 68 hours of exposure to the flue gas desulfurization simulant. Visual ranking in Table 3 was used to qualitatively differentiate alloy performance after the crevice assembly immersion test.

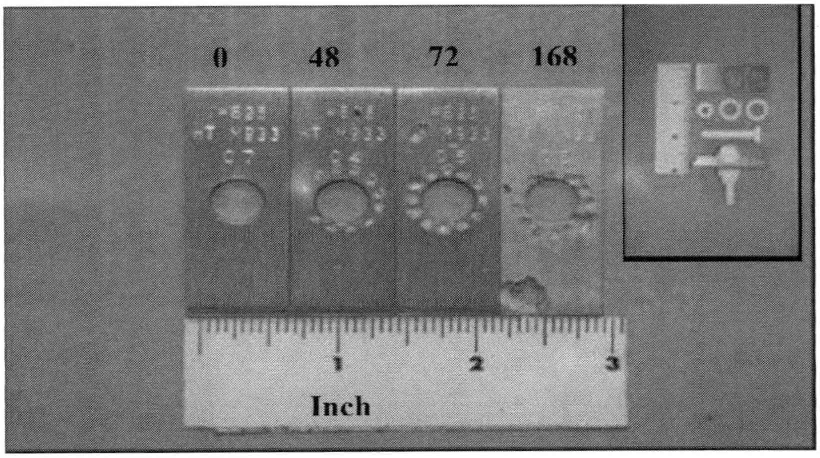

Figure 4. Alloy 625 in flue gas simulant at 75°C for 0, 48, 72 and 168 hours with a PTFE artificial crevice former (insert at the upper right shows the assembly). Note deep pits where the crevice former device contacted the sample and the through thickness pitting away from the crevice former device (far right).

The metal corrosion rate can be converted into a penetration rate per year using the measured metal weight loss of multiple crevice assembly coupons results, as shown in Table 3. This quantity is typically expressed in mils (1mil = 0.001 inch) penetration per year (mpy; 1 mpy equals 25.4 μm per year) and can be calculated from

$$MPY = 534W/D \cdot A \cdot T \qquad (1)$$

where W is weight loss in milligrams, D is metal density in grams per cubic centimeter, A is area in square inches, T is time in exposure hours and "534" converts units from (gm, cm, square inches and hour) to mil per year.[16] This equation was used to convert experimental weight loss data into corrosion rates (mpy) in Table 3.

Alloys C22 and I686 exhibited outstanding corrosion resistance (<1 mpy); C276 and CW6M exhibited excellent corrosion resistance (~ 1-5 mpy). For corrosion resistant alloys, corrosion rates greater than 5–20 mpy are considered to be excessive.[16] Much higher corrosion rates (> 100 mpy) were observed in Alloy 625 to 316L in Table 3. For corrosion resistant alloys to serve in these industrial exhaust systems, uniform corrosion is not the main failure mechanism; instead, localized corrosion is the main cause of failure. Typical localized corrosion attack examples are pitting and crevice corrosion. These constitute the most undesirable failure mechanisms as time-to-failure is difficult to predict. As shown in Table 1, vapor in these industrial exhaust systems contain strong acids with Cl^- and F^- ions which are associated with severe localized corrosion.

Table 3. Initial screening results based on flue gas desulfurization simulant.

Alloy		Flue gas desulfurization simulant				
		Crevice Corrosion Assembly Corrosion Rate¶		E_{rp}-E_{corr}	Hysteresis Loop	Visual§ Ranking
Common Designation	UNS #	MPY	mm/yr	mV vs. SSC¥		
C22	N06022	0.31	0.008	0.4178	Positive	1
I686	N06686	0.28	0.007	0.4332	Positive	1
C276	N10276	2.52	0.064	0.4223	Positive	1
CW6M	N30107	3.62	0.092	0.3796	Positive	2
625	N06625	227	5.77	0.4351	Positive	3
825	N08825	385	9.79	0.3864	Negative	5
20Cb-3	N08020	182	4.63	Negative	Negative	7
2205	S31803	292	7.41	0.5427	Positive	4
904L	N08904	112	2.85	0.3165	Negative	6
316L	S31603	168	4.27	Negative	Negative	8

#Corrosion rate, mil/yr, by weight loss — ¶ at 75°C

¥Silver-silver chloride (SSC) reference electrode

§The visual ranking in (lab exposures) assigned "1" to those samples that had not obviously failed (pitting/cracking) and qualitatively ranked the remaining coupons (all deemed to have failed by virtue of having pitted) from least to most severely pitted.

3.2 Electrochemical Measurements

Candidate materials were also subjected to cyclic potentiodynamic polarization (CP) scans to probe their susceptibility to localized corrosion. The preferred criteria for metals to resist localized corrosion from CP results include positive hysteresis loop, high breakdown potential (E_{bd}), high repassivation potential (E_{rp}) and high value of E_{rp}-E_{corr} (corrosion potential) . CP scans of 316L in the flue gas simulant indicated a negative hysteresis loop, as displayed in Figure 5 (a). This means that the protective oxide film on the surface of 316L is not stable and 316L is thus susceptible to pitting/crevice corrosion in the simulant solution. On the contrary, positive hysteresis loops, from CP scans, were observed for C276 in the simulant, as shown in Figure 5 (b). This indicates that the oxide film on the C276 surface is passive (stable) and C276 is not susceptible to pitting/crevice corrosion in this environment.

Based on the results in Table 3, the best performance with respect to pitting/crevice corrosion is expected to be Alloy C22, I686, C276, and CW6M (in this order). Actual field tests were subsequently performed to further evaluate alloy performance under real service environments.

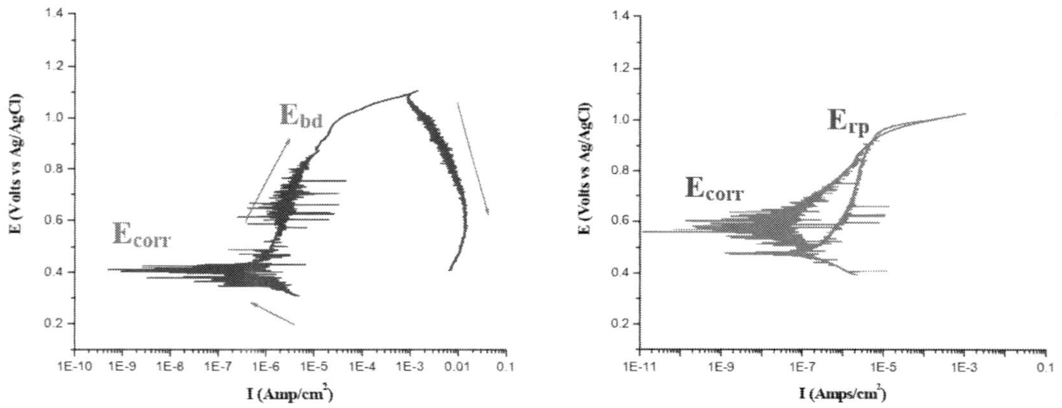

(a) **(b)**

Figure 5. (a): Cyclic polarization scan of 316L with negative hysteresis loop in flue gas simulant solution. (b): Cyclic polarization scan for C276 alloy with positive hysteresis loop in flue gas simulant solution.

3.3 Field Coupon Test Racks

Uncoated coupons

Test racks for uncoated (bare) coupons were used at three test sites for field evaluation. They were installed inside the exhaust duct system as shown in Figure 3 (a). After retrieval from the test sites, the racks were returned for further evaluation. Metal coupons were removed from each rack, cleaned, weighed, and evaluated under optical microscopes and by scanning electron microscopy (SEM).

In the sulfuric acid pickling environment, all four alloys showed very low corrosion rates (< 1 mpy), as summarized in Table 4. Figures 6 (a)-(d) display SEM pictures of the tested coupons with polishing grooves still visible on the metal surface. Only CW6M experienced minor pitting corrosion; a small pit is observable near the center-right of Figure 6 (d). The results indicated that the oxide films of C22, I686 and C276 were passive and that these alloys can survive long-term exposure in the sulfuric acid pickling environment.

In the HF/HNO_3 environment, pitting corrosion was apparent for all four alloys as displayed in Figure 6 (e)-(h). Most of the coupons' surfaces have reacted with the environment, and only C22 still has polishing grooves visible on its surface. The corrosion rates were very high for C276 and CW6M, up to 125 mpy and 108 mpy, respectively as shown in Table 4. In addition, C22 and I686 experienced 27.5 mpy and 58.2 mpy corrosion rates, respectively, in the HF/HNO_3 environment.

In general, the corrosion rate differences for similar environments are likely to be caused by the different processing temperature for each facility, with a lower processing temperature corresponding to a lower corrosion rate. The HF/HNO_3 pickling bath for location three was at ambient temperature instead of 60°C or 82°C for the other two locations, and most coupons experienced a lower corrosion rate at this site.

Table 4. Field exposure testing results for uncoated coupons.

Alloy	Steel Pickling Operation Environment											
	H_2SO_4, 96°C[1] 2 weeks		HF/HNO_3, 60°C[1] 2 weeks		H_2SO_4, 93°C[2] 4 weeks		HF/HNO_3, 82°C[2] 4 weeks		H_2SO_4, 52°C[3] 10 weeks		HF/HNO_3, ambient temp[3] 10 weeks	
	mpy[#]	SEM	mpy[#]	SEM	mpy[#]	SEM	mpy[#]	SEM	mpy	SEM	mpy[#]	SEM
C22	0.30	a	15.9	b	0.45	a	27.5	b	0.039	a	3.81	b
I686	0.30	a	16.7	b, d	0.80	a	58.2	b, d	0.047	a	5.22	b
C276	0.71	a	65.0	b	1.61	a	125	b, c	0.093	a	16.68	b
CW6M	0.63	b	32.1	b	3.84	b	108	b, c	0.257	a	35.58	b

[1]Location one, [2] Location two, [3] Location three

[#]Uniform corrosion rate, mil/yr, by weight loss

(a) Uniform corrosion, (b) Pitting corrosion, (c) Stress corrosion cracking, (d) Intergranular corrosion

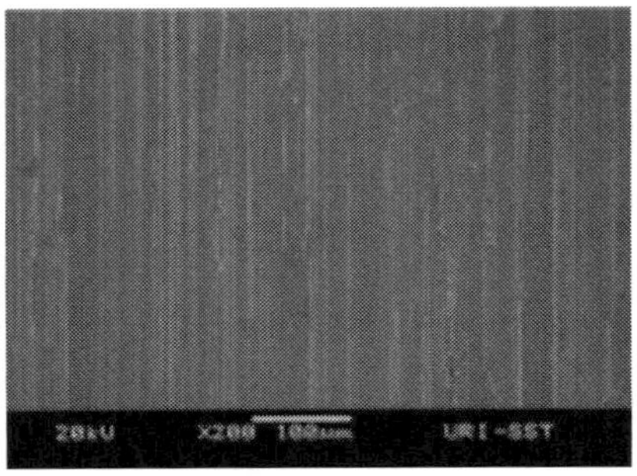

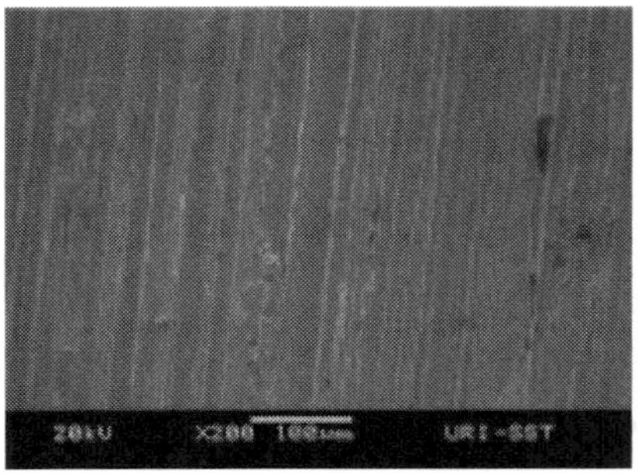

(a) C22

(b) I686

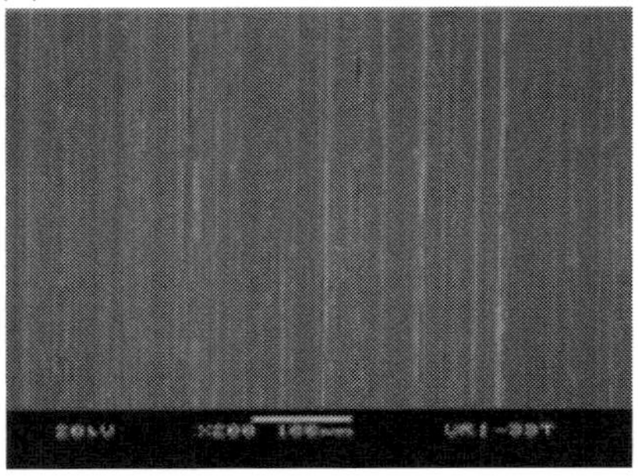

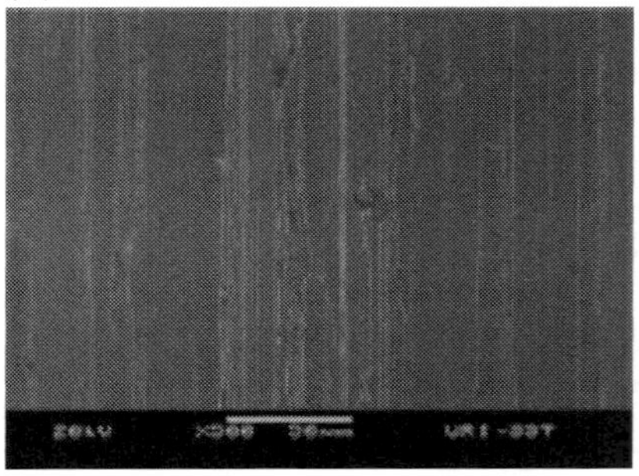

(c) C276

(d) CW6M

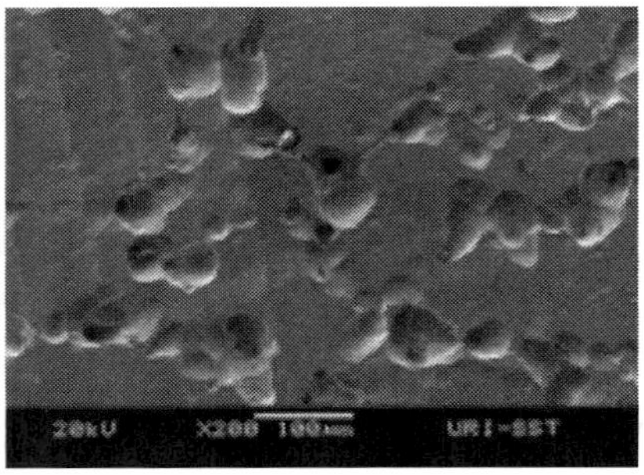

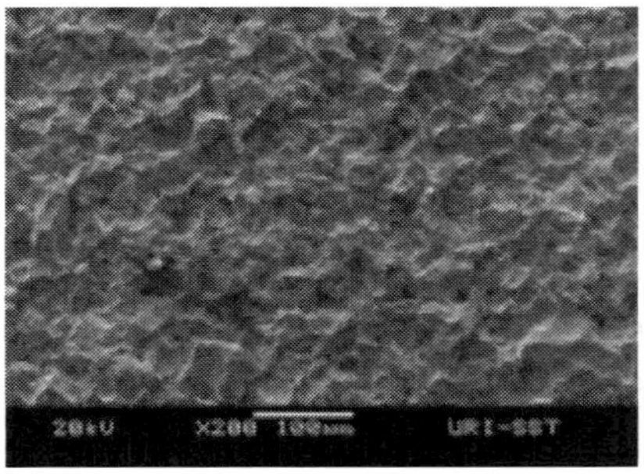

(e) C22

(f) I686

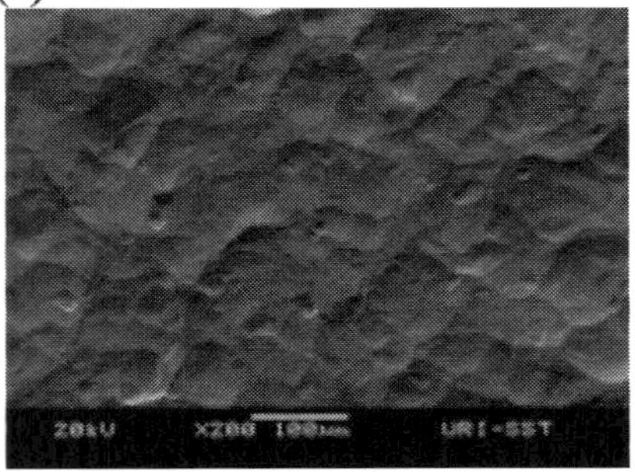

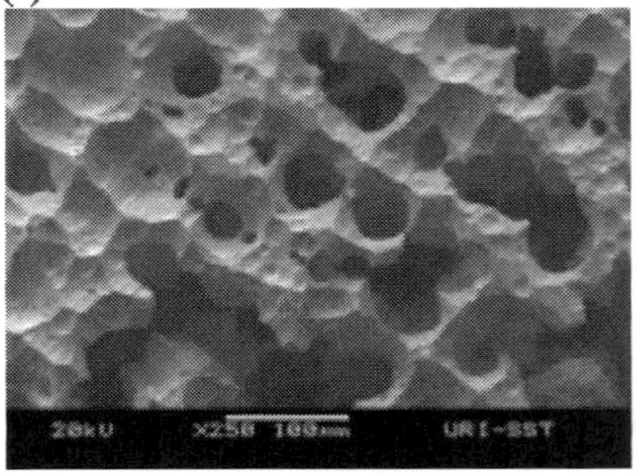

(g) C276

(h) CW6M

Figure 6. Representative SEM images of coupons exposed to environments of H_2SO_4 for two weeks - (a) to (d), and HF/HNO_3 for 4 weeks - (e) to (h). (a) C22, no corrosion. (b) I686, no corrosion. (c) C276, no corrosion. (d) CW6M, minor pitting corrosion with a pit visible near center-right of the picture. (e) C22, pitting corrosion. (f) I686, pitting corrosion. (g) C276, high corrosion rate with shallow pits. (h) CW6M, severe pitting corrosion.

Coated coupons

To further reduce metals' corrosion in the HF/HNO_3 environment, four coating systems were selected to coat the metal substrate based on literature reviews[17,18,19] including ethylene-chlorotrifluoroethylene (ECTFE, or Halar®), ethylene-tetrafluoroethylene (ETFE), perfluoralkoxy (PFA) and epoxy, and they were applied to the best performing alloy determined by the corrosion tests described above.

Two test racks were installed in the acid pickling ductwork at location three for approximately three months. The coupons are shown in Figure 7 (a). No visible coating degradation can be observed for coupons exposed to the sulfuric acid environment. Another test rack was exposed to a mixed acids pickling tub with 0.5-4% hydrofluoric acid plus 6-18% nitric acid (HF/HNO_3) at near ambient temperature for three months, with the results shown in Figure 7 (b). Visible blisters are present on the PFA and epoxy coated coupons. (Some of the coating on the PFA coated coupon has been removed, exposing the underlying metal.)

Figures 8 (a) to (d) display representative SEM images of the coated coupons' metal substrate exposed to the HF/HNO_3 pickling environment. No metal corrosion was found on the sandblasted metal surface of the ECTFE and ETFE coated coupons, as shown in Figures 8 (a) & (b). However, corrosion was observed on coupons with PFA and epoxy coatings because of coating delamination (blistering), as shown in Figure 8 (c) & (d), where the sandblasted marks on the metal surface were partially corroded away. Table 5 summarizes the field testing results. All four coating systems display good environmental resistance to the sulfuric acid (H_2SO_4) pickling atmosphere, with no visible delamination (blistering) of the coatings or corrosion of the metal substrate. Blistering of the coating normally indicates that the aggressive fluid has penetrated through it and accumulated underneath it; subsequently, this may cause the metal substrate to corrode locally.

Table 5. Field testing results for coated coupons.

		Exposure Environment					
		H_2SO_4, 52°C¥ 3 months		HF/HNO_3, Ambient Temp¥ 3 months		HF/HNO_3, 71°C§ 3 months	
Coating	Thk*	Coating	SEM	Coating	SEM	Coating	SEM
ETFE	40.4	a	c	a	c	a	c
PFA	21.8	a	c	b	d	b	d
ECTFE	31.8	a	c	a	c	a	c
Epoxy	27.1	a	c	b	d	d	d

*Average Coating Thickness (mils)
(a): intact coating; (b): blistering, coating delamination
(c): no substrate corrosion; (d): substrate corrosion--pitting corrosion
¥ Location one § Location three

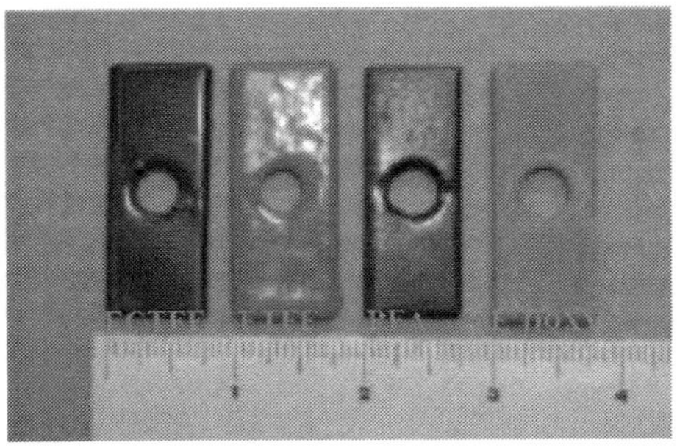

(a) (b)

Figure 7. Photographs of coupons from test racks installed at location three testing site. (a): Photograph of coated coupons installed in the H_2SO_4 pickling environment; no coating degradation. (b): Photograph of coated coupons removed from field test rack installed in the HF/HNO_3 pickling environment; visible blisters for PFA (coating shown partially removed) and epoxy coatings.

(a) (b)

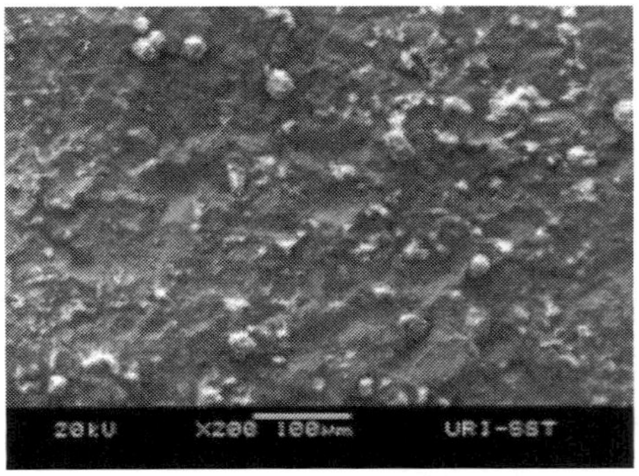

(c) (d)

Figure 8. Representative SEM images of coupons exposed to HF/HNO_3 environment for three months at location three testing site. (a): ECTFE coated substrate; no corrosion observed on the sandblasted metal surface. (b): ETFE coated substrate; no corrosion observed on the sandblasted metal surface. (c): PFA coated substrate under the blisters; corroded surface. (d): Epoxy coated substrate under the blisters; corroded surface.

4. CONCLUSIONS

Based on laboratory and field testing results, high nickel alloys such as Alloy C22 demonstrated good corrosion resistance in environments like industrial flue gas desulfurization simulant and sulfuric acid exhaust fumes, but not as well in the nitric/hydrofluoric acid processing environments. In addition, fluoropolymer type coatings such as ECTFE and ETFE can effectively protect the underlying alloy substrate from corrosion in flue gas desulfurization simulant, sulfuric acid exhaust fumes and nitric/hydrofluoric acid exhaust fumes. These test results provide the basis for the development of fire protection hardware for extremely corrosive industrial exhaust duct environments.

5. ACKNOWLEDGEMENTS

The authors thank J.C. Harrington, David Fuller, Jon Barna, Mitchell Bachuchin, Scott Baker, Larry Moore, Terry Cooper, John Sharland, Harold Moore, Thomas McCarty, and Blair Swinnerton of FM Global for their contributions to this work.

6. REFERENCES

1. FM Global Loss Prevention Data Sheet, 7-78, "Industrial Exhaust System", 1-14, September 2000.
2. Agarwal, D. C., "Alloy Selection Methodology and Experiences of the FGD Industry in Solving Complex Corrosion Problems: The Last 25 Years", Corrosion 96, NACE P 00447, 1996.
3. Metals Handbook, 9th ed., Vol. 13, 641-657, ASM International, Metals Park, OH, 1987.
4. Understanding the Hazard, "Fire in Industrial Exhaust Systems", FM Global, Johnston, RI, 2006.
5. Understanding the Hazard, "Plastics in Construction", FM Global, Johnston, RI, 2004.
6. Understanding the Hazard, "Heated Plastic Tanks", FM Global, Johnston, RI, 2004.
7. Kehler, B.A, Ilevbare, G.O. and Scully, J.R., "Crevice Corrosion Stabilization and Repassivation Behavior of Alloy 625 and Alloy 22", Corrosion, Vol. 57, No. 12, 1042-1065, 2001.
8. Pitt, W. G. and Andersen, T. N., "Corrosion of Alloys in Simulated Smelter FGD Scrubber Solutions", Materials Performance, No. 5, 26-29, 1982.
9. Phull, B. S., Mathay, W. L. and Ross, R.W. Ross, "Corrosion Resistance of Duplex and 4-6% Mo-Containing Stainless Steels in FGD Scrubber Absorber Slurry Environments", Corrosion 2000, NACE P 00578, 2000.
10. Sridhar, N., Wu J. B. C. and Corey, S. M., "The Effect of Acid Mixtures on Corrosion of Nickel-Base Alloys", Materials Performance, Vol. 26, No. 10, 17-23, 1987.
11. Agarwal, D. C., Herda, W. R. and Berry, R. W., "Reliability/Corrosion Problems of FGD Industry Cost Effective Solutions by Ni-Cr-Mo Alloys & an Advanced 6 Mo Alloy 31", Corrosion 2000, NACE P 00574, 2000.
12. ASTM G48-00, "Standard Test Methods for Pitting and Crevice Corrosion Resistance of Stainless Steels and Related Alloys by Use of Ferric Chloride and Solution", American Society for Testing and Materials, Philadelphia, PA, 2004.
13. ASTM G61-86, "Standard Test Method for Conducting Cyclic Potentiodynamic Polarization Measurements for Localized Corrosion Susceptibility of Iron-, Nickel-, or Cobalt-Based Alloys", American Society for Testing and Materials, Philadelphia, PA, 2004.
14. Dunn, D. S., Cragnolina, G. A. and Sridhar, N., "An Electrochemical Approach to Predicting Long-Term Localized Corrosion of Corrosion-Resistant High-Level Waste Container Materials", 90-104, Corrosion 56, 2000.
15. Su, P. C. and Devereux, O. F., "Alternating and Direct Current Electrochemical Studies of a Wool Wax-Based Corrosion Preventive Coating on Aluminum Alloy 2024", 419-427, Corrosion 54 1998.
16. Fontana, M. G., Corrosion Engineering, 3rd Ed., McGraw-Hill, New York, 171-173, 1986.

17. Schweitzer, P. A. editor, "Corrosion and Corrosion Protection Handbook", Marcel Dekker, New York, 538-545, 1989.
18. Metals Handbook, Vol. 13A ,814-850, ASM International, Metals Park, OH, 2003.
19. Stratmann, M., "Corrosion Stability of Polymer-Coated Metals – New Concepts Based on Fundamental Understanding", 1115-1126, Corrosion 61, 2005.

Prediction of Source Location and Rate of a Chemical Release for Emergency Response: Back Trajectory Approach

Shahryar Khajeh Najafi
Camarillo, California
shah@safersystem.com

Suneetha Burla
Camarillo, California
sburla@safersystem.com

ABSTRACT

With the ever-growing chemical industry, chemical storage and transportation may pose a threat to neighboring communities should there be an accidental release. In order to safeguard the workers and communities, agencies such as the National Institute for Occupational Safety and Health (NIOSH) and the Environmental Protection Agency (EPA) have made it mandatory for all industrial facilities to comply with their listed rules and regulations. The Emergency Planning and Community Right-to-know Act (EPCRA) requires that the public be given access to information on chemicals at individual facilities for emergency planning in case of episodic releases at the site. A large number of state-of-the-art modeling technologies and engineering tools have been developed and are readily available to successfully predict and track the cloud or plume of a chemical during an accidental release using only information on the release scenario and the meteorological data at the time of release. These models provide an assessment of the impact of a release and are essential and extremely useful for emergency planning in order to save lives and resources in the event of a release.

After the events of September 11, 2001, the paradigm has shifted to predicting an intentional release of a toxic chemical, which has potentially high impact if the targets are public locations such as football stadiums, shopping malls, and army bases located either domestically or internationally. In such cases, identification of the source location and the rate of release becomes a big challenge for hazmat personnel tasked with responding to the chemical attack. This is in great contrast to situations involving an accidental chemical release where the source of a release is known, or, in most cases, can be quickly identified.

In an attempt to find a solution to this problem, a method is proposed for locating the source of a chemical using only the basic knowledge of meteorological information (wind speed, direction and stability) and two or more concentration measurements of the chemical released. The technique used in this approach is called the reverse corridor or back trajectory method. This method uses the horizontal dispersion information obtained from surface meteorological data at receptor locations to narrow the search domain for the release location, thus reducing the computational cost. The effects of the main parameters of dispersion, such as wind meander and fluctuations in sensor concentration on the sensitivity of the model, are discussed. This model establishes a confidence level for prediction of a release location under varying circumstances.

The data regarding prediction of the source location and its associated confidence level are valuable pieces of information for emergency responders when seconds count.

1. INTRODUCTION

Source characterization is very critical for emergency response when there is an accidental release or leak of a toxic or hazardous chemical in a plant site, when a chemical is being transported to a different location, and when there is a terrorist attack in a public location such as a football stadium, a shopping mall, or a railway station. In order to identify and characterize the source parameters of a toxic chemical release several research groups have developed and tested various models and methods for source location and the release rate estimation based on the downwind concentration measurements (Shah and Gilbert 2003, Padmanathan et al 2004, Sue and Young 2008 and Rao 2007). Such techniques include Back Trajectories and Cluster Analysis, Bayesian approaches, Variational and Adjoint methods, and Genetic Algorithms. These techniques can be primarily categorized into forward techniques which predict source characteristics from the source to the receptor (sensor) and backward models which invert from the receptor to the source.

This paper mainly focuses on investigating the feasibility and accuracy of locating the source of a chemical release using only basic information on the meteorological data and two or more concentration measurements of the chemical released. The reverse corridor or the back trajectory technique is used to narrow the space domain for the source location and a Gaussian model is used to calculate the concentration at points of interest.

Section 2 presents the methodology for constructing back trajectory from horizontal dispersion parameters obtained from surface meteorological data. The criteria for identifying the possible zone for a release location are discussed. Finally, a method is proposed for locating the release source by searching for the global minimum error between the measured and predicted concentrations.

Section 3 describes the experimental set-up used for current study. The results predicted by the model under several test conditions are presented in section 4. First, the conventional straight-line Gaussian diffusion model is tested under constant meteorological conditions. Considering the non-stationary diffusion conditions, the model is then tested by introducing some random noise to the measured data. The Gaussian model is used for forward and backward calculation of concentration measurements. Section 4.1 includes testing of the model without considering any fluctuation at the measuring points, section 4.2 describes the effect of sensor position on the model sensitivity and section 4.3 presents the results when the model is tested with different levels of noise due to non-stationary nature of meteorological data. Finally in section 5, the conclusions of the work and the usefulness and robustness of the current model are summarized.

2. METHODOLOGY

2.1 Constructing the Back Trajectory

Back trajectory formulation is based on surface meteorological data. Figure 1 represents the methodology for constructing the corridor. The horizontal plume dispersion parameter σ_y, based on ten minute averaging, is obtained from the equation of the form

$$\sigma_y = ax^b \tag{1}$$

Where x is the distance, and a, b are the constants determined by the Pasquill stability condition. The values of a are 0.4, 0.36, 0.34, 0.32, 0.32 and 0.31 for stability classes ranging from A through F respectively. The corresponding values of b are 0.91, 0.86, 0.82, 0.78, 0.75 and 0.71. The horizontal stability class is determined by the azimuthal component of wind direction. The standard deviation of the wind direction fluctuations, based on ten minute running average, is used to derive the horizontal stability. Tables related to the standard deviation of azimuthal fluctuation of wind direction to atmospheric stability are published by the EPA.

The reverse corridor is a wedge originating from the sensor location pointing in the mean wind direction as measured at the meteorological tower. If the wind is blowing from the west, the reverse corridor is drawn from the east. To calculate the wedge angle, a downward distance of x meter is assumed from the receptor against the wind direction (OA). A line segment of length $2.14\sigma_y$, where σ_y is the horizontal dispersion parameter found from equation (1) using an appropriate stability curve, is drawn perpendicular to OA at A. The wedge is then constructed by connecting the other end of the line segment to the sensor location. The factor of 2.14 sets the cloud width at which the concentration falls to about one percent of the centerline value.

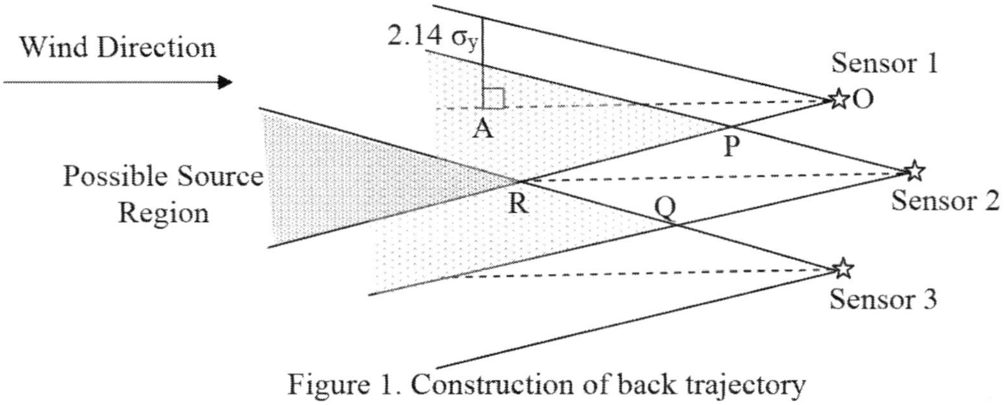

Figure 1. Construction of back trajectory

2.2 Narrowing down the Search Zone

The number of deployed sensors and their distribution play an important role in targeting the location of a chemical release. Rather than searching for an infinite space upwind of the sensors, one can utilize the sensors' distribution to narrow down the search zone.

Assume a case with three sensors: sensor 1, sensor 2 and sensor 3, as shown in Figure 1. A reverse corridor is drawn from each sensor as described in the previous section. The wedge drawn for each sensor encompasses the area where the release could have occurred. The intersection of two wedges provides a region where any release initiated within that region would impact the two sensors. The wedges drawn from each sensor intersect with each other at points P for sensors 1 and 2, Q for sensors 2 and 3, and R for sensors 1 and 3 as shown in the Figure 1. The shaded region formed by the intersection of all of the wedges from sensors 1, 2, and 3 is called the common wedge, which would contain the most probable area of the release source location.

2.3 Locating the Source

A mesh or grid that contains square cells with user definable dimensions is overlaid onto the common wedge as shown in Figure 2. The nodes on each cell define the locus of points used as possible source locations. The algorithm for source location is based on two loops, an outer loop where the rate is incremented, and an inner loop where the domain for possible source location is swept from point R up to a predefined distance in the direction opposite to the wind.

For each trial rate and trial source location, a concentration value is predicted by the Gaussian model for each impacted sensor. Actual measured concentration at each sensor is compared against the model prediction, and the error between the two values is recorded. The minimum recorded value obtained from the inner loop is labeled a local minimum. For each trial rate the procedure is repeated until the absolute minimum, called global minimum, is achieved. This is a very intensive computing process. In order to simplify the process, only the ground level release is studied and the Gaussian model was selected due to its convenience and rapid response. As an alternative, any other model can also be used at increased computational costs.

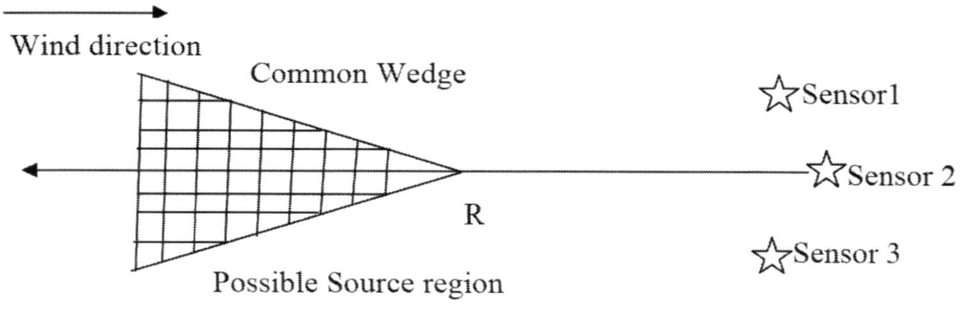

Figure 2. Source Location by back trajectory

3. EXPERIMENTAL SET-UP

To test the feasibility and accuracy of the source location model, simulated releases are carried out using the Gaussian plume model. This is done with the understanding that using the Gaussian model in both forward simulation and the reverse method biases the result in favor of the reverse method. A Cartesian co-ordinate system (x, y) (z=0, for ground release), where x represents the downwind distance and y represents the cross-wind distance from the source location, is used with the x-axis oriented in the mean wind direction and the y-axis in the horizontal cross-wind direction.

The forward Gaussian model is used to compute the concentration of a chemical release at each sensor position from a known location and for a known release rate. This concentration value is called as "measured" concentration. A wide range of possible values for the release rates are chosen, ranging from 0.01 through 10 kg/s, with an increment of 0.01 Kg/s. The search domain is limited to a distance of 10 km from the common wedge apex. A total of 1000 different release rates are considered for the test. For each rate, the entire possible source domain is swept to find the local average minimum error between the predicted and the measured values for all sensors. The absolute global minimum obtained from the simulation of 1000 rate changes yields the most possible source locations and their corresponding release rates.

Three sensors S_1, S_2 and S_3 are located within the plume at a distance (x_1, y_1), (x_2, y_2) and (x_3, y_3) respectively from the source O (0, 0), as shown in the Figure 3, such that they are impacted by the chemical.

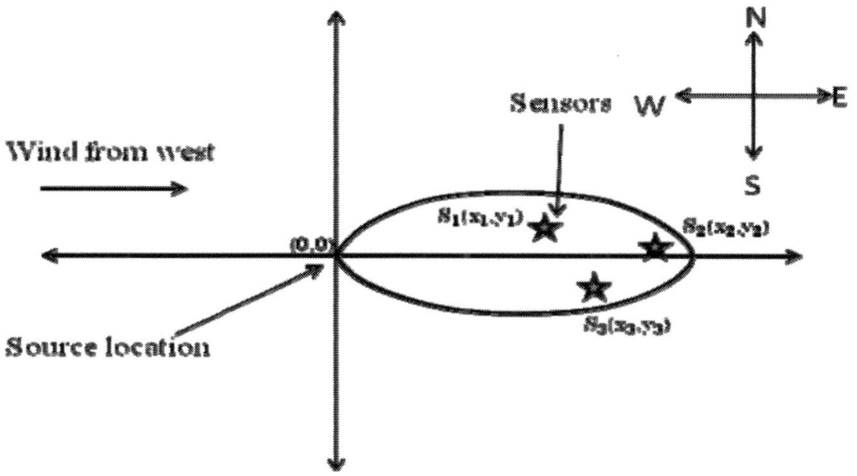

Figure 3. Plume/Sensor impact

The concentrations at each sensor location, $C_1(x_1, y_1)$, $C_2(x_2, y_2)$ and $C_3(x_3, y_3)$ can be calculated using the Gaussian form of plume equation:

$$\langle C\rangle(x, y, z) = \frac{Q_m}{2\pi\sigma_y\sigma_z u} \exp\left(-\frac{y^2}{2\sigma_y^2}\right) \times \left\{ \exp\left[-\frac{(z-H_r)^2}{2\sigma_z^2}\right] + \exp\left[-\frac{(z+H_r)^2}{2\sigma_z^2}\right] \right\}$$

(2)

$\langle C\rangle(x, y, z)$ = Average concentration (15 min)
Q_m = Release rate (Kg/s)
σ_y, σ_z = Dispersion coefficients = f (stability class, downwind distance)
u = Wind speed (m/s)
y, z = Coordinates (m)
H_r = Release height (m)
For the case in study, z = 0 and H = 0, hence the equation (2) can be simplified to

$$(C)(x, y, z) = \frac{Q_m}{2\pi\sigma_y\sigma_z u}\exp\left(-\frac{y^2}{2\sigma_y^2}\right)$$

(3)

4. TESTING MODEL VALIDITY

In order to test the validity of the model we present some numerical calculations and compare them with the model prediction. The model is tested for the following conditions:

1) Wind speed (1.5m/s, 3m/s and 6m/s)
2) Wind direction from west
3) Stability Classes 1 through 6 (A through F)
4) Different levels of concentration fluctuations (40% and 80% noisy data)

4.1 Case Study (1): Ideal case scenario with no sensor data fluctuations

Three sensors located at (100, 21), (300, 49) and (500, 50) from the actual source location are chosen and the model is tested for the release rates of 0.01, 0.1, 1.0 and 10.0 Kg/s.

This is an ideal case where we do not expect any inherent fluctuations in the sensor data due to turbulence and the sensor data is free of any noise. Under these conditions, the concentration measured by the sensors is the same as that calculated using the forward Gaussian model. This test enables us to evaluate the model alone because several sources of error are eliminated from consideration, knowing that the sensor data is exact. The predicted source locations for various stability classes from A through F, for wind blowing from the west at a speed of 1.5m/s, are plotted as shown in Figure 4 and the results are also tabulated in Table 8 in the Appendix.

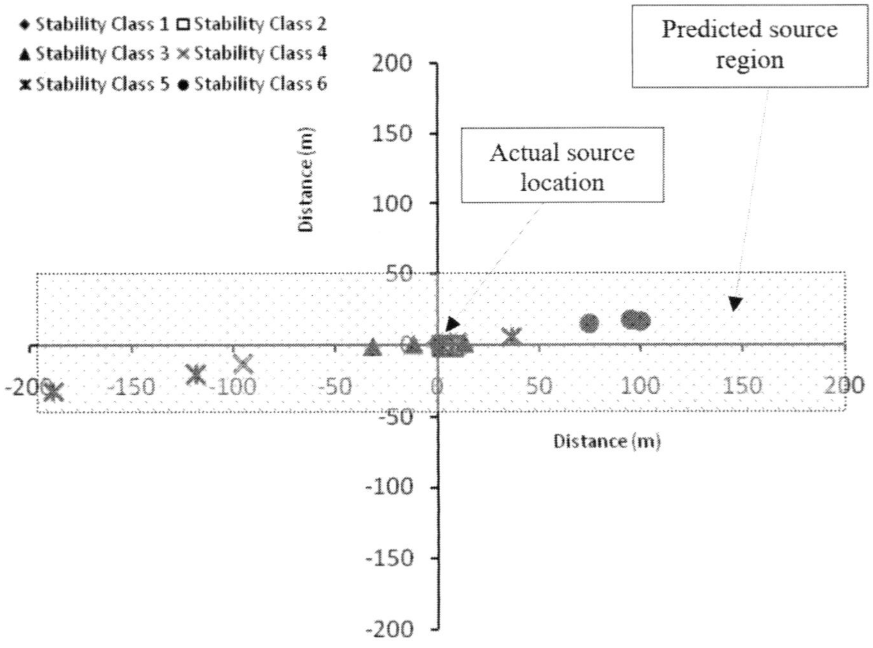

Figure 4. Comparison of the model predicted source location against the actual source location for different stability classes and release rates at wind speed of 1.5m/s

The data points for each stability class correspond to four different release rates viz. 0.01, 0.1, 1.0 and 10.0 kg/s. From Figure 4, it can be seen that the model predictions for the source location values are very close to the actual source location (0, 0). For a given release rate, the accuracy with which the model predicts the source location decreases with the increase in stability class. For lower stability classes, the predicted source locations are concentrated around the origin (0, 0) (actual source location) and as we move to higher stability classes, the predicted source locations are more dispersed and farther from the actual source location. This is especially true in cases with low release rates of 0.01Kg/s and 0.1 Kg/s and under highly stable conditions (5 and 6), where the predicted source location values are far from the actual source location values.

The release rates predicted by the model are also compared against the actual release rates for various stability classes in Table 1. It can be seen that, except for the higher stability classes (5 and 6), the release rates predicted by the model match well with the actual release rates.

Table 1. Comparison of the model predicted release rates against the actual release rates for different stability classes at wind speed of 1.5m/s

Actual Source location	Release rate (Kg/s)	Predicted rate (Kg/s)					
		Stability Class					
		1	2	3	4	5	6
(0,0)	0.01	0.01	0.01	0.01	0.01	0.04	0.01
(0,0)	0.1	0.1	0.1	0.11	0.16	0.25	0.01
(0,0)	1.0	1.0	0.98	0.96	0.94	0.98	0.11
(0,0)	10.0	10.1	9.8	9.6	9.8	7.6	1.6

Similar results are obtained when the model is tested for the wind speeds of 3m/s and 6 m/s blowing from the west. The results for the model predictions for the source location at a wind speed of 3m/s and 6m/s are plotted in Figure 5 and Figure 6 respectively. For higher stability classes (5 and 6) and at low release rates (0.01 and 0.1 Kg/s), the predicted source location values are more dispersed and farther from the actual source location for higher wind speeds of 3m/s and 6m/s. Nonetheless, the predicted source location values are still at a radius of 200m from the actual source, which is a reasonable distance limit to locate the source. A box is drawn on the graph, enclosing all the data points for source location predictions under all stability conditions. Thus the model is tested under several conditions of different wind directions, wind speeds and other sensor locations (sensor locations chosen such that they are neither too saturated nor negligibly impacted), and for all the conditions it has been observed that the source can be located within the area enclosed by the box. Hence, without loss of generality, it can be concluded that under ideal conditions of no sensor fluctuations, the model can predict the source location within a 0-50m radius from the actual source location for a wide range of wind speeds (1.5 to 6m/s), except under conditions of very high stability classes(5 an 6) and very low release rates (0.01 and 0.1Kg/s), in which the source location can be predicted within 200m radius from the actual source. The values for the predicted source locations for various stability classes ranging from A through F, with wind blowing from the west at a speed of 1.5m/s, 3m/s and 6m/s are also tabulated in Table 8, Table 9, and Table 10, respectively, in the Appendix.

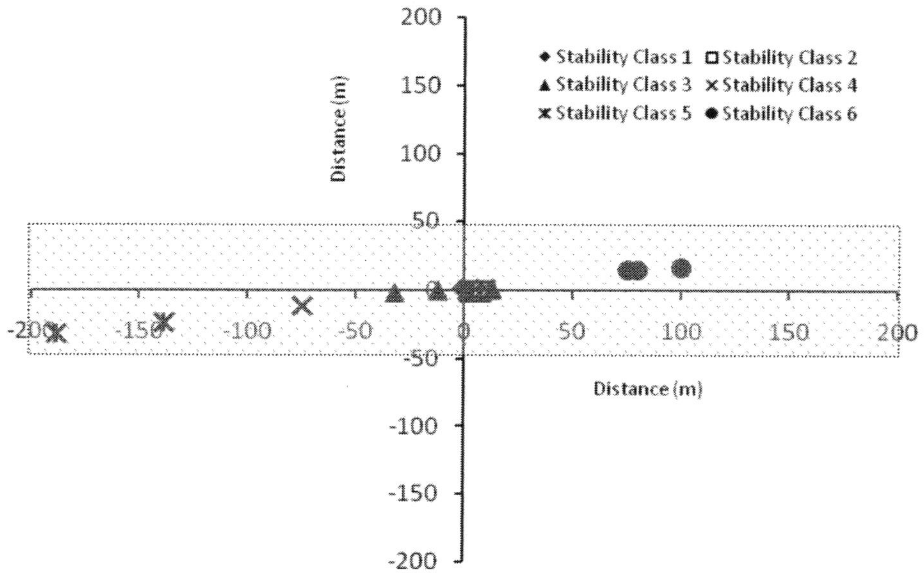

Figure 5. Comparison of the model predicted source location against the actual source location for different stability classes and release rates at wind speed of 3m/s

The predicted release rates for wind speeds 3m/s and 6m/s are tabulated in Table 2 and Table 3 respectively. The results for release rate predictions at wind speeds 3m/s and 6m/s are very similar to the results at wind speed 1.5m/s. Except for the higher stability classes (5 and 6), the model predictions for release rates match well with the actual release rates for all the stability classes.

Table 2. Comparison of the model predicted release rates against the actual release rates for different stability classes at wind speed of 3m/s

Actual Source location	Release rate (Kg/s)	Predicted rate (Kg/s)					
		Stability Class					
		1	2	3	4	5	6
(0,0)	0.01	0.01	0.01	0.01	0.01	0.04	0.01
(0,0)	0.1	0.1	0.1	0.11	0.15	0.29	0.01
(0,0)	1.0	1.0	0.98	0.96	0.94	0.98	0.11
(0,0)	10.0	10.1	9.8	9.6	9.8	9.8	1.6

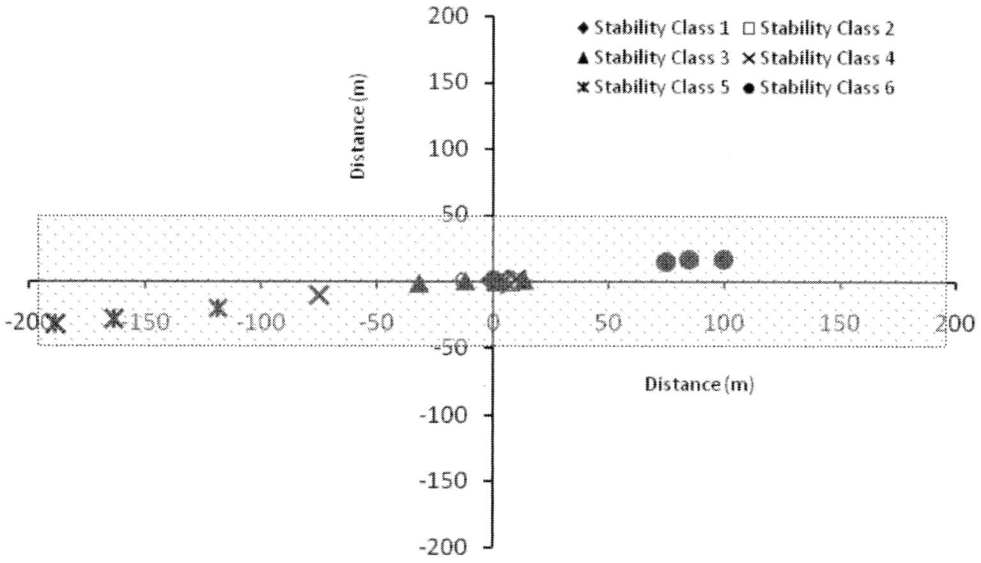

Figure 6. Comparison of the model predicted source location against the actual source location for different stability classes and release rates at wind speed of 6m/s

Similar results for the source location and the release rate predictions at all three different wind speeds (1.5m/s, 3m/s and 6m/s) indicates that the model predictions are independent of the wind speeds, in the range 1.5-6m/s.

Table 3. Comparison of the model predicted release rates against actual release rates for different stability classes at wind speed of 6m/s

Actual Source location	Release rate (Kg/s)	Predicted rate (Kg/s)					
		Stability Class					
		1	2	3	4	5	6
(0,0)	0.01	0.01	0.01	0.01	0.01	0.04	0.01
(0,0)	0.1	0.1	0.1	0.11	0.15	0.34	0.01
(0,0)	1.0	0.99	0.98	0.96	0.94	2.5	0.12
(0,0)	10.0	10.1	9.8	9.6	9.4	9.8	0.98

4.2 Case Study (2): Sensitivity to sensor location

To test the model's sensitivity to the sensor location, different locations other than those mentioned for the previous case study are selected viz. (45, 5), (70, -7) and (100, 11) for the three sensors in the study. A wind speed of 3m/s is randomly selected because we have already seen from previous test results that the model predictions are not very sensitive to the wind speed. The test results are as shown in the following Table 4.

Table 4. Comparison of the model predicted source location against the actual source location for different stability classes and release rates at wind speed of 3m/s

Actual Source location	Release rate (Kg/s)	Predicted Source Location (m, m)					
		Stability Class					
		1	2	3	4	5	6
(0,0)	0.01	(1,-1)	(-1,0)	(2,0)	(13,0)	(-10,0)	(-315,-14)
(0,0)	0.1	(1,-1)	(-1,0)	(-8,0)	(13,0)	(-10,0)	(-45,0)
(0,0)	1.0	(1,-1)	(4,0)	(2,0)	(13,0)	(-10,0)	(-45,0)
(0,0)	10.0	(1,0)	(-1,0)	(2,0)	(7,0)	(0,0)	(-45,0)

From Table 4, it can be seen that the model predictions for the source location using the current set of sensor locations are very different from the model prediction results generated using a different set of sensor locations in the previous Case study (1),especially for cases under high stability classes of 5 and 6 at very low release rates. The predicted source location is about 300m away from the actual source under stability class 6 at very low release rate of 0.01 kg/s.

Table 5. Comparison of the model predicted release rates against the actual release rates for different stability classes at wind speed of 3m/s

Actual Source location	Release rate (Kg/s)	Predicted rate (Kg/s)					
		Stability Class					
		1	2	3	4	5	6
(0,0)	0.01	0.01	0.01	0.01	0.01	0.01	0.01
(0,0)	0.1	0.1	0.1	0.12	0.1	0.1	0.03
(0,0)	1.0	0.98	0.98	0.96	0.94	0.99	0.39
(0,0)	10.0	9.8	9.8	9.6	9.6	9.6	5.8

Similarly, from Table 5 it can be seen that the model predicted values for the release rates under higher stability classes 5 and 6 have changed significantly. This is possible when the concentration values of the released chemical at the sensor location are beyond the measurable concentration range of the sensors. This indicates that the model predictions are highly sensitive to the sensor positions in the field. Hence, it is highly important to take multiple sensor readings at various random locations and care should be taken when selecting the sensors for study. The sensors are chosen such that they are neither too saturated nor negligibly impacted. Before making the decision about the final predicted source location and release rate, the model results for various sensors located at multiple locations are weighed against each other and the readings from negligibly impacted and overly saturated sensors are eliminated from consideration.

4.3 Case Study (3): Sensitivity to sensor data

The same sensor locations, as mentioned for case study 4.1, are selected viz. (100, 21), (300, 49) and (500, 50) for the three sensors in this study.

In this case, some noise is introduced into the sensor data in order to take into account the fluctuations in the sensor data due to turbulence and wind meander. The concentration data measured at the sensor locations using the Gaussian equation are corrupted by adding different levels of noise. Two cases, one with 40% noise in the concentration data and the other with 80% noise, are examined. To study the effect of 40% noise in the concentration data, for each sensor a random concentration value between $0.6C_m$ and $1.4C_m$ is chosen, where C_m is the measured concentration calculated using a forward Gaussian model for the corresponding sensor, and then the simulation was run. This accounts for cases when both the concentration is overestimated and is underestimated due to meteorological data fluctuations. Thus for this case, several random concentration values within the mentioned limit are created for every sensor under consideration. These tests allow us to evaluate the sensitivity of the model to different concentration fluctuations that are inevitable in the real world sensor data.

With the introduction of 40% noise in the sensor data, it can be seen that the model predictions for the source location values are very far from the actual source location (0, 0), for stability classes 5 and 6 at very low release rates of 0.01kg/s and 0.1kg/s. These results when compared with the ideal case study 4.1, with no concentration fluctuations and with all the other conditions the same, show that the model predictions are not significantly affected by the introduction of 40% noisy data.

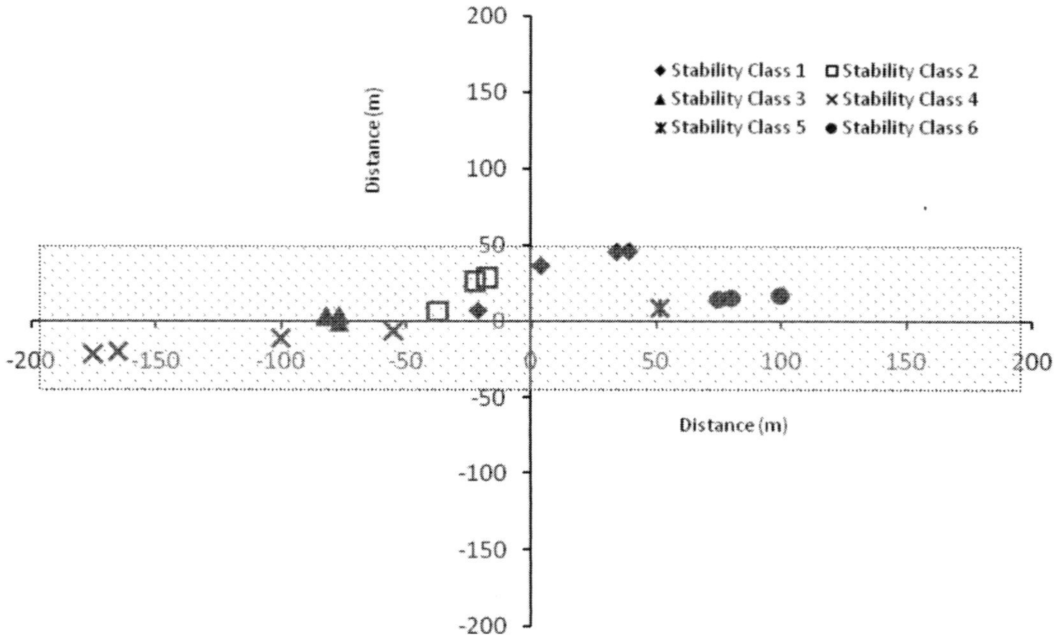

Figure 7. Comparison of the model predicted source location against the actual source location for different stability classes and release rates at wind speed of 3m/s for 40% noisy data

Table 6 shows the predicted release rates with the introduction of 40% noise into the sensor concentration readings. These results when compared against the data with no sensor fluctuations show that the predicted release rates are very different from the actual release rates and the accuracy of the release rate prediction does not follow any trend with the stability class.

Table 6. Comparison of the model predicted release rates against the actual release rates for different stability classes at wind speed of 3m/s for 40% noisy data

Actual Source location	Release rate (Kg/s)	Predicted rate (Kg/s)					
		Stability Class					
		1	2	3	4	5	6
(0,0)	0.01	0.01	0.01	0.01	0.01	0.07	9.8
(0,0)	0.1	0.04	0.04	0.08	0.11	0.24	0.04
(0,0)	1.0	0.62	0.63	0.97	1.6	10.9	0.56
(0,0)	10.0	5	5.2	6.6	11	11	8.1

Similarly, for a case with 80% noise in the concentration data, a random concentration value between $0.2C_m$ and $1.8C_m$ is chosen, and then the simulation is run. The predicted source locations and the release rates are tabulated for different levels of noise. With the increase in the noise in the sensor data to 80%, similar results were obtained for the source location and the release rate predictions.

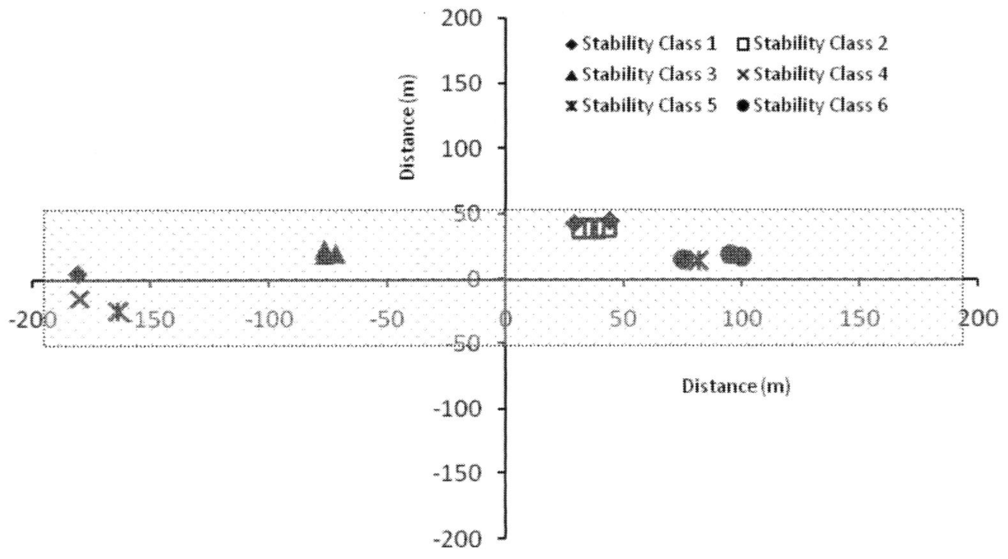

Figure 8.Comparison of the model predicted source location against the actual source location for different stability classes and release rates at wind speed of 3m/s for 80% noisy data

Table 7 shows the predicted release rates with the introduction of 80% noise in the sensor concentration readings. These results match well with the predicted release rate data with no sensor fluctuations. The results when compared with those with the 40% noisy data show that the release rate predictions with the introduction of 80% noise are much better than those with 40% noisy data.

Table 7. Comparison of the model predicted release rates against the actual release rates for different stability classes at wind speed of 3m/s for 80% noisy data

Actual Source location	Release rate (Kg/s)	Predicted rate (Kg/s)					
		Stability Class					
		1	2	3	4	5	6
(0,0)	0.01	0.02	0.01	0.02	0.05	10.9	0.95
(0,0)	0.1	0.3	0.36	0.66	5.8	11	0.04
(0,0)	1.0	0.87	0.99	1.3	1.5	1.3	0.59
(0,0)	10.0	10.4	10.5	9.9	11	10.8	7.3

Nevertheless, it can be concluded that even with a noisy sensor data, the model predicts the source parameters with a decent level of confidence, except under high stability conditions (5 and 6) and at low release rates (0.01 and 0.1Kg/s).

The values for the predicted source locations for various stability classes from A through F for 40% and 80% noisy data are also tabulated in Table 11 and Table 12 respectively in the Appendix.

5. CONCLUSION

In this paper, we explored the robustness of a relatively simple back trajectory model for constructing the reverse corridors in order to trace the source characteristics such as the release location and the release rate of a toxic chemical. Two different cases were studied in order to test the validity of the model. The first test involved an ideal case scenario where exact concentrations were provided using the forward Gaussian model. The second case was such that fluctuation due to wind meander and atmospheric turbulence at measuring points are taken into consideration. The latter was simulated by adding random noise to the measured data. In both cases, the model results match well with the actual data, in their ability to locate the source and the release rate of a chemical release.

The model was tested for different wind speeds, viz. 1.5m/s, 3m/s and 6 m/s, and the results showed that the model predictions are independent of wind speed for a given stability class and they match well with the actual data. The model was also tested for varying stability classes, A through F (1 through 6), and the results for source location showed that the accuracy of the model prediction decreases with the increase in stability class. This is particularly true under higher stability classes 5 and 6 at very low release rates.

Different prediction results for two different sets of sensor positions indicate that the model is sensitive to the chosen sensor location. The results for the source location and the release rates, when the model was simulated with the noisy sensor data, viz. 40% and 80%, matched very well with the true data (except under highly stable conditions at low release rates), indicating that the back trajectory method coupled with a simple Gaussian plume model is capable of providing a reasonable prediction of source variables.

6. REFERENCES

1. Khajeh Najafi , Shahryar, and Gilbert, Ernest J., *Use of real-time measurement for estimating release rate,* Institute of Chemical Engineers Symposium Series, American Institute of Chemical Engineers, 2003
2. Kathirgamanathan, Padmanathan, McKibbin, Robert, and McLachlan Robert I., *Source release-rate estimation of atmospheric pollution from a non-steady point source at a known location,* Netherlands: Environmental Modeling and Assessment, 2004

3. Haupt, Sue Ellen, and Young, George S., *Paradigms for source characterization*, 15th Joint Conference on the applications of air pollution meteorology with the A&WMA, New Orleans: American Meteorological Society, 2008
4. Rao, Shankar K., *Source estimation methods for atmospheric dispersion*, Atmospheric Environment, Science Direct, 2007

7. APPENDIX

Table 8.Comparison of the model predicted source location against the actual source location for different stability classes and release rates at wind speed of 1.5m/s

Actual Source location	Release rate (Kg/s)	Predicted Source Location (m, m) Stability Class					
		1	2	3	4	5	6
(0,0)	0.01	(-1,1)	(2,0)	(-12,0)	(5,1)	(-190,-32)	(100,17)
(0,0)	0.1	(-1,1)	(2,0)	(-32,-2)	(-95,-14)	(-120,-20)	(75,15)
(0,0)	1.0	(-1,1)	(8,-1)	(13,1)	(10,1)	(1,0)	(75,15)
(0,0)	10.0	(-1,0)	(8,-1)	(13,1)	(5,1)	(36,6)	(95,17)

Table 9.Comparison of the model predicted source location against the actual source location for different stability classes and release rates at wind speed of 3m/s

Actual Source location	Release rate (Kg/s)	Predicted Source Location (m, m) Stability Class					
		1	2	3	4	5	6
(0,0)	0.01	(-1,1)	(2,0)	(-12,0)	(5,1)	(-190,-32)	(100,17)
(0,0)	0.1	(-1,1)	(2,0)	(-32,-2)	(-75,-10)	(-140,-24)	(75,15)
(0,0)	1.0	(-1,1)	(8,-1)	(13,1)	(10,1)	(1,0)	(75,15)
(0,0)	10.0	(-1,0)	(8,-1)	(13,1)	(5,1)	(1,0)	(80,15)

Table 10.Comparison of the model predicted source location against the actual source location for different stability classes and release rates at wind speed of 6m/s

Actual Source location	Release rate (Kg/s)	Predicted Source Location (m, m) Stability Class					
		1	2	3	4	5	6
(0,0)	0.01	(-1,1)	(2,0)	(-12,0)	(5,1)	(-190,-32)	(100,17)
(0,0)	0.1	(-1,1)	(2,0)	(-32,-2)	(-75,-10)	(-164,-28)	(75,15)
(0,0)	1.0	(4,-2)	(8,-1)	(13,1)	(10,1)	(-120,-20)	(75,15)
(0,0)	10.0	(-1,0)	(8,-1)	(13,1)	(10,1)	(1,0)	(85,17)

Table 11.Comparison of the model predicted source location against the actual source location for different stability classes and release rates at wind speed of 3m/s for 40% noisy data

Actual Source location	Release rate (Kg/s)	Predicted Source Location (m, m) Stability Class					
		1	2	3	4	5	6
(0,0)	0.01	(-21,7)	(-37,6)	(-77,-1)	(-100,-10)	(-230,-36)	(100,17)
(0,0)	0.1	(4,36)	(-20,30)	(-82,3)	(-175,-22)	(-243,-40)	(75,15)
(0,0)	1.0	(34,45)	(-23,26)	(-82,3)	(-165,-20)	(-210,-33)	(75,15)
(0,0)	10.0	(40,46)	(-23,26)	(-77,4)	(-55,-7)	(50,9)	(80,15)

Table 12.Comparison of the model predicted source location against the actual source location for different stability classes and release rates at wind speed of 3m/s for 80% noisy data

Actual Source location	Release rate (Kg/s)	Predicted Source Location (m, m)					
		Stability Class					
		1	2	3	4	5	6
(0,0)	0.01	(-180,4)	(-230,-7)	(-380,-20)	(-122,2)	(-285,-35)	(100,17)
(0,0)	0.1	(45,45)	(32,39)	(-77,23)	(-300,44)	(-265,-26)	(75,15)
(0,0)	1.0	(30,43)	(38,38)	(-77,20)	(-12,-1)	(-245,-24)	(75,15)
(0,0)	10.0	(45,45)	(43,40)	(-72,20)	(-2,0)	(-180,-15)	(95,15)

Lessons from Safety Culture Evaluations at Process Facilities

Steve Arendt, ABS Consulting, 16855 Northchase Drive, Houston, TX 77060

Interest sparked by the BP-Baker Panel review caused many companies to undertake evaluations of their safety culture. This paper will compare the results from safety culture evaluations involving six companies and 15+ facilities from several process industries. These results were developed using a performance assurance review approach that mapped process safety performance outcomes to their culture causal factors.

The evaluations used a variety of culture data collection approaches, including surveys, interviews, and work observations. All of the studies involved the use of CCPS's essential features of a good process safety culture as the common diagnostic framework.

The lessons presented will be of use to all companies seeking to build and nurture better process safety cultures at operating facilities.

Spot The Hazard!
A Cultural Extension of Hazard Identification Training

Robert F. Wasileski
Sr. Corporate Process Safety Engineer
NOVA Chemicals, Inc.
U.S. Operating Center
Moon Township, Pennsylvania, USA 15108
wasiler@novachem.com

A 10th PPSS Anniversary Paper

ABSTRACT

Hazard Identification Training (HIT) programs have been used in the chemical processing industry to raise awareness with employees on what constitutes a process hazard and techniques for identifying these hazards in the workplace. Quite often these HIT programs will deliver the course material to the participants as classroom instruction, computer-based training (CBT), organized field reviews or a combination of these formats. These presentations tend to be one-time efforts, or cyclic programs that reach the target audience on some periodic basis (e.g., triennially). While the content of the course material may be educational, these programs often have limited success due to their inability to make a lasting impression with the participants. Organizational and personnel changes, employee turnover, vanishing corporate memory and shifting priorities can all have an undermining effect on the well-intended efforts of traditional HIT programs. The challenge thus becomes finding ways to keep the knowledge both relevant and current, in order to effect lasting cultural change in a dynamic environment.

This paper presents a unique extension to HIT known as "Spot The Hazard", which uses the facility's intranet to reinforce the concepts taught in the classroom. Spot The Hazard blends photographic examples and technical knowledge with aspects of behavior modification. The photographs that are presented have not been staged and thus, are germane to the employees. The inherent on-going nature of the program ensures the knowledge remains current, while at the same time it provides a platform for making positive cultural change.

The concepts of Spot The Hazard are explained in greater detail in the current paper, along with a representative number of actual examples from the past year.

. INTRODUCTION

This paper presents a creative and unique cultural extension to Hazard Identification Training (HIT) known as "Spot The Hazard!". The "Spot The Hazard!" Program (hereafter also referred to simply as the "Program") is a web-based monthly contest that is operated on the facility's intranet.

The Program began following a two-year HIT initiative to educate operating, maintenance and technical staff on hazard awareness and identification [1]. This initiative covered 12 operating facilities within the Corporation. Senior process safety and loss prevention professionals delivered the course material, in an interactive session lasting 6 to 7 hours. The curriculum included subject matter on fire science; sources of flammables and combustibles; sources of ignition; and principles of prevention and protection. Employees were taught the science of combustion and technical concepts that influence the combustion process (e.g., flash point temperature, auto-ignition temperature, inert atmospheres, etc.). In addition, photographs were taken at each operating facility and incorporated into the curriculum to illustrate where combustibles are found, where ignition sources may be found, and examples of prevention and protection facilities.

II. CONCEPTS AND APPROACH

The focus of the aforementioned HIT initiative was oriented towards the science of fire and preventing uncontrolled process fires in the operating facilities, as evidenced by the curriculum. However, the fundamental objective of the course was common to any other HIT program used in similar industries: to modify the thought processes and behavior of the course participants.

It is necessary to comment on the difference between "education" and "training". The name "Hazard Identification Training" is actually a misnomer, yet it is routinely used in the chemical processing industries to describe this type of initiative. Training aims "to impart a set of established facts and skills and to obtain a uniform predictable behavior from the trainees without the necessity of their understanding why they should act in the prescribed manner" [2]. Education, on the other hand, "seeks to have the student learn skills and to understand why actions are taken or not," [2] and to "observe, analyze and question, to formulate hypotheses and make conclusions and then to act, live and modify their actions according to these conclusions" [2].

The significant difference between the two terms, therefore, lies in the concept of *understanding*. Understanding the principles presented in any educational program is absolutely necessary for application of the knowledge. In particular with hazard recognition, it is not sufficient to simply train employees to recognize the same hazards in the field as those presented in the classroom. Employees must understand the principles being presented, and then be able to translate that knowledge to the field to identify hazards in new and unique situations based on the learned principles. It is only when employees *understand* the fundamental principles that they can begin to apply the knowledge to different situations [3]. In turn, the behaviors and thought processes of the employees begin to change. When the collective organization inherits this knowledge and consistently applies it to identifying and evaluating hazards, true cultural change can occur.

III. SAFETY CULTURE

Two sources frequently cited for definitions of "safety culture" are the UK Health and Safety Executive and the Center for Chemical Process Safety (CCPS). The UK Health and Safety Executive definition is "…the product of the individual and group values, attitudes, competencies and patterns of behavior that determine the commitment to, and the style and proficiency of, an organization's health and safety programs" [4]. The CCPS suggests a direct and more succinct definition: "Safety culture is how the organization behaves when no one is watching" [5]. These definitions are specific to *safety* culture, and while they are somewhat different, they share the common notion that organizational culture is a reflection of the shared common values that drive organizational performance [6]. In effect, the process safety performance of an organization is a function of the degree to which the collective organization embraces the process safety policies, programs and infrastructure in place.

Leadership is an essential component of managing culture, and plays a significant role in influencing cultural change. The CCPS [5] recognizes that leadership cannot effect cultural change with mandates and edicts. Rather, positive movement comes from "leadership by example", where the desired values and behaviors are advanced from being vision and set into action when leaders *model* the vision. Successful leaders inspire, enable and challenge the organization towards improvement. It is no coincidence then that the model for transformational leadership [6] includes these concepts in its four dimensions: Charisma, Inspiration, Individual Attention and Intellectual Challenge. Transformational leadership applies these characteristics – charisma, inspiration, individual attention and intellectual challenge – consistently and frequently to a model, in order to gradually move the values of the collective whole in a positive direction.

The key contributors to successful transformational leadership lie in *consistency* and *frequency*. The vision and the message must be consistent. The communication and positive reinforcement must be frequent. This combination yields change. In the context of Hazard Identification Training (HIT), this framework translates to a curriculum of hazard identification principles and techniques that are consistently communicated, and reinforced on a frequent basis with employees. However, this is often where traditional HIT programs fall short in their ability to effect cultural change. HIT tends to be either a one-time initiative, or one of many required training courses that is scheduled on a periodic basis (e.g., triennially) for employees. The course material is simply delivered too infrequently to keep the organization focused. Furthermore, the effectiveness of HIT programs can be compromised by organizational and personnel changes, employee turnover, vanishing corporate memory and shifting business priorities. These factors create significant challenges for process safety professionals, and often limit the success of HIT programs in terms of shifting the culture of the organization.

IV. PROGRAM DESCRIPTION

"Spot The Hazard!" places photographs of actual hazardous conditions found in the process on a web page of the intranet. Employees are given the opportunity to "spot the hazard" in the photograph, and submit their answers electronically through the integrated e-mail system. At the end of the contest period, one winner is randomly selected from among the correct responses to receive a reward. The winner is announced to the entire facility through e-mail, along with a detailed explanation of the process hazard, the reason why it is a hazard, and recommended actions to correct the concern. Every employee that submitted a correct response is also recognized in a positive manner in the winning e-mail announcement.

The curriculum from the previously discussed HIT initiative forms the foundation for the "Spot The Hazard!" Program. The photographs generally aim to capture potential sources of combustible materials, ignition, hazardous exposures and the like. The e-mail communications are used to announce the contest winner, but more importantly reiterate the technical principles and knowledge that support the identification and evaluation of process hazards.

With each new contest, there is an opportunity to *consistently* demonstrate how the knowledge has been applied to yet another unique and different situation. Furthermore, the key principles and relevant knowledge are *frequently* reinforced with employees, to maintain a collective focus on hazard identification and risk assessment (HIRA).

Figure 1. Example Screen Shot of Spot The Hazard! Web Page

CORE COMPONENTS

The core elements of the "Spot The Hazard!" Program include:

1. *Hazard Identification*. Photographs of process hazards identified at the facility form the foundation of the Program. The photographs are never 'staged' or simulated. Actual conditions are photographed as they are found. Generally the Site Process Safety Engineer will identify these conditions during Planned General Inspections (PGI) or ad-hoc surveys. However, all employees are encouraged to submit findings for future contests, and this does in fact occur.
2. *Risk Assessment*. Once a hazardous condition has been identified, Operations is consulted for a simple risk assessment. This serves two purposes. First, the risk assessment determines if the concern is legitimate and the acceptability (or unacceptability) of the risk. Second, it ensures buy-in from the asset owner and key stakeholders before the photograph is used on the intranet.
3. *Problem Solving*. The risk assessment discussion will, quite often, prompt recommendations to correct the deficiency. The asset owner is responsible for implementing corrective action(s), and given ample time to do so before the photograph is used on the intranet.
4. *Employee Participation*. Employee Participation is one of the fundamental elements of any process safety program, and the "Spot The Hazard!" web page provides an additional means to accomplishing this. The web page is accessible to all Site employees, and everyone is encouraged to submit a response through the integrated e-mail system.
5. *Process Safety Education*. At the end of each monthly contest, the winner is announced in a site-wide e-mail. This communication is the vehicle for continuing education. Here, details are provided to employees regarding what the process hazard was, why it is a hazard, and how the hazard can be eliminated or how the risk can be mitigated. This stage provides the educational component to the Program, to **consistently and frequently connect with employees**.

LEADERSHIP AND CULTURAL PARALLELS

Aside from the obvious benefits, such as gains in employee participation and education on hazard identification, the Program has progressively demonstrated a number of 'soft benefits'. These less tangible, but equally important, benefits include an increase in social capital, new leadership platforms and positive cultural shift.

1. *Social Capital*. The term "social capital" refers to a core concept in organizational behavior, and may be described as the existence of a specific set of informal values or norms shared among members of a group that permits cooperation among them. Social capital is indicated by consensus, collective action, and the amount of trust demonstrated in a community or among individuals. Unlike other forms of capital that are depleted by use, social capital is depleted by non-use. Harvard political scientist Robert D. Putman [7] describes two types of social capital: 'bonding social capital' and 'bridging social capital'. Bonding social capital refers to the value of social networks within homogenous groups, whereas bridging social capital refers to that value between socially heterogeneous groups. Bonding social capital explains why some groups become isolated and disenfranchised from the rest of society (or an organization). Bridging social

capital, on the other hand, works across horizontal networks of individuals and groups to enhance productivity and consensus, and thus is seen as a positive asset. The "Spot The Hazard!" Program fosters development of these horizontal networks by communicating with teams across the organization, and allowing process safety professionals the opportunity to become engaged with the rest of the facility in a meaningful fashion.

2. *Leadership Platforms.* Seven definable leadership practices have been connected to the development and support of high-performing cultures [6]: vision, credibility, collaboration, feedback and recognition, accountability, communication, and action-orientation. The "Spot The Hazard!" Program applies each of these practices in the following ways:
    i. Vision. For every hazardous condition shown, employees are provided with recommendations to correct the deficiency or mitigate the risk. In other words, a vision is provided for the 'desired state' of operation.
    ii. Credibility and Collaboration. These two practices work in tandem, and are accomplished by discussing the hazards with the asset owners. Buy-in from the asset owners ensures credibility with the various stakeholders, and satisfies the need for collaboration.
    iii. Feedback and Recognition. Participants who submit a correct response to the monthly contest are recognized in the winning e-mail announcement. Furthermore, the contest winner is recognized with a reward and given positive feedback for their accomplishment.
    iv. Accountability. The Program uses photographs taken from all areas of the facility. No single operating area is excluded nor singled-out from being used in the Program. This has the effect of reminding employees that process safety is the responsibility of every employee at each level of the organization.
    v. Communication and Action-Orientation. The Program intends to be proactive, by identifying hazardous conditions before they result in undesirable consequences. These conditions are communicated to the asset owners and resolved in a timely manner. By its very nature, the process is tuned towards timely communication and results-oriented action.

3. *Cultural Shift.* A positive shift in the process safety culture is clearly one of the desired outcomes of the Program. Technical principles are orchestrated with concepts of leadership and education to produce the desired step change. The e-mail communications consistently reinforce the technical aspects and connect them to principles of process safety and loss prevention engineering.

V. ANALYSIS

At the time this document was prepared, the "Spot The Hazard!" Program had been in operation for approximately one year. The following subjects have been featured in the Program thus far:

- Photo Contest #1 – Energized Abandoned Conduit
- Photo Contest #2 – Impaired Fire Protection System
- Photo Contest #3 – Free Standing Compressed Gas Cylinder
- Photo Contest #4 – Disabled Pressure / Vacuum Vent
- Photo Contest #5 – Unsupported Abandoned Piping
- Photo Contest #6 – Unsafe Venting of Pressure Relief Valve
- Photo Contest #7 – Dangerous Hose Connections
- Photo Contest #8 – Hydrochloric Acid Piping Deadleg
- Photo Contest #9 – Unrestricted Access to Electrical Substation
- Photo Contest #10 – Unprotected Combustible Trash Receptacles
- Photo Contest #11 – Uncontained Acid Piping Drains
- Photo Contest #12 – Open Ended Piping System
- Photo Contest #13 – Horizontal Hookup on Flexible Hose
- Photo Contest #14 – Abandoned Wood Pipe Support
- Photo Contest #15 – Unsafe Pipe and Hose Supports

Figure 2 illustrates the total number of responses received to each contest, and the corresponding number of correct entries. *(Note: the total number of employees at the facility is approximately 200 persons.)*

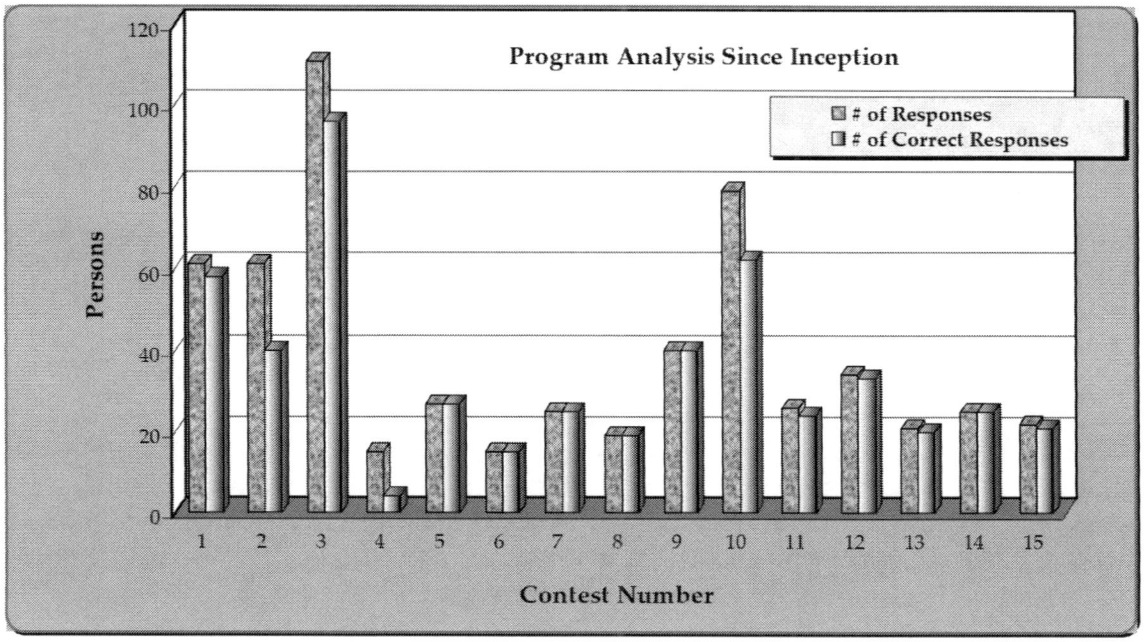

Figure 2. Program Analysis: Total Responses Received and Number of Correct Responses

VI. CASE EXAMPLES

Several examples have been selected from the Program for presentation in the current section. These examples are representative of the typical content included in the e-mail communications to the facility following each contest, with the exception of names and identities, which have been removed.

Contest #4: Disabled Pressure / Vacuum Vent

The Photo: Photo #4 featured a pair of Pressure/Vacuum (P/V) Vents on the roof of the #2 Unit. This particular model of P/V Conservation Vent uses a liquid-loaded flexible diaphragm for inbreathing and outbreathing operation, rather than the traditional weighted or spring-loaded pallets typically found in a conservation vent. The advantage of the flexible diaphragm is that it will not freeze in cold weather or stick closed in condensing services (e.g., polymerizing vapors).

The Process Hazard: These P/V Vents require a liquid glycol solution inside the pressure and vacuum chambers to operate properly. The close-up photograph revealed that three of the four chambers were empty, as evidenced by the empty sight glasses on top of the vents. Only one sight glass had a dipstick visible, indicating the chamber had been filled with glycol. In addition, several employees noted that the shutoff valves between the conservation vents and the process were closed, when in fact one would expect them to be (car-sealed) open. (*Note: In the actual contest, two photographs were used. One provided a close-up image and the other, not shown here, displayed the vents along with their ancillary equipment.*)

The Reason this is Hazardous: Pressure/Vacuum Vents are often installed to protect equipment, such as tanks, by allowing the process to breathe in and out. However, if the pressure and vacuum chambers have not been filled with glycol, the vents cannot operate as designed. Similarly, the valves between the vents and the process must be open to allow the vents to function. In this instance, the conservation vents shown in the photograph were identified during a pre-startup safety review. This reinforces the hazard identification process, the value of pre-startup safety reviews and the need to understand relatively new technologies.

What Should be Done: Prior to Startup, the pressure and vacuum chambers should be filled with the correct quantity of glycol solution, corresponding to the designed settings of the vents. The shutoff valves must also be car-sealed open, since these valves are critical to the operation of a process safety device.

Contest #7: Dangerous Hose Connections

The Photo: Photo #7 featured a simple hose station with two capped connections. This particular hose station is located near the Site's barge unloading dock, and used for liquid water-treatment chemicals.

The Process Hazard: This photograph showed several hazards, including a manual block valve in the open position immediately upstream of the hose connection. In addition, the locking mechanisms (sometimes referred to as "dog ears") on the caps were not secured in place. Several other employees further noted the deteriorated condition of the pipe support and the absence of line labels.

The Reason this is Hazardous: The potential for both loss of process containment and personnel exposure exists in this situation. With the block valve in the open position, it is possible that liquid inventory is trapped behind the cap, and may be under substantial pressure. This places employees at risk of being exposed when it comes time to remove the cap and use the hose connection. Furthermore, because the locking mechanisms are not secured with wire or a band, they could unexpectedly come off, especially if pressure builds behind the cap.

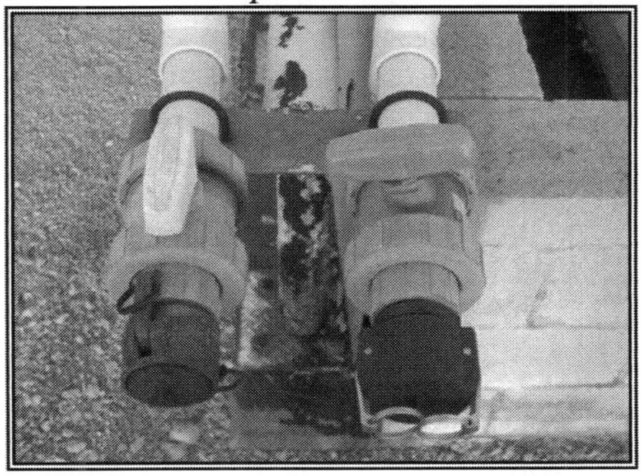

What Should be Done: The manual block valves on this, and all hose stations, should be closed when they are not in use. The locking mechanisms on the caps should also be secured in place with Velcro bands, wire ties, or a similar method. Quite often drain valves will also be installed between the block valve and end cap, so the line can be checked for inventory and pressure before the cap is removed.

Contest #13: Horizontal Hookup on Flexible Hose

The Photo: Contest #13 featured a photograph of a black flexible hose on the third floor of the Screening Building.

The Process Hazard: The hose in the photograph was connected to the piping on the horizontal, and the weight of the coiled hose had caused the hose to kink (partially collapse) at the connection point. This condition is indicative of flexure beyond the minimum bend radius of the hose assembly.

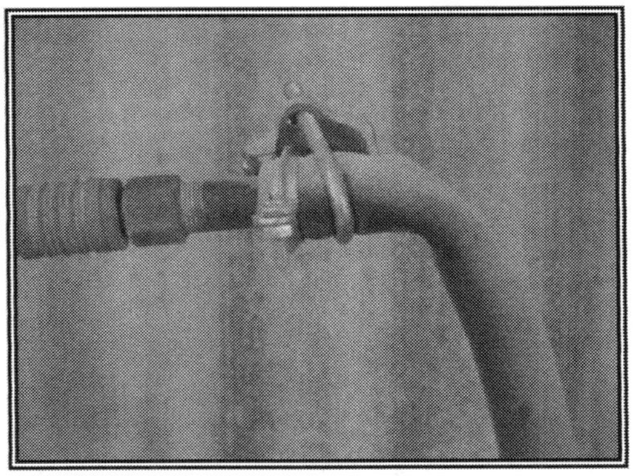

The Reason this is Hazardous: Hoses and flexible connections are often the least considered, least regulated component in either utility or process applications. Yet, they are expected to function under a wider variety of physical and chemical stresses than any other mechanical component. As a result, hose assemblies and flexible connectors are highly susceptible to failure.

In this particular instance, the horizontal hookup of the hose has resulted in severe bending at the pipe connection, which may contribute to premature failure of the hose assembly. In addition, the hose clamps do not meet specification, and the piping transition is not an engineered design.

What Should be Done: Piping systems installed for hose connections should have the pipe outlet directed downward, so the hose is connected in the vertical plane. This will protect employees from exposure in case of accidental rupture, slow the aging process of the hose by "straightening" the flow path, and reduce bend stresses at the end of the hose assembly. Also, proper engineering designs should be used to connect hoses to piping systems, making use of the correct hose clamps.

VII. CONCLUSIONS

In some respects the field of process safety management has taken a step change in new directions over the last few years. NASA's experiences with the shuttles *Challenger* [8] and *Columbia* [9] have prompted our industry to re-examine tragic events, such as the explosion at Longford [10], the Piper Alpha oil platform accident [11], and most recently the BP Texas City refinery explosion [12] for contributing cultural causes. Similarly, the CCPS has devoted an entire chapter to the subject of safety culture in its Guidelines for Risk Based Process Safety [13], indicating the necessity of a strong safety culture for strong process safety performance.

The CCPS and others [14] have developed guidance for creating and fostering a strong process safety culture. This guidance typically includes the development of strategic plans, tools and communication devices to *sustain* the culture, once it has been developed. The "Spot The Hazard!" Program is an illustration of how the process safety professional can extend a traditional instrument (i.e., Hazard Identification Training) to a new format for sustainability.

The "Spot The Hazard!" Program has been in operation for approximately one year, yet indications of its success are already evident. For example, it's quite common for participants to raise attention to hazards and concerns beyond the main focus of the photograph. In fact, individual entries have often contained several responses, speaking to both the acceptance and commitment of the employees to the Program.

Another indication of the success of the Program has been seen in the number of ideas submitted by employees, as they look forward to the continuation and expansion of the Program. Initially, the author photographed various conditions around the facility to kickoff the Program and build momentum. Over the course of the past year, however, specific suggestions have come from all levels of the organization, including plant management. This type of participation is particularly interesting, in that it demonstrates an elevated level of "hazard recognition" among the plant population.

Perhaps the most noteworthy indication of success is in the fact that the Program is still in operation, even after the author's departure from the facility. After moving on to another role in the company, the "Spot The Hazard!" Program was voluntarily adopted by a production supervisor who has committed to provide the leadership necessary to maintain the Program. This is testament to the value that plant supervision sees in the Program.

VIII. RECOMMENDATIONS

The future of the "Spot The Hazard!" Program lies in integration. This begins when the Program is pollinated into the other manufacturing locations throughout the company. By having the Program in place at all facilities, the benefits can be compounded many times over, as knowledge is transferred between all facilities.

Furthermore, there is substantial opportunity to collect and analyze data from the Program. For example, it would be beneficial to understand the types and number of process hazards that are being identified. Hazards could potentially be categorized by type (e.g., fire/explosion, toxicity, electrical shock, etc.) and compared across the organization and within a facility. In addition, participation rates and incident trends may be relevant measures to assess the effectiveness of the Program on influencing process safety culture.

The key measure of success, however, will be integrating the Program on a much larger scale, in such a way that the impact to the organization is greatest. The maximum program benefit will only be realized when all stakeholders are involved. The greatest yield will be when process improvement becomes the product of shared input by operations, maintenance, engineering, training and management. When this occurs, integration has blossomed and the benefits are magnified.

On a broader scale, the "Spot The Hazard!" Program can be readily adapted to similar industries, and tailored to the types of hazards present in that particular industry, while still achieving the same desired results. A better-educated workforce with a mindset of hazard awareness and recognition is, after all, in the common interest of all of us.

IX. REFERENCES

[1] **Phillips, K. Gerry, Nibber, Norman, and McPhail, M. Reid**, "Hazard Awareness Training In Managing Operational Risks", *51st Canadian Chemical Engineering Conference, Process Safety & Loss Management Symposium*, Halifax, Nova Scotia, October 2001.

[2] **Carney, A.L.**, "Factors In Instructional Design: Training versus Education", *Instructional Technology Lab Seminar*, University of Illinois, Chicago, March 2003.

[3] **Louvar, Joseph, and Hendershot, Dennis**, "Education Materials for Universities and Industry", *Process Safety Progress*, **26**(2), pp. 85-89, 2007.

[4] **UK Health and Safety Executive**, "Safety Culture: A Review of the Literature", HSL/2002/25, 2002.

[5] "Building Process Safety Culture: Tools to Enhance Process Safety Performance", New York: Center for Chemical Process Safety of the American Institute of Chemical Engineers, 2005.

[6] **Killimett, Patrick**, "Organizational Factors That Influence Safety", *Process Safety Progress,* **25**(2), pp. 94-97, 2006.

[7] **Putnam, Robert**, *Bowling Alone: The Collapse and Revival of American Community,* Simon and Shuster, 2000.

[8] Report of the Presidential Commission on the Space Shuttle Challenger Accident, National Aeronautics and Space Administration, Washington, DC, June 6, 1986.

[9] Columbia Accident Investigation Board, Report Volume 1, United States Government Printing Office, Washington, DC, 2003.

[10] **Hopkins, Andrew**, *Lessons From Longford – The Esso Gas Plant Explosion*, CCH Australia Limited, 2000.

[11] UK Department of Energy, *The Public Inquiry Into the Piper Alpha Disaster*, London: HMSO, 1990.

[12] The Report of the BP U.S. Refineries Independent Safety Panel, January 2007, (www.bp.com/bakerpanelreport).

[13] Center for Chemical Process Safety, *Guidelines For Risk Based Process Safety,* Hoboken: John Wiley & Sons, Inc., 2007.

[14] **Kadri, Shakeel H., and Jones, David W.**, "Nurturing a Strong Process Safety Culture", *Process Safety Progress,* 25(1), pp. 16-20, 2006.

Establishing a Personal Philosophy for Promoting Safety, Health and Environmental Excellence

James Gary, Shaw E&I, Inc., Newport Chemical Depot, Newport, IN
Michele Summers, Purdue University, College of Technology, Lafayette, IN
Lee Gary, Strategic Management Services, New Orleans, LA

A 10th PPSS Anniversary Paper

Successful managers have long recognized Safety, Health and Environmental (SHE) performance as an essential factor for outstanding industrial operations. In recent years, however, the requirement for a personal commitment to SHE by front-line managers has taken on the added dimension of representing how they are perceived by wider stakeholder audiences, including employees, family members, communities, news media, governmental agencies and – with increasing scrutiny – financial markets. Additionally, overall SHE performance has become a key determinant in whether an organization – public or private – even continues to exist or not.

Programs designed and implemented to achieve new and/or expanded SHE success can strengthen organizational performance – most notably, financial results. At the new heart of SHE continuous improvement is the need for operating managers to develop and adopt a Personal Philosophy for SHE excellence.

Based on lessons we have learned through the years, essential components of a SHE Personal Philosophy should, at a minimum, include:

- Delivering outstanding performance to our Stakeholder---Family, Community, Co-Workers, Regulators, Customers, Ourselves---is everyone's **Responsibility.**
- Attaining SHE excellence improves **financial** performance.
- Meeting or exceeding SHE goals satisfies the **Permissive** to operate.

In short, line operating managers can no longer simply delegate SHE responsibility to staff and back office administrators. Lip service is unacceptable. The consequence of any commitment less than passionately promoting a Personal Philosophy can become career limiting for line managers – and undesirable for the overall organization.

A question that runs through the minds of individuals is…How will I be remembered? What will others think of me both during my lifetime and after? What should I be doing now to influence what those thoughts will be? The same questions are also contemplated by the leaders of industrial organizations—or at least the good ones. The answers for industrial organizations, such as manufacturing, distribution, or waste treatment facilities, are important because it will drive how their leaders will plan and execute their strategies.

Traditionally, industrial organizations have "made their mark" in many different ways and are usually remembered after years of operation—or even after the end of operations—for one or more particular reasons. One of the more common reasons is that it has become a Benchmark and other organizations measure themselves against it. Setting records is another reason for being remembered. Such records are usually associated with first time accomplishments, and could be in the technology, financial, or productivity areas. While the preceding two reasons are most often associated with positive achievements, industrial organizations can also be remembered for negative accomplishments and attaining an infamous reputation, usually involving improper or illegal activities associated with financial, personnel or community relations matters. A 2002 Time story headlined "Inside the growing Enron scandal: how evidence was shredded and top executives fished for a bailout as the company imploded" hi-lighted just such a negative situation. Further, even though there is a long and varied list of things that industrial organization can and will be remembered for, today the most likely topic is SHE performance-- and it will have significant impact whether it is good or bad. For example, almost nine months after an explosion at the British Petroleum refinery in Texas City, TX killed 15 workers, a November 24, 2005 headline in the Houston Chronicle stated "B P's Own Audit Finds Big Lapses in Management." SHE performance is applauded when it is positive and has a beneficial effect on communities, but poor performance will undoubtedly result in very negative community feelings, often to the point of people wanting someone to go to jail. Therefore, every effort must be made to achieve positive results.

So, recognizing that SHE performance will be remembered whether it is good or bad, most industrial organizations will focus on achieving good performance. The truly successful ones, however, will strive to become great because they want to be as far as possible away from mediocre or, worse yet, poor performance. Furthermore, most have learned that great SHE performance will also positively influence activities in other areas of their operations and definitely improve overall performance. There are many strategies to be pursued when striving for greatness in SHE and the following identifies some of the more successful ones:

- *Involve everyone in defining SHE Goals*
- *Conduct Training, training, and more training on SHE Programs and Processes*
- *Establish a positive atmosphere where Operating Procedures can be questioned*
- *Maintain excellent 2-way communications with Regulatory Oversight Agencies*
- *Provide incentives and rewards for continuously improving SHE Performance*

There are also many writings on the subject and countless seminars and continuing education programs designed to produce SHE programs which will yield outstanding results. As a line manager, one can become overwhelmed by the shear volume of information available on the subject. For example, the Occupational Safety and Health Administration publication entitled "Voluntary Protection Programs: An

OSHA Cooperative Program," outlines ways to achieve model Safety and Health programs and performance and does provide a good roadmap for developing master Health and Safety plans, but implementation requires an inordinate amount of time and detailed knowledge of numerous technical principles routinely articulated by SHE professionals and regulators. Furthermore, experienced SHE staff members are also more than willing to lecture and pontificate on what line managers should be doing to lead the effort in their organizations. As a division manager, I reached the point in my career where I decided to develop my own personal philosophy on managing SHE performance and, to this day, it still serves me very well. By combining the salient points of SHE theories with my own hands-on experience and judgment from years of industrial operations, I have been able to produce a simple yet effective philosophy for plant managers to use as a guide in developing their own personal philosophy. While I appreciate all the volumes of information on the subject, the key point for me has always been a personal commitment to the belief that outstanding SHE performance is critical to the overall performance of an industrial organization. Doing well in SHE sets the tone for duplicating outstanding performance in other functions, including financial performance. So, by focusing on SHE performance, an industrial organization can develop a culture of excellence that extends to all facets of the operation and results in overall outstanding performance.

While most line managers generally strive to simplify systems and streamline them to the bare essentials, it is necessary that there be some program in place to memorialize and communicate the leader's Vision. Years ago I worked for a plant manager that based his entire Quality Program on the saying "One gets what is **Inspected** and not what is **Expected."** Every one in the facility knew he meant to get away from your desk and get out into the plant and to observe and to talk with people to know what was really going on to assess the true quality status of the site, regardless of what the Quality Department was also doing. The simplicity of that statement returned to me as I set out to outline my SHE personal philosophy. So, seeking to maximize impact as well as maintain simplicity, I developed my own personal philosophy and it is based on the fact that managing SHE is just like managing any other function - only much more important and it is as follows:

- Delivering outstanding performance to our Stakeholders---Family, Community, Co-Workers, Regulators, Customers, Ourselves---is everyone's **Responsibility.**
- Attaining SHE excellence improves **financial** performance.
- Meeting or exceeding SHE goals satisfies the **Permissive** to operate.

Let's look a little deeper into each of the above statements.

First, in today's world, anyone operating in an industrial environment has a responsibility to do so in a safe and environmentally compliant manner. We are accountable to many stakeholders who care very much about the outcome of our actions. We owe it to our families to preserve our opportunities to make a living and provide for their needs, but also to protect our personal health and be there for them when needed. We must protect the well-being of the community by ensuring that nothing harmful threatens its safe and healthy existence. Our co-workers need to be assured that we are as interested in their personal safety as we are in our own and will do nothing to risk their security. Governmental regulators must feel confident that we are doing the right things and not taking short cuts in executing our jobs. Also, regulators need to know that our facilities are completely open and accessible to them at any time, so trust and partnering can be fully developed. Further, we must continually ensure to our customers that we are doing nothing to jeopardize the goods or services that they are counting on us to deliver. Lastly, we owe it to ourselves to protect and maintain our ability to remain on the job and satisfy all stakeholders by not compromising performance in Safety, Health and Environmental matters. Meeting our responsibilities to all of these stakeholders will provide the basis for our operations remaining open and our being allowed to pursue productivity and financial goals.

Next, financial performance will definitely be enhanced by strong SHE performance because there are potential significant penalties associated with unsafe or non-compliant operations execution. While personnel injuries and equipment damage incur direct costs as a result of emergency response activities and follow-up medical or maintenance expenses, the impact on increased insurance rates can be staggering. Conversely, the effect of good performance can also significantly reduce insurance rates and other loss prevention costs. For example, while managing a $30 million operation, reducing loss experiences and claims resulted in a reduction of approximately $0.5 million in the corporate allocation of insurance premiums to the site profit/loss statement. This represented an increase of about 17% in the site profitability because it went straight to the bottom line. Additionally, the medical and plant maintenance and operating costs associated with responses to injuries and equipment damage are also eliminated with the reduction of such events. So, improving SHE performance simply makes sense financially and the results are directly observable for all to see. Some managers feel it is not appropriate to speak about SHE performance in financial terms because it is an emotional issue and could be misinterpreted as insensitive on the part of the company. While it is necessary to continually remind everyone that the primary purpose of strong Health and Safety programs is to avoid personal pain and suffering, it is also quite appropriate to point out that there is a significant cost associated with poor performance and that could eventually threaten the continuation of the operation. Conversely, good performance will make a site financially stronger while competing with other operations for continued existence and even growth—and that is good for all employees at the site. For example, if a particular location is selected as the site for a new capital expansion program, higher paying supervisory jobs will be added and will provide financial growth opportunities for many employees, or special bonuses may be given to all employees in recognition of outstanding performance. With respect to environmental performance, poor or non-compliant operations could result in enforcement actions from regulatory agencies, leading ultimately to fines and even closure of a facility. The facility mentioned above that received a reduction of $0.5 million in insurance premiums also received a $1 million enforcement action from its state environmental oversight agency. While the final penalty was reduced to about $500,000 it still did a lot to reduce the savings that had come from the improvement in Health and Safety performance. Luckily, the enforcement action was a one-time occurrence and the facility has since become a model environmental site, avoiding reoccurrence of similar negative activities, while maintaining the positive Health and Safety performance and associated financial benefits. The lessons learned from these improvements were also shared across the company and has been credited with preventing similar occurrences at other sites utilizing comparable processes and treatment technologies. While these other sites may be located in different states or countries, the operating practices are just as applicable because they are based on sound principles.

Lastly, it is necessary to meet all SHE requirements to obtain the "permissive" to actually operate. In this context, permissive means permission or authorization to proceed with operations and is analogous to an instrument control system allowing functions to proceed only after all prerequisite conditions have been satisfied. At no time can any facility be allowed to operate if minimum standards for safe and compliant activities are not being met. It simply is unacceptable to by-pass, disregard or override any requirement established for minimum safe and compliant operation. To help get this concept across to everyone on site at the particular facility mentioned in the preceding paragraph, a "Green Light" program was established. Think of a traffic signal light, and when it is Green, facility operations are allowed to proceed freely and without restrictions. If the light turns to Yellow, based on anyone having questions or concerns about any phase of operations, a Pause Work status is implemented and activities proceed under extreme caution and usually with reduced rates while the concerns are thoroughly evaluated. The light remains Yellow as long as any concerns exist. Just as it is unacceptable to proceed thru a yellow or caution signal on a traffic light, it is also unacceptable to disregard a yellow or caution signal in an operating facility, and serious consequences result for anyone doing so intentionally. If the light turns Red at any time, operations are immediately terminated and all efforts are directed toward getting the light back to Green. For example, if an explosion meter reaches an Lower Explosive Limit (LEL) of 10%, goes into alarm, but then drops back down below the 10% alarm set point, a "Yellow Light" condition exists. All work in the area must be paused until the reason for the alarm is investigated. Only after the cause for the alarm is determined and corrected can work be safely resumed. If causes can not be definitively determined, then a "Red Light" condition is declared and work is completely and permanently stopped until the condition is corrected and "Green Light" status re-established. There simply can be no deviation allowed in administering this Green Light Program and utilizing it completely will result in outstanding SHE performance. It may take a while for all employees to truly believe that when the light goes Yellow or Red that production must be reduced or terminated and all efforts directed toward correcting the deviant situation, but once fully accepted and religiously followed, the resulting performance will provide exceptional results.

The three principles contained in our personal SHE philosophy are very simple, easy to communicate and understand, and most importantly, deliver outstanding results. Also, they are just as applicable today as they were when I developed them as a younger manager learning how to keep a struggling facility open and competing with more seasoned and profitable operations. In fact, the greatest competitive challenge to the site referenced earlier came from another facility within our own company operating in the same state and initially much more profitably than the one I was managing. While the two sites had many similarities, there was one significant difference and that eventually led to the demise of one and the success of the other. We'll learn more about the story of these two facilities using the following Case Study.

CASE STUDY

AN ANALYSIS OF TWO COMMERCIAL HAZARDOUS WASTE INCINERATOR FACILITIES

This case study involves the evaluation of the SHE programs and performance of two separate but similar facilities located within the same Midwestern state and of how their eventual fates differed so dramatically. The lessons learned from this case study will be helpful in understanding the importance of a strong and effective SHE Program- and line management ownership of it- and how it can affect the financial performance of a business enterprise and possibly even its existence.

The first facility, let's call it Northern Incineration, was an established and extremely profitable enterprise that was the flagship operation for the parent corporation, a Fortune 200 company, involved in various service operations. The second facility, we'll name it Southern Incineration was a fledging operation that had only flirted with breakeven financial performance. Southern Incineration had much smaller existing processing capabilities but was planning for major expansion in future years. Being located in the same state, although at opposite ends and approximately 300 miles between them, both facilities were regulated by the same state agencies in SHE matters and their surrounding communities were almost identical in make-up. Each facility was located in close proximity to a major urban area and very visible in local media eyes. The actual processing techniques used at each facility were very similar and, being part of the same parent company, utilized the same personnel procedures and practices with respect to Human Resource activities, such as hiring, pay, benefits and other similar programs. Therefore, the basic skills and qualifications of the staffs at each facility were equally capable and similarly trained. Corporate oversight for all SHE programs were identical and represented the very latest in available technology and the very best available professional staffs to support the two facilities. The future was bright for both facilities since they were part of a growing market and it appeared they would continue to operate and expand their business for several years in to the future. Only one facility, however, did grow and today continues as an extremely viable and profitable enterprise. The other has long since ceased operations and several of the former employees convicted of illegal SHE activities, fined and some received jail sentences. How could that happen? How could so similar facilities, with the same parent organization, end up at such diverse points?

Northern Incineration was a facility that appeared on the surface to be well run and committed to sound SHE performance. During the everyday operation, however, of the facility two separate groups existed—one running the "front office" and one "running the plant." Theoretically, the site manager, who was located in the front office, was directing all activities for the entire facility, but only occasionally entered the actual operating facility, and then usually only because he had to do so for some customer reason or special plant tour. The general feeling in the plant was that everything important happening did so in the Front Office and was associated with profit and loss. Running the plant was a manager that was generally believed to have little respect for SHE performance and even less for actual compliance. He had mastered the art of "looking the other way" when necessary to meet production records. When corporate oversight groups indicated there could be underlying compliance problems in SHE areas, those warnings were not aggressively heeded. As rumors began to spread about possible compliance violations, the surrounding communities began to become interested in the safety of their families and possible negative implications to property values. Since Northern Incineration had not established a good relationship with their community during the good times, when it would have been an easier initiative, the facility was now engaged in all-out warfare with its neighbors. Customers began to defect rather than do business with an organization of questionable reputation. As politicians inevitably do, agency investigations soon followed and increased significantly as the site's actual SHE performance began to falter, eventually culminating in an explosion that shut the facility down for several months for repair - at enormous costs. It never restarted.

Southern Incineration was a very small upstart facility that was struggling to establish itself and overcome some of the equipment and technology problems it had lived with since beginning operations. The management and operating staffs were truly one team and even occupied the same office facilities—with little to no distinction between front office and operating plant. The managers "walked the cow path: every day with the operators and mechanics and had a good appreciation for the SHE compliance challenges that existed. While there was tremendous pressure to improve financial performance to the level of its sister facility to the north, the mangers were committed to instilling in the entire organization the importance of strong SHE performance based on line management responsibility and individual accountability. Strong relationships were established with the community that have lasted until the present, and were a major part of how the facility was able to overcome the initial negative impact that resulted from an enforcement action that arose out of misunderstanding with the sate regulatory agency over the interpretation of a particular rule. In fact, subsequent to that incident, the agency has actually commended the facility for its on-going excellent compliance programs. The site has long since overcome its early technology deficiencies, experienced significant expansion of capacity, and has grown to be a major player in the incineration services industry. It has continually reduced operating costs by strong involvement of all employees in improvement programs and experienced significant reductions in insurance premiums as a result of lower claims. Customers began to recommend the facility to others. In keeping with the commitment to SHE excellence, the facility has set several SHE records and received their company President's award for various aspects of SHE performance. Its future is still bright.

DISCUSSION

1. What is the role of line management in establishing the SHE performance of an enterprise?
2. How do employees know and feel when management is truly committed to SHE?
3. In what ways can improved SHE performance lower operating costs, increase sales and raise profitability?
4. How would you contrast the ways the two facilities dealt with agency enforcement actions? How do the lessons learned in this case study relate to your business?

After reviewing the Case Study and discussion questions, it becomes obvious why Southern Incineration overcame its early financial challenges and went on to become the flagship operation for the corporation. The commitment to SHE excellence not only provided the basis for remaining open and in operation during challenging times, but also served as the model for attaining excellence in other areas, such as productivity, customer service and financial performance. So, since all organizations will be remembered in some form or fashion, leaders should make every effort to be remembered positively and one of the best ways to accomplish that is to attain outstanding performance in Safety, Health and Environmental areas. It is important for the boss to develop a simple and straight forward personal philosophy and articulate it to everyone in the organization. Once success is achieved and sustained in Safety, Health and Environmental performance, it will carry over to all other facets of the operation, and remembrances will be favorable and long lasting.

Rev 5
14Feb08

Retool Your Process Safety Program for Efficiency and to Meet Multiple Management System Requirements

Philip M. Myers
Advantage Risk Solutions, Inc., Sunbury, OH 43074
Ph: (740) 965-6304, Email: pmyers@ARiskSolution.com

A 10th PPSS Anniversary Paper

ABSTRACT

Since the initial development of most process safety management programs, there have been many advances in process safety. This may be a good time to restructure and retool process safety programs as formal management systems, designed for efficiency in managing chemical process safety and extracting the greatest organizational value.

This paper examines how process safety management programs can be redesigned and retooled to actively extract improved efficiencies while meeting regulatory requirements and those of multiple management system standards. The British Standards Institute recently published new guidance in PAS99 to provide standardization for the management system integration process.

Process safety program efficiencies can be gained by considering the overall management system and all of the program elements and activities as processes that can be mapped and improved. Efficiencies can also be gained in terms of considering process safety activities and their impacts on the business in terms of the risks, re-directing resources from low risk areas to those posing the greatest risks and potential organizational benefits through management control. Finally technology and new tools can be utilized to extract additional value for the business. This paper captures current thinking and trends in the redesign of process safety management systems.

Keywords: Process Safety, PSM, Management System, Efficiency, Technology

INTRODUCTION

At the time of initial implementation of process safety, many companies developed "programs" (either as internal initiatives or in response to external regulatory compliance requirements) rather than complete, formal management systems. Management systems are defined by the British Standards Institute (BSI) as "system(s) to establish policy and objectives and to achieve those objectives" [1] and by the Center for Chemical Process Safety (CCPS) as "a formally established set of activities designed to produce specific results in a consistent manner on a sustainable basis [2]."

So what value do companies see in establishing formal process safety management systems? The American Chemistry Council (ACC), through its Responsible Care® Management System initiative identifies some of these considerations [3] as: management systems are seen as key performance drivers, many businesses already use management systems to organize activities and deliver improved performance, regulatory (state and federal) excellence programs recognize companies that implement management systems, and management systems can be used to demonstrate commitment to various stakeholders. Most importantly, management systems are used for better management and control of key processes, activities, and identified risks to business objectives - while also prioritizing allocation of resources. In addition, industry associations such as the American Chemistry Council (ACC) and the Synthetic Organic Chemical Manufacturers Association (SOCMA) now require member companies to implement formal management systems (at various levels) as a condition of continued membership. This can, in turn, change the industry standard thereby impacting other companies in the chemical process industries as well. In short, the trend is to develop and implement formal management systems including process safety, and the resultant management systems are viewed as good for business.

There have been numerous advances in process safety over the years, incorporating various learnings, new approaches, development of best practices, advancement of new techniques and improvements to others, implementation of various electronic systems, as well as advances in business and management systems themselves. As a result, redesign of overall programs and further development of formal process safety management systems may be desirable and beneficial. A formal review of existing programs and activities should include the overall program or management system, as well as individual elements or sub-processes. It can be helpful to view process safety systems and activities as work processes. Process maps and/or flow diagrams can be used to gain an overall picture and more detailed insights to the current situation.

The design and development of management systems must consider and fit the overall company culture, as well as available resources that can be sustained. At the center of most modern management systems is a risk based approach, focusing efforts and allocating resources according to the risks posed to the business and its objectives.

CHARACTERISTICS OF MANAGEMENT SYTEMS

While there are many reference materials regarding management systems, CCPS [4] provides useful key characteristics. Characteristics of management systems are categorized as General, Organizing, Implementing, and Controlling, as follows:

General -

- Explicit Goals and Objectives
- Well-Defined Scope
- Clear-Cut Desired Outputs
- Consideration of Alternative Achievement Mechanisms
- Well-defined Inputs and Resource Requirements
- Identification of Needed Tools and Training

Organizing –

- Strong Sponsorship
- Clear Lines of Authority
- Explicit Assignments of Roles and Responsibilities

- Formal Procedures
- Internal Coordination and Communication

Implementing –
- Detailed Work Plans
- Specific Milestones for Accomplishments
- Initiating Mechanisms

Controlling –
- Performance Standards and Measurement Methods
- Checks and Balances
- Performance Measurement and Reporting
- Internal Reviews
- Variance Procedure
- Audit Mechanisms
- Corrective Action Mechanisms
- Procedure Renewal and Re-Authorization

In the recent *Guidelines for Risk Based Process Safety* [2] CCPS identifies a number of additional factors to address in the design, development and implementation of process safety management systems. These include the purpose and scope, personnel roles and responsibilities, tasks and procedures, inputs, outputs and desired results or work products, personnel qualification and training, activity triggers, schedules and deadlines, resources and tools, metrics, continuous improvement, management reviews, and auditing.

These management system characteristics and key factors can be used together with company and site specific information to design effective overall management systems and/or to improve individual process safety management elements.

INDUSTRY ASSOCIATION REQUIRED MANAGEMENT SYSTEMS

Some industry associations now require the development and implementation of specific formal management systems, including the American Chemistry Council (ACC) and its Responsible Care® Management System (RCMS) [3] and the Synthetic Organic Chemical Manufacturers Association (SOCMA) ChemStewards® Management System [8]. Each of these management systems is based on a continuous improvement approach that follows the basic PLAN – DO – CHECK – ACT philosophy. ACC required certification of all current Responsible Care® companies by the end of 2007 to either the RCMS or RC14001 [9] management system. The latter of these combines RCMS and ISO14001 certification into a single process. The SOCMA required ChemStewards® Management System is based on three tiers that companies may be required to or choose to meet and is accompanied by various levels of verification. The three tiers are Tier 1 Fundamentals, Tier 2 Enhanced Performance, and Tier 3 Excellence. SOCMA members are required to achieve certification within a period of three years. For existing SOCMA members, onsite management system verification is required by the end of 2008. Both

ACC and SOCMA require onsite management system verifications every three years after initial certification.

As an example, an outline of the SOCMA ChemStewards® Management System is given here, as follows.

Planning (**PLAN**)
- o Policy
- o *Hazards and Risks*
- o *Legal and Other EHS&S Requirements*
- o *Objectives and Targets*
- o Action Plans

Implementation, Operation and Accountability (**DO**)
- o Responsibilities and Accountabilities
- o *Training, Awareness, and Competency*
- o Communication
- o Management System Documentation
- o Operational Control
- o Emergency Preparedness and Response

Performance Measurement and Corrective Action (**CHECK**)
- o Monitoring & Measurement
- o *Non-Conformance and Corrective/Preventative Actions*
- o Records
- o *Internal Management System Audits*

Management System Review (**ACT**)
- o Management System Review

The minimum management system requirements are shown above in italics for Tier 1 companies. Tier 2 companies must meet all of the above listed requirements. SOCMA members can elect to pursue Tier 3 verification only after verification that the Tier 2 requirements are met. Note that the setting of objectives and targets is required for even Tier 1, as a fundamental aspect of all management systems. It should also be noted that when planning for industry association required management systems (such as ChemStewards®), regulations and other company commitments may result in additional management system requirements.

INTEGRATION OF MULTIPLE MANAGEMENT SYSTEMS

As management standards have been developed, companies have often sought to integrate management systems to avoid duplicity and to obtain gains in efficiency. Some of the candidates for integration with process safety management systems include quality [5], environment [6], safety [7], and security.

CCPS [10] outlines an approach to integrate process safety management, environment, safety, health, and quality management systems. However, new guidance from BSI was published in 2006 (Publicly Available Specification PAS99:2006) [1] to standardize integration of multiple management systems. This guidance builds upon recent experience in integrating additional management systems. The guidance is intended to improve the business focus, utilize a holistic approach to managing risks, result in a coordinated approach to systems to avoid duplicity and gaps, and to provide an efficient management system. The general approach is to pull out commonalities from each management system that is to be integrated and form the foundation of the integrated system. Built on top of that are aspects to address specific individual management system requirements, as shown in *Figure 1*. The integrated management system uses the PLAN – DO – CHECK – ACT philosophy. The overall management system structure is shown in *Figure 2*. Key components of the overall integrated management system include policy, planning, implementation and operation, performance assessment, improvement, and management review.

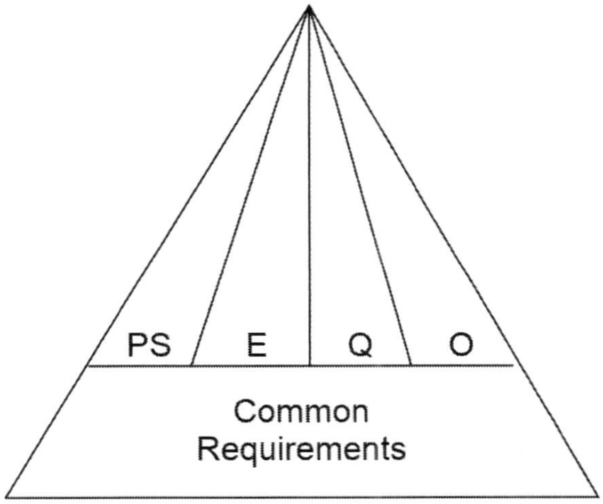

PS – Process Safety
E – Environment
Q – Quality
O – Other (Safety, Security, etc.)

Figure 1. Integration of Multiple Management Systems

As is the case for any integrated (or "independent") management system, it is a good idea to utilize normal business processes and activities to the extent possible. In this way the process safety management system becomes a normal, recognized part of doing business – and is not viewed by the company as a sideline activity that is "the process safety leader's job". Integrating process safety management with normal business processes is an effective strategy for a company of any size. However, this integration can be extremely helpful to smaller companies with more limited resources - to ensure that activities are carried out and built into an integrated formal process (such as together with an existing quality management system).

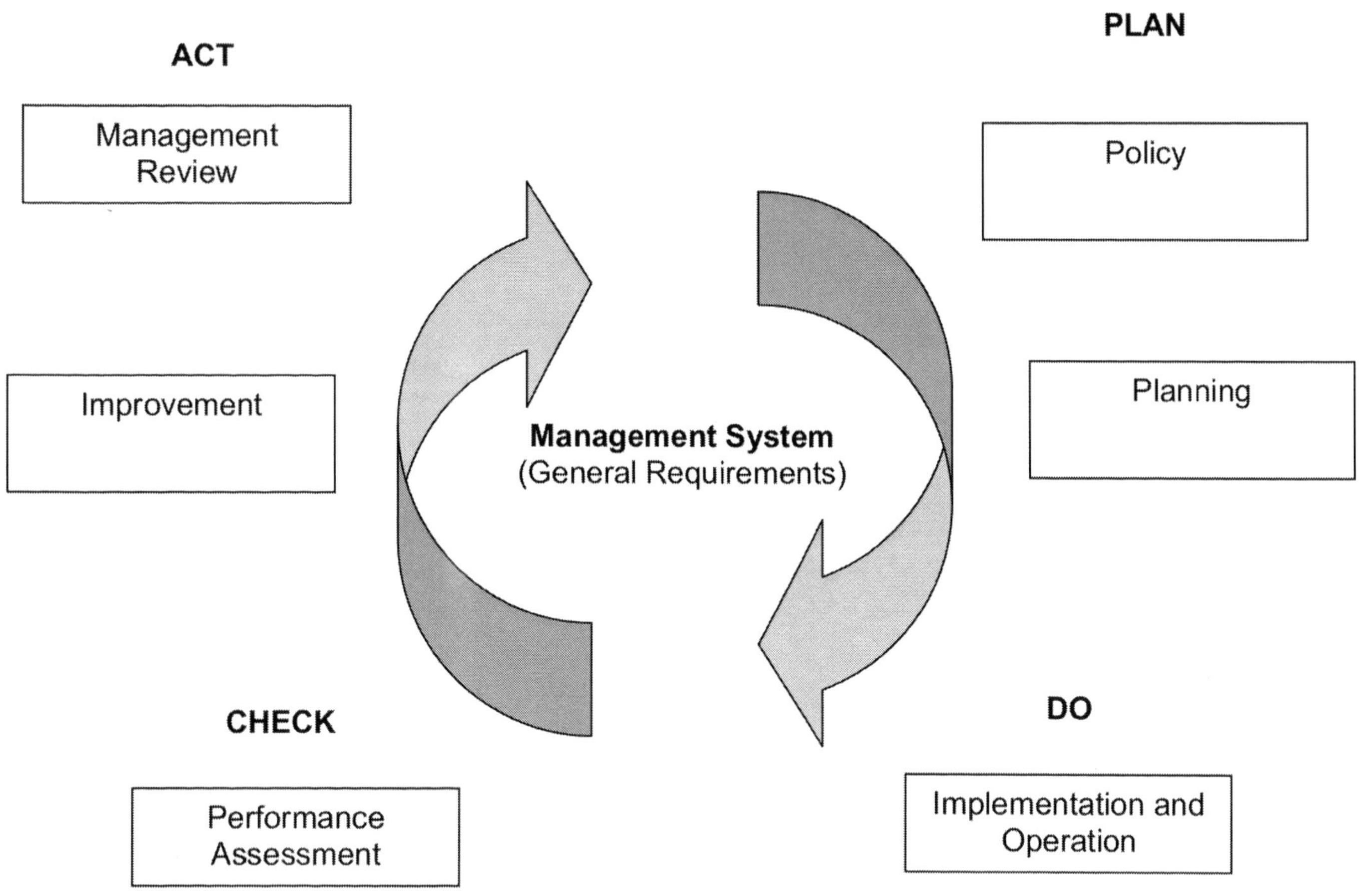

Figure 2. Common Management System Structure

IDENTIFYING AND MAXIMIZING EFFICIENCIES

It is important for companies to begin the redesign, development, and retooling of process safety management systems as a formal process, using a framework for reviews and changes. A key part of the formal process is the identification and maximization of solutions yielding improved efficiencies. At the same time it is important to balance those potential efficiencies against PSM performance and overall effectiveness. CCPS [2] refers to effectiveness as a function of both PSM performance and efficiency. As changes to the process safety management system are considered, it is also important that they are evaluated within the overall context of the company or site safety culture – and should be consistent with the larger view of supporting (maintaining and/or developing) a positive process safety culture.

General Approach and Methods

Whenever possible, the focus is first on increasing process safety management system efficiencies where it can be done without compromising performance. This is essentially the elimination of excess or waste that in reality does not add value to the overall process safety performance. When companies first conduct a critical review of their management system, there typically is some "low hanging fruit" where changes can be readily identified to improve efficiency without negative impacts on performance. Also keep in mind that waste can be in terms of very inefficient use of resources due to the lack of a formalized system, guidance, or aids – not simply from doing more than is needed or justified. Following this first pass in the formal process, a number of possible changes identified to increase efficiency may also have a "down side" – that is a potential for somewhat negative impacts on performance. It then becomes a balancing act, trying to extract efficiencies and productivity while maintaining an appropriate level of process safety assurance.

Focus Areas

When designing or improving a process safety management system, potential efficiencies can be identified for the *overall management system* and for various *elements or sub-processes*. It is import to identify the scope of the formal review at the onset of the project. Sometimes during the course of the formal review process it is acknowledged that formal processes may not even exist, and often process gaps are identified. These are areas that can be explored for gaining efficiencies. To be efficient in process safety management a formal overall management system is needed. Key parts of the overall management system that are sometimes missing or "dated" are current, relevant objectives (and targets); *real* continuous improvement; and periodic "investigative" reviews of non-conformities as well as management reviews.

Besides review of the overall management system and individual elements, efficiencies can be identified by considering various aspects of integration of the elements, and integration with other management and plant systems. It is especially important to evaluate integration issues (to the extent reasonably practical) when the scope of the formal review project is focused on a particular element or sub-process.

What to Look For

There are a number of things to look for in the conduct of formal management system reviews. A logical place to begin is to look for waste in the use of time or resources generally. This involves the elimination of activities that are not needed or are somewhat redundant. This often evolves into a discussion of "quality" and level of detail. Judgment is necessary to sort out the needed outputs and the appropriate level of detail. Some individuals may attempt to justify excessive detail or additional outputs as necessary for "quality". The formal review process will need to identify what is necessary and/or prudent and minimize or eliminate additional activities that do not add value. Again, care must be taken not to compromise process safety or to undermine a positive process safety culture. However, it is necessary to evaluate the value of activities in order to focus resources and gain process safety efficiencies. Another area of consideration is how to handle non-regulatory areas. In some cases efficiencies can be gained by handling non-regulated areas differently, while in other cases it may actually be more efficient to simply apply the same approach for all plant areas. This is highly dependent upon the activity and associated level of effort, as well as being subject to overall company philosophies. Management systems are used to apply an appropriate level of effort to each issue or problem, and the continuous improvement component involves finding new or better ways to do what is needed.

Following is a list of some general things to look for when developing more efficient systems:

- Excess or waste
- Ways to reduce effort
- Missing formal systems (or identifiable gaps)
- Systems that are "overkill" for the need
- Lack of templates
- Insufficient guidance and/or training
- Paper-based systems
- Multiple systems (paper or electronic)
- Segregated information/data/systems (not generally accessible)
- Management system steps with no "processing"
- Work flows that simply follow the company organizational chart
- Bottlenecks
- High demand/high activity areas
- High resource areas
- Repetitive activities
- Infrequent, but repetitive activities

Another helpful approach may be to view the management system(s) in terms of the following categories to identify efficiencies: people, things, and systems. A significant effort in this type of formal review addresses the systems and there are many aspects to consider, including:

- Providing new or improved tools
- Streamlining

- Reducing unnecessary reviews
- Reducing cycles
- Measuring and evaluating activities
- Limiting the amount of time a single person spends on a specific activity
- Utilization of personnel with specialized skills or abilities – including internal and external consultants
- Resource allocation – financial, personnel
- Interfaces and integration
- External influences and effects

In addition to these general areas where efficiencies can be uncovered, CCPS member companies have identified a number of specific efficiency considerations for overall process safety management systems and for individual elements [2]. These specific examples can provide a starting point for brainstorming and addressing a company's or plant site's specific situation and needs.

Risk-Based Approach

A risk-based approach is now at the heart of most modern management systems. Companies generally view any occurrence that impedes or prevents the company from achieving its stated objectives and targets as a risk. While there are many diverse risks to businesses from inside and external to the organization, the management system is used to focus resources on those potential events or occurrences that can have the most significant negative impacts, as well as on processes and opportunities that positively add to profitability. This same approach can be used in applying risk-based concepts to process safety management systems themselves. Process safety leaders and practitioners are typically familiar with risk-based techniques and applications to chemical process systems. However, most process safety professionals are much less familiar with the inward application of similar concepts and approaches to the process safety management system(s).

Application of risk-based approaches to process safety management systems is a case of "doctor heal thyself" – using existing or similar practices, techniques and approaches normally applied externally to the "internal" management system itself. Some of the same questions typically asked may be pertinent in reviewing management systems and changes to them in a risk-based context, such as: What can go wrong?, What is the consequence or impact if it does go wrong?, How likely is it to happen?, and So what does that mean to the organization and decisions to be made? Analogies can be drawn for the use of basic brainstorming and various simplified hazard identification techniques (what-if, what-if/checklist), likelihood evaluation techniques (event trees) and risk evaluation (risk matrices) related to management systems. Simplifications and modifications of other useful approaches such as for accident investigation (applied to non-conformities), and management of change (including continuous improvement), and Layer of Protection Analysis (LOPA) (layers of review/review cycles) provide useful analogies to stir thoughts regarding applications to the management processes as well. Of course, the applications are significantly modified and the level of effort applied with any of these techniques should be justified by the importance of or level of risk associated with the subject area(s) of interest.

Similar to the review of chemical process risks, it is helpful to have a graphical depiction of the process safety management system – the overall system and its sub-processes. This can be accomplished through various graphical techniques such as using process mapping, flowcharting, or fishbone diagrams for example. It is also helpful to have a certain amount of available data about the management system in terms of inputs, outputs, process safety management performance, efficiency and overall effectiveness. However, it is reasonable to use whatever data is currently available to begin the formal review process, using both quantitative and qualitative information, without waiting to invest significant resources in development of additional metrics and to gather that information over an extended period of time. (Going forward, however, development of additional metrics to address specific areas can be very beneficial.) For significant management system changes, it may be helpful to utilize a defined a pilot testing phase to identify problems or areas for improvement prior to final incorporation in the overall management systems and "roll out" to the entire company or plant site.

In applying a risk based approach to management systems, it may be helpful to use a project leader that:

- is outside of the day-to-day process safety activities
- has a strong business focus
- clearly understands all aspects of process safety and management systems
- has a significant amount of risk evaluation experience and readily recognizes risk trade-offs
- is comfortable and effective in communicating and working with senior management
- is willing to ask all of the "dumb" questions that can lead to significant improvements.

Example Efficiency Opportunities

A number of companies have process safety programs that may not formally incorporate some aspects of management systems. Common areas for improvement exist in terms of setting and maintaining relevant objectives (and annual targets), in planning, and in actually reviewing and addressing non-conformities in the management system.

There is much to be gained in terms of explicitly considering efficiencies as a constructive part of process safety management systems. Other typical areas of opportunity exist in terms of:

- Further development of electronic systems (especially for smaller companies)
- Retooling of combined Management of Change (MOC)/Pre-Startup Safety Review (PSSR) processes
- Handling of recommendations/actions/findings – including writing, tracking, resolution, closeout and documentation (for PHA, MOC, PSSR, audit, incident investigation, and management review action items)
- Optimal use of personnel resources – addressing issues of leadership; training; reviews (series, parallel, team-based); too many, too few, wrong people for the task, level of senior management interest and direction

- Hazard identification and associated techniques and analysis (considering rigor, cost, risk, need, and issues of misapplication)
- Training – use of Learning Management Systems [10] [11] and various electronic solutions
- Use of new technologies and tools (e.g. wikis for collaboration)
- Organizational issues (especially following mergers, acquisitions, or divestitures)

SUMMARY

While process safety management programs have been in existence for many years, clear benefits can be achieved by revisiting these programs and conducting formal reviews of them. Businesses clearly recognize the benefits of strong formal management systems to maximize the use of resources, focusing them on the most significant areas to mitigate "down side" risks while maximizing the "up side" opportunities and profitability. Process safety management systems can benefit from utilization of the same approach, developing formal overall management systems to focus process safety resources while achieving improvements in efficiency. At the heart of most current management systems is the application of a risk-based approach to the management systems and associated activities. The chemical process industries can take the next step forward to advance process safety using risk-based concepts and techniques to develop and/or formally review process safety management systems. These systems can then be redesigned or improved to provide an appropriate level of process safety performance while systematically extracting efficiencies and providing maximum value to the organization.

REFERENCES

1. BSI PAS99:2006, *Specification of Common Management System Requirements as a Framework for Integration*, London: British Standards Institution, 2006.
2. CCPS, *Guidelines for Risk Based Process Safety*, Hoboken: John Wiley & Sons, Inc., 2007.
3. ACC, RCMS:2005, *Responsible Care® Management System*, Arlington: American Chemistry Council, 2005.
4. CCPS, *Guidelines for Technical Management of Chemical Process Safety*, New York: American Institute of Chemical Engineers, Center for Chemical Process Safety, 1989.
5. BS EN ISO 9001:2000, *Quality Management Systems – Requirements*, 2000.
6. BS EN ISO 14001:2004, *Environmental Management Systems – Requirements with Guidance for Use*, 2004.
7. OHSAS 18001:2007, *Occupational Health & Safety Management Systems – Specification*, 2007.
8. SOCMA, *ChemStewards® Management System*, Washington: Synthetic Organic Chemical Manufacturers Association, 2004.
9. ACC, RC14001:2005, Arlington: American Chemistry Council, 2005.
10. Myers, Philip M., Watson, Brent, and Watson, Molly. "Create and Deliver Effective Training Programs Using Instructional Systems Design and E-Learning, Traditional and Blended Learning", Proceedings of the 9th Process Plant Safety Symposium and 3rd Global Congress on Process Safety, 2007.
11. Myers, Philip M., Watson, Brent, and Watson, Molly. "Effective Training Programs Using Instructional Systems Design and E-Learning", *Process Safety Progress*, Published online 12 December 2007 in Wiley InterScience (www.interscience.wiley.com). DOI 10.1002/prs. 10245, ©2007 American Institute of Chemical Engineers.

Improved Process Safety Management and Simple Metrics

Jack Chosnek
Knowledge*One*
Houston, Texas
jc@knowledge1.net

Ricky Clifton
Gulf Coast Waste Disposal Authority
Houston, Texas
rclifton@gcwda.com

A 10th PPSS Anniversary Paper

ABSTRACT

A Process Safety Management (PSM) system was implemented at Gulf Coast Waste Disposal Authority (GCWDA) to enable management of the increased risks brought about by the introduction of large quantities of oxygen in the processing of chemical waste and the destruction of emissions through Regenerative Thermal Oxidizers. This was a voluntary effort by GCWDA since they do not fall under the OSHA PSM Standard (29 CFR 1910.119). A tailored system was developed that could comply with the OSHA Standard if necessary, but included additional elements that improved on it and that are practiced by the U.S. leading chemical companies.

These elements include Management Commitment, Management Review, Conduct of Operations and Continuous Improvement. In order to allow for continuous and proactive management review simple metrics were created for each of the PSM elements rather than depending upon a periodic audit or the common use of incident statistics.

This paper will deal with the implementation of the PSM system from both the perspectives of the consultant that helped in the implementation and the facility management that had to adopt it. The successes and some of the barriers for implementation in a small company will also be discussed as well as the management success in the year after its implementation.

Introduction

When Gulf Coast Waste Disposal Authority (GCWDA) decided to increase their level of process safety due to increased risks, the challenge was how to devise a system that could be managed and sustained by a small company with limited resources.

GCWDA was ready to bring in large quantities of oxygen into their Bayport plant in order to increase the treatment efficiency, and thus the capacity, of its waste digestion tanks. In addition, these large atmospheric aeration tanks (110 ft in diameter) were going to be covered and their off-gas sent to newly installed regenerative thermal oxidizers in order to significantly reduce emissions from their operation. These changes would introduce significant higher risks to the plant than what had been common. Because GCWDA provides chemical waste disposal services to more than sixty petrochemical facilities located in the Bayport industrial park southeast of Houston, Texas, its safe continuous operation is critical to the smooth and environmentally-compliant operation of the petrochemical industry in the area. Thus, implementation of a process safety system that would help ensure uninterrupted operation became a necessity.

Need for a PSM System

Although GCWDA had safely operated its Bayport facility for many years, the number of customers had been steadily increasing and the nature of the incoming wastes was changing and its variability was becoming more pronounced. This required increased monitoring and added instrumentation, but it was being handled as the plant had little turnover and the operating personnel had become very experienced. This had lead to a dearth of written standard operating procedures (SOPs) in both the operations and maintenance areas. Because of the low organic concentration of the incoming waste, the operation was perceived as low risk and the emphasis had been on occupational safety rather than process safety. Flammability issues were limited to the main lift (pumping) station which had been mitigated by increasing communication with all the waste discharging facilities to avoid similar incidents.

With the proposed expansion of the facility and the introduction of oxygen and new more complex equipment, a need for a more structured operating environment was recognized. A Process Hazards Analysis (PHA) of the project was conducted with the help of personnel from the client industries and this PHA unveiled potentially significant risks. It became evident to GCWDA that an improved process safety system would be needed and Knowledge*One* was engaged to help in this endeavor.

Deciding on a PSM System

Because GCWDA was not subject to OSHA's Process Safety Standard (29 CFR 1910.119) given the very limited quantity of organics in the plant and their great dilution, a tailored system could be devised. The system would have to be suitable for a small facility, capable of being monitored with minimal resources, and of course, most important, able to ensure a high level of safety. A framework similar to OHSAS 18001[1] was adopted since it includes important elements such as management commitment, management review, and a continuous improvement cycle and documentation control, which OSHA 1910.119 does not offer. All the elements of OSHA 1910.119 were

[1] OHSAS 18001 is an international occupational health and safety management system specification in the same vein as ISO 9001 and ISO 14001.

included, because they do provide a structure for maintaining a safe operating environment. The elements were not copied exactly as written in 1910.119 but tailored for the company and the facility. With a little tweaking this management system would be totally compliant with the OSHA Process Safety Standard. In addition, other elements that are practiced in leading petrochemical companies, such as Conduct of Operations (sometimes called Operational Discipline) were included. As opposed to OSHA's, which is a performance-based standard, this system was designed with specific requirements of well defined scope.

Design and Implementation of the PSM System

A team was formed in order to review in depth the policies of procedures as were being developed. To assure that a workable system would be developed and buy in from top management would be obtained, the team included the Manager of Operations, Facility Manager, the authority's Technical Director, Director of projects, Operations Supervisor, and the plant's Compliance Coordinator. As needed, the Maintenance Supervisor was called in for the meetings. During the development of the system a full time Safety/Process Safety professional was hired and he became an integral part of the team. The first decision of the team was that approval of any policy was to come from the authority's General Manager.

As each policy was developed and reviewed, all personnel were trained on the content of the policy, and those affected on all the new aspects of their work as required by the policy. During the process a risk matrix was developed for the company, with the participation of the top management of the company.

The first policy that was develop was an overall policy named Facility Safety and Health Policy that described the framework, set the management expectations, described the elements that would be considered and established overall metrics. All policies developed under this policy would have, besides its requirements, an objective, a scope, a periodic review of the policy and a person responsible for that review. The main component of this master policy was a continuous improvement cycle, where management commitment and goals were set, a plan was made, the plan executed, the results were assessed, and corrective action would be taken if the results did not match the plan. The cycle would then begin again with new goals with inherent improvement in those goals.

The other important component of the master policy was dictating periodic management review of the process safety management system. In order to facilitate this review, besides the overall metrics, each procedure was to have measurable actions or results that could either be tracked or compared to a standard. The intent was not to rely solely on an audit that would be conducted every few years or on lagging indicators such as incidents.

The overall metrics that were set are:

- Number and trend of incidents and near misses per hours worked (monthly).
- Number and trend of open, temporary, and past due changes under Management of Change (MOCs), (monthly).
- Percent of employees attending required safety meetings or training (monthly).

- Number of client slugs and spills incidents requiring formal enforcement[2,3].
- Number of LEL (Lower Explosive Limit) required control actions on First Step tanks.

Although the last metric could have been included in the first one, it was separated to indicate its importance to the facility.

The second policy that was developed was a Data and Document Control policy, which was formalizing the existing record keeping, necessary as part of the environmental compliance required by the Environmental Protection Agency (EPA) and Texas Commission on Environmental Quality (TCEQ). The Safety and Health (S&H) documents were included in this controlled record keeping and the opportunity was used to move from a purely paper system to computerized document control. This turned out to be a great benefit during the development of the PSM system.

This policy was immediately followed by an S&H Documentation policy which defined which documents were to be kept and what data was to be maintained. This policy has very similar requirements to those of OSHA's Process Safety Information. The records, S&H Documentation, and the policies themselves were placed under Management of Change (MOC), which was the next policy to be developed. Description of the policies and the metrics that were set follows with emphasis on the differences with the OSHA Process Safety Standard.

Elements of the PSM System

For the most part the Safety/Process Safety Coordinator was designated to either review or initiate the periodic review of the policies, and to collect the information required by the policies and present it to the Facility Manager at the periods dictated by the policies.

The S&H policies that were established are:

- **Data and Document Control**. Provides for a header in which the frequency of review of the document is stated and its retention date. Designates the Records Coordinator as responsible for control of documents.
- **Safety and Health Documentation**. Described above, provides also for maintaining a list of all S&H documents. It explicitly includes documents related to the PSM System itself such as training records, records of management reviews, records required by laws and regulations, etc.
- **Management of Change (MOC)**. Similar to OSHA's MOC. Requires monthly reporting on open, temporary and past due MOCs. The system is managed with the help of a database that automatically collects these metrics, has e-mail requests and notifications, and secure electronic approvals (there are no paper

---

[2] Slugs refer to discharges of highly concentrated organics in a short time period.

[3] GCWDA as an Authority has legislative power and is required to enforce the environmental permits made with the companies (clients) that discharge waste into the facility.

MOCs). Although not mandated by the policy the Facility Manager reviews all the MOCs that are written.

- **Conduct of Operations**. Establishes the structure of the operations organization and responsibilities for maintaining competent and sufficient operations personnel. It describes the format and content of Standard Operating Procedures (SOPs). It sets standards of operation which include having a list of critical SOPs to be performed as checklists, a list of equipment whose malfunction initiates immediate action to correct, maximum times to acknowledge alarms and take action on them, allowance for process shutdown on failure of critical interlocks, operator communications and expected behavior. The metrics are the monthly number of inoperable interlocks and alarms with duration of inoperability, the percent of SOPs reviewed during the quarter and the percent of operators that have had refresher training during the semester.
- **Hazards Identification and Risk Control**. This policy dictates the use of Process Hazards Analyses (PHAs) similar to what OSHA has, but adds the requirement to use a risk-based approach to hazard elimination or reduction. In addition it identifies process areas where modifications require a PHA and specifically dictates that an operator be a member of the PHA team. It also describes other mechanisms for hazard identification besides PHAs, such as work permits, MOCs, work orders, pre-startup safety reviews, etc.
- **Pre-Startup Safety Review (PSSR)**. Besides the requirements of verifying that everything has been done in accordance with specifications, that all necessary procedures are in place, that personnel has been trained, and all PHA recommendations have been resolved, the Facility Manager has to sign the PSSR, and any items that are not in compliance with the requirements have to be signed off by the Facility Manager and the Process Engineer responsible for the startup, and that a plan of completion with dates be made.
- **Mechanical Integrity**. It calls for similar requirements to those of OSHA, but it requires considering risk for deciding on the frequency of inspections. All maintenance specific procedures are developed and captured through this section. Semi-annual maintenance training statistics is reviewed in this required metric.
- **Emergency Planning and Response**. This procedure provides the most probable scenarios that can affect the facility and requires that they be addressed in the Emergency Plan. It spells out areas that need to be covered in the Emergency Plan such as Roles and Responsibilities, Emergency Communications, On-Site Personnel Safety, Emergency Plant Operation, Securing of Physical Assets and Personnel Evacuation. It also requires that the Emergency Plan contain a specific response to events such as a hurricane.
- **Contractors**. This policy spells out in detail the facility's and the Contractor's responsibilities much as OSHA does but adds detail to Contractor security and accountability inside the facility. The required metrics are the monthly number of incidents in which contractors were directly or indirectly involved, classified by contractor company, and a yearly evaluation of contractor training records.
- **Training and Competence**. Besides the OSHA requirements of being trained on the process and on the safety of job tasks, the policy requires the Facility Manager to have a competent work force that is also trained on all the aspects of the PSM

System. The metrics are quarterly statistics on operator training, semi-annual statistics on maintenance training, and annual statistics on training of the workforce.

- **Safe Work Practices**. OSHA's Hot Work section is substituted with this policy that requires development of procedures for safe work in the areas of Hot Work, Lockout/Tagout, Personal Protective Equipment (PPE), Confined Space Entry and Elevated Work. Except for PPE, a permitting system was developed for all these areas. The metrics for this policy are the annual number of permits given in each category.
- **Incident Investigation**. This policy defines the timing and type of the investigation after an incident, as well as the format and data requirements of the incident report. Incidents are divided into two types, depending on the severity of the incident and the investigation and report are commensurate with the type. Metrics are the monthly number of incidents and near misses involving employees or contractors, and a monthly report on the number of incident investigations where reports have not been completed.
- **Employee Participation**. Sets the requirements that employees of all levels will participate in all aspects of the PSM System. Specifically, the opportunity for feedback on all S&H policies needs to exist, review of SOPs by operators before approval is required, and at least one experienced operator is required to be part of the team conducting a PHA, a PSSR, a high-severity incident investigation or a formal internal audit.
- **Management System Audits**. This policy requires evaluating the performance of the PSM system by reviewing the facility's compliance with all the S&H policies and procedures. A protocol for the audit is defined as well as the format of the report and the timing for corrective actions, if any. Initially a yearly frequency was set for this audit.
- **Safety and Health Management Review**. This policy determines the manner and timing of the management reviews. In essence, it dictates a monthly review by the Facility Manager and staff of all the metrics that were due for that month as mandated in each policy.

Integration of the PSM System with Plant Operations

Although initially resisted because it represented a significant change in the way that things were done, the PSM System was well integrated into plant operations and became a way of life. Among other items, plans were made to include the MOC number in work orders and purchase orders related to an MOC, in order to verify that all potential changes were recognized and managed accordingly; once a month part of the weekly staff meeting was spent in reviewing the required system metrics; the corporate training tracking database was used to satisfy the training documentation requirements; and a contractor package meeting all the policy requirements, to be handed out to all contractors, was prepared by the authority's Technical Services Group.

Management's Perspective of the PSM System

Management has been extremely supportive of the PSM system from the design phase through its implementation, and its integration to the facility's operation. It is management's belief that the system has not only provided a high degree of safety, but has provided structure and stability to plant operations that have spilled over to other areas of the company. The framework created for the system has been organizationally useful for improved training, company knowledge retention and safety and environmental compliance.

Follow-up

Audits were conducted by the consultant at the end of the initial implementation and a year later. The audits did not reveal any systemic deficiencies—findings were mainly missed actions due to lack of familiarity or insufficient experience on performing required action. All in all, the system is working well and personnel are satisfied that it introduces a higher level of safety to their work environment.

Strengthening Process Safety Requirements in HSE Management System – an NOC's Experience

Morris Kho Kee Wee

PETRONAS
Kuala Lumpur, Malaysia
morris@petronas.com.my

Busari Jabar

PETRONAS
Kuala Lumpur, Malaysia
busari_jabar@petronas.com.my

A 10th PPSS Anniversary Paper

ABSTRACT

PETRONAS, the national oil company of Malaysia was formed in 1974. Since then, it has transformed itself into a fully integrated multi-national Oil and Gas company which is operating in 33 countries.

A structured HSE Management System based on risk and quality management approach had been established and implemented across the Group in 1999. HSE Management System audits have been conducted since 2002 and the results show that the management system has been put in place in most of its operating companies. The audits' findings also highlighted various weaknesses in the management system especially with regard to the non-specific and non-prescriptive requirements of process safety.

A task group was formed in 2006 to formulate the implementation strategy and detailed work plans to enhance existing HSE Management System incorporating detailed process safety requirements. Eight technical standards which set minimum standards on the following process safety aspects were developed - Management of Change; Operating Procedures; Mechanical Integrity; Process Hazard Analysis; Process Safety Information; Pre-Activity Safety Review; Design Integrity and Proprietary Licensed Technology Assessment. Standard training modules were also developed for the eight process safety aspects and training sessions conducted for all operating companies. Subsequently, on-site situational assessments were conducted to identify the gaps and develop the implementation plan for each operating company.

The ultimate aim is to strengthen the process safety requirements and full implementation of the enhanced HSE Management System by the end of 2008.

1. INTRODUCTION

PETRONAS was incorporated in 1974 as the National Oil Company of Malaysia. Starting as a national oil company involved in exploration and production activities locally, it has transformed into a multi-national oil company now operating in 33 countries in both upstream and downstream oil and gas businesses. To sustain its business in the challenging global operating environment, a structured Health, Safety and Environment (HSE) Management System (HSEMS) was established in 1999 based on risk and quality management approach for Group-wide implementation.

Based on HSE Management System audits conducted at its operating companies, it was found that most companies have put in place its HSE management system with some of them going a step further in integrating the management system with other international standards such as ISO 14001 and OHSAS 18001. However, the findings from the HSE audits also highlighted various process safety concerns, most of which related to weaknesses in the HSE management system or work processes. Examples are: outdated or unavailability of process safety information or operating procedures; poor management of changes; inadequate process hazards review and other process safety related issues.

The audits' findings, coupled with numerous internal process related incidents such as fires, spillages and failure of safety critical equipment, has propelled the Management to issue a directive to all operating companies to give more emphasis and focus on Process Safety Management (PSM).

At the Group level, a Process Safety Workgroup was set up in January 2006 to develop and implement appropriate strategy and detailed work plan for Group-wide process safety implementation. The effort was championed by the Managing Director/Chief Executive Officer of a major refinery and sponsored by the Senior General Manager of the Group HSE Division. The Process Safety Work Group is comprised of representatives drawn from various operating companies and the scope of work covers the following:

- Develop process safety implementation strategy and detailed work plan covering roles and responsibilities, project milestones and targets
- Develop specific and prescriptive technical documents or guidelines including audit checklists to set minimum standards and capture best practices for the relevant process safety aspects
- Develop process safety training modules and implement "train-the-trainers" and other process safety training programs
- Facilitate and guide process safety implementation in all operating companies
- Monitor and report on the implementation progress to the Management.

2. IMPLEMENTATION APPROACH

The approach adopted in process safety implementation is through integration of the process safety requirements, in the forms of minimum standards and best practices, into the existing HSE management system. Through mapping of the elements of the HSEMS with those of PSM, it was found that majority of the process safety elements or requirements have already been addressed in the HSEMS. Those which were missing or inadequately defined were incorporated as enhancement in the existing HSEMS. In addition, eight (8) new standards or guidelines have also been developed to set minimum standards and document the best practices in the following aspects:

- Management of Change
- Process Hazard Analysis
- Operating Procedures
- Pre-Activity Safety Review
- Process Safety Information
- Mechanical Integrity
- Design Integrity
- Proprietary and Licensed Technology Assessment

In developing the above documents, references were made to established process safety standards and requirements used by the industries such as the Guide Books of Centre for Chemical Process Safety, Occupational Safety and Health Acts and Regulations of the United States, Guide Books of Health and Safety Executive of the United Kingdom and the Control of Industrial Major Accident Hazards Regulations of Malaysia. The linkage and integration of the above process safety aspects with the HSEMS is illustrated in Figure 1.

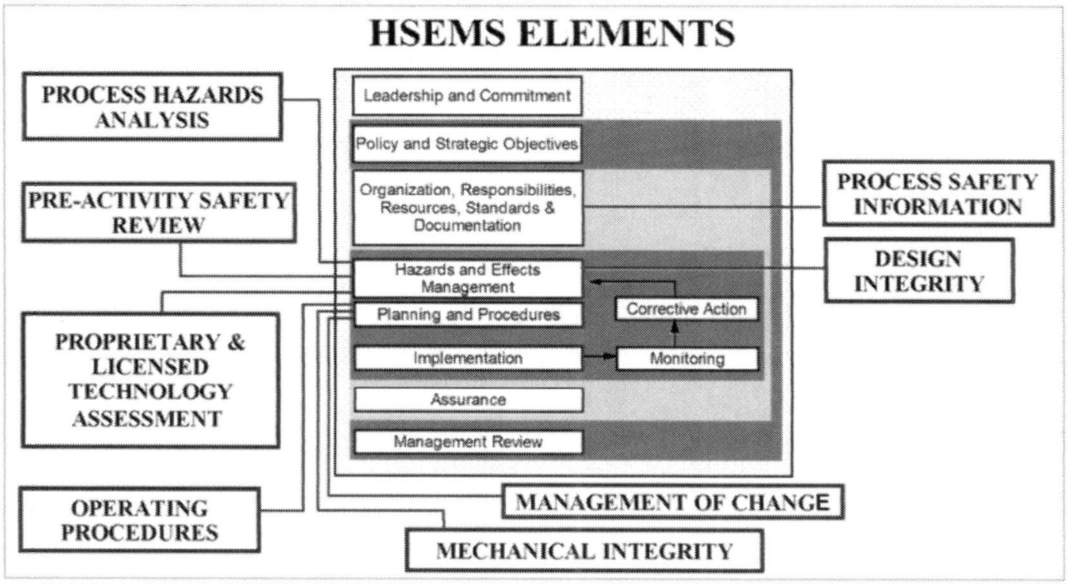

Figure 1 : Integration of Process Safety Aspects with the HSEMS

To ensure adequate and consistent understanding on the requirements and expectation of the Group's standards and requirements on process safety, training modules have also been developed and training programs organized at the Group level. They include:

- Process Safety Workshop for top management and focal persons of the operating companies to instill awareness on process safety and provide opportunity for them to identify and discuss implementation issues and challenges

- Train-the-Trainer Training on each process safety aspects for process safety focal persons of the operating companies.

Promotion and communication on process safety implementation have been carried out through the quarterly Group HSE Bulletin, regular management briefings and presentation to various levels of personnel within the organization.

2.1 Site Implementation

To assist operating companies in the implementation, a generic process safety implementation framework has been established as summarized in Figure 2 below. In general, the implementation framework is divided into five (5) major phases i.e. Planning and Organisation; Site Assessment and Development of Gap Closure Plan; System Development; System Implementation; and Monitoring and Measurement. Depending on the size and complexity of the operating companies and gaps identified, the implementation time frame could vary from 12 to 60 months.

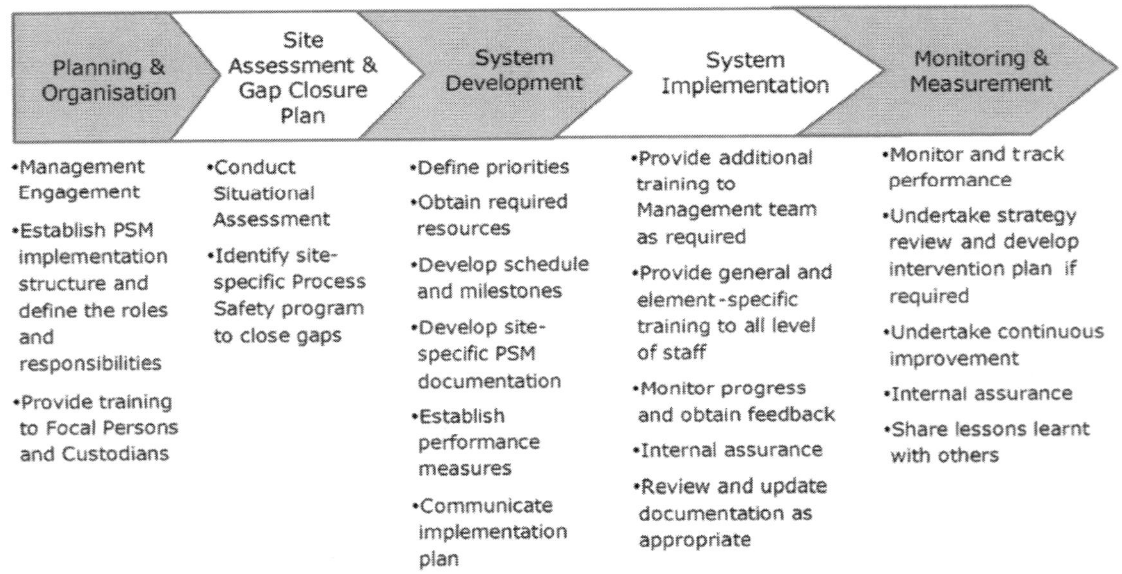

Figure 2 : Process Safety Implementation Framework for the Operating Companies

Apart from training for relevant personnel as discussed earlier, one of the main activities in process safety implementation is a situational assessment carried out to assess the adequacy of current systems and arrangements at the sites, to identify gaps and develop improvement plans towards meeting the Group's process safety requirements. It was noted that the adequacy and compliance status of some process safety requirements have also been addressed in a number of other management system audits carried out in the Group such as the HSE Management System, Reliability and Integrity Management Systems and Plant Operational Capability Standard audit/assessments. To minimize duplication of efforts and findings already been addressed in the above audits, information from various audit reports, if any, were referred to while undertaking the process safety situational assessments. A multi-disciplinary team with representatives from across the operating companies was mobilized for the assessment to optimize resources as well as to promote sharing of practices and lessons learnt amongst the various operating companies.

The process safety situational assessments results would provide information and guidance to the operating companies in the development of their strategies and road-map for process safety implementation at their sites. Based on findings from the assessment, necessary resources (internal or external e.g. consultant) were identified and allocated in closing the gaps such as development or updating of site specific systems, procedures/instructions; training; updating of process safety information; conducting process hazards analysis etc.

2.2 Monitoring and Measurement

A Process Safety Community of Practice (CoP) has also been formed as part of technical networking groups in PETRONAS with the following objectives:

- Networking amongst process safety practitioners
- Sharing of best practices and lessons learnt
- Performance measurement and benchmarking
- Knowledge enhancement and innovation in areas related to process safety

To assist in the Group-wide implementation of process safety, the CoP initially focuses on implementation issues such as training, situational assessment, progress monitoring and sharing of best practices. Areas of concern and lessons learnt pertaining to the implementation are also discussed amongst the CoP members. Sub-workgroups have also been formed to work on various initiatives such as development of key performance indicators, training requirements, competency development programs and others relating to process safety. Progress on process safety implementation are monitored and discussed through the CoP and reported to the Management on periodic basis.

To ensure process safety requirements are adequately addressed and effectively implemented in all the operating companies, regular audits are carried out within the Group at three different levels as follows:

- Tier-1 Audit - carried out at the plant or site location to assess the level of compliance to the process safety procedures and requirements. This audit is conducted at a periodic interval by the relevant personnel within the operating company

- Tier-2 Audit - carried out at the company level to assess and assure adequacy and effectiveness of the systems and procedures in place in managing process safety. This audit activity is led by the relevant management personnel within the company and supported by technical specialists

- Tier 3 Audit - carried out to assess the company's overall performance in managing process safety as part of the wider HSE Management System and is normally carried out by independent parties external to the organization e.g. Group HSE Division with support of technical specialists

Even though the target for full implementation of process safety as part of enhancement of existing HSE Management System is set at December 2008, many challenges are faced by operating companies in meeting the deadline especially with regards to shortage of competent process safety personnel and other competing priorities within the organization.

3. REFERENCES

1. PETRONAS Technical Standards : *PTS 60.0101 Health, Safety and Environmental Management System*, PETRONAS

American Institute of Chemical Engineers
4th Global Congress on Process Safety
April 6-10, 2008
New Orleans, LA

Implementation of DHS Chemical Facility Anti-Terrorism Standards

David A. Moore, PE, CSP
AcuTech Consulting Group
2001 North Beauregard Street
Alexandria, VA 22311
www.acutech-consulting.com

A 10th PPSS Anniversary Paper

Abstract

Section 550 of the Homeland Security Appropriations Act of 2007 ("Section 550"), enacted on October 4, 2006, provided the Department of Homeland Security (DHS) with authority to promulgate "interim final regulations" for the security of certain high risk chemical facilities in the United States. The Chemical Facility Anti-Terrorism Standards (CFATS) regulation (6 CFR Part 27) is risk-based and performance-based, which makes it both particularly progressive and flexible and yet challenging due to the nature of the rule.

The asset owners of high risk chemical facilities must interpret regulatory requirements and DHS guidance, and then stand up a set of security measures that achieve risk-based performance standards depending on their Tier level. Congress prohibited the DHS from requiring specific security measures while giving them the authority to approve site security plans. DHS must then interpret the security posture established by the asset owner and evaluate the proposed measures against the risk-based performance standards to judge the level of compliance. Two key issues emerge – the need to define adequate security and compliance in a completely performance-based regulation.

This paper overviews the requirements and describes some of the more difficult aspects. Experience with the regulation has proven that, for the higher tier sites particularly, critical cutting edge issues in homeland security are being addressed including the need to deal with a number of highly challenging attack scenarios. The more difficult risk-based performance standards include:

- Deter, Detect, and Delay - Deter, detect, and delay an attack, creating sufficient time between detection of an attack and the point at which the attack becomes successful;
- Restrict Area Perimeter - Secure and monitor the perimeter of the facility.

REFERENCES

US Code of Federal Regulations, 6 CFR Part 27.

KEYWORDS

Chemical security; risk-based performance standards; U.S. chemical security regulations; Chemical Security Anti-terrorism Standards (CFATS); security vulnerability analysis (SVA), site security plans (SSP).

INTRODUCTION

Until October 4, 2006, when the President signed the 'Department of Homeland Security Appropriations Act of 2007' (the Act), the Department of Homeland Security (DHS) lacked the authority to regulate the security of high-risk chemical facilities. This gave the Department of Homeland Security no later than 6 months after the date of enactment of the Act to issue interim final regulations establishing risk-based performance standards for security of chemical facilities and requiring vulnerability assessments and the development and implementation of site security plans for chemical facilities. DHS is now implementing an appropriate regulatory program under Section 550 of that Act. The interim final rule was published in April, 2007, as 6 CFR Part 27, and became effective in June, 2007.

The legislative language is very brief - about 740 words – so DHS has extensive flexibility to define the program requirements. Some of the relevant Section 550 language to this paper includes the following requirements:

"Such regulations shall permit each such facility, in developing and implementing site security plans, to select layered security measures that, in combination, appropriately address the vulnerability assessment and the risk-based performance standards for security for the facility. This provision allows that a site has a high degree of flexibility in choosing from various typical ways to manage to the performance standards DHS adopts. Provided further, that the Secretary may not disapprove a site security plan submitted under this section based on the presence or absence of a particular security measure, but the Secretary may disapprove a site security plan if the plan fails to satisfy the risk-based performance standards established by this section. Provided further, that the Secretary may approve alternative security programs established by private sector entities, Federal, State, or local authorities, or other applicable laws if the Secretary determines that the requirements of such programs meet the requirements of this section and the interim regulations. This means that some existing programs may be found as equivalent, but that depends on the fact that the regulations are no more significant that any existing industry program, which is unlikely for at least the highest risk sites.

ISSUES OF CONCERN AT HIGH RISK FACILITIES

DHS is collecting information and assessing facilities based on several security issues or groups of issues:

A. Risk to public health and safety from a release of:
1. Toxic chemicals;
2. Flammable chemicals;
3. Explosive chemicals.

B. Risk to public health and safety from theft or diversion:
1. Chemical weapons and precursors with "bathtub chemistry" potential;
2. Weapons of mass effect (toxic inhalation gases that can be used as weapons directly);
3. Improvised Explosive Device (IED) Precursors.

C. Risk to public health and safety from sabotage or contamination of materials that could release toxic inhalation gases if exposed to water.

DHS is also collecting information on two additional security issues, but will not tier facilities based on this information at this time:

D. Risks to the ability of the government to provide critical services in the event of an emergency:
1. Public health
2. Potable drinking water and electric power
3. National defense

E. Risks to the national or regional economy

APPENDIX A – CHEMICALS OF INTEREST

When DHS released Appendix A on November 20, 2007[1], the implementation phase of CFATS was able to take place. Appendix A contained 325 individual chemicals with 81 of these chemicals listed for multiple security issues. For example, Titanium Tetrachloride is considered a security issue for Release Toxics, Theft Weapons of Mass Effect and for Sabotage/ Contamination. The final breakout by security issue was:

1. Release – 187 chemicals
    - Toxics
    - Flammables
    - Explosives
2. Theft and Diversion – 162 Chemicals
    - Chemical Weapons (CW)/ Chemical Weapon Precursors (CWP)
    - Weapons of Mass Effect (WME)
    - Explosives (EXP)/Improvised Explosive Device Precursor (IEDP)
3. Sabotage/Contamination – 60 chemicals

[1] 6 CFP Part 27, Appendix to Chemical Facility Anti-Terrorism Standards; Final Rule – issued on November 20, 2007.

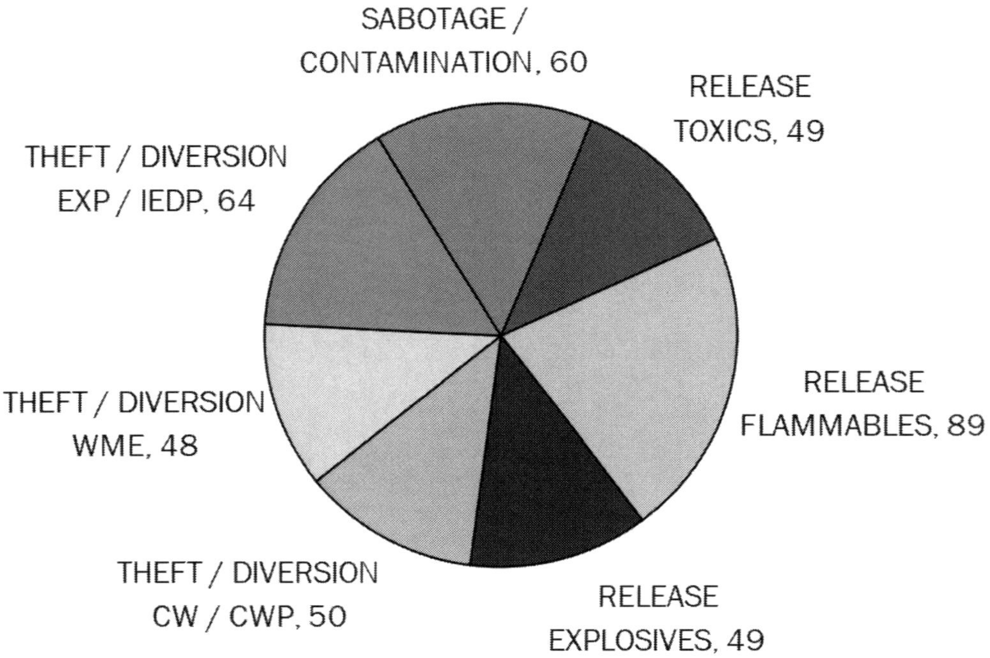

REGULATORY PROCESS

DHS is addressing chemical facilities with the most significant risk profiles as early in the program as possible. For each phase, the program would contain several basic steps:

- Chemical facilities fitting certain risk profiles would complete a "Top-screen" risk assessment methodology accessible through a secure Department website. The Department would use this methodology to determine if a chemical facility "present[s] a high level of security risk" and should be covered by this program.

- If the Department determines that a chemical facility qualifies as "high risk," the Department would require the facility to prepare and submit a Security Vulnerability Assessment (SVA) and Site Security Plan (SSP), and would provide technical assistance to the facility as appropriate. Following a facility's submission of these materials, the Department would review the submissions for compliance with risk-based performance standards. The Department (or when appropriate, a DHS-certified third-party auditor) would follow up with a site inspection and audit.

- The facility's SVA or SSP may be found deficient, in which case it must consult, object, or appeal depending on the stage of the process. If the SVA and/or SSP are ultimately disapproved, the covered facility would be required to revise its plan and resubmit the materials to meet the Department's performance standards, or face the penalties and other remedies set forth in the statute. Once approved, the Department would notify the facility of its continuing obligations—based on its level of risk—to maintain and periodically update its SVA and SSP, which are required to be submitted.

The general sequence of the regulatory scheme is illustrated in Figure 1.

Figure 1 - Strategic Approach CFATS

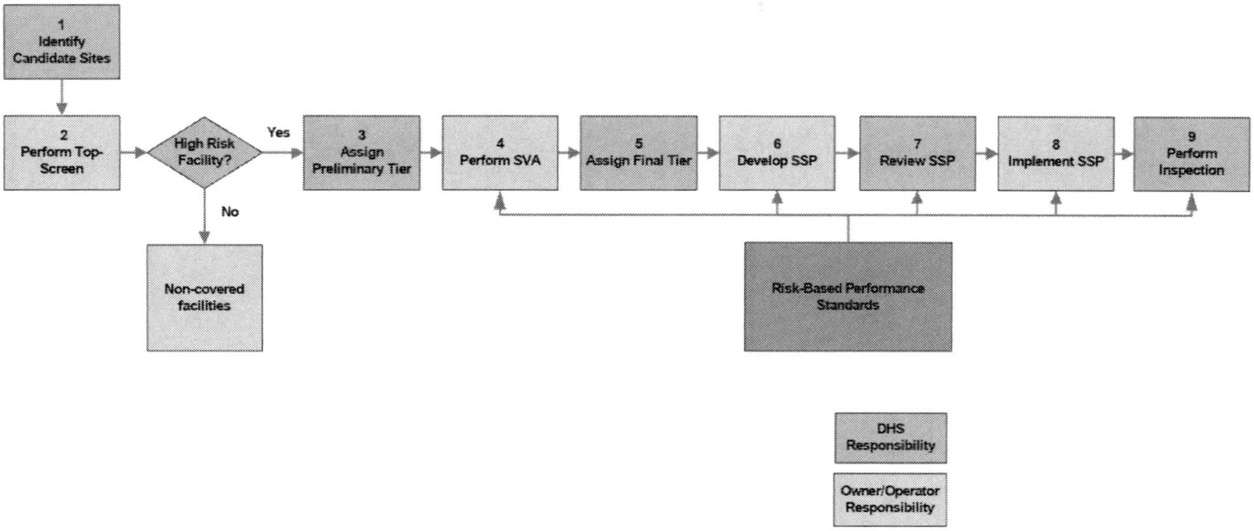

After a review of the Site Security Plan, the Department will preliminarily approve it or disapprove it. In the case of a preliminary approval, the Department will issue a Letter of Authorization to the covered facility. After preliminarily approving a Site Security Plan, the Department will inspect each facility in order to determine compliance with the requirements of this part. (The inspection provisions are discussed more fully above). After issuing a Letter of Authorization, the Department will schedule an inspection of the facility. After the inspection, if the Department concludes that the Site Security Plan addresses the vulnerabilities identified in the Vulnerability Assessment, satisfies the risk-based performance standards, and has been satisfactorily implemented, the Department will issue a Letter of Approval to the covered facility.

LIKELY IMPACTS

This is very significant for the chemical and petroleum sector in that up to this point there was no Federal regulation that defined security expectations for fixed chemical facilities except the U.S. Coast Guard Maritime Security Regulations[2] or to a limited extent the Department of Transportation regulations for transportation security found in HM-232.. MTSA did not explicitly define risk-based performance standards or tiers. While the MTSA regulations were performance-based for the most part, the USCG has authority to specify security measures, whereas DHS does not.

The Interim Final Rule[3] 'Chemical Facility Anti-Terrorism Regulations' in April, 2007, has requirements for up to 50,000 or more 'chemical' sites to address the following key areas:

[2] Pub. L. No. 107-295, Nov 25, 2002

[3] Federal Register: December 28, 2006, Volume 71, Number 249, Proposed Rules, Pages 78275-78332.

1. Risk-Based Performance Standards (RBPS)
2. Vulnerability Assessment (VA)
3. Site Security Plans (SSP)

VULNERABILITY ASSESSMENT METHODOLOGY

DHS intends to require a two-part process to perform both a preliminary "consequence" analysis and a more thorough vulnerability assessment on chemical facilities. The first segment of the methodology is a screening tool known as the "Top-screen," and is designed to be used through a secure Department website. For chemical facilities, the Top-screen solicits answers to a series of questions intended to assess the level of damage that could result from a terrorist incident at the facility. The Top-screen process primarily asks questions to determine whether the facility handles chemicals of interest in excess of the Screening Threshold Quantities but also has other questions related to manufacturing capacity and importance. For example, questions from the Top Screen User Manual (January 2008) include the following:

> **Refinery Capacity**
>
> **Enter the production capacity of the refinery in barrels per day. The production capacity, also known as the nameplate capacity, is the product output under conditions optimized for maximum quantity for the production facility, as demonstrated by one or more test runs.**
>
> **Typical Operating Capacity (bpd)**
> [Q:1.5-386]
>
> **Enter the design capacity, or theoretically calculated product output, of the refinery in barrels per day. The design capacity of an operable petroleum refinery is expressed in terms of barrels per day of crude capacity, cracking capacity, desulphurization, or amounts of products by grade.**
>
> **Maximum Design Capacity (bpd)**
> [Q:1.5-387]

o

The Top-screen also includes queries regarding manufacture and storage of explosives materials, and seeks information on quantities of chemical substances and precursors addressed by the Chemical Weapons Convention. See 22 U.S.C. 6701. The Top-screen process is intended to gather information both to evaluate the consequences of a catastrophic explosion or release and to assess the possible danger if dangerous chemicals are stolen. A more detailed description of the Top-screen process is available as Appendix A.

The second segment of the assessment process provides the tools to conduct a thorough facility Vulnerability Assessment and could also be utilized via a secure website. It has three fundamental steps, each with detailed instructions:

1. Identify the assets on the facility;
2. Apply specified threat scenarios to each asset to quantify the resulting consequences if an attack succeeded; and

3. Apply the threat scenarios to each asset in light of the security measures in place and evaluate the likelihood and the degree to which the attack could succeed.

LIMITATIONS OF DHS TO SPECIFY SECURITY MEASURES

The limitation on DHS to define any particular security measure is particularly significant. It allows industry the flexibility of defining security measures that achieve a certain level of performance while requiring DHS to define a risk-based performance standards process that adequately achieves a given security level under those constraints.

This is a performance-based and risk-based regulation, fortunately, since it is inconceivable that a regulation could be developed that prescribes a security regime that would be appropriate for all of the possible applications across the industry. This is especially true in the case of the broadly defined 'chemical sector' over which DHS has authority.

The Act restrains DHS from requiring any specific measures and, in particular, any inherently safer technology (IST). Earlier proposed bills had mentioned IST directly, but this legislation is silent on IST, and expectations for inherent safety are not likely to be included in the final regulation. This is consistent with testimony given before the Senate Environment and Public Works Committee in 2006 when they were considering the feasibility of including inherently safer provisions in Congressional legislation[4], as well as other references on the topic.[5]

RISK-BASED PERFORMANCE STANDARDS PER TIER LEVEL

The Department believes that the RBPS and the Section 550 Program should incorporate risk-based tiering. Section 550 provides the Department with authority to regulate those chemical facilities "that, in the discretion of the Secretary, present high levels of security risk." Thus, the risk-based tiers would differentiate and create tiers among those facilities that, as described above, qualify as presenting "high levels of security risk" and are thus "covered facilities."

Depending on the facility's tier level on a scale of 1-4, facilities will be subject to the application of Risk Based Performance Standards (RBPS) at each level, and a range of protective security measures will follow. The sites will have the flexibility to select layered security measures that, in combination appropriately address the vulnerability assessment and the risk-based performance standards for security for the facility.

DHS will determine that the specific options selected by the site owner/operator protect against the various potential attack scenarios in the VA and adequately address the RBPS. The security measures chosen by the site owner/operator shall address:

---

[4] Testimonies of Moore, David A., and Hendershot, Dennis C., before the Senate Committee On Environment And Public Works, Full Committee Hearing on Inherently Safer Technology in the Context of Chemical Site Security , Wednesday, June 21, 2006

[5] Mannan, Sam, White Paper – "Challenges in Implementing Inherent Safety Principles in New and Existing Chemical Processes", Texas A&M University, 2002

- the overall security requirements along strategic lines of prevent, detect, delay, and response;
- the vulnerabilities identified in the security vulnerability assessment;
- every RBPS associated with each tier level as applicable;
- the security objectives for anti-terrorism relevant to the site which include the prevention of:
  1. intentional release of a regulated chemical substance from a covered facility with the intent to cause harm to the public or employees;
  2. theft or diversion of a regulated chemical substance;
  3. contamination of a reactive regulated chemical substance stored in a transportation container; and
  4. damage or disruption to a critical chemical process with the intent to degrade the production capacity or availability

The interim final rule included the following requirements for Risk-Based Performance Standards (RBPS):

Section 27.230 Risk-Based Performance Standards

(1) Restrict Area Perimeter. Secure and monitor the perimeter of the facility;

(2) Secure Site Assets. Secure and monitor restricted areas or potentially critical targets within the facility;

(3) Screen and Control Access. Control access to the facility and to restricted areas within the facility by screening and/or inspecting individuals and vehicles as they enter, including,
(i) Measures to deter the unauthorized introduction of dangerous substances and devices that may facilitate an attack or actions having serious negative consequences for the population surrounding the facility; and
(ii) Measures implementing a regularly updated identification system that checks the identification of facility personnel and other persons seeking access to the facility and that discourages abuse through established disciplinary measures;

(4) Deter, Detect, and Delay. Deter, detect, and delay an attack, creating sufficient time between detection of an attack and the point at which the attack becomes successful, including measures to:
(i) Deter vehicles from penetrating the facility perimeter, gaining unauthorized access to restricted areas or otherwise presenting a hazard to potentially critical targets;
(ii) Deter attacks through visible, professional, well maintained security measures and systems, including security personnel, detection systems, barriers and barricades, and hardened or reduced value targets;
(iii) Detect attacks at early stages, through counter-surveillance, frustration of opportunity to observe potential targets, surveillance and sensing systems, and barriers and barricades; and

(iv) Delay an attack for a sufficient period of time so to allow appropriate response through on-site security response, barriers and barricades, hardened targets, and well-coordinated response planning.

(5) Shipping, Receipt, and Storage. Secure and monitor the shipping, receipt, and storage of hazardous materials for the facility;

(6) Theft and Diversion. Deter theft or diversion of potentially dangerous chemicals;

(7) Sabotage. Deter insider sabotage;

(8) Cyber. Deter cyber sabotage, including by preventing unauthorized onsite or remote access to critical process controls, such as Supervisory Control And Data Acquisition (SCADA) systems, Distributed Control Systems (DCS), Process Control Systems (PCS), Industrial Control Systems (ICS); critical business systems; and other sensitive computerized systems;

(9) Response. Develop and exercise an emergency plan to respond to security incidents internally and with assistance of local law enforcement and first responders;

(10) Monitoring. Maintain effective monitoring, communications and warning systems, including
(i) Measures designed to ensure that security systems and equipment are in good working order and inspected, tested, calibrated, and otherwise maintained;
(ii) Measures designed to regularly test security systems, note deficiencies, correct for detected deficiencies, and record results so that they are available for inspection by the Department; and
(iii) Measures to allow the facility to promptly identify and respond to security system and equipment failures or malfunctions;

(11) Training. Ensure proper security training, exercises, and drills of facility personnel;

(12) Personnel Surety. Perform appropriate background checks on and ensure appropriate credentials for facility personnel, and as appropriate, for unescorted visitors with access to restricted areas or critical assets, including,
(i) Measures designed to verify and validate identity;
(ii) Measures designed to check criminal history;
(iii) Measures designed to verify and validate legal authorization to work; and
(iv) Measures designed to identify people with terrorist ties;

(13) Elevated Threats. Escalate the level of protective measures for periods of elevated threat;

(14) Specific Threats, Vulnerabilities, or Risks. Address specific threats, vulnerabilities or risks identified by the Assistant Secretary for the particular facility at issue;

(15) Reporting of Significant Security Incidents. Report significant security incidents to the Department and to local law enforcement officials;

(16) Significant Security Incidents and Suspicious Activities. Identify, investigate, report, and maintain records of significant security incidents and suspicious activities in or near the site;

(17) Officials and Organization. Establish official(s) and an organization responsible for security and for compliance with these standards;

(18) Records. Maintain appropriate records; and

(19) Address any additional performance standards the Assistant Secretary may specify.

Of all of the performance standards expected of industry, the more challenging ones may include, for the higher tier sites particularly, the need to address cutting edge issues in homeland security that include the need to deal with a number of highly challenging attack scenarios. The more difficult risk-based performance standards include:

- Deter, Detect, and Delay - Deter, detect, and delay an attack, creating sufficient time between detection of an attack and the point at which the attack becomes successful;
- Restrict Area Perimeter - Secure and monitor the perimeter of the facility.

Guidance will be issued by DHS to help clarify their expectations, but even so the real issue is that there may be expectations well beyond the current capabilities of the regulated sites to defend against such attacks. This will be, in the final analysis, the heart of CFATS – how secure is secure enough?

SITE SECURITY PLANS

Under Section 550, the Department must also require that "high risk" chemical facilities develop and implement "Site Security Plans." The statute specifies that the Department "shall permit each facility, in developing and implementing Site Security Plans, to select layered security measures that, in combination, appropriately address the Vulnerability Assessment [for the facility] and the risk-based performance standards for security for the facility."

A Site Security Plan must address both the "Vulnerability Assessment" for the covered facility and the applicable "risk-based performance standards." To address the Vulnerability Assessment, the plan must identify and describe the function of the measures the covered facility will employ to address each of the facility's vulnerable areas.

Focusing on those vulnerable areas, the Site Security Plan must then address specific modes of potential terrorist attack and how each would be deterred or otherwise addressed. For example, a facility must select, develop and describe security measures intended to address potential attacks involving: (1) a VBIED (vehicle borne improvised explosive device); (2) a water-borne explosive device (if applicable); (3) an assault team; (4) individual(s) on the premises with explosives or a firearm, or (5) theft of certain chemicals; and (6) the possibility of insider or cyber sabotage.

* "In addition, a covered facility's Site Security Plan must identify how the layered security measures selected by the covered facility meet the Department's risk-based performance standards.

The types and intensity of measures necessary to satisfy these standards will of course depend on the risk-based tier of the covered facility at issue. Covered facilities will also have a continuing obligation, which will vary based on their risk-based tier, to maintain and periodically update their Site Security Plan. Further, the statute specifically prohibits the Department from rejecting a Site Security Plan, simply because it does not incorporate a **specific** type of security measure: "[T]he Secretary may not disapprove a Site Security Plan submitted under this section based on the presence or absence of a particular security measure."

APPROVAL OF VULNERABILITY ASSESSMENTS AND SITE SECURITY PLANS

Section 550 states that "the Secretary shall review and approve each vulnerability assessment and site security plan required under this section." See Pub. L. No. 109-295, sec. 550(a). To implement this provision of the statute, and consistent with the implementation plan discussed herein, the Department will require all covered facilities to submit Vulnerability Assessments and Site Security Plans to the Department. The Department will review and approve or disapprove each Vulnerability Assessments in accordance with proposed § 27.215. If the Department approves the Vulnerability Assessment, the Department will issue a letter to the covered facility so stating.

After a review of the Site Security Plan, the Department will preliminarily approve it or disapprove it. In the case of a preliminary approval, the Department will issue a Letter of Authorization to the covered facility. After preliminarily approving a Site Security Plan, the Department will inspect each facility in order to determine compliance with the requirements of this part. (The inspection provisions are discussed more fully above). After issuing a Letter of Authorization, the Department will schedule an inspection of the facility. After the inspection, if the Department concludes that the Site Security Plan addresses the vulnerabilities identified in the Vulnerability Assessment, satisfies the risk-based performance standards, and has been satisfactorily implemented, the Department will issue a Letter of Approval to the covered facility.

The Secretary may disapprove a site security plan if the plan fails to satisfy the risk-based performance standards established by this section. If the Department concludes that the Site Security Plan has not been satisfactorily implemented, the Department will consult with the covered facility as provided in proposed 27.240(b) and schedule a second inspection.

CONCLUSION

The Department of Homeland Security has proposed a regulatory framework for enhancing the security of the chemical sector with respect to acts of terrorism based upon the following principles:

- Not all chemical facilities present the same level of risk, with the greatest scrutiny focused on those that, if attacked, could endanger the greatest number of lives, have the greatest economic impact, or present other significant risks.
- The concept is to base facility security on enforceable performance standards tied to the types and severity of potential risks posed by terrorists, but allowing facilities the flexibility to select among appropriate site-specific security measures that will most effectively address those risks.
- In creating this regulatory framework, DHS should recognize the progress many responsible companies have made to date and should build on that progress.
- Interpreting the risk based performance standards will pose significant challenges due to the completeness of them as well as interpreting them across a wide variety of facilities handling chemicals.

There is little doubt that the new regulation will generate significant discussion and potential controversy with regard to the threats to be evaluated, the handling of sensitive information, and conflict with State-based security initiatives applicable to the chemical industry. However, once these hurdles have been overcome this initiative will significantly improve the overall security posture of high risk chemical facilities across the nation.

Deploy Six Sigma to Improve Process Safety

R.Muthukrishnan
Reliance Industries Limited, Mumbai, India

A 10th PPSS Anniversary Paper

ABSTRACT

Unlike other areas of business, in the field of Process Safety by sharing knowledge everyone gains and no one loses. Knowledge on Process Safety is abundantly available in the net and for free. In spite of adequate knowledge available on the subject, the chemical process industry continues to suffer from frequent high profile incidents such as BP Texas refinery, Buncefield, UK oil storage explosion, etc.

CCPS has recently published a comprehensive book on the subject: "Guidelines for Risk based Process Safety" that recommends risk based strategy for Process Safety management. The book identifies several management challenges in implementing such risk based strategy.

It is observed that weak safety management systems are the major obstacle in achieving excellence in Process Safety. Companies the world over have leapfrogged their financial performance by applying proven business process improvement methodologies such as TQM and Six Sigma. The same methodologies can be effectively deployed to improve Process Safety as well.

A modified Six Sigma methodology is proposed for improving the Process Safety management systems. Application of this methodology is explained using "Operational Readiness Check" – one of the 20 risk based Process Safety elements.

1. Introduction:

Process Safety deals with the potential release of hazardous chemical substances from chemical process plants resulting in fires, explosions and toxic effects and injury to people and damage to the assets and the environment. Most process incidents, however small, can escalate into large scale releases with devastating consequences unless preventive and responsive measures are planned and pursued early. During the past decade alone, process accidents have been responsible for hundreds of injuries and fatalities as well as billions of dollars in property damage and lost profits. With the trend towards larger plants and higher output, large scale process incidents are far more likely today than previously.

Process Safety practices and formal safety management systems have been in place in many companies for many years. PSM is widely credited for reductions in major accident risk and in improved chemical industry performance. Nevertheless, many organizations continue to be challenged by inadequate management system performance, resource pressures and stagnant Process Safety results. Incident investigations continue to identify inadequate management system performance as a key contributor to the incident. And audits reveal a history of repeat findings indicating chronic problems whose symptoms are fixed again and again without effectively addressing the technical and cultural root causes.

2. Process Safety Management has evolved over the years

Monitoring safety performance has traditionally been reactive, driven by injury statistics and focused on the worker to make improvements. In the early industrial era, accidents were attributed to the unsafe act of workers. This placed the focus of safety interventions on the worker.

Traditional safety utilizes programs, work rules, training, meetings, committees, signage, behavior modification, consequence management, coaching and, ultimately, punishment to manage the worker into safe practices and actions.

Soon it was realized that the employee is only a small part of the bigger organizational system. It takes the worker, equipment, processes, procedures, planning (both tactical and strategic), supervision and management, leadership, vision and values to achieve injury and incident free conduct of the business.

As a first step, companies began using Technical Standards and Codes to manage Process Safety. Technical standards have continued to evolve throughout the 20th century and they remain one of the foundations of Process Safety management till date. These standards included both internal company practices and external consensus based standards, such as those of ANSI, API, ASME and NFPA, etc.

Major accidents like the Bhopal disaster, Flixborough etc that occurred in the 70s and 80s, demonstrated that technical standards alone were insufficient to prevent accidents. In reaction to public concerns about the actual and potential effects of these major accidents, government agencies issued regulations to define minimum levels of accident prevention activity for the protection of workers, the pubic and environment. Many accident prevention regulations including OSHA's PSM standard in 1992 and Seveso Directives in the Europe in 1982 are performance based regulations that allow companies some degree of latitude to tailor their Process Safety activities to the appropriate level of risk.

Regulations establish only minimum requirements that in some situations may not be enough to adequately manage risk while in other cases, may force companies to overwork Process Safety issues. The PSM and RMP regulations prompted many companies to implement new activities that have achieved Process Safety performance improvements. Although this catalyzed many positive results these regulatory compliance drivers have also forced some companies adopting a compliance only mindset. A compliance based approach has not resulted in zero accidents at the workplace.

With the growth of various TQM programs, many companies applied the approach of continuous improvement, a hall mark of a quality focused operation, to Process Safety programs: "let us learn from our experience" and "keep raising the bar". However this strategy failed in the case of Safety Management as in many cases, the frequency of accidents was very low though with high impact on consequence. The absence of loss events does not reliably indicate that a Process Safety management system is working well. Several years later we still see serious accidents continue to occur.

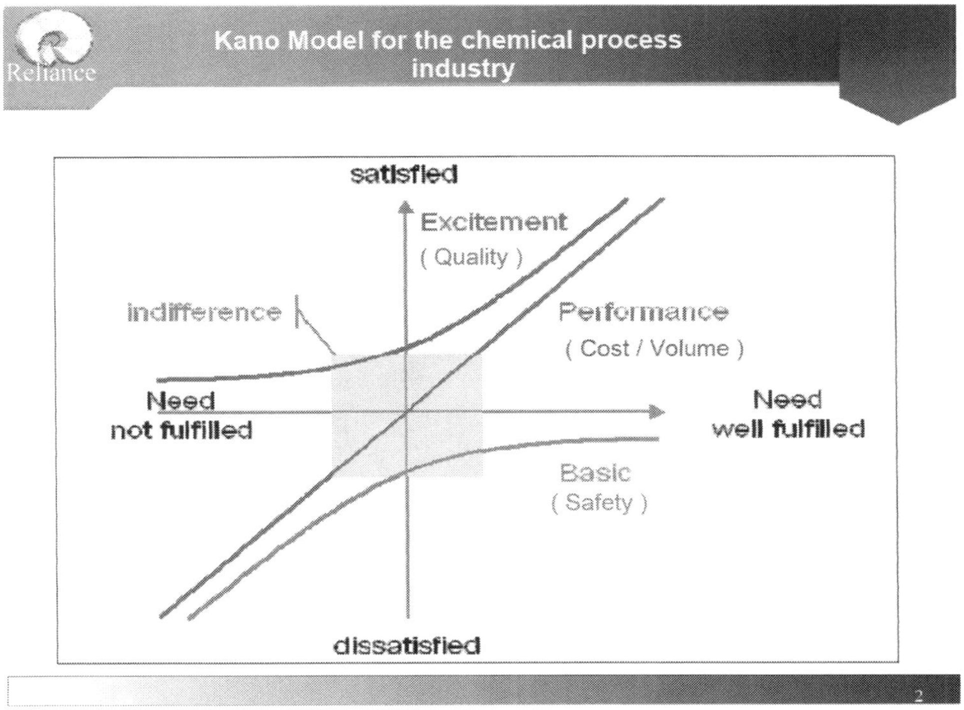

Fig 1. Kano Model of the Chemical Process Industry

The Kano model captures the dilemma of the chemical process industry very well. The four key drivers for the industry are cost, volume, quality and safety. Reducing costs and increasing volumes drives the business; this is represented by the blue ‘Performance’ line. Supplying differentiated products that exceed customer expectations delights the customers and the business alike and this is represented by the red ‘Excitement’ line. Maintaining a good safety performance is given; while any decrease in performance is not tolerated, an improved performance does not bring excitement. All companies try to figure out the optimum investment of resources. This is represented by the ‘Green’ line.

3. Need for applying modern business process improvement methodologies

If you type Chemical Process Safety, Google search returns more than 4 million websites. Many of these sites share their information and knowledge on Process Safety and accident case histories and provide good networking opportunities. Noteworthy are CCPS, MKOPSC at Texas A&M, OSHA, USA and HSE, UK. Unlike other areas of business, in the field of Process Safety by sharing knowledge everyone gains and no one loses. Knowledge on Process Safety is abundantly available in the net and for free.

In spite of adequate knowledge available on the subject, the chemical process industry continues to suffer from frequent high profile incidences such as BP Texas refinery, Buncefield, UK oil storage explosion, etc. It is not due to lack of subject matter expertise; people in the industry know the importance of Process Safety; it is just that they need to know how to reach there. It is not that the companies do not want to deploy resources to ensure Process Safety. When the reputation and public image are at risk, companies do beyond what is required by statutory requirements.

What is required is development and implementation of good management systems that would build upon the Process Safety knowledge and skills. Companies the world over have leapfrogged their financial performance by applying proven business process improvement methodologies such as TQM and Six Sigma. The same modern business process improvement systems that successful companies world over deployed to improve their bottom line could be used to achieve excellence in Process Safety management systems.

To state simply,

Process Safety expertise + Business process management systems and tools = Excellence in Process Safety

4. Six Sigma as a methodology for improving Process Safety Management Systems

Six Sigma Process Improvement Methodology is a rigorous approach to improving business processes by addressing the underlying causes of variation. The early exponents were Motorola and GE in the 1980s. Since then, many organizations whether in manufacturing or service, have successfully deployed Six Sigma to achieve significant improvements in cost, quality and cycle time.

Applying Six Sigma thinking to safety has tremendous possibility and potential. The same organizational systems that drive efficiency and quality are applied to the safety process. In Six Sigma deployment, safety can be addressed in tactical as well as strategic planning. Safety goals can be aligned with business objectives, thereby creating a linkage between resource needs and allocation.

The Six Sigma methodology derives its strength from the following characteristics:

4.1 Process focus:

Six sigma focuses on processes. Good processes deliver consistent, quality outputs in a timely, efficient and effective manner. The outputs are only a result of the activities in the process (what we do) and the inputs (the information, materials and services we use). Variation in the output (measured as cost, quality and time) is directly related to the cumulative variation of the elements of the process. Six Sigma concentrates on understanding the relationships between the inputs, the activities in the process and the outputs, so that we can change the important variables to deliver the 'best' result to the 'customer'.

4.2 Project approach:

Six Sigma utilizes project based approach for improvement of business processes. A Six Sigma project seeks to improve the performance by getting to the root cause and fixing it permanently, rather than applying quick fixes that often only address the symptoms. Projects typically go for four to six months. However, initial 'learning' projects or those where data is difficult to collect could take longer.

4.3 Team based:

Process knowledge often resides with the people involved in the activities in the process and any changes to the process will affect them directly. Six Sigma teams aim to get many perspectives on the problem under review by involving different disciplines, suppliers, customers and 'process owners' and the subject matter experts. Core teams are often around five to seven in number with other relevant inputs at appropriate times as required. The project leaders called black or green belts will have been trained in the tools and techniques of the methodology.

4.4 Rigorous phase gated methodology:

Six Sigma Process Improvement projects use a rigorous, phase gated methodology with the following phases:

- **Define** – to scope, establish the process, its inputs and outputs, set up the team and plan the project.
- **Measure** – to gather data to baseline the current 'as is' state and identify the possible causes of variation and waste.
- **Analyze** – to understand the relationship between the inputs and the outputs of the process under review so that the 'vital few' causes of variation and waste can be identified.
- **Improve** – to identify potential solutions, select the best one and implement the new process incorporating the improvements.
- **Control** – to put in place the controls and monitors that will sustain the improvements over time.

This methodology ensures that goals are finalized before the project is started, ensures that solutions obtained are not quick-fix type; have been evolved after considering all factors that contribute to the problem and the chosen solution is implemented to demonstrate the benefits and finally measures are identified that will ensure that the solution will not degrade to its old state of affairs with passage of time.

4.5 Variety of Tools and Techniques:

The Six Sigma tool kit has a number of tools and techniques that help teams work through the methodology. Tools range from very simple but effective scoping documents through to quite complex statistical analysis. The tools typically fall into four groups:

- **Process management tools** – e.g. process mapping, waste identification, bottleneck reduction
- **Project management tools** – e.g. Gantt charts, project plans, resource plans
- **Data management tools** – e.g. Pareto charts, regression, correlation, control charts, hypothesis testing, designed experimentation, using Minitab software

- **Change management tools** – e.g. brainstorming, stakeholder management, communication planning.

5. Key features of the proposed DMDIC methodology

Six Sigma projects are executed by two different methodologies - one for improvement of existing processes and another for launching new processes. The new processes could also be in place of existing processes if marginal improvements may not give the desired result.

The improvement methodology consists of five phases, known as DMAIC. DMAIC stands for Define, Measure, Analyze, Improve and Control. The steps have been explained earlier. (See Section 4.4)

For new processes the methodology is called Design for Six Sigma (DFSS) and the cycle is known as DMADV - Define, Measure, Analyze, Design and Verify. There are many variations to the above methodology. For example the methodology IDDOV stands for Identify, Define, Develop, Optimize, and Verify.

In the case of Process Safety management systems, CCPS has come out with the application of new Process Safety management systems called Risk based Process Safety management. The details of the processes are covered well in the book, "Guidelines for Risk Based Process Safety" (RBPS) published by CCPS.

This approach is different from the previous approaches of continuous improvement or compliance based. The approach recognizes varying risks that exist in different companies. The risks may vary due to the size of the company, resources availability, hazardous nature of the chemicals or processes handled, and above all the safety culture at the company. This approach also calls for monitoring the leading indicators as well as performance indicators and taking corrective action before the performance deteriorates.

This approach is a quantum jump over the previous approaches and incorporates many of the proven business process management tools, such as:

- Monitoring and controlling the leading indicators
- Management review
- Stakeholder engagement
- Resources commitment appropriate to the risk level, etc.

Since capturing all the above call for installing a new process, a DFSS methodology, namely DMDIC is proposed for implementation of Process Safety elements. DMDIC stands for Define, Measure, Develop, Implement and Control.

- **Define** – the scope of the project, do a macro mapping of the process, set up the team, plan the project schedules and tolgate reviews.
- **Measure** – develop metrics for measuring the performance of the new process in terms of leading and lagging indicators as well as to measure its efficiency and effectiveness. Agree on the metrics that will be tracked to gain insight into the process.
- **Develop** – the final process that addresses all failure modes and causes that can lead to unacceptable performance of the process.
- **Implement** – the process by way of training, communication and change management techniques.
- **Control** – verify the performance of the process in actual situation and put in place additional controls and monitors that will sustain the benefits of the new process

It is the sequential execution of the above phases with application of necessary tools and techniques that ensure that the newly installed processes would continue to give the desired performance.

The execution methodology and a brief description of the tools and techniques for each phase are described in detail below.

6. Define phase – Tools and deliverables

6.1 Define a clear Project Charter

Six Sigma utilizes the project-by-project approach and is executed through cross-function teams using project management methodology. "Project management is the application of knowledge, skills, tools, and techniques to a broad range of activities to meet project requirements," according to the Project Management Institute. Work breakdown analysis, schedule development, risk analysis, scope definition, status reporting and cost budgeting are common processes that project managers use to plan, execute, control and close projects. These processes and associated tools that work for transactional projects and manufacturing projects also work for executing the RBPS elements in an organization.

A project charter is the first step in the Define step. The charter helps to specify necessary resources and boundaries for the project.

It includes a clear definition of accountability and team roles and responsibilities. The charter is signed off by the project champion and the project manager. The project champion is a higher level executive and by approving the document, the project champion shows his commitment to the project and agrees to ensure that the organization assigns the appropriate resources.

Below is a detailed view of various elements in a typical charter:

Business Case explains why there is a need to undertake the project and how it will support organizational objectives.

Problem Statement provides a more specific and focused description of the symptoms arising from the problem the team will address.

Goal Statement describes results anticipated from the project. The goals state specific project targets to achieve the desired project purpose. The targets are stated as measurements. A goal statement is made using the principles of ***SMART.***

Project Scope identifies boundaries for the project and describes the specific project scope. What is not part of the project is also clarified to bring in clear focus to the project.

Role of Team Members identifies who are the core members for the project. The role of other support members who contribute to the success of the project is captured through a ***RACI diagram***.

Project Milestones define time-bound clear milestones in the life of the project. This includes tollgate reviews and management reviews to monitor the progress of the project.

State ***the Financial or Soft Benefits*** that is expected to be realized from the project. In the case of Process Safety management systems the soft benefits of changing the organization culture and realizing the goal of zero Process Safety related accidents are the benefits that far outweigh any financial benefits.

At the completion of every phase of the project, a ***tollgate review*** is conducted by the project champion to ensure all the deliverables are in place and to proceed to the next phase.

An example of a project charter for installing a "Management system for Operational Readiness Check" is shown in Figure 2. Operational Readiness Check is one of the 20 RBPS elements in the Process Safety Management Systems.

The SMART goal statement and RACI diagram need a little more explanation.

6.2 Make a SMART goal statement

The goal statement generally takes the form of performance objectives and the goal(s) are specific, measurable, attainable, relevant and timely (SMART).

The goal statement for the project is: "The average no. of incidents recorded during the last two years is xx with a severity level of 2 or more. It is planned to reduce such incidence to a max of 2 for all startups that occur at the site in any year."

Note the SMART elements in this statement:

Specific - The project aims to decrease the number of incidences to 2 that have severity level of more than 2.
Measurable - The targets are measurable and mechanism exists on how to measure them.
Attainable - Should be reasonable, within the ability of the team and at the same time challenging enough.
Relevant – The objective directly aligns well with the business objectives of the organization.
Timely - A deadline for the project exists and owned by all team members.

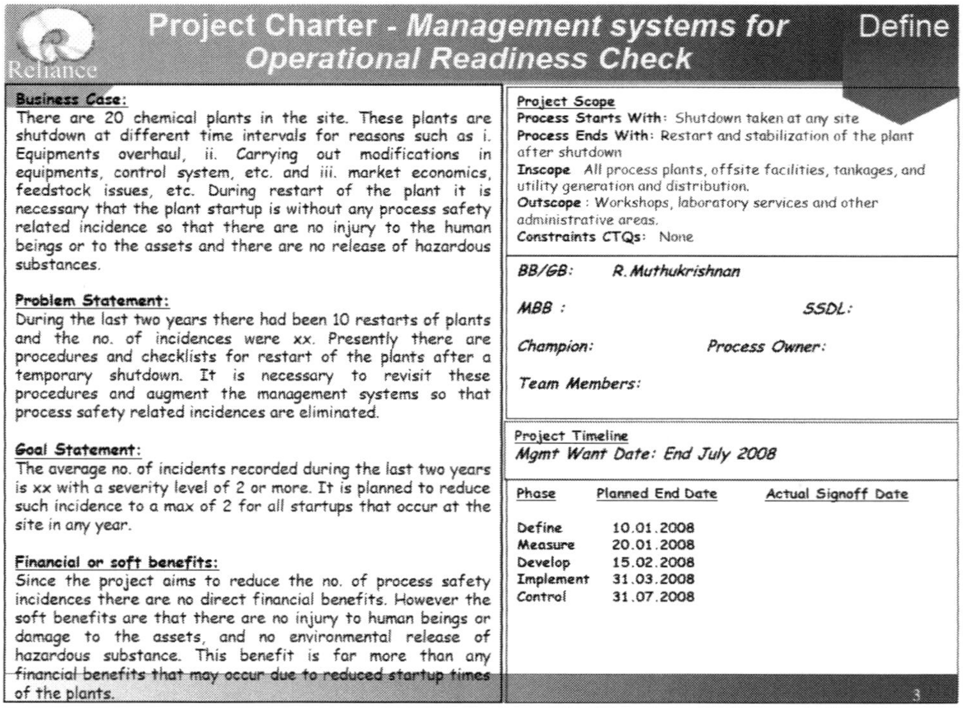

Fig 2. Project Charter for Operational Readiness Check

6.3 Prepare a RACI Diagram

A RACI diagram is a graphical display that denotes the roles and responsibilities of key people in a project. The RACI diagram assigns tasks into four responsibility types. The first letter of the role types make up the acronym RACI:

Responsible: Those who do the actual work.

Accountable: Those who are ultimately accountable for the completion of the work.

Consulted: Those who provide inputs to the development and implementation of the project solutions. If any of the *consulted* need to spend considerable time on the project, it may be a good idea to induct him as a *responsible* member rather than *consulted*. (Consider this as two-way communication.)

Informed: Those who need to be kept up to date on the progress of the particular phase. (Consider this as one-way communication.)

Below is a RACI diagram filled in for the project on Operational Readiness Check process implementation.

Reliance

RACI Diagram - Operational Readiness Check

Define

The RACI Diagram

Functional Role	Project Role	Define	Measure	Develop	Implement	Control
Plant Operations Manager	Project Manager	R	R	R	R	R
Plant Operations Engineer	Team Member	R	R	R	R	R
Process & Technology Services Engineer	Team Member	R	R	R	R	R
Safety Officer	Team Member	R	R	R	R	R
Site Operations Manager	Champion	A	A	A	A	A
Site Head	Business Owner	C	I	I	I	A
Maintenance Manager	Team Member	R	R	R	R	R
Software Engineer	Consultant	I	I	C	C	I
Business Process Improvement Manager	Consultant	C	C	C	C	C

4

Fig 3: RACI Diagram for the Project - Operational Readiness Check

It is essential to have a good mix of skill sets, roles and personality types on the project. The rank or position of the team members also plays a key part. If you have too many high level executives they may have little time available to participate in meetings let alone perform the project tasks. Besides involving the Process Safety experts in the project and the line function people, a business process consultant conversant with six sigma methodology is required as a consultant for successful achievement of the goals.

7. Measure phase – Tools and Deliverables

In the Measure phase, key measures are identified, data collection is planned and executed, process variation is displayed, performance baseline is established and sigma level of the process is calculated.

The following are the key activities that are relevant to the Measure phase:

1. Create detailed process map
2. Identify input, output & process Measures
3. Agree on the key measures to track and monitor
4. Establish data collection plan
5. Develop dashboards with specification limits & targets

7.1 Prepare a detailed process map

Process management involves defining macro and micro processes, assigning ownership, and measuring the performance of each process.

A SIPOC diagram is a tool typically employed in the Measure phase to identify all relevant elements of the project.

The tool name prompts the team to consider the Suppliers (the 'S' in SIPOC) of your process, who provide the Inputs (the 'I') to the process, the detailed Process (the 'P') your team is improving, the Outputs (the 'O') of the process, and the Customers (the 'C') that receive the process outputs. In some cases, Requirements of the Customers can be appended to the end of the SIPOC for further detail.

In some instances it may be better to map the SIPOC for the current process and the SIPOC for the desired improved process.

A SIPOC is a simple process mapping tool used to map the entire chain of events from trigger to delivery of the target process.

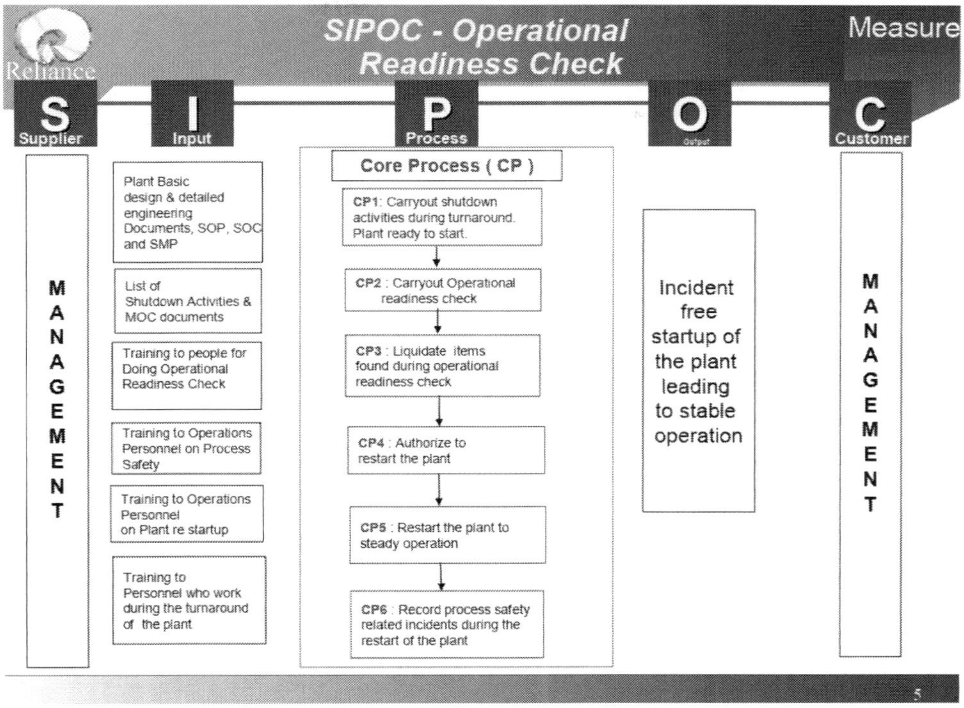

Figure 4: Process Map for Operational Readiness Check

7.2 Identify right metrics for the project

"How's it going?"

How many times have you been asked that question about your business? How do you answer that question quantitatively and completely?

Most of us can answer the question qualitatively, but ... is that a complete answer? Or is it just a partial or superficial response? Can the answer be supported by examples?

Just being able to answer the question is important. It means that you, in fact, do know how well you're doing, and where your project is headed. On the converse, if you can't answer the question, that may be a precursor to dark times ahead for you and your project.

Metrics selected should be simple and straightforward and most of all relevant. Typically there is no one metric that fits all the requirements for a particular situation. Metrics should be based on what, in fact, needs to be measured to improve the process, rather than use one that is available in the current measurement system.

Many Six Sigma professionals advocate the use of a Balanced Scorecard type of approach for the selection of project metrics. The Balanced Scorecard approach includes both financial and non-financial metrics, as well as lagging and leading measures across the four areas or perspectives: Financial, Customer, Internal Processes, and Employee Learning and Growth.

Metrics generally fall into two categories:

1. Performance and output metrics, also called lagging metrics and
2. Input and process metrics, also called leading metrics

Performance metrics are output measures of the process; that is, they assess the overall performance of the process. They are external in nature and are most closely tied to outputs, customer requirements, and business needs for the process.

Input and Process Metrics are measures that ascertain why a process is not performing up to expectations. They are usually associated with internal process steps and inputs to the process.

Brainstorming and affinity diagrams are found to be particularly effective tools for identifying and finalizing metrics for tracking.

While selecting the final metrics for tracking performance, make sure you understand whether we are tracking efficiency metrics or effectiveness metrics. Efficiency is usually a measure of speed and cost; improvement in efficiency is achieved by eliminating waste in time or resources. Effectiveness is usually a measure of how well you are meeting the requirements of the customer and the overall objectives of the process. If too much emphasis is placed upon speed and resource utilization, and cost in safety processes, effectiveness and quality would suffer.

In transactional processes such as safety systems, we should not place too much emphasis on speed, resource utilization and costs as otherwise the outcome and effectiveness of the process would suffer.

Once the right metrics for tracking performance has been identified the next step is the data collection plan that involves setting the specification limits and capturing the current performance. Typically you'll use run charts (performance over time) to see how the project goals are achieved.

8. Develop phase – Tools and Deliverables

In this phase, the final process that addresses all the failure modes and the causes that can lead to unacceptable performance of the process is developed by the team using the tools like cause and effect diagram and FMEA.

The steps involved are:

1. Identify causes that result in defective performance of the process using C & E diagram, 5 whys and FMEA and through brainstorming.

2. Develop procedures that address deficiencies identified in the above two steps.

3. Prepare the list of actions that were identified in FMEA for implementation.

8.1 Cause-and-Effect Diagrams

When utilizing a team approach to problem solving, there are often many opinions as to the problem's root cause. One way to capture these different ideas and stimulate the team's brainstorming is the cause and effect diagram, commonly called a fishbone diagram. The fishbone will help to visually display the many potential causes for a specific problem or effect. It is particularly useful in a group setting and for situations in which little quantitative data is available for analysis.

To construct a fishbone, start with stating the problem in the form of a question, such as 'Why is the no. of Process Safety incidences high during plat startup?' Framing it as a 'why' question helps in brainstorming, as each root cause idea should answer the question. The question is placed at the 'head' of the fishbone.

The rest of the fishbone then consists of one line drawn across the page, attached to the problem statement, and several lines, or 'bones,' coming out vertically from the main line. These branches are labeled with different categories.

8.1.1 Five Whys

The 5 Whys is a technique used in the Develop phase of the Six Sigma methodology. It's a great Six Sigma tool that doesn't involve data segmentation, hypothesis testing, regression or other advanced statistical tools, and in many cases can be completed without a data collection plan.

By repeatedly asking the question, "Why" (five is a good rule of thumb), you can peel away the layers of symptoms which can lead to the root cause of a problem. Very often the ostensible reason for a problem will lead you to another question. Although this technique is called "5 Whys," you may find that you will need to ask the question fewer or more times than five before you find the issue related to a problem.

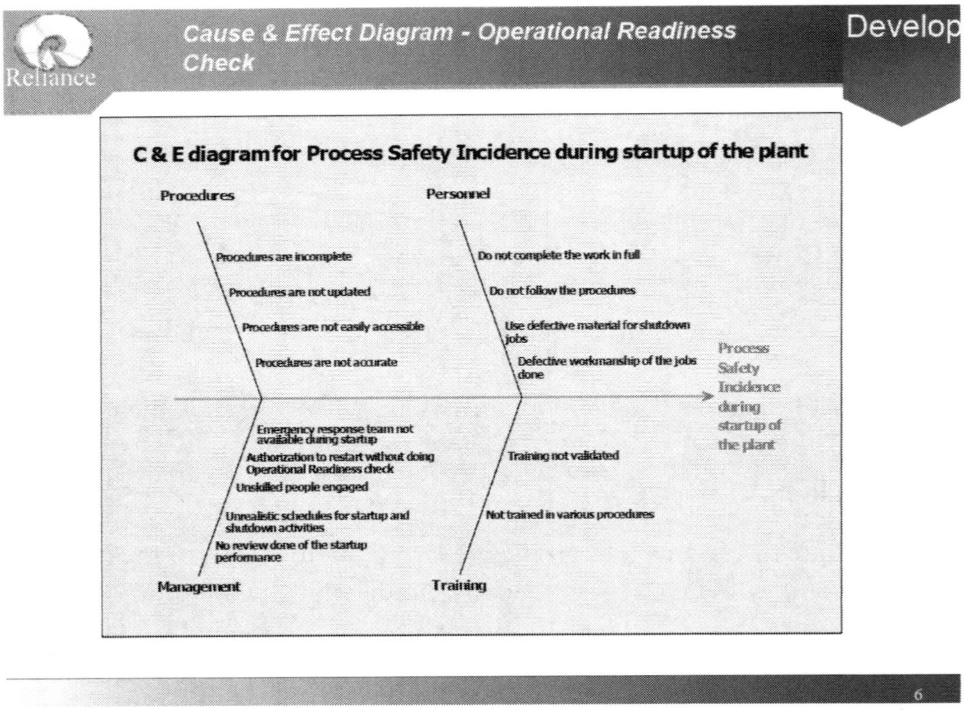

Fig 5. Cause and Effect Diagram for Process Safety Incident during startup

8.2 FMEA

Failure Mode and Effects Analysis called FMEA is a proactive tool that enables the identification and prevention of process or product errors before they occur. It is a systematic way to examine a process for possible ways in which failure can occur and then to redesign the processes so that the new process eliminates the possibility of failure.

The FMEA team should not exceed 6 to 10 people, although this may depend on the process complexity. Each team should have a leader and/or facilitator, record keeper or scribe, time keeper and a champion. During the FMEA meeting, the team must have members knowledgeable about the process or subject matter. It is advisable to include facilitators with skills in team dynamics and rapid decision-making. Ground rules help define the scope and provide parameters to work within. Typically the process owner leads the FMEA team and keeps it updated.

The idea is to focus improvement efforts on the failures that have the biggest impact on customers. The highest scoring failure modes are those that happen a lot many times, that are bad when they happen, and/or that are unlikely to be detected. Difficult-to-detect errors obviously are more likely to lead to undesirable outcomes.

The team then completes the FMEA analysis for the highest-scoring failure modes and for any that get the highest severity scores, even if they do not score that high overall. Obviously, a business wants to make sure any possible disaster is prevented, even if it is unlikely to occur. "Completing the analysis" means looking at the potential causes for the error mode, identifying ways to detect the problem, developing recommended actions, and assigning responsibility for monitoring the process and taking action when warranted.

FMEA can be used at either end of a DMAIC or Design for Six Sigma project: At the beginning of a project, FMEA can help a team better scope the opportunity by defining the types of failures and narrowing the focus to a specific type of problem. In the Implement phase, FMEA can help uncover potential problems (especially unintended consequences) with suggested solutions, thereby allowing timely adjustments. In the Control phase, FMEA helps identify what measures need to be in place to make sure that failures will not happen in the future.

FMEA works best in a team environment with cross-functional representation. Subject matter expertise is critical.

8.3 Document the procedure

Documenting the procedures can be done as per the guidelines given in the 8th RBPSM element "Operating Procedures" in the book Guidelines for Risk Based Process Safety.

Again this phase is completed with the toll gate review by the project champion.

9. Implement – Tools and Deliverables

This is where a strategy is developed and implemented. This phase prepares the target group (stakeholders) to adapt to the newly developed management systems and procedures. Implementation is achieved through training, communication and change management techniques.

9.1 Change management

Doing the same old thing and expecting different results is the definition of insanity. As much as new tools and a new roadmap empowers people to do things differently, systemic constraints -- be it organizational structure or reward system -- if not addressed adequately can seriously damage the credibility of the effort and make cynics out of employees.

While Six Sigma is based on solid principles and well-founded data, without departmental or organizational acceptance of change, Six Sigma solutions and tools may not be effective. Actions to confront big roadblocks early in the deployment phase can do the magic of boosting morale and providing momentum to overcome psychological hurdles throughout the organization.

Teams may decide to use change management tools such as CAP (Change Acceleration Process) to help build support and facilitate rapid improvement. Careful planning, communication, participation and ensuring that senior leaders are well-informed throughout the process will greatly increase the chance for a smoother implementation.

9.1.1 Creating a Powerful Vision

Developing a clear vision is important in making a culture change a reality. Having a shared vision and communicating it well are essential in galvanizing a workforce to come together and stay together, during the process of change.

It requires a shared vision. It is one, in which everybody has a role to play, everybody clearly understands his or her role, and everyone knows "what's in it for me?"

9.1.2 E- enablement of the Process Safety management systems

One of the ways to bring changes to existing system is to e-enable all the functions of the new process. Employing web based automation, companies have achieved near perfect compliance to procedures and improved the performance of their core business processes in human resources, procurement, finance, customer relations and logistics. Similarly the core processes of Process Safety systems also can be web automated and made available to all the users. Thus scheduling operational readiness check, authorization for restart of the plant, recording the audit findings, recording the process safety incidences during the restart, etc. can all be executed in the web based systems. Install the software and allow access through the company intranet systems. This makes sure people don't do things in the old way.

If the core processes in safety have not been earlier e-enabled, it may be a good idea, to include one software specialist in the project as a team member. Bringing the software engineer early in the project helps to understand the requirements better and develop the software to meet the requirements of the project owner.

9.1.3 Using Influencing Strategy

If the political, technical or cultural attitudes of key stakeholders are identified (and verified), a strategy may be worked out for influencing them to strengthen, or as a minimum, maintain their level of support for the change. This simple tool helps the team assess stakeholder issues and concerns, identifying and creating a strategy for those who must be "moved" to a higher level of support.

Reliance **Influencing Strategy -Operational Readiness Check** Implement

Key Stakeholder	Type of resistance	Underlying issue	Strategy
Ram	Technical	Feeling of inadequacy or stupidity	Education, involvement and information
Goyal	Political	Loss	Stress what is gained through implementing solutions
Naresh	Organizational	Control	Give control or credit for the solutions
Jay	Individual	Emotional paralysis	Doing less over a period of time

7

Fig 6. Influencing Strategy

9.1.4 <u>Other methods to overcome resistance</u>

When you are implementing a new process or management systems, there may be resistance visible or hidden from several quarters. Here are several approaches that can be used to counter these occurrences.

- Provide examples that the new process is ingrained with other existing business processes.
- Make it impossible for anyone to ignore the process. This can be accomplished by tying the success of the process to personal training plans, annual review, and other well-established business processes.
- Provide training to ensure people adapt to a new process, especially when they have become accustomed and experienced in another process. Training takes away the technical resistance to the new process.
- The validity of the change can be strengthened by recommendations from independent sources. Evidence that other companies or departments have been successful with similar efforts can assist in gaining acceptance. It becomes difficult to argue against change when there is a proven track record in place by others. Before

rolling out the new process, do your research and be prepared to cite examples of success stories.

- Another common point of resistance is to cite the inflexibility with the software tool or application that supports the process. Ensure there is a feedback loop for the new software. This allows for user suggestions and provides for continuous improvements. Also it builds a sense of ownership and contribution becomes apparent.
- Do not grant exceptions to the newly installed process. Allowing exceptions to the change in process typically is a way to ensure failure. This has the potential to spiral out of control as others attempt to be granted the same exceptions. Worse yet, others will attempt to expand exceptions. Soon the process is left dangling as an exception; in essence the exceptions become the process. Additionally, once exceptions are granted, many business case assumptions are thrown into disarray. In most instances the business case assumptions are based on full usage of the process improvement, not exceptions. Once exceptions are allowed less than 100 percent benefit potential is realized.
- Others attempt to delay the implementation with suggestions that this is not the right time for change. Honestly, there is never a right time for change. There will always be internal and external factors that will have competing influence with the process improvement.

9.2 Communicate, Communicate & Communicate

Making the decision to implement a new Process Safety management program into your organization is just the first step on a long journey. The best-laid plans may go awry if the focus is solely on the technical side, without considering the cultural and communication aspects of the equation.

Since implementing Process Safety usually involves changing human behavior, it is critical to include a communication plan that identifies and addresses human concerns. Initiating transformation of any magnitude within the organization requires meaningful dialogue with engineers, front line supervisors, maintenance staff, contract workmen and other key stakeholders. The leadership team must communicate early and often -- clearly conveying the vision, strategies and benefits for all concerned.

Remember to communicate up, down and sideways across the organization. Also keep in mind that communication is an ongoing process - not a one-time event. Consider the variation in your audiences and the methods that will be most appropriate for them. If there are many people in the organization without access to email or computers, for example, reliance on web-based communications alone may not be a viable option. Also consider the messages that will be shared at different stages of deployment and establish a regular rhythm for communicating through a variety of channels.

All organizations know that communication takes time and effort – but the investment is worthwhile. It is critical for people to be reminded of the vision but also how far they have come. This helps maintain morale and belief in the change process. Positive evidence that things are changing will combat any cynics.

When you are implementing a change process, communicate about 10 times more frequently than you think is necessary.

10. Control Phase – Tools and deliverables

In the **Control** phase, the performance of the process is verified in actual situation and necessary controls are put in place to sustain the benefits of the new process

The control phase consists of executing the following activities:

- Control methods and
- Control plan

There are several ***control methods*** that ensure the gain of the project stays even with passage of time. The only way to control a process is to apply control methods on Xs, the inputs and not on the Ys, the outputs. Inputs for identification of control methods come from the following:

- Final process map
- Updated FMEA
- Process SOPs and SMPs
- Training materials

The Japanese Poka-Yoke method called Mistake Proofing works towards eliminating an error condition from occurring. Usually through brainstorming Poka Yoke measures can be discovered. It advocates designing or developing tools, techniques and processes such that it is impossible or very difficult for people to bypass system or make mistakes. If eliminating error is not possible, provide warning signals in the system so that these can help to make corrections before the error escalates to a bad performance.

Statistical process control charts serve to identify variations due to common causes and assignable causes. While SPC charts are essential they do not by themselves constitute as good control methods. It is the corrective measures that constitute the control methods. If you use Poka Yoke as the control methods for all the inputs and the process parameters, then there is no need for SPC charts because the input variables should not be able to misbehave.

Having standard operating procedures is essential but are not sufficient to achieve consistent performance. Right half of the FMEA provides a good deal of control methods.

Human errors can be prevented by building good human systems such as:

- Training
- Discipline
- Skill building
- SOPs
- Standardization, etc.

Control plan documents all the control methods and actions to control the process.

A control plan should typically contain the following elements: The first column contains all the variables that need to be controlled. The second column lists what changes were made in order to control those key variables at the target values. The third tells what signals would indicate an adverse change in the variable and the last column gives the correction method.

The control plan reflects the long term control strategy to ensure improvements remain effective. Documenting the control procedures helps institutionalize the control process and extend the gains of the project to other areas. The control methods and control plans should be signed off by the business owner and project champions.

11. Roadmap for implementing the 20 RBPSM elements

One of the major performance measures of Process Safety excellence is the Safety Culture. In the Six Sigma parlance it is called Y and all other PSM elements are 'x's. Y is a function of 'x's.

$$Y = f(x1, x2, x3,,,,, xn)$$

You cannot control Y; you can only monitor Y. If you want to modify or improve Y it can be done only by tweaking the 'x's that contribute to Y. So no Six Sigma project need to be done to improve Y. It would automatically improve once all other elements are improved.

The only quantitative way to monitor Safety Culture is through surveys. Feedback from surveys should be taken seriously and the improvements liquidated before embarking on another surveys. Culture cannot be changed overnight. People have a tendency of going back to old ways unless there is a strong reason for them to stick to the new ways (What's in it for me!). So there should be adequate time before another survey is initiated.

One of the RBPSM elements is Metrics and Measurements; however this is part of all other PSM elements and gets developed during implementation of other PSM elements.

Effective processes for the remaining 18 Process Safety management elements could be implemented using the above DMDIC approach. Simultaneous implementation of all the 18 elements is strongly recommended as many elements have strong interaction with each other and the progress of one depends strongly on the other. For example training and Management of Change procedures are intimately connected with 'Operational Readiness'. If simultaneous implementation is not possible grouping of the elements could be worked out for sequential implementation.

Also investigation of Process Safety incidence should be done using the well established DMAIC methodology to ensure that we get to the root cause and minimize repetitive incidences of the same kind.

Continuous improvement of all the elements should be done using the following tools:

- Audits
- Management reviews
- Communications
- Change management strategy
- Review of the relevance of metrics and performance tracking

As we progress with the newly installed procedures, we would have plenty of data on the performance of the 20 elements. Further strengthening of the 20 elements could be done using the DMAIC methodology.

12. Conclusion

CCPS has recognized the following major challenges in moving to the Risk Based Process Safety management systems:

1. Developing an adequately detailed and accurate understanding of risk
2. Managing the initial difficulty in selecting the appropriate performance metrics
3. Acquiring the discipline required to maintain these performance metrics
4. Developing organizational trust and integrity to candidly review performance and

5. Overcoming resistance to making management system changes based on the predictive metrics.

Many of the above challenges could be overcome by installing processes based on the DMDIC methodology.

The cross functional team that is entrusted with the implementation of the RBPSM elements develops an adequately detailed and accurate understanding of the risk and they also develop appropriate performance and process metrics during different phases of the DMDIC project. Well prepared C& E diagram and the FMEA reflect the current state of safety culture at the site and they provide a basis for developing desired management systems and procedures for the chosen element. Change management issues are addressed during the implementation phase. In the control phase the team develops the discipline required to maintain the performance

It should be noted that the procedures developed at one site would not be applicable at another site as the size and hazardous nature of operations vary besides the safety culture. However teams working at another site could use the documents for working out procedures applicable to their site.

13. References

1. Breyfogle III, Forrest W, *Implementing Six Sigma, Smarter Solutions Using statistical Methods*, John Wiley & Sons Inc, 2003.
2. Doble, Mukesh, *Six Sigma and chemical process safety*, Intl. J. Six Sigma and Competitive Advantage, Vol. 1, No.2, 2005
3. Eckes, George, *Six Sigma for everyone*, John Wiley & Sons, New Jersey
4. *Guidelines for Risk Based Process Safety*, AIChE Center for Chemical Process Safety, New York, NY, 2007.
5. *Guidelines for performing Pre-startup Safety Reviews*, AIChE Center for Chemical Process Safety, Wiley, New York, NY, 2007.
6. http://www.isixsigma.com/
7. http://www.pmi.org

Quality Assurance and Instrumented Safety Systems

Angela E. Summers, PhD, PE, SIS-TECH Solutions, LP
12621 Featherwood Drive, Suite 120, Houston, TX 77034

A 10th PPSS Anniversary Paper

ABSTRACT

A perfect process would have no hazards, but perfection is impossible in the real world. Latent conditions exist in the equipment, procedures, and personnel training, eventually presenting challenges to safe operation when enough accumulate. Safe operation is maintained through implementation of a risk reduction strategy relying on a wide variety of safety equipment practices to prevent releases of highly hazardous chemicals. Quality design and management are absolutely essential if real risk reduction and incident prevention is to be achieved – not just calculated risk reduction. This paper uses the Shewhart Cycle to introduce the various activities involved in achieving safe operation using instrumented safety systems (ISS). The paper follows the Plan, Do, Check, and Act process to discuss quality assurance and its application to ISS.

INTRODUCTION

Accidents continue to happen because many owner/operators use injuries and fatalities as the predominant metric to demonstrate safe operation. A focus on direct impact can lead to a normalization of loss of containment events and a tolerance for events which increase the likelihood of impact. Further, injuries and fatalities should occur so infrequently that impact data is meaningless for monitoring performance.

Accidents often occur when safety equipment is not properly designed, installed, operated, tested and maintained. Industry has adequate theory and standards to ensure process equipment can be operated safely. Preventing errors and improving safety requires a systems approach that reduces the conditions contributing to error. The problem is not bad people and lack of competency; the problem is that the systems governing equipment are not rigorous enough to ensure reliability.

A quality management system must be used to sustain the rigor necessary to maintain equipment reliability; otherwise, accidents will still occur when enough latent conditions accumulate. A proactive approach monitors for behaviors, errors, and failures that lead to releases. Identifying improvement opportunities is essential to counter this accumulation and minimize risk.

PLAN

Deming believed that 85% of a worker's effectiveness is determined by the system he works within, only 15% by his own skill [1]. Planning ensures that work processes yield equipment that operates consistently in a safe manner, fulfills government and jurisdictional requirements, and meets recognized good engineering practices. The output of planning is a management system of policies, practices, and procedures that seek to identify and control releases of highly hazardous chemicals. Recommended work practices and activities are provided for instrumented protective systems in the CCPS book, Guidelines for Safe and Reliable Instrumented Protective Systems [2] and for safety instrumented systems in ANSI/ISA 84.00.01-2004 [3].

There is no substitute for knowledge. [4] Only a small amount of knowledge can save many hours of work or prevent mistakes leading to process hazards. Many owner/operators are losing process knowledge and history as operators and technical staffs retire or simply leave for better jobs. Errors accumulate unless there is continuous analysis and improvement of safety practices to counteract loss of expertise through retirements and downsizing and equipment degradation through age and obsolescence. Process knowledge is sustained by a foundation of written process safety information (PSI), covering the process hazards, technology, and equipment.

Knowledge grows when hazard evaluation procedures [5] are used throughout the equipment life to identify and evaluate significant events involving process operation. The event risk is analyzed qualitatively or quantitatively to determine the causes and potential frequency of occurrence. Independent protection layers are implemented to ensure that single failures or errors do not compromise safe operation. When the residual risk exceeds the owner/operator risk criteria, additional administrative and engineered safeguards are recommended and implemented to reduce the risk below the criteria. Knowledge also evolves over time as research and development yields operational facilities. Loss of containment events identify weaknesses in the risk reduction strategy leading to the implementation of more safeguards.

Personnel should be trained in the process hazards associated with their work activities. Personnel must have the skills and knowledge necessary to perform their work with the desired quality, so minimum job entry skills and knowledge should be specified. When on-the-job training is required, the training program should address how the skills and knowledge are developed in a timely and safe manner and how progress is measured [2].

A written design basis should define the PSI for the safety equipment and should be traceable to the process hazards analysis. For safety instrumented systems, the design basis is the hardware and software safety requirements specification. [3] The design basis should be maintained under revision control for the life of the equipment and should ensure the following:

- The safety equipment has been proven to work in the operating environment with the required performance.
- The system is designed to exceed the risk reduction specified in the hazards analysis.

However, the most important things are unknown and unknowable." [4] So, even when good people apply adequate theory and standards, there are always lessons to be learned. To counteract the unknown, owners/operators rely on a defense-in-depth strategy using multiple independent protection layers to lower operational risk [6]. These layers include instrumented safety systems (ISS) that achieve or maintain a safe state when unacceptable process conditions occur. Defense-in-depth also minimizes of common cause, common mode and systematic errors that cause multiple layers to fail [7,8].

Finally, planning must consider security and management of change. Physical and cyber access to the ISS should be restricted using administrative procedures and physical means [2]. Independence assessments should consider data communication and human interface failures. Written procedures should address how to initiate, document, review, and approve changes to ISS other than replacement in kind. Any change to the process and its equipment should be evaluated through a management of change process to determine the impact on the ISS requirements.

DO

The Do phase involves the implementation of the various policies, practices, and procedures. From a project implementation perspective, the detailed engineering is completed yielding an ISS installation that operating according to its design basis and exceeds the required performance. Detailed engineering includes sufficient information to ensure ISSs are properly specified, constructed, installed, commissioned, operated, and maintained. Equipment installed in ISS should be proven to provide the required performance in similar operating environments.

Equipment classification also considers the core attributes of protection layers, namely independence, functionality, integrity, reliability, auditability, management of change, and access security. Detailed design should provide an ISS equipment list identifying the equipment by a unique designation (e.g., the tag number) and required inspection and proof test interval.

Validation activities include an input to output test of each new or modified ISS to demonstrate and document that the equipment is installed according to the specification and operates as intended for each operating mode. Validation should be satisfactorily completed prior to the initiation of any operating mode where a hazardous event could occur that requires the operation of a new or modified safety system.

Proof tests are periodically conducted using a written procedure to demonstrate the successful operation of the ISS and to identify and correct deviations from the design basis and equipment specification. Maintenance personnel are trained on the procedures as necessary to ensure equipment is maintained in its "as good as new" condition. The proof test interval is chosen based on the relevant regulatory or insurance requirements, equipment history in a similar operating environment, manufacturer's recommendations, and risk reduction requirements.

Operating plans should consider the inspection and preventive maintenance requirements necessary to maintain the equipment in the "as good as new" condition. ISS proof tests should demonstrate that the mechanical integrity program is maintaining the required equipment performance. Review of mechanical integrity program records and trends is fed forward into the check phase of the quality cycle. Operating procedures should cover the safe and approved methods for interacting with the safety equipment, such as bypassing, manual initiation, and reset. Operations personnel should be trained and tested on the procedures as necessary to ensure correct actions are taken. Operator actions in response to abnormal operation should be recorded and periodically assessed.

CHECK

By what method? Only the method counts. [4] The check phase applies metrics to the work processes to provide a standard means to assess performance against requirements. Sustainable operation is achieved by focusing on metrics which provide performance indication on a real-time basis. Example metrics are provided in Table 1 for the ISS. Additional recommended metrics have been published by CCPS [9].

The most important things cannot be measured. [1] Selecting appropriate metrics to track can seem like an overwhelming task. Sometimes, technical personnel want to measure everything just because they can. Metrics should be carefully chosen, so that just the right amount of meaningful data is collected. Good metrics drive personnel to do the right thing. It is unfortunate, but true, that personnel will behave contrary to reason and to the best interest of the company if necessary to "make their numbers."

On the other hand, the more that is known about the equipment and what is affecting its operation, the better the risk can be managed. For safety equipment, the quality of the installed equipment is limited by the rigor, timeliness and repeatability of mechanical integrity activities. Near miss and incident investigations should evaluate any identified ISS inadequacy or failure. Spurious trips and process demands should be tracked and compared with expectations in the hazard analysis. Management of change processes should be used to resolve performance gaps.

In the real world, many owner/operators are essentially following the old adage: "Measure with a caliper. Mark it with a scribe. Cut it with a chain saw." Their process hazards analysis is becoming increasingly quantitative with more factors and modifiers; the verification of risk reduction uses multiple significant digits; and the mechanical integrity record simple states "passed."

The real world must come into balance, because the risk reduction strategy is proven by mechanical integrity data. The risk reduction provided by the ISS is the inverse of its probability of failure on demand (PFD) that is calculated as the number of devices failed dangerously divided by the total number of devices tested. Using probabilistic techniques, the PFDs of specific equipment can be used to determine the system performance for comparison to the design basis assumptions [7].

Repeated failures indicate that the mechanical integrity program is inadequate. Failure tracking is essential to close the safety lifecycle. Root cause analysis is used to determine why metrics are trending in the wrong direction, so that actions plans can be implemented to improve the management system, equipment, procedures, and personnel training.

Continuous improvement is incorporated in PSM through a concept often called "grandfathering," where the owner/operator determines and documents that the existing equipment is designed, maintained, inspected, tested, and operated in a safe manner. This requires an assessment of the existing design and management practices against current good engineering practices and process requirements. The review should determine whether the existing ISS is operating per the design basis and the current management system is sufficient to yield the required risk reduction.

ACT

"What is a system? A system is a network of interdependent components that work together to try to accomplish the aim of the system. A system must have an aim. Without an aim, there is no system. The aim of the system must be clear to everyone in the system. The aim must include plans for the future. The aim is a value judgment [4].

The act phase involves the actions taken in response to trends in metrics and to continuous improvement opportunities. It is the opportunity for the owner/operator's safety culture to shine and for risk to be driven as low as reasonably practicable. Action plans should define a path forward, milestones, and timelines. Plans should be periodically assessed to determine whether there is a need to accelerate the schedule or broaden the plan objectives. For example, a planned ISS upgrade may be accelerated when the manufacturer withdraws support for the equipment. To be successful, action plans should be communicated to affected personnel so they understand and commit to it.

Continuous improvement is necessary to stay on top of latent conditions that present potential safety challenges and weaken protection layers. Implementing upgrades aimed at improving long-term operational effectiveness takes time to complete, depending on the complexity and degree of change involved. As the ISS is changed, operating plans and targets should consider any additional risk to be managed during the transition. The ISS operating and mechanical integrity basis should be reviewed and needed revisions implemented to ensure equipment, procedures, and personnel training remain in sync with modifications.

SUMMARY

Deming believed that experience by itself teaches nothing and that data without context is meaningless. Information gained from experience must be interpreted against a framework of expected behavior, equipment design and operating performance. But, experience is not always the best teacher. Without an understanding of the underlying root causes, raw data can be misinterpreted creating a flawed view of reality. Only data understood within its proper context provides a solid foundation for safe operation.

Accidents are prevented when safety issues are approached from a quality perspective. The plan, do, check, and act phases are essential to maintaining safe and reliable operation. A management system supported with metrics should be used to establish targets and monitor performance against policies, practices, and procedures. Periodic gap analysis should used to verify that actual performance exceeds expectations established in the hazard analysis and design basis. Performance gaps should be closed with action plans that reduce risk and prevent accidents.

REFERENCES

1. Deming, W. Edwards, *Out of Crisis,* MIT Press, (1986).
2. *Guidelines for Safe and Reliable Instrumented Protective Systems,* American Institute of Chemical Engineers, NY, (2007).
3. ANSI/ISA 84.00.01-2004, *Functional Safety: Safety Instrumented Systems for the Process Industry Sector*, Instrumentation, Systems, and Automation Society, NC, (2004).
4. Deming, W. Edwards, *The New Economics for Industry, Government, Education,* 2nd Edition, MIT Press, (2000).
5. *Guidelines for Hazard Evaluation Procedures, Second Edition with Worked Examples*, American Institute of Chemical Engineers, NY, (1992).
6. *Layer of Protection Analysis: A Simplified Risk Assessment Approach*, American Institute of Chemical Engineers, NY, (2001).
7. ISA TR84.00.02, *Safety Instrumented Functions (SIF) – Safety Integrity Level (SIL) Evaluation Techniques*, Instrumentation, Systems, and Automation Society, NC (2002).
8. ISA TR84.00.04, *Guidelines for the Implementation of ANSI/ISA 84.00.01-2004 (IEC 61511)*, Instrumentation, Systems, and Automation Society, NC (2005).
9. CCPS/AICHE, "Process Safety Leading and Lagging Metrics," proposed metrics for review published on AICHE website (Jan 2008).

Table 1 Example Metrics Related to Instrumented Safety Systems

Lifecycle Step	Example Metric
Hazard Analysis	Total number of hazard and risk analysis scheduled during defined interval
	▪ Number completed
	▪ Number behind schedule
	▪ Percentage hazard and risk analysis behind schedule
	▪ For those behind schedule, total number of days behind schedule
Design Basis	Total number of safety equipment
	▪ Number of equipment with as-built documentation
	▪ Number of equipment with redlined or missing documentation
	▪ Percentage of equipment with redlined or missing documentation
Mechanical Integrity	Total number of inspections scheduled during defined interval
	▪ Number of inspections on-schedule but incomplete
	▪ Number of inspections completed
	▪ Number of inspection behind schedule
	▪ Percent on-time for inspection
	▪ Percent behind schedule for inspection
	▪ For completed, number of successful inspections
	▪ Percentage of successful inspections
	Total number of equipment tests scheduled during defined interval
	▪ Number of tests incomplete
	▪ Number of tests completed
	▪ Number of test behind schedule
	▪ Percent on-schedule for test
	▪ Percent behind schedule for test
	For completed tests:
	▪ Number of tests with "as found" within equipment specification
	▪ Number of tests with "as found" outside equipment specification (i.e., failed dangerously, failed safe, or degraded state)
	▪ Percentage of tests within equipment specification
	▪ Percentage of tests outside equipment specification
	Total number of safety equipment
	▪ Total number of failures found by diagnostics
	▪ Total number of failures found by inspection and testing
	▪ Total number of failures requiring equipment repair or replacement
	▪ Total number of failed equipment returned to service within allowable repair time
	▪ Percentage of equipment returned to service within the allowable repair time

Lifecycle Step	Example Metric
Degraded Operation	Total number of safety equipment that are out of service (bypassed, disabled, overridden, or under test/repair) during process operation for defined interval
	▪ Total number hours that safety equipment is out of service
	▪ Number of safety equipment that are returned to service within the allowable repair time
	▪ Number of safety equipment that are out of service, but covered under MOC
Process Performance	Percentage of start-ups involving abnormal or emergency operation
	Total number of process shutdowns during defined interval
	▪ Number due to spurious operation of safety equipment
	▪ Number due to abnormal or emergency operation
	Total number of safety alarms during defined interval
	▪ Number of standing or nuisance alarms
	▪ Number of safety alarms requiring response

Safe Management of Unforeseen Delays in Mechanical Integrity Inspection Schedules

James H. Willis
Air Products & Chemicals, Inc
Allentown, PA 18195
willisjh@airproducts.com

Harold W Thomas
Air Products & Chemicals, Inc
Allentown, PA 18195
thomashw@airproducts.com

Brian R. Dunbobbin
Air Products & Chemicals, Inc
Allentown, PA 18195
dunbobbr@airproducts.com

A 10th PPSS Anniversary Paper

ABSTRACT

The days of "frequent" annual maintenance outages where mechanical integrity testing and inspections can be performed are gone. The new reality of modern industrial gas and chemical plant operation includes expectations of ever increasing durations between scheduled maintenance down time that have sometimes reached 5 years and beyond. In addition, in the interest of keeping initial capital cost under control, there is constant pressure to restrict redundant architectures and systems intended to allow on-line maintenance.

Required safety systems that have the potential for unrevealed failures lead to a requirement for plant down time to properly test and inspect those systems. Because our customers expect high reliability with long runs between maintenance outages it is prudent to design our facilities and mechanical integrity (M.I.) plans to meet the agreed commercial strategy. However, unforeseen events can cause delays in maintenance outages and thus our window to complete mandatory non-deferrable M.I. activities. Air Products' process for reviewing the consequence and likelihood of these delays allows appropriate action to be taken to ensure continued safe operation while complying with customer needs to reschedule downtime windows to the greatest extent possible.

This paper outlines the policy used to review and manage the effects of inspection and test delays on facility risk. The hazard review process is administered and documented through the company's existing Management of Change process. The review itself is handled using a team with diverse technical expertise that can uncover and address key factors affecting safety. Levels of required management approval and Process Safety representation varies based on the expected length of the planned delay. Duration of the extension dictates the level of technical reviews and provides guidance as to the whether qualitative or quantitative techniques are appropriate. In either case, mitigation techniques may be used to temporarily manage the increased level of risk associated with the inspection or test delay.

Implementation is challenging as the plant culture migrates from one of local decisions and control to one of understanding the multidisciplinary impact these delays create. The preferred solution is always to find a way to perform the task as planned or by using alternate "on stream" inspection methods without requiring an extension review.

1. INTRODUCTION

Maintaining safety in industrial facilities goes far beyond worker safety concerns such as slips, trips and falls. The industry recognizes that control and containment of the process is integral to protecting people and the environment. Mechanical Integrity (M.I.) programs are implemented for a variety of disciplines such as Safety Instrumented Systems, Relief Devices and Pressurized Equipment to achieve this control of the process energy and fluids.

In order to meet commercial demands, delays to plant outages are sometimes necessary. Because delays in safety related inspections and tests can increase the likelihood of an undesirable event, these delays are only permitted in exceptional circumstances. Technical review and approval of the delay is required to assure consistent Mechanical Integrity at the site.

Much of the discussion included here is from the viewpoint of a large company operating a large number of small to medium sized facilities where leadership and technical resources are often regional, covering a significant number of sites. Large, self-contained facilities may find some differences in their own support organization that eliminates some of the variables and concerns contained herein.

2. CONDITIONS, HISTORY & REQUIREMENTS

In the past, outages on a one or two year basis were more commonplace in industry. These routine "frequent" maintenance outages made it relatively simple to schedule and complete inspections and tests such as safety instrumented system/function revalidation and pressure relief device proof testing. As customer requirements changed, required intervals between scheduled maintenance outages changed as well. With some intervals between scheduled maintenance downtime stretching up to 5 years and beyond, scheduling of down time inspections became more difficult and unexpected delays and schedule changes have now become an ongoing concern.

It was once common and acceptable for decisions concerning inspection timing and delays to be made at the site level by local leadership. Matching our downtime requirements for inspection and testing with a customer, say a large oil refinery, can cause problems in completing the required tasks, as in many cases, if our plant goes down, the refinery can not operate. The primary variables considered in the decision were safety and environmental, process plant availability and cost. Site leadership did not always have an in depth understanding of the rate of degradation of equipment and how inspection intervals affected the likelihood of an undesirable event. As M.I. philosophy took hold, understanding that limitations do exist in the equipment and that periodic inspection, based on knowledge of specific equipment degradation, is essential to safe operation. A review of root cause investigations from various industry incidents shows that abnormal conditions and temporary procedures are often to blame for hazardous releases. It is imperative that temporary activities and delays in required inspections and tests be properly reviewed, managed and integrated with the Management of Change process.

Authorization of M.I. delays must start with a thorough understanding of the process conditions, degradation mechanisms, maintenance history and proposed duration of the delay. It is impossible to properly assess likelihood of failure of a component without knowing its design life based on real conditions and failure mechanisms. One cannot rely only on generic or manufacturer's recommended design life or maintenance intervals. All normal and transient operating conditions must be considered as well as local environmental effects on the equipment. In fact, experience shows that transient conditions are often the "worst case".

3. THOROUGH TECHNICAL REVIEWS

Although the facility's goal is to gain approval to safely defer the inspection/test requirement, the preferred option is always to find some way to perform the inspection/test activity. Prior to making the request for extension, the facility manager will hold a small "pre-screen" meeting with the facility personnel and appropriate support engineers. The goal of this session is two fold:

1) Validate the need for extension for each item. If the inspection or test can be completed through alternate methods or by operating redundant equipment temporarily, the request for extension will not be needed.
2) If the extension request will proceed, the team will assign action items to collect pertinent data required by the technical review team.

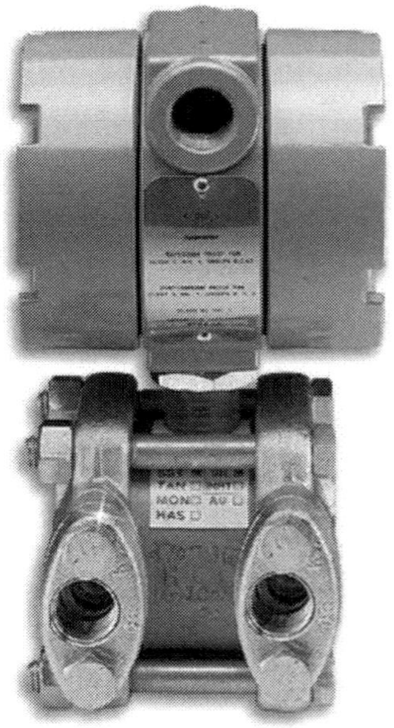

In order for the pre-screen meeting to be productive, the following supporting documentation should be available, such as Piping and Instrument Diagrams (P&ID), equipment specifications, electrical schematics, process parameters and inspection/test requirements for each item to be extended.

Provided that the need for deferral is validated, the extension review process can proceed. To make meaningful connections between the remaining life of the component and the operating conditions under which it is subjected, a diverse technical team is assembled for each review session. The involvement from other disciplines may be a limitation from a resource management point of view[1] but input from this diverse team is very important as it is fundamental to the effectiveness in this process. Involving this varied team allows the capture of important details from each technical discipline:

- Facility Manager
- Facility Operator (specific unit experience required)
- Process Engineer (specific unit or technology experience required)
- Maintenance Engineer/Technician/Leader (specific equipment experience required)
- Engineer for discipline/technology being reviewed
- Process Safety Engineer

The ensuing technical review can be qualitative or quantitative, based upon recommendations from the Process Safety Engineer (not required for temporary extensions less than 90 days). The depth of the technical review will be based on the complexity of the system under review.

For example: Discussion concerning single component inspections may only involve a simple review of facts while changes affecting broad systems may involve a detailed Quantitative Risk Assessment (QRA). This review may address the following issues, based on system complexity and technical discipline:

1) Plant components are designed, operated and inspected/tested through the M.I. Program in a specific manner to achieve and maintain certain levels of risk reduction. The assumed levels of risk reduction in the design, operation and M.I. program must be documented in order to assure through the technical review that the new test frequencies still meet the risk target.
2) Remaining useful life for critical equipment is documented and the proof test interval is no longer than one half of the remaining useful life as determined by fitness for service.
3) Each layer of protection operating in stand by mode (i.e. PSV, SIF, etc.) has auditable documented risk reduction performance criteria necessary to achieve the defined hazard rate target.
4) Temporary mitigating activities or conditions that can be put in place to ensure continued safe operation while the Mechanical Integrity task is under an extension.

The technical team shall consider normal and transient operating conditions, past maintenance and operating history of the component and expected degradation mechanisms.

The approach used is similar to Risk Based Inspection (RBI) techniques. RBI studies are known to have advantages in evaluation of the operational risks associated with extending time between unit shutdowns[1]. Historical evidence is an important part of a successful review. The more information that is available for the system, the easier these reviews can typically be. As such, a reliable system of archiving maintenance history is quite valuable to the overall review process. In fact, collection and intelligent evaluation of such data can also be used to review and validate the baseline M.I. inspection and test program.

"We gain substantial insight into the relationship between inspection and risk by recognizing which component of risk a particular inspection activity is intended to reduce.[2]" Since the natural result of deferring inspection and test activity is an increase in likelihood of a failure, mitigation is a highly regarded method in this process for temporarily adding other safety measures to counteract the basic increase in failure likelihood for the equipment in question. The object can either be to reduce the likelihood of a demand on the component or reduce the consequence should a demand occur. Below are examples of possible mitigation steps for pressure relief devices.

Likelihood Reduction:

- Sources of overpressure positively isolated or adequately limited for duration of extension.
- Lock open of path to second relief device for duration of temporary extension period.
- Additional means of overpressure protection implemented for the period of the extension on pressure system.
- Installation of mechanical stops.

Consequence Reduction:

- Deny or limit access to area of potential hazard during extension period.
- Temporary containment methods to capture liquid leaks (environmental concerns)
- Temporary reduction in hazardous material inventory such as reducing operating levels in tanks.
- Alter operating conditions to limit demand cases.

4. STANDARDIZED WORK PROCESS FOR CONSISTENT RESULTS

Discrete facilities with dedicated support staff benefit from assigned personnel completing each review, using their experience with the process to produce consistent, repeatable decisions. This is an ideal situation for yielding consistent results. However, consistency can be a challenge for companies which operate a large number of facilities where multiple groups of technical support personnel are often regionally based to better serve a number of facilities. Centralized technical resources are less common, so clear procedures are required to drive consistency rather than by using a single group of resources to cover all extension reviews. Each team may contain members who are new to the process so it is vital that the procedure be clear and easy to follow.

As you might expect, different individuals will have varying preferences and beliefs which can be detrimental to consistency when the reviews performed are often qualitative. Our solution involved a basic set of boundaries and rules combined with logic flow diagrams to guide the process. For illustration, Figure 1 shows the basic steps involved in the review of a pressure relieving device.

Figure 1: Pressure Relief Device Work Flow Diagram Example

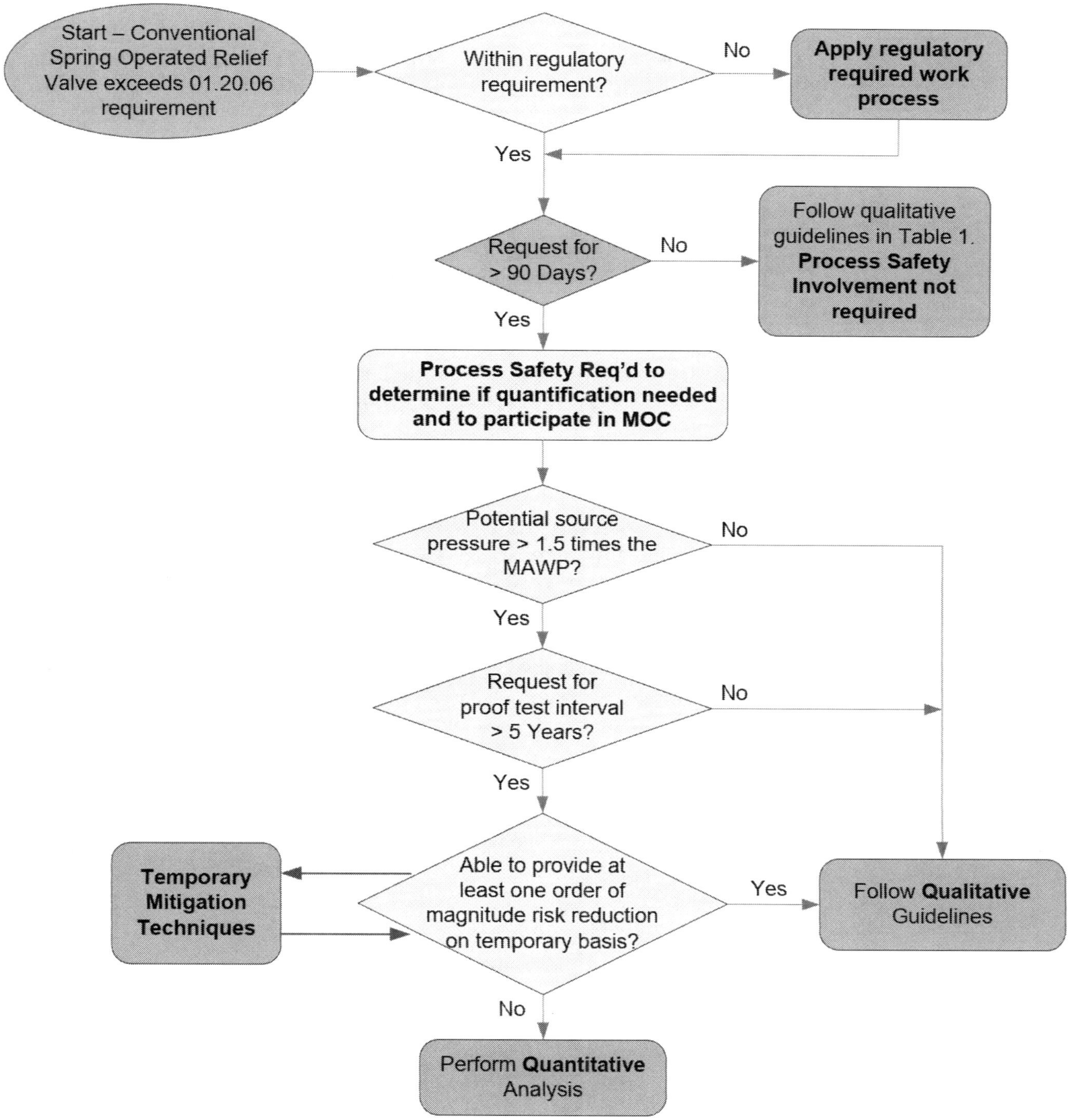

5. LEADERSHIP TEAM AWARENESS AND APPROVAL

The technical review and approval is an important part of the process but does not result in the final decision. In fact, the technical team recommendations are used as a decision making tool by the ultimate approval authority which varies by the length of the requested deferral. Table 1 shows the stepwise increase in management level as the deferral interval increases:

Table 1: Temporary Inspection or Test Frequency Deviation Approval Levels

In Compliance	Facility Manager	Senior Line Manager	Senior Line Manager and Process Safety	Operating Group V.P. and Process Safety
≤ 30 days beyond test or inspection due date	31-60 days beyond inspection or test due date	61-90 days beyond inspection or test due date	90-180 days beyond inspection or test due date.	>180 days beyond test or inspection due date.

The process of elevating the management level for approval as risk increases is in contrast to the old protocol of local unit managers making most of their own decisions despite having possible gaps in their knowledge of technical details. The delays and thus the resulting risk modifications are now visible to leadership teams. Through key process indicators (KPI's) and knowledge of plant activities, managers pay closer attention to the level of process safety at facilities and drive company wide improvements.

Once the technical review team has reviewed the facts and made a recommendation, the approver must make a final decision on whether to grant a deferral by taking the technical team results under advisement. It is possible for a senior level manager to override the technical team recommendation. The implication is that the approving manager takes responsibility for the safety and environmental risk posed by the deferral.

6. BARRIERS TO IMPLEMENTATION

The mechanical integrity aspect of process safety in facilities has vastly improved as compared to just over 5 years ago. Many unit managers viewed inspections and tests as a maintenance chore in the facilities and not something that had a significant affect on safety or environmental quality. Specific equipment inspections might have been recognized as a benefit to reliability but not plant EH&S.

The gap was due to lack of understanding as to what factors go into a hazard rate. Although process safety professionals clearly understand the importance of designing inspection and test programs to avoid the "one in a million" event, local unit leaders do not have reason to think in those terms. A typical facility might operate for up to 50 years making the "one in a million" event unlikely to occur during the unit's lifetime. When those numbers are observed over 500+ facilities, the possible events become more real and inspection/test programs must address the possibility. When looking at the data, this effort makes perfect sense to more senior management looking at 500+ facilities but may look superfluous to an individual responsible for a single site.

As a result, decisions concerning inspections have in some cases been made at the local level and not all the factors necessary for decision making may have been considered when delaying an M.I. activity by months or even years. Implementing a program for elevating approval of deferrals removes that power from the local leaders and places it higher in the management ranks. Clearly one could expect resistance to this change as power is transferred.

Those not intimately tied to mechanical integrity may not understand the direct link between component remaining life, operations practices and the inspections performed to "catch" issues before they become loss of containment. This can develop into an attitude that these maintenance items are a low priority and should not affect plant run time.

7. Technical Guidance Example

When performing technical analyses to determine if a temporary extension is permissible, it is important to understand the basis of the equipment's failure rate and how the extension would decrease its likelihood of performing successfully. There are two aspects of the problem. The first is to know that the equipment continues to operate in its useful life. The second is to determine whether the decrease in performance due to a longer proof test interval results in sufficient loss of protection that the hazard rate target is no longer met. In many cases, the initial scope coupled with a standard design, operation, and mechanical integrity program results in a level of performance that exceeds requirements. As long as the expected risk reduction for the protection layer is documented, this determination is fairly straight forward as long as the equipment can be assumed to be operating within its useful lifetime. The information below provides some insight regarding conventional pressure relief devices when considering the question of useful life and probability of failing to function on demand.

Figure 2

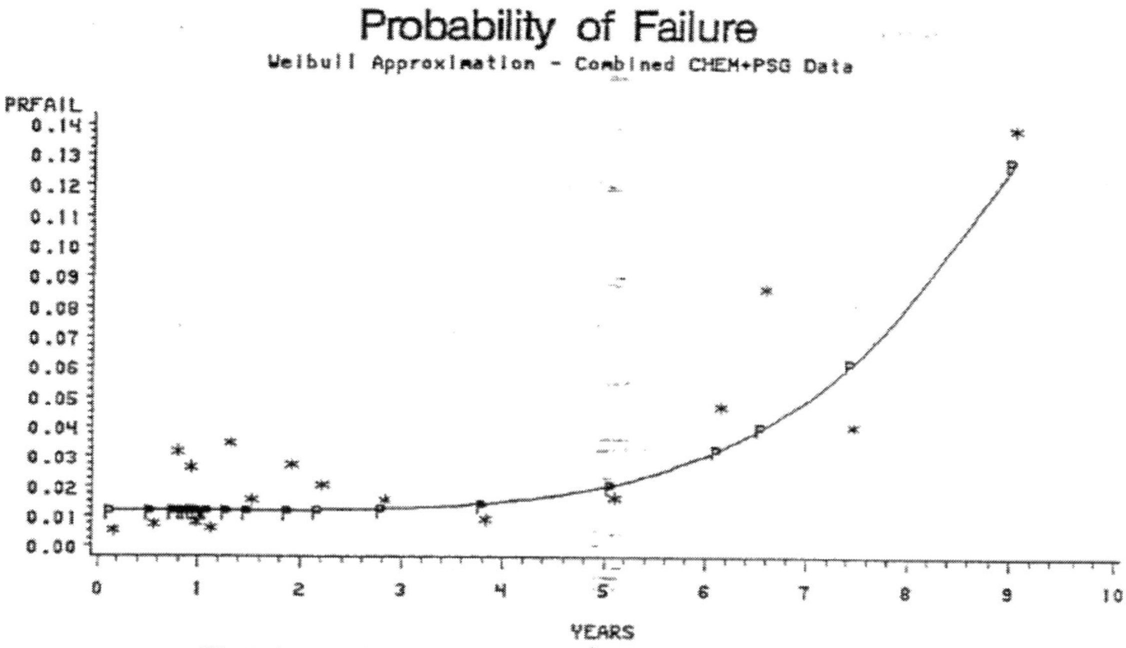

Fig 3. Probability of Failure versus Time for CHEM+PSG Data

One means of estimating useful life is data analysis. The basis of the graph below is predominantly an analysis of typical Air Products spring operated relief valves in air separation plants with some hydrogen/carbon monoxide plant data. The **failure mode = relief valve fails to open**. Fail to open is defined as "proof test report indicated valve did not open" OR "the valve did not lift at a test pressure greater than or equal to 1.5 times its set pressure". The data was analyzed using the quantal response method[3].

This data compares favorably to others[4] who have engaged in testing designed to produce proof test data that corresponds to good experimental design technique and analysis programs.

Just as important is a qualitative review of the specific application to determine if the valve in question is more likely to exhibit known failure mechanisms. For instance, Gross and Harris[4] have determined that one failure mechanism is corrosion of carbon steel due to atmospheric exposure, even though the internal wetted parts were stainless steel. They also found this to be true when the valve was stainless steel, but had carbon steel springs and spring washers. From a qualitative perspective, exposure to the elements coupled with material of construction will have an effect.

8. CONCLUSION

The deferral of Mechanical Integrity activities must involve the expertise of a wide variety of unit knowledgeable personnel. Without this varied input into the discussion, key parameters can be missed. In reviewing industry incidents, we see that temporary conditions are often identified as initiators of an event. This work process assures through thorough technical reviews that all relevant factors are considered and ultimately that temporary exceptions to normal Mechanical Integrity activities still meet the unit's established hazard rate targets.

In order to operate the safest possible facility, systems and programs that are sustainable must be created that are palatable to the employees responsible for their use. It goes without saying that both Process Safety and Mechanical Integrity are cornerstones of the safety structure in an operating facility. However, strict and rigid standards and practices can be unproductive when attempting to enforce them on multiple global facilities. Support and participation improves when fundamentally sound work processes are developed with appropriate flexibility.

9. REFERENCES

1. Valbuena, Ricardo, *Implementing and Evergreening RBI in Process Plants*, Materials Technology Institute, 2005.
2. American Petroleum Institute, R*isk Based Inspection Base Resource Document: Publication 581,* API Publishing Services, Washington DC: 2000
3. Sheesley, J.H., Thomas, H.W, Valenzuela, C.A.; *Quantal Response Analysis of Relief Valve Test Data*, ASQC 49th Annual Quality Congress, Cincinnati, Ohio, May 1995.
4. Gross, R.E., Harris, S.P.; *Analysis of Safety Relief Valve Proof Test Data to Optimize Lifecycle Maintenance Costs*, RAMS, 2008.

RECENT DEVELOPMENTS AND TECHNOLOGY IMPROVEMENTS IN API RISK-BASED INSPECTION PLANNING TECHNOLOGY

P. A. Henry, P.E.
The Equity Engineering Group, Inc.
Shaker Heights, Ohio USA

D. A. Osage, P.E.
The Equity Engineering Group, Inc.
Shaker Heights, Ohio USA

A 10th PPSS Anniversary Paper

ABSTRACT

A Joint Industry Project for Risk-Based Inspection (RBI JIP) was initiated and managed by API within the refining and petrochemical industry in 1994. The work of this JIP resulted in two publications, API 580 Risk-based Inspection and API 581 Base Resource Document – Risk-Based Inspection. The concept behind these publications was for API RP 580 to introduce the principles and present minimum general guidelines for RBI while API 581 provides quantitative RBI methods. The API RBI JIP has made major advancements since the publication of these two documents and is working on the second edition of API 581. The second edition will be published in a three volume set, Part 1 – Inspection Planning Using API RBI Technology, Part 2 – Probability of Failure in API RBI, and Part 3 – Consequence Modeling in API RBI.

This paper provides an overview of this new three volume set, describes the advances made in risk-based inspection technology, provides an overview of a new consequence modeler designed to evaluate complex fluids, and explains a new approach that utilizes API 579-1/ASME FFS-1 Fitness-For-Service (FFS) models to determine the probability of failure component of risk. This new approach reinforces the complementary nature of FFS and RBI for asset management and ensures that standard models and assessment techniques are used.

1. INTRODUCTION

The API Risk-Based Inspection (API RBI) methodology may be used to manage the overall risk of a plant by focusing inspection efforts on the process equipment with the highest risk. API RBI provides the basis for making informed decisions on inspection frequency, the extent of inspection, and the most suitable type of Non-Destructive

Examination (NDE). In most processing plants, a large percent of the total unit risk will be concentrated in a relatively small percent of the equipment items. These potential high-risk components may require greater attention, perhaps through a revised inspection plan. The cost of the increased inspection effort may sometimes be offset by reducing excessive inspection efforts in the areas identified as having lower risk.

API 580 [1] provides guidance on developing a risk-based inspection program for fixed equipment and piping in the hydrocarbon and chemical process industries. API 581 provides quantitative RBI methods to establish an inspection program. A complete re-write of API 581 [2] as an API Recommended Practice is currently being prepared to capture new technology that has been developed by the API JIP, to provide a logical step-by-step procedure that will enable practitioners to better understand the methodology, and to make the technology clear to facilitate an effective peer review. The title of the document is being changed to *API RBI Technology* to emphasize the fact that significant technology, including industry best practices, is required to effectively implement an RBI program. The second edition of API 581 is targeted for publication during 2008 and will present the API RBI methodology in a three part volume:

- Part 1 – Inspection Planning Using API RBI Technology
- Part 2 – Determination of Probability of Failure in an API RBI Assessment
- Part 3 – Consequence Analysis in an API RBI Assessment

The calculation of risk in API RBI involves the determination of a probability of failure combined with the consequence of failure. Failure in API RBI is defined as a loss of containment from the pressure boundary resulting in leakage to the atmosphere or rupture of a pressurized component. As damage accumulates in a pressurized component during in-service operation the risk increases. At some point, a risk tolerance or risk target is exceeded and an inspection is recommended of sufficient effectiveness to better quantify the damage state of the component. The inspection action itself does not reduce the risk; however, it does reduce uncertainty thereby allowing better quantification of the damage present in the component.

In API RBI, a component is defined as any part that is designed and fabricated to a recognized code or standard. For example a pressure boundary may consist of multiple components (cylindrical shell sections, formed heads, nozzles, tank shell courses, tank bottom plate, etc.). Equipment is defined as an assemblage of components.

An overall event tree for the API RBI methodology is presented in Figure 1. This figure illustrates how the probability of loss of containment from in-service process equipment and the consequence analysis portions of API RBI fits within the overall RBI methodology. Probabilities of loss of containment are a function of the generic failure frequencies for particular components and the calculated damage state (damage factors) of the component being evaluated.

2. PROBABILITY OF FAILURE

The probability of failure used in API RBI is computed from Equation (1).

$$P_f(t) = gff \cdot D_f(t) \cdot F_{MS} \tag{1}$$

In this equation, the probability of failure, $P_f(t)$, is determined as the product of a generic failure frequency , gff , a damage factor, $D_f(t)$, and a management systems factor, F_{MS} .

The generic failure frequency for different component types was set at a value representative of the refining and petrochemical industry's failure data. The generic failure frequency is intended to be the failure frequency prior to any specific damage occurring from exposure to the operating environment, and are provided for several discrete hole sizes for various types of processing equipment (i.e. process vessels, drums, towers, piping systems, tankage, etc.). Discrete hole sizes and an associated failure frequency are introduced into the assessment to model release scenarios. API RBI uses four hole sizes to model the release scenarios covering a full range of events (i.e. small leak to rupture).

Adjustment factors are applied to the generic failure frequencies to reflect departures from the industry data to account for damage mechanisms specific to the components' operating environment and to account for reliability management practices within a plant. The damage factor is determined based on the applicable damage mechanisms (local and general corrosion, cracking, creep, etc.) relevant to the materials of construction and the process service, the physical condition of the component, and the inspection techniques used to quantify damage. The damage factor modifies the industry generic failure frequency and makes it specific to the component under evaluation.

The management systems adjustment factor, F_{MS}, accounts for the influence of the facility's management system on the mechanical integrity of the plant equipment. This factor accounts for the probability that accumulating damage which results in loss of containment will be discovered in time and is directly proportional to the quality of a facility's mechanical integrity program. This factor is derived from the results of an evaluation of a facilities or operating unit's management systems that affect plant risk.

The damage factor is applied on a component and damage mechanism specific basis while the management systems factor is applied equally to all equipment within a plant. Damage factors with a value greater than 1.0 will increase the probability of failure, and those with a value less than 1.0 will decrease it. Both adjustment factors are always positive numbers.

Damage factors do not provide a definitive Fitness-For-Service assessment of the component. The basic function of the damage factor is to statistically evaluate the amount of damage that may be present as a function of time in service and the effectiveness of an inspection activity to quantify that damage.

Methods for determining damage factors appear in API 581 for the following damage mechanisms:

- Thinning – D_f^{thin}
- Component Linings – D_f^{elin}
- External Damage (corrosion and stress corrosion cracking) – D_f^{extd}
- Stress Corrosion Cracking (internal) – D_f^{scc}
- High Temperature Hydrogen Attack – D_f^{htha}

- Mechanical Fatigue (Piping Only) – D_f^{mfat}
- Brittle Fracture – D_f^{brit}

If more than one damage mechanism is present, then the principal of superposition, with a special modification for general thinning and external damage, and component linings, is used to determine a total damage factor in accordance with Equation (2).

$$D_{f-total} = \min\left[D_f^{thin},\ D_f^{elin}\right] + D_f^{extd} + D_f^{scc} + D_f^{htha} + D_f^{brit} + D_f^{mfat} \tag{2}$$

3. CONSEQUENCE OF FAILURE

3.1 General

Loss of containment of hazardous fluids from pressurized processing equipment may result in damage to surrounding equipment, serious injury to personnel, production losses, and undesirable environmental impacts. In API RBI, the consequences of loss of containment are determined using well established consequence analysis techniques [3], [4], [5] and are expressed as an affected impact area or in financial terms. Impact areas from such event outcomes as pool fires, flash fires, fireballs, jet fires, and vapor cloud explosion (VCEs) are quantified based on the effects of thermal radiation and overpressure on surrounding equipment and personnel. Additionally, cloud dispersion analysis methods are used to quantify the magnitude of flammable releases and to determine the extent and duration of personnel exposure to toxic releases. Event trees are utilized to assess the probability of each of the various event outcomes and to provide a mechanism for probability-weighting the loss of containment consequences.

An overview of the API RBI consequence analysis methodology is provided in Figure 2.

Methodologies for two levels of consequence analysis are provided in API RBI. A Level 1 consequence analysis provides a simplistic method to estimate the consequence area based on lookup tables for a limited number of generic or reference hazardous fluids. A Level 2 consequence analysis methodology has been added to API 581 that is more rigorous in that it incorporates a detailed calculation procedure that can be applied to a wider range of hazardous fluids.

3.2 Level 1 Consequence Analysis

API 581 [2] provides a simplistic method for evaluating the consequences of hazardous releases for a limited number of reference fluids. The available reference fluids are shown in Table 1. A reference fluid is selected from Table 1 that closely matches the normal boiling point (NBP) and molecular weight (MW) of the fluid contained within the process equipment. The consequence area is then determined from a simple polynomial expression that is a function of the release magnitude.

For each discrete hole size, release rates are calculated based on the phase of the fluid. For liquid phase Equation (3) is used, and for vapor phase Equation (4) or (5) is used. Equations (4) and (5) are based on critical or subcritical flow through an orifice or hole.

$$W_n = C_d K_v \rho_l \left(\frac{A_n}{144}\right)\sqrt{\frac{2 \cdot g_c \cdot (P_s - P_a)}{\rho_l}} \tag{3}$$

$$W_n = C_d P_s A_n \sqrt{\left(\frac{k \cdot MW \cdot g_c}{RT_s}\right)\left(\frac{2}{k+1}\right)^{\frac{k+1}{k-1}}} \quad \textit{for sonic velocity} \tag{4}$$

$$W_n = C_d P_s A_n \sqrt{\left(\frac{MW \cdot g_c}{RT_s}\right)\left(\frac{2k}{k-1}\right)^{\frac{2}{k}}\left(1-\left(\frac{P_a}{P_s}\right)^{\frac{k}{1-1}}\right)} \quad \textit{for subsonic velocity} \tag{5}$$

For the Level 1 analysis, a series of consequence analyses were performed to generate consequence areas as a function of the reference fluid and release magnitude. In these analyses, the major consequences were associated with pool fires for liquid releases and VCEs for vapor releases. Probabilities of ignition, probabilities of delayed ignition, and other probabilities in the Level 1 event tree were selected based on expert opinion for each of the reference fluids and release types (i.e. continuous or instantaneous). These probabilities were constant and independent of release rate or mass. These analyses resulted in consequence area equations of the form shown in Equation (6).

$$CA = a \cdot X^b \tag{6}$$

The variables a and b in Equation (6) are provided for some of the reference fluids in Tables 2 and 3. If the release is steady state and continuous such as the case for small hole sizes, then the release rate is substituted into Equation (6) for X. If the release is considered instantaneous, for example, as a result of a vessel or pipe rupture, then the release mass is substituted into Equation (6) for X. The transition between a continuous release and an instantaneous release in API RBI is defined as a release where more than 4500 kg (10,000 lbs) of fluid mass escapes in less than 3 minutes.

The final flammable consequence areas are determined as a probability-weighted average of the individual consequence areas calculated for each release hole size. This is performed for both the equipment damage and the personnel injury consequence areas. The probability weighting utilizes the hole size distribution and the generic frequencies of the release hole sizes selected. The equation for probability weighting of the consequence areas is given by Equation (7).

$$CA^{flam} = \left(\frac{\sum_{n=1}^{4} gff_n \cdot CA_n^{flam}}{gff_{total}}\right) \tag{7}$$

The total generic failure frequency, gff_{total}, in the above equation is determined using Equation (8).

$$gff_{total} = \sum_{n=1}^{4} gff_n \tag{8}$$

The Level 1 consequence analysis procedure is a simplistic method for approximating the consequence area of a hazardous release. The only inputs required are basic fluid properties (such as MW, density and ideal gas specific heat ratio, k) and operating conditions. A calculation of the release rate or the available mass in the inventory group (i.e. the inventory of attached equipment that contributes fluid mass to a leaking equipment item) is also required. Once these terms are known, the consequence area is determined from Equations (6) and (7).

A similar procedure is used for determining the consequences associated with releases of toxic chemicals such as H_2S, ammonia or chlorine. Toxic impact areas are based on probit equations and can be assessed whether the stream is pure or a percentage of a hydrocarbon stream.

One of the main limitations of the Level 1 consequence analysis is that it can only be used in cases where the fluid in the component can be represented by one of the reference fluids. The Level 1 consequence analysis has been used in the refining industry over the past 10 years. However, as international interest has grown in API RBI in the refining and petrochemical industries, as well as in the chemical industries, it became apparent that the limited number of reference fluids available in the consequence area tables was not sufficient. As a result, the Level 2 analysis was developed to calculate consequence areas for releases of hazardous fluids using a more rigorous approach. The Level 2 analysis also resolves inconsistencies in the Level 1 analysis related to release type and event probabilities.

3.3 Level 2 Consequence Analysis

A detailed calculation procedure is provided for determining the consequences of loss of containment of hazardous fluids from pressurized equipment. The Level 2 consequence analysis was developed as a tool to use where the assumptions of the simplified Level 1 consequence analysis were not valid. Examples of where the more rigorous Level 2 calculations may be necessary are cited below:

1. The specific fluid is not represented adequately within the list of reference fluid provided in Table 1, including cases where the fluid is a wide-range boiling mixture or where the fluids toxic consequences are not represented adequately by any of the reference fluids.
2. The stored fluid is close to its critical point, in which case, the ideal gas assumptions for the vapor release equations are invalid.
3. The effects of two-phase releases, including liquid jet entrainment as well as rainout need to be included in the assessment.
4. The effects of BLEVEs and non-flammable explosions are to be included in the assessment (not included in the Level 1 Assessment).

5. The effects of pressurized non-flammable explosions, such as are possible when non-flammable pressurized gases (e.g. air or nitrogen) are released during a vessel rupture, are to be included in the assessment (not included in the Level 1 Assessment).
6. The meteorological assumptions used in the dispersion calculations that form the basis for the Level 1 consequence analysis table lookups do not represent the site data.

The Level 2 consequence analysis procedures presented in API 581 provide equations and background information necessary to rigorously calculate consequence areas for several flammable and toxic event outcomes. A summary of these events are provided in Table 4.

To perform Level 2 consequence analysis calculations, the actual composition of the fluid stored in the equipment is modeled. Fluid property solvers are available which allow the analyst to calculate fluid physical properties more accurately. The fluid solver will also provide the ability to perform flash calculations to better determine the release phase of the fluid and to account for two-phase releases. In many of the consequence calculations, physical properties of the released fluid are required at storage conditions as well as conditions after release to the atmosphere.

A cloud dispersion analysis must be performed as part of a Level 2 consequence analysis to assess the quantity of flammable material or toxic concentration throughout vapor clouds that are generated after a release of volatile material. Modeling a release depends on the source conditions, the atmospheric conditions, the release surroundings, and the hazard being evaluated. Employment of many commercially available models, including SLAB or DEGADIS [6], account for these important factors and will produce the desired data for a Level 2 API RBI consequence analysis.

The event trees used in the Level 2 consequence analysis are shown in Figures 3 and 4. Significant improvement in the calculations of the probabilities on the event trees have been made in the Level 2 analysis procedure. Unlike the Level 1 analysis, the probabilities of ignition on the event tree are not constant with release magnitude. Consistent with work of Cox, Lee and Ang [7], the Level 2 event tree ignition probabilities are directly proportional to the release rate. The probabilities of ignition are also now a strong function of the MW of the fluid. The probability that an ignition will be a delayed ignition is also now a function of the release magnitude and how close the operating temperature is to the auto-ignition temperature (AIT) of the fluid. These improvements to the event tree will result in consequence impact areas that are more strongly dependent on the size of release and the flammability and reactivity properties of the fluid being released.

4. RISK ANALYSIS

4.1 Determination of Risk

The calculation of risk can be determined as a function of time in accordance with Equation (9). This equation combines the probability of failure and the consequence of failure as described earlier in this paper.

$$R(t) = P_f(t) \cdot C(t) \tag{9}$$

Note that the probability of failure, $P_f(t)$, is a function of time since the damage factor as shown in Equation (1) increases as the damage in component due to thinning, cracking or other damage mechanisms accumulate with time. Figure 5 illustrates that the risk associated with individual damage mechanisms can be added together by superposition to provide the overall risk as a function of time.

In API RBI, the consequence of failure, $C(t)$, is assumed to be invariant with time. Therefore, Equation (9) can be rewritten as shown in Equations (10) and (11) depending on whether the risk is expressed as an impact area or in financial terms.

$$R(t) = P_f(t) \cdot CA \qquad \textit{for Area-Based Risk} \tag{10}$$

$$R(t) = P_f(t) \cdot FC \qquad \textit{for Financial-Based Risk} \tag{11}$$

In these equations, CA is the consequence impact area expressed in units of area and FC is the financial consequence expressed in economic terms. Note that in Equations (10) and (11), the risk is varying with time since the probability of failure is a function of time.

4.2 Risk Matrix

Presenting the results in a risk matrix is an effective way of showing the distribution of risks throughout a plant or process unit without numerical values. The risk matrix used in API RBI is shown in Figure 6. In this figure, the consequence and probability categories are arranged such that the highest risk ranking is toward the upper right-hand corner. Numerical values associated with consequence and probability categories are also shown in Figure 6.

Risk categories (i.e. High, Medium High, Medium, and Low) are assigned to the boxes on the risk matrix. In API RBI the risk categories are asymmetrical to indicate that the consequence category is given higher weighting than the probability category.

Component items residing towards the upper right-hand corner of the risk matrix will most likely take priority for inspection planning because these items have the highest risk. Similarly, items residing toward the lower left-hand corner of the risk matrix tend to take lower priority because these items have the lowest risk. Once the plots have been completed, the risk matrix can then be used as a screening tool during the prioritization process.

5. INSPECTION PLANNING

5.1 Overview

The premise of inspection planning using API RBI is based on the fact that at some point in time, the risk as defined in Equations (10) and (11) will reach a specified risk target. The risk target is established by the Owner-User. When or before the risk target is reached, an inspection of the equipment is recommended based on a ranking of the component damage mechanisms that have the highest calculated damage factors. Although the act of inspecting a piece of equipment does not necessarily reduce the inherent risk associated with that piece of equipment, inspection does provide knowledge of the damage state of the vessel and reduces uncertainty. As a result, the probability that loss of containment will occur is directly related to the amount of information that is available from inspection and the ability to quantify that damage.

In API RBI, reduction in uncertainty is a function of the effectiveness of the inspection in identifying and quantifying the type and extent of the damage. Some inspection techniques are better, for example, in detecting thinning (general corrosion) damage than others. On the other hand, an inspection technique appropriate for general corrosion may not be very effective in detecting and quantifying damage due to local thinning or cracking.

From this discussion, the calculated risk as performed in API RBI is not only a function of time but it is also a function of the knowledge obtained on the condition or damage state of the component determined in an effective inspection program. When inspection effectiveness is introduced into the risk Equations (10) and (11), the equations can be rewritten as Equations (12) and (13):

$$R(t,I_E) = P_f(t,I_E)\cdot CA \qquad \textit{for Area-Based Risk} \qquad (12)$$

$$R(t,I_E) = P_f(t,I_E)\cdot FC \qquad \textit{for Financial-Based Risk} \qquad (13)$$

5.2 Inspection Effectiveness

In API RBI, the inspection effectiveness is graded A through E, with an A inspection providing the most effective inspection available (90% effective) and E representing no inspection. A description of the inspection effective levels for general thinning damage is given in Table 5.

To illustrate the method in which different inspection levels effect the damage factor and probability of failure, consider the example of the general thinning damage mechanism (procedures for modifying damage factors based on inspection effectiveness are provided in API 581 for all damage mechanisms). For general thinning, API RBI utilizes an approach based on a metal loss parameter, A_{rt}. The damage factor is calculated as a function of this parameter and is based on the premise that as a pressure vessel or piping wall corrodes below the construction code minimum wall thickness plus the specified corrosion allowance, the damage factor will increase. An inspection program for general thinning will result in a reduction of the damage factor based on the effectiveness of the inspection to quantify the corrosion rate. As an example, the general thinning damage factor, D_f^{thin}, for a component with an A_{rt} equal to 0.5 is 1200 if there is no inspection (i.e. Inspection Effectiveness is E) as shown in Table 6. If a *B level* inspection is performed, the damage factor is reduced to 600. If two *B level* inspections have been completed, the damage factor is further reduced to 200. When these damage factors are substituted into Equation (1), it becomes apparent that an effective inspection program can reduce the probability of failure of a component and the risk of loss of containment.

5.3 Inspection Planning

In planning inspections using API RBI, a plan date is typically chosen far enough out into the future to include a time period covering one or several future maintenance turnarounds. Within this period, three cases are possible based on predicted risk and the specified risk target.

1. *Case 1 – Risk target is exceeded at a point in the future prior to the inspection plan date* – This is the classical case and is represented in Figure 7. In this case, the results of an inspection plan will be the number of inspections required, as well as the type or inspection effectiveness required, to reduce the risk at the future plan date down below the risk target. The target date is the date where the risk target is expected to be reached and is the date of the recommended inspection.
2. *Case 2 – Risk already exceeds the risk target at the time the RBI analysis is performed* – This case is shown in Figure 8 and indicates that the current risk at the time of the RBI analysis exceeds the risk target. An immediate inspection will be recommended at a level sufficient to reduce the risk at the future plan date down below the risk target.
3. *Case 3 – Risk at the future plan date does not exceed the risk target* – This case is shown in Figure 9 and indicates the predicted future risk at the plan date will not exceed the risk target and therefore, no inspection is recommended during the plan period.

The concept of how the different inspection techniques with different effectiveness levels can reduce risk is shown in Figure 7. In the example shown, a *B Level* inspection was recommended at the target date. This inspection level was sufficient since the risk predicted after the inspection was performed was determined to be below the risk target at the plan date. Note that in Figure 7, a *D Level* inspection at the target date would not have been sufficient to satisfy the risk target criteria. The projected risk at the plan date would have exceeded the risk target.

6. API RBI AND FFS – COMPLEMENTARY TECHNOLOGIES

Integration of the FFS assessment procedures with the in-service inspection codes also facilitates effective use of the emerging API RBI technologies to maximize equipment availability and to improve sustainable maintenance performance. The FFS and RBI relationship depends upon the type of RBI study. In a RBI study using a qualitative evaluation, FFS assessment procedures can be used to alter the risk-ranking of equipment based on the level of damage and the results of the assessment. In a RBI study using a quantitative evaluation, the FFS assessment procedures provide a model for flaw and damage analysis that can be used to establish a probability of failure. The probability of failure can be combined with a consequence of failure model to produce risk that can be subsequently utilized in the RBI study. Work is underway to identify areas to improve API 579-1/ASME FFS-1 [8] to facilitate use with the API in-service inspection codes, API 510 [9], API 570 [10], and API 653 [11]. When work is complete, API 581 can be used to set the scope, method of inspection, and inspection interval for a piece of equipment. If damage is found, the inspection interval can be modified based on the results of a FFS assessment performed in accordance with API 579-1/ASME FFS-1.

7. FUTURE DEVELOPMENT WORK

7.1 Overview

The technology embedded within API RBI is continuing to progress. As the technology is improved, the methodology will be modified and revisions to API 581 will be issued. Highlights of the known technological improvements planned for the probability of failure, consequence analysis and inspection planning methods in API RBI are provided below.

7.2 Probability of Failure

Currently, Probability Of Failure (POF) is based on superposition of applicable damage factors calculated for each of the damage mechanisms present in equipment. An alternative methodology for calculating probability of failure is being developed based on Fitness-For-Service (FFS) damage models as described by Osage [8]. Inputs to these structural reliability models will include statistical continuous distributions for material properties, physical dimensions, applied loadings, inspection effectiveness, and metal loss. The result of this approach will be a probabilistic calculation of the POF. Initial efforts using the new probabilistic approach will be in evaluating the damage from general and local metal loss. Other areas to be examined will be brittle fracture, environmental cracking, high temperature hydrogen attack, and creep damage.

Currently, API RBI utilizes four discrete hole sizes with a hole size probability distribution based on generic failure frequencies established after review of industry failure data. These hole size failure distributions are applied to all damage mechanisms covered within API RBI. Intuitively, the hole sizes used in the risk assessment should be strongly dependent on the active damage mechanisms. A case in point is stress corrosion cracking or any other damage mechanism likely to result in a crack. A crack size distribution as well as calculation procedures to determine the crack opening area would be more appropriate than using a discrete set of holes sizes and release areas. It is also possible that a single hole size may be developed based on damage mechanism. Development work is planned for this area.

Another planned development activity is to perform a more detailed assessment of the industry failure data to re-evaluate the generic failure frequencies. Inherently, the industry failure data is not generic in nature, that is, the data includes the effects of in-service damage accumulated during operation. The concept of generic failure frequency is that the failure data should be independent of service life or damage mechanism. It is believed that the generic failure frequencies currently used are too conservative.

Finally, the management systems factor, F_{MS}, used to quantify the impact that the plant mechanical integrity program has on the probability of failure should be further evaluated.

7.3 Consequence Analysis

The consequence analysis procedures in API RBI utilize a dense gas (heavier than air) cloud dispersion modeler to evaluate the impact areas associated with flammable and toxic releases. This is an adequate assumption for most refinery and petrochemical process fluids. However, for light gas releases, such as hydrogen and methane, the use of a dense gas model may be too conservative. Future modifications to the Level 2 Consequence analysis in API RBI will incorporate the use of a neutrally buoyant cloud dispersion model.

There is also some room for improvement in the calculation of the event tree probabilities. Research has recently been completed by the Energy Institute [13], including a review of ignition probabilities and correlations. This work needs to be reviewed to determine the need to modify the correlations currently being used in API RBI. There is also the need to improve the methods for determining the probability that a delayed ignition will result in either a flash fire or VCE. The probability that a delayed ignition will result in a VCE can easily be tied to the NFPA reactivity value of the fluid, which is a qualitative measure of a fluid's propensity to detonate.

A technology of interest to the petrochemical industry is the ability to analyze the consequences associate with dust explosions. Currently, API RBI does not provide a method for evaluating the consequences of solid (powder) releases.

7.4 Inspection Planning

One of the most common questions asked when performing an API RBI analysis is related to risk targets. Some Owner-Users have internal corporate guidelines for risk tolerance; however, using risk tolerance guidelines to develop risk targets for use in an RBI analysis is not straightforward. In many cases, Owner-Users have no internal guidelines on risk targets. Future work to develop consistent guidelines for determining both area-based and financial-based risk targets is needed.

The current algorithm used for inspection planning, whereby inspection recommendations to reduce overall risk are based on targeting the damage mechanism that has the highest calculated damage factor needs improvement. Currently, there is no economic evaluation made to determine the most cost effective inspection plan for a particular combination of damage mechanisms. An optimization methodology that considers all damage mechanisms, inspection costs, and inspection effectiveness needs to be developed.

8. NOMENCLATURE

age	is the time since the last thickness reading
A_n	is the cross sectional hole area associated with the nth release hole size
A_{rt}	is the metal loss parameter
C_a	is the corrosion allowance
C_r	is the corrosion rate
C_d	is the coefficient of discharge
$C(t)$	is the consequence of failure as a function of time
CA	is the consequence impact area
CA_{cmd}	is the final component damage consequence area
CA_{inj}	is the final personnel injury consequence area
$D_{f-total}$	is the total damage factor
D_f^{thin}	is the damage factor for thinning
D_f^{elin}	is the damage factor for component linings
D_f^{extd}	is the damage factor for external damage, i.e. corrosion under insulation
D_f^{scc}	is the damage factor for stress corrosion cracking
D_f^{htha}	is the damage factor for high temperature hydrogen attack
D_f^{mfat}	is the damage factor for mechanical fatigue
D_f^{brit}	is the damage factor for brittle fracture
$D_f(t)$	is the damage factor as a function of time, equal to $D_{f-total}$ evaluated at a specific time
F_{MS}	is the management systems factor
FC	is the financial consequence
g_c	is the gravitational constant
gff	is the generic failure frequency
k	is the release fluid ideal gas specific heat capacity ratio
K_v	is the viscosity correction factor
MW	is the release fluid molecular weight
$pfbii_{2r}$	is the probability of a fireball given an immediate ignition of a rupture release of a two-phase fluid
poi_{ll}	is the probability of ignition given a leakage release of a flammable liquid
poi_{vl}	is the probability of ignition given a leakage release of a flammable vapor
poi_{2l}	is the probability of ignition given a leakage release of a flammable two-phase fluid

poi_{lr}	is the probability of ignition given a rupture release of a flammable liquid
poi_{vr}	is the probability of ignition given a rupture release of a flammable vapor
poi_{2r}	is the probability of ignition given a rupture release of a flammable two-phase fluid
$poii_{ll}$	is the probability of immediate ignition given ignition of a leakage release of a flammable liquid
$poii_{vl}$	is the probability of immediate ignition given ignition of a leakage release of a flammable vapor
$poii_{2l}$	is the probability of immediate ignition given ignition of a leakage release of a flammable two-phase fluid
$poii_{lr}$	is the probability of immediate ignition given ignition of a rupture release of a flammable liquid
$poii_{vr}$	is the probability of immediate ignition given ignition of a rupture release of a flammable vapor
$poii_{2r}$	is the probability of immediate ignition given ignition of a rupture release of a flammable two-phase fluid
$pvcedi_{ll}$	is the probability of VCE given a delayed ignition of a leakage release of a flammable liquid
$pvcedi_{vl}$	is the probability of VCE given a delayed ignition of a leakage release of a flammable vapor
$pvcedi_{2l}$	is the probability of VCE given a delayed ignition of a leakage release of a flammable two-phase fluid
$pvcedi_{lr}$	is the probability of VCE given a delayed ignition of a rupture release of a flammable liquid
$pvcedi_{vr}$	is the probability of VCE given a delayed ignition of a rupture release of a flammable vapor
$pvcedi_{2r}$	is the probability of VCE given a delayed ignition of a rupture release of a flammable two-phase fluid
P_a	is the atmospheric pressure
P_s	is the storage or normal operating pressure
$P_f(t)$	is the probability of failure as a function of time
$P_f(t, I_E)$	is the probability of failure as a function of time and inspection effectiveness
POL	is the probability of leakage
POR	is the probability of rupture
ρ_l	is the liquid density
R	is the universal gas constant
$R(t)$	is the risk as a function of time

$R(t, I_E)$	is the risk as a function of time and inspection effectiveness
t_{min}	is the minimum required thickness for the component established using the applicable construction code
t_{rd}	is the thickness reading
T_s	is the storage or normal operating temperature
W_n	is the release rate associated with the nth release hole size

9. REFERENCES

[1] API, API RP 580 *Recommended Practice for Risk-Based Inspection*, American Petroleum Institute, Washington, D.C.

[2] API, API RP 581 *API RBI Technology*, American Petroleum Institute, Washington, D.C, In Preparation.

[3] CCPS, *Guidelines for Consequence Analysis of Chemical Releases*, ISBN 0-8169-0786-2, published by the Center for Chemical Process Safety of the American Institute of Chemical Engineers, 1999.

[4] van den Bosch, C.J.H., and Weterings, R.A.P.M., Ed., *Methods for Calculation of Physical Effects* (TNO Yellow Book, Third Edition), Chapter 6: Heat Flux from Fires, CPR 14E (ISSN 0921-9633/2.10.014/9110), Servicecentrum, The Hague, 1997."

[5] CCPS, *Guidelines for Evaluating the Characteristics of Vapor Cloud Explosions, Flash Fires, and BLEVEs*, ISBN 0-8169-0474-X, published by the Center for Chemical Process Safety of the American Institute of Chemical Engineers, 1994.

[6] OFCM , *Directory of Atmospheric Transport and Diffusion Consequence Assessment Models* (FC-I3-1999), *published* by the Office of the Federal Coordinator for Meteorological Services and Supporting Research (OFCM) with the assistance of SCAPA members, the document is available at http://www.ofcm.gov/atd_dir/pdf/frontpage.htm.

[7] Cox, A.W., Lees, F. P., and Ang, M.L., *Classification of Hazardous Locations*, Rugby: Instn Chem. Engrs., 1990.

[8] Osage, D.A., "API 579-1/ASME FFS-1 2006 – A Joint API/ASME Fitness-For-Service Standard For Pressurized Equipment", ESOPE Conference, Paris, France, 2007.

[9] API, API 510 *Pressure Vessel Inspection Code: Maintenance Inspection, Rerating, Repair and Alteration*, American Petroleum Institute, Washington, D.C., 1999.

[10] API, API 570 *Piping Inspection Code: Inspection, Repair, Alteration, and Rerating of In-Service Piping Systems*, American Petroleum Institute, Washington, D.C., 1998.

[11] API, API 653 *Tank Inspection, Repair, Alteration, and Reconstruction*, American Petroleum Institute, Washington, D.C., 1998.

[12] Valbuena, R.R, *Implementing and Evergreening RBI in Process Plants*, ISBN #1-57698-040-5, published by the Materials Technology Institute of the Chemical Process Industries, Inc., 2005.

[13] EI Publ., *Ignition Probability Review, Model Development and Look-up Correlations*, First Edition, Energy Institute, London, ISBN 978-0-85293-454-8, 2006.

[14] NFPA, *SPFE Handbook for Fire Protection*, Society of Fire Protection Engineering and the National Fire Protection Association, ISBN 0-87765-354-2 (NFPA No. HFPE-95), 1995.

[15] Lees, F.P., *Loss Prevention in the Process Industries*, Butterworths, London, 1986

[16] CCPS, *Estimating the Flammable Mass of a Vapor Cloud*, ISBN 0-8169-0778-1, published by the Center for Chemical Process Safety of the American Institute of Chemical Engineers, 1999.

Table 1 – List of Reference Fluids Available for Level 1 Consequence Analysis

Reference Fluid	Examples of Applicable Materials
C_1 - C_2	methane, ethane, ethylene, LNG, fuel gas
C_3 - C_4	propane, butane, isobutane, LPG
C_5	pentane
C_6 - C_8	gasoline, naphtha, light straight run, heptane
C_9 - C_{12}	diesel, kerosene
C_{13} - C_{16}	jet fuel, kerosene, atmospheric gas oil
C_{17} - C_{25}	gas oil, typical crude
C_{25+}	residuum, heavy crude, lube oil, seal oil
H_2	hydrogen only
H_2S	hydrogen sulfide only
HF	hydrogen fluoride
Water	Water
Steam	Steam
Acid	Acid, caustic
Aromatics	Benzene, Toluene, Xylene, Cumene
Ammonia	Ammonia
Chlorine	Chlorine

Table 2 – Equipment Damage Flammable Consequence Area Equation Constants

Reference Fluid	Continuous Releases Constants								Instantaneous Releases Constants							
	Auto-Ignition Not Likely (CAINL)				Auto-Ignition Likely (CAIL)				Auto-Ignition Not Likely (IAINL)				Auto-Ignition Likely (IAIL)			
	Gas		Liquid		Gas		Liquid		Gas		Liquid		Gas		Liquid	
	a	*b*	*a*	*b*	*a*	*b*	*a*	*b*	*a*	*b*	*a*	*b*	*a*	*b*	*a*	*b*
C_1-C_2	43.0	0.98			280.0	0.95			41.0	0.67			1079	0.62		
C_3-C_4	49.48	1.00			313.6	1.00			27.96	0.72			522.9	0.63		
C_5	25.17	0.99	536.0	0.89	304.7	1.00			13.38	0.73	1.49	0.85	275.0	0.61		
C_6-C_8	29.0	0.98	182.0	0.89	312.4	1.00	525.0	0.95	13.98	0.66	4.35	0.78	275.7	0.61	57.0	0.55
C_9-C_{12}	12.0	0.98	130.0	0.90	391.0	0.95	560.0	0.95	7.1	0.66	3.3	0.76	281.0	0.61	6.0	0.53
C_{13}-C_{16}			64.0	0.90			1023	0.92			0.46	0.88			9.2	0.88
C_{17}-C_{25}			20.0	0.90			861.0	0.92			0.11	0.91			5.6	0.91
C_{25} +			11.0	0.91			544.0	0.90			0.03	0.99			1.4	0.99
H_2	64.5	0.992			420.0	1.00			61.5	0.657			1430	0.618		
H_2S	32.0	1.00			203.0	0.89			148.0	0.63			357.0	0.61		
HF																
Aromatics	17.87	1.097	103.0	1.00	374.5	1.055			11.46	0.667	70.12	1.00	512.6	0.713	701.2	1.00

Table 3 – Personnel Injury Flammable Consequence Area Equation Constants

Reference Fluid	Continuous Releases Constants								Instantaneous Releases Constants							
	Auto-Ignition Not Likely (CAINL)				Auto-Ignition Likely (CAIL)				Auto-Ignition Not Likely (IAINL)				Auto-Ignition Likely (IAIL)			
	Gas		Liquid		Gas		Liquid		Gas		Liquid		Gas		Liquid	
	a	*b*	*a*	*b*	*a*	*b*	*a*	*b*	*a*	*b*	*a*	*b*	*a*	*b*	*a*	*b*
C_1-C_2	110.0	0.96			745.0	0.92			79.0	0.67			3100	0.63		
C_3-C_4	125.2	1.00			836.7	1.00			57.72	0.75			1769	0.63		
C_5	62.05	1.00	1545	0.89	811.0	1.00			28.45	0.76	4.34	0.85	959.6	0.63		
C_6-C_8	68.0	0.96	516.0	0.89	828.7	1.00	1315	0.92	26.72	0.67	12.7	0.78	962.8	0.63	224.0	0.54
C_9-C_{12}	29.0	0.96	373.0	0.89	981.0	0.92	1401	0.92	13.0	0.66	9.5	0.76	988.0	0.63	20.0	0.54
C_{13}-C_{16}			183.0	0.89			2850	0.90			1.3	0.88			26.0	0.88
C_{17}-C_{25}			57.0	0.89			2420	0.90			0.32	0.91			16.0	0.91
C_{25} +			33.0	0.89			1604	0.90			0.081	0.99			4.1	0.99
H_2	165.0	0.933			1117	1.00			118.5	0.652			4193	0.621		
H_2S	52.0	1.00			375.0	0.94			271.0	0.63			1253	0.63		
HF																
Aromatics	64.14	0.963	353.5	0.883	1344	0.937	487.7	0.268	18.08	0.686	0.14	0.935	512.6	0.713	1.404	0.935

Table 4 – Consequence Event Outcome for Level 2 Analysis

<table>
<tr><th>Event Outcome</th><th>Description</th><th>General Procedure</th></tr>
<tr><td>Pool Fire [3], [4], [7]</td><td>Occurs as a result of immediate ignition of a flammable liquid from a pressurized process vessel or pipe that leaks or ruptures.</td><td>1. Determine pool fire size
2. Calculate burning rate
3. Calculate flame length and tilt
4. Determine radiant energy emitted
5. Determine energy received at distant points (need view factor and atmospheric transmissivity)
6. Calculate safe distance</td></tr>
<tr><td>Jet Fires [3]</td><td>Occurs as a result of immediate ignition of a flammable vapor or two-phase jet release from a pressurized process vessel or pipe that develops a hole.</td><td>1. Calculate flame length
2. Determine radiant energy emitted
3. Determine energy received at distant points (need view factor and atmospheric transmissivity)
4. Calculate safe distance</td></tr>
<tr><td>Fireball [3]</td><td>Occurs as result of the immediate ignition of a flammable, superheated liquid/vapor released due to a vessel or pipe rupture. Fireballs always occur in combination with a physical explosion or a BLEVE.</td><td>1. Determine available flammable mass
2. Determine fireball diameter, height and duration
3. Determine radiant energy emitted
4. Determine energy received at distant points (need view factor and atmospheric transmissivity)</td></tr>
<tr><td>Flash Fire [3], [5]</td><td rowspan="2">Occurs as a result of a delayed ignition of a vapor cloud. The source of the vapor cloud could either be from a vapor or two-phase jet release or evaporation off the surface of an un-ignited liquid flammable pool.</td><td>1. Determine if cloud source is continuous (plume) or instantaneous (puff)
2. Utilize cloud dispersion model to determine the amount of flammable material (greater than LFL) that is in the source cloud</td></tr>
<tr><td>Vapor Cloud Explosion[3], [5], [15], [16]</td><td>1. Determine if cloud source is continuous (plume) or instantaneous (puff)
2. Utilize cloud dispersion model to determine the amount of flammable material (between LFL and UFL) that is in the source cloud
3. Determine equivalent amount of TNT
4. Calculate overpressure as a function of distance</td></tr>
<tr><td>BLEVE [5]</td><td>Occurs upon rupture of a vessel containing a superheated but pressurized liquid that flashes to vapor upon release to atmosphere</td><td>1. Determine equivalent amount of TNT which is a function of the storage pressure and the amount of fluid that flashes
2. Calculate overpressure as a function of distance.</td></tr>
<tr><td>Physical Explosion [5]</td><td>Occurs upon rupture of a vessel containing a pressurized flammable or non-flammable vapor</td><td>1. Determine equivalent amount of TNT which is a function of the storage pressure and volume of vapor
2. Calculate overpressure as a function of distance.</td></tr>
<tr><td>Toxic Release</td><td>Occurs upon release of toxic fluid to the atmosphere through a hole or due to a rupture</td><td>1. Determine if cloud source is continuous (plume) or instantaneous (puff)
2. Utilize cloud dispersion model to determine the portion of the cloud that exceeds the toxic limit (concentration and duration) of the fluid</td></tr>
</table>

Table 5 – Guidelines for Assigning Inspection Effectiveness – General Thinning

Inspection Category	Inspection Effectiveness Category	Intrusive Inspection Example	Non-intrusive Inspection Example
A	Highly Effective	50 to 100% examination of the surface (partial internals removed), and accompanied by thickness measurements	50 to 100% ultrasonic scanning coverage (automated or manual) or profile radiography
B	Usually Effective	Nominally 20% examination (no internals removed), and spot external ultrasonic thickness measurements	Nominally 20% ultrasonic scanning coverage (automated or manual), or profile radiography, or external spot thickness (statistically validated)
C	Fairly Effective	Visual examination without thickness measurements	2 to 3% examination, spot external ultrasonic thickness measurements, and little or no internal visual examination
D	Poorly Effective	External spot thickness readings only	Several thickness measurements, and a documented inspection planning system
E	Ineffective	No inspection	Several thickness measurements taken only externally, and a poorly documented inspection planning system

Table 6 – Thinning Damage Factors

A_{rt}	Inspection Effectiveness												
	E	1 Inspection				2 Inspections				3 Inspections			
		D	C	B	A	D	C	B	A	D	C	B	A
0.02	1	1	1	1	1	1	1	1	1	1	1	1	1
0.04	1	1	1	1	1	1	1	1	1	1	1	1	1
0.06	1	1	1	1	1	1	1	1	1	1	1	1	1
0.08	1	1	1	1	1	1	1	1	1	1	1	1	1
0.10	2	2	1	1	1	1	1	1	1	1	1	1	1
0.12	6	5	3	2	1	4	2	1	1	3	1	1	1
0.14	20	17	10	6	1	13	6	1	1	10	3	1	1
0.16	90	70	50	20	3	50	20	4	1	40	10	1	1
0.18	250	200	130	70	7	170	70	10	1	130	35	3	1
0.20	400	300	210	110	15	290	120	20	1	260	60	5	1
0.25	520	450	290	150	20	350	170	30	2	240	80	6	1
0.30	650	550	400	200	30	400	200	40	4	320	110	9	2
0.35	750	650	550	300	80	600	300	80	10	540	150	20	5
0.40	900	800	700	400	130	700	400	120	30	600	200	50	10
0.45	1050	900	810	500	200	800	500	160	40	700	270	60	20
0.50	1200	1100	970	600	270	1000	600	200	60	900	360	80	40
0.55	1350	1200	1130	700	350	1100	750	300	100	1000	500	130	90
0.60	1500	1400	1250	850	500	1300	900	400	230	1200	620	250	210
0.65	1900	1700	1400	1000	700	1600	1105	670	530	1300	880	550	500

A_{rt}	Inspection Effectiveness												
	E	4 Inspections				5 Inspections				6 Inspections			
		D	C	B	A	D	C	B	A	D	C	B	A
0.02	1	1	1	1	1	1	1	1	1	1	1	1	1
0.04	1	1	1	1	1	1	1	1	1	1	1	1	1
0.06	1	1	1	1	1	1	1	1	1	1	1	1	1
0.08	1	1	1	1	1	1	1	1	1	1	1	1	1
0.10	2	1	1	1	1	1	1	1	1	1	1	1	1
0.12	6	2	1	1	1	2	1	1	1	1	1	1	1
0.14	20	7	2	1	1	5	1	1	1	4	1	1	1
0.16	90	30	5	1	1	20	2	1	1	14	1	1	1
0.18	250	100	15	1	1	70	7	1	1	50	3	1	1
0.20	400	180	20	2	1	120	10	1	1	100	6	1	1
0.25	520	200	30	2	1	150	15	2	1	120	7	1	1
0.30	650	240	50	4	2	180	25	3	2	150	10	2	2
0.35	750	440	90	10	4	350	70	6	4	280	40	5	4
0.40	900	500	140	20	8	400	110	10	8	350	90	9	8
0.45	1050	600	200	30	15	500	160	20	15	400	130	20	15
0.50	1200	800	270	50	40	700	210	40	40	600	180	40	40
0.55	1350	900	350	100	90	800	260	90	90	700	240	90	90
0.60	1500	1000	450	220	210	900	360	210	210	800	300	210	210
0.65	1900	1200	700	530	500	1100	640	500	500	1000	600	500	500

Notes: Determine the row based on the calculated A_{rt} parameter shown below. Then determine the thinning damage factor based on the number and category of highest effective inspection. Interpolation may be used for intermediate values.

$$A_{rt} = \max\left[\left(1 - \frac{t_{rd} - C_r \cdot age}{t_{\min} + C_a}\right),\ 0.0\right]$$

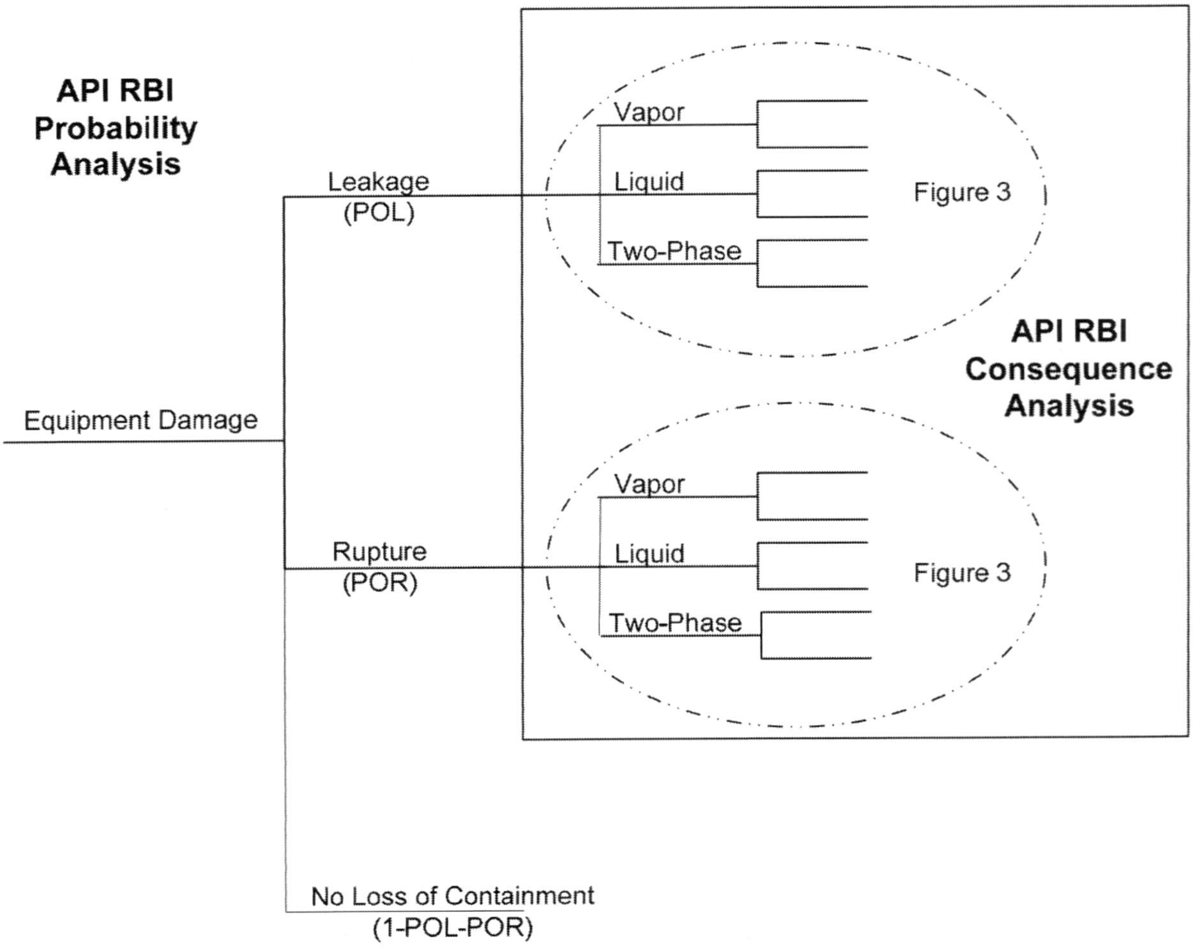

Figure 1 – Overall Event Tree Used in API RBI

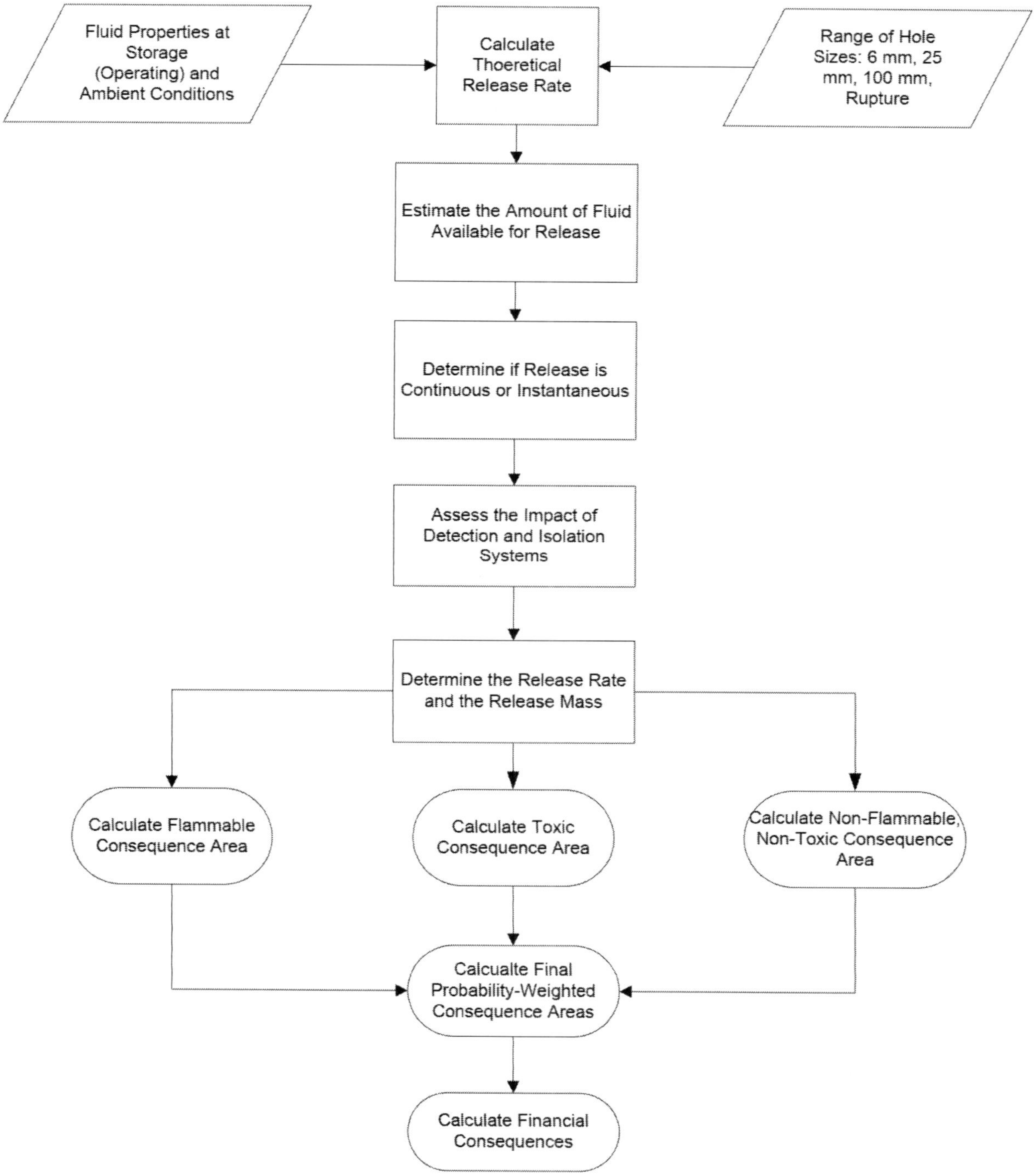

Figure 2 – Consequence Analysis Procedure in API RBI

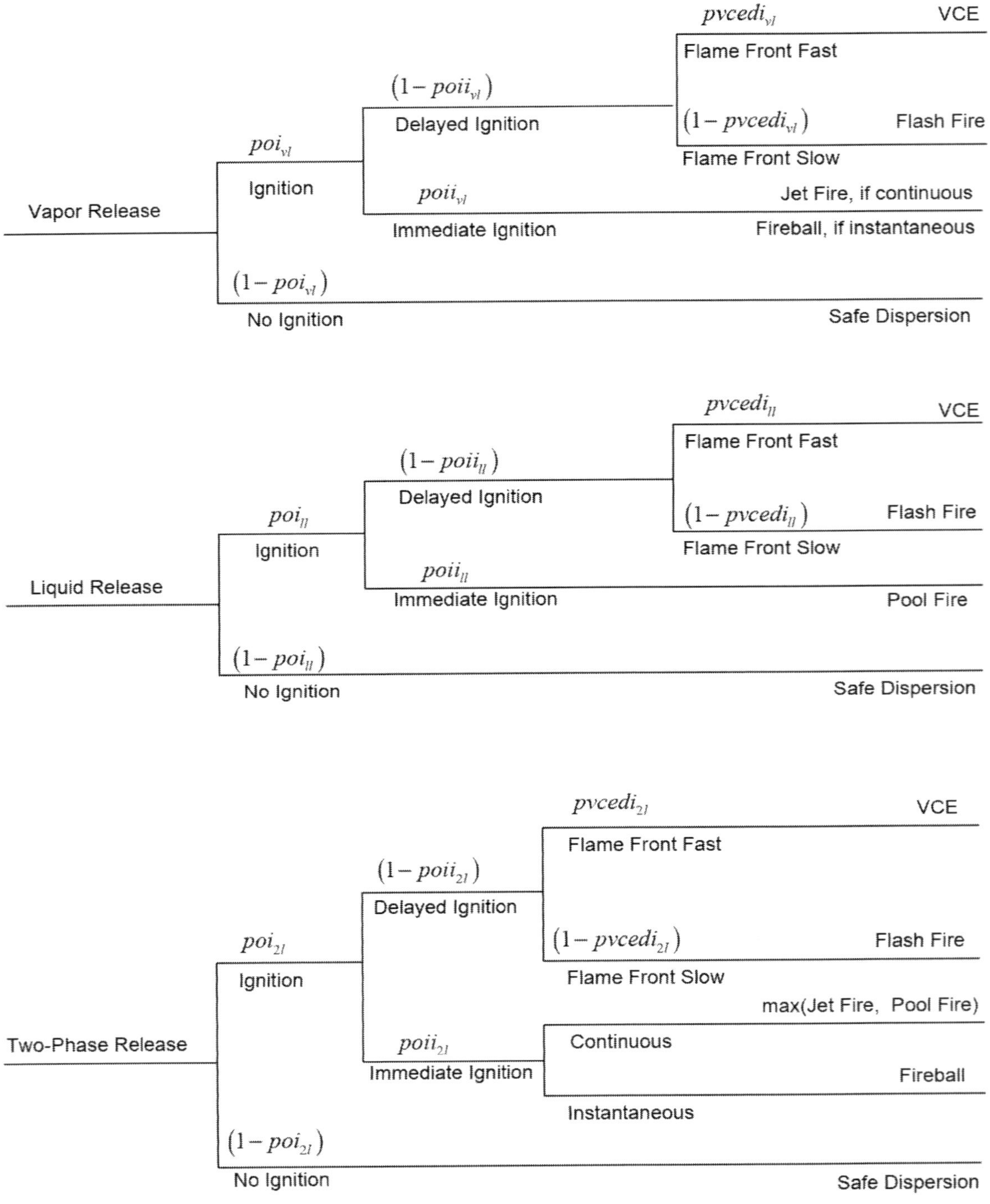

Figure 3 – Consequence Analysis Event Tree for Leakage (Small Hole) Case in API RBI

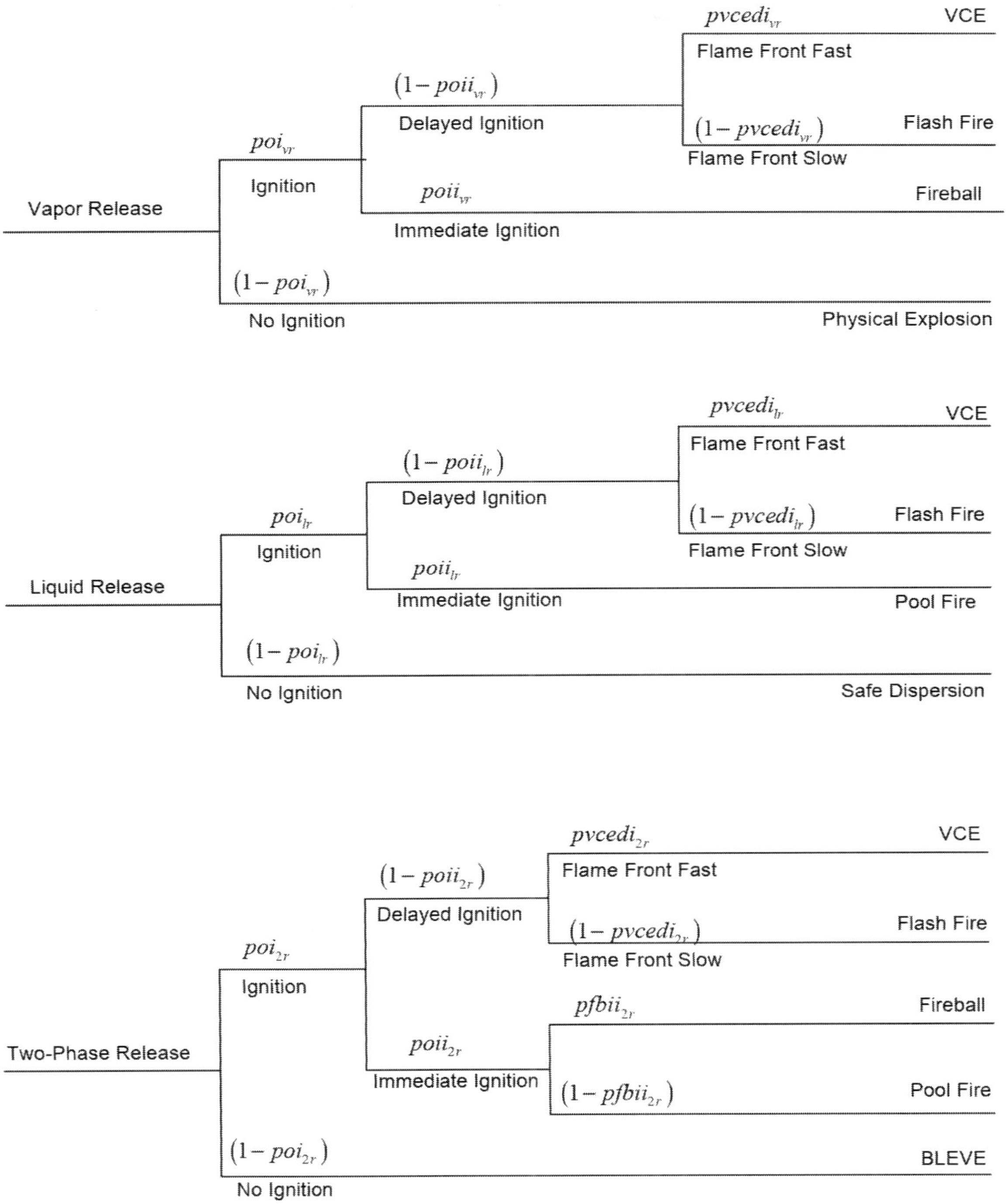

Figure 4 – Consequence Analysis Event Tree for Rupture Case in API RBI

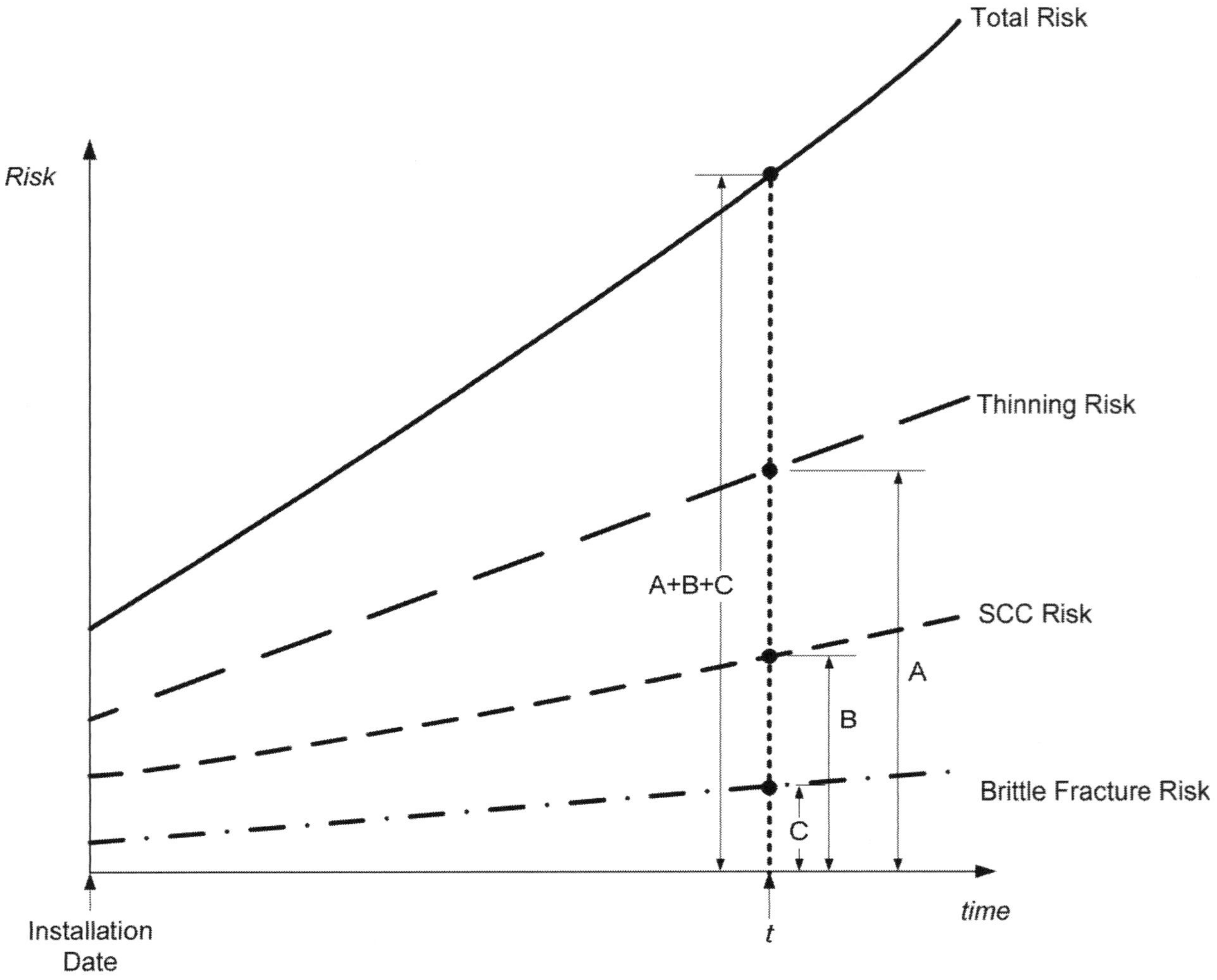

Figure 5 – Superposition Principle for the Calculation of Risk in API RBI

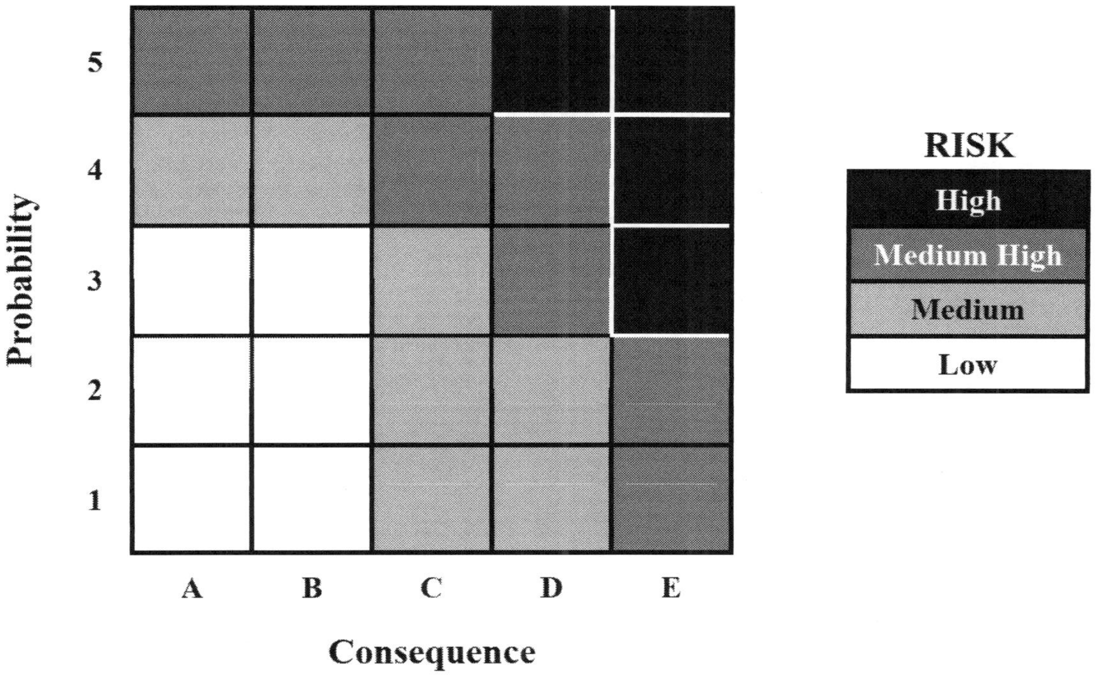

Numerical Values Associated with Consequence and Probability Categories in API RBI

Probability Category (1)		Consequence Category (2)	
Category	**Range**	**Category**	**Range**
1	$D_{f-total} \leq 2$	**A**	$CA \leq 100$
2	$2 < D_{f-total} \leq 20$	**B**	$100 < CA \leq 1000$
3	$20 < D_{f-total} \leq 100$	**C**	$1000 < CA \leq 3000$
4	$100 < D_{f-total} \leq 1000$	**D**	$3000 < CA \leq 10000$
5	$D_{f-total} > 1000$	**E**	$CA > 10000$

Notes:

1. **In terms of the total damage factor.**
2. **In terms of maximum component consequence area.**

$$CA = \max\left[CA_{cmd},\ CA_{inj}\right]$$

Figure 6 – Risk Matrix

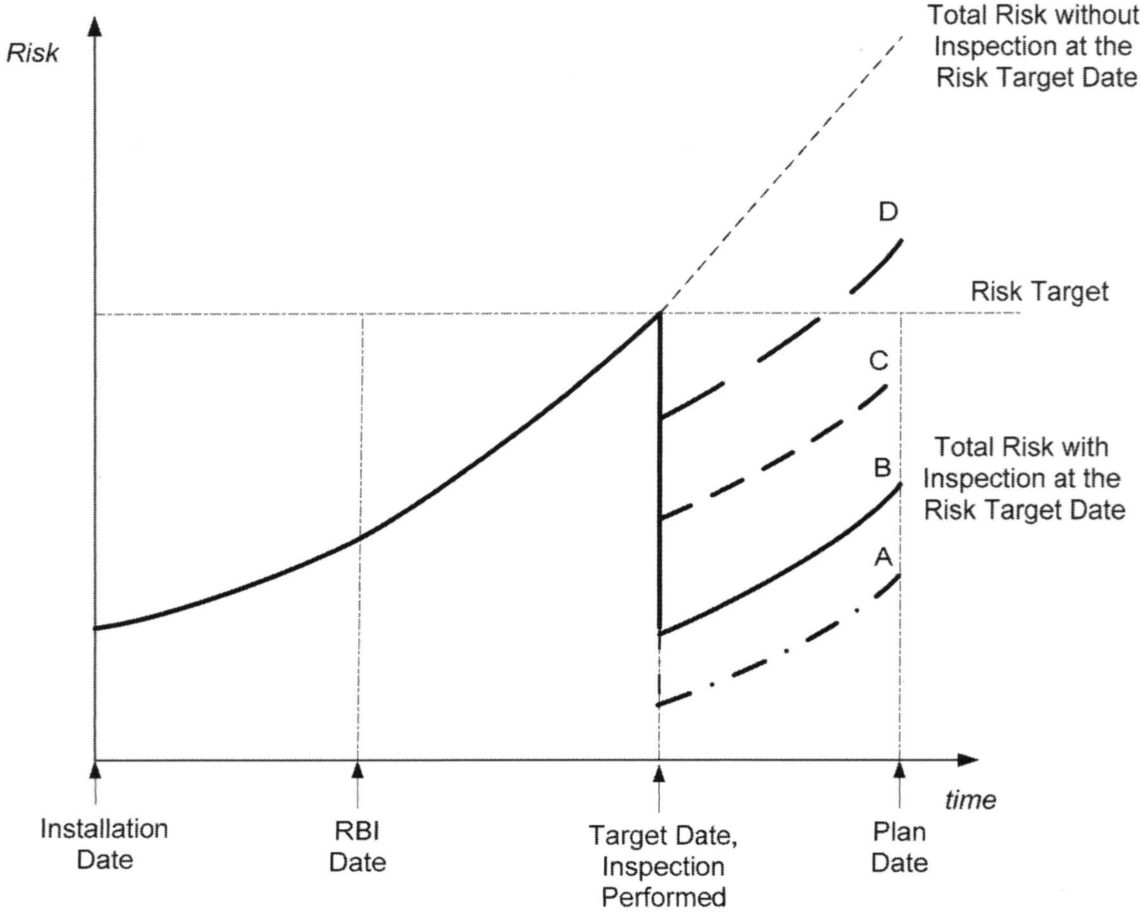

Figure 7 – Case 1: Inspection Planning When the Risk Target is Exceeded Between the RBI Date and the Plan Date

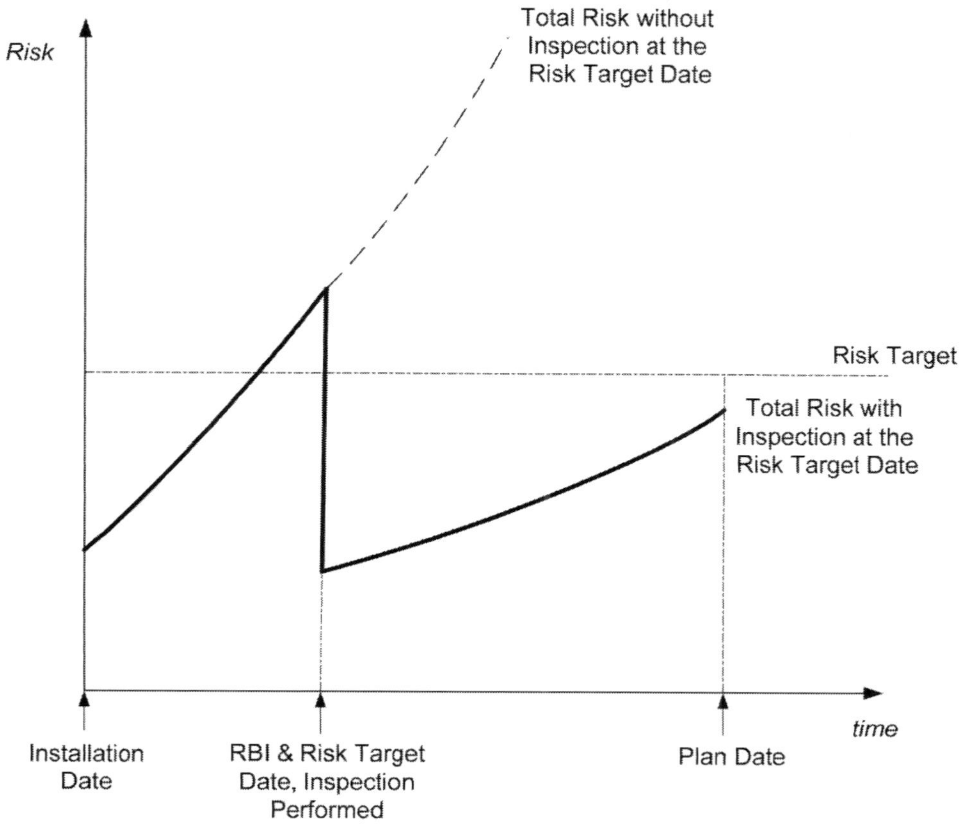

Figure 8 – Case 2: Inspection Planning When the Risk Target has been Exceeded Prior to the RBI Date

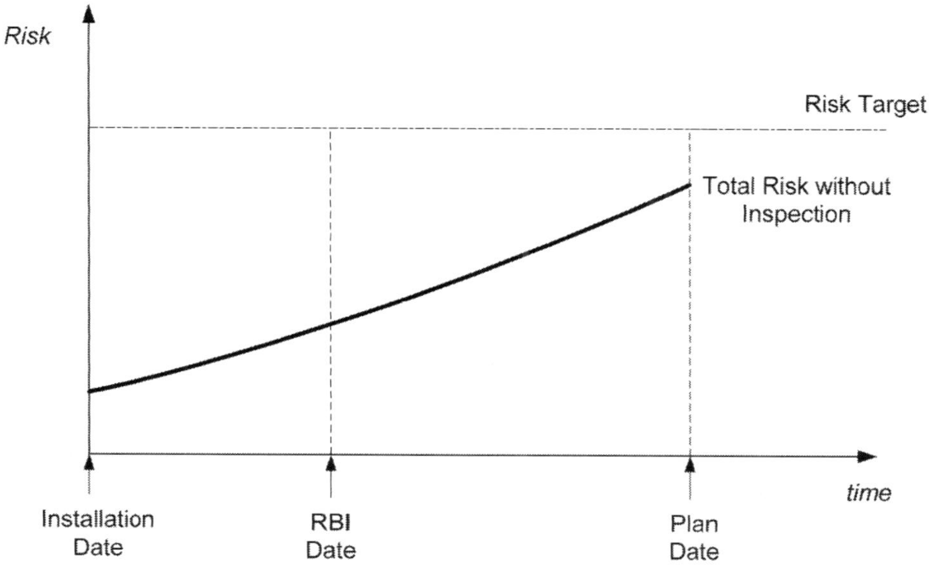

Figure 9 – Case 3: Inspection Planning When Risk Target is Not Exceeded Prior to the Plan Date

The Family Tree of Protective Requirements Specifications

Kevin L. Klein
Celanese Corp.
Houston, TX
Kevin.klein@celanese.com

David Dietrich
Solutia, Inc.
Alvin, TX
djdiet@solutia.com

A 10th PPSS Anniversary Paper

ABSTRACT

The Instrumented Protective System (IPS) lifecycle requires the development of a Protective Requirements Specification which is the design basis for IPS. This document can take many forms; a collection of documents, a single document, or in a database. This paper investigates what is takes to document the Protective Requirements Specification through a relational database. At first glance, this looks like a simple task with functions, systems, and components, but deep down there are some family secrets that need to be uncovered to better understand those relationships.

1. INTRODUCTION

So why do we even need the Protective Requirements Specification (PRS) and what is required? ANSI/ISA 84.00.01-2004 and IEC 61511 both require in clause 10 the development of a PRS as early as the feasibility stage of the project design, and to continue PRS use throughout the IPS lifecycle. The PRS contains three major sections; process information, hardware specifications, and software specifications. The process information is used to communicate why the IPS is needed, process conditions, environmental conditions, environmental conditions, and process dynamics. The hardware section specifies the Instrumented Protective Function (IPF) design such as fault tolerance, inputs and sensors, logic solvers, actions and final elements, and support systems. Finally where a PES is used a software specification is requires with information such as program logic and human interface requirements.

The PRS is the key document to understanding the design, management, and operation of an Instrumented Protective Function. These specifications can be called by various names depending on your perspective. For instance in ISA they are called Safety Requirements Specifications, while in the CCPS book Safe and Reliable Instrumented Protective Systems it is called a design basis. In this paper the term PRS will be used. Whatever you call it the purpose of this document is two fold during the development of an instrumented protective function. First, it is used to communicate the information about the process, the hazards, and the design between the process, maintenance, and engineering. Second, once the function is installed, the PRS is used to define how the function operates and should be managed, how it was designed, and why the function exists and is used during Management of Change and training. So this document needs to be kept up to date all the time. In order to make sure this is done, the document needs to useful to many different people.

So how do we as engineers make the PRS records of real value to the process engineer, operators, manufacturing engineers, maintenance engineers, maintenance technicians, and supervision? Will everyone need to see the same information? If you develop this document as a paper document, will you have to make different versions for each person? How much work will this be for everyone to keep up to date? Is there a way to present the information differently to each of these different groups without having to duplicate all the information many times? Do you put all the information in one place? Can you reference some of the data? How do you find the information? Many questions like these led us to decide to develop a database to track and record all the information required for an IPS.

In this day of electronic records we have many different options to choose from when documenting something such as written text, spreadsheets, and relational databases. But what is the best platform for recording the PRS information? This will depend on how you will be using the information. If you just want to document the information and not use it as a reference, then using a word document is acceptable. Spreadsheets are great at compiling lists of information, but you can not show relationships between the information. Finally the relational database allows you to identify the different relationships between information and reduce the duplication of that information. This is important with a PRS, since the information for one IPF may have many different relationships with another IPF. If you use a spreadsheet or a text file you will need to type in the information many times in some cases. In addition, when you change a piece of information, you need to track down the direct source of all that information. This can lead to much additional work as well as inaccurate records. This is why we chose to utilize a relational database for the PRS documentation.

2. IDENTIFY THE DATABASE PLATFORM

There are many options for the programmer when developing a database for the PRS. Some of the factors you need to consider are size of the application, how many locations will the software support, how do you want to access the information, and where will the software reside to name a few. These types of decisions will determine what the database will look like. Will this be an application just for the engineers, or will it be used by the operators and maintenance? How many users will access the data? Will it be accessed by printed reports or available to them electronically? Will each site maintain individual databases or is a centralized database a better fit? What security and auditing functions need to exist? Does a platform already exist in the organization that could support the database? These questions need to be asked of both the users and the developers of the database. One must also identify support roles for the database application. Who will develop it? How and where will the data be warehoused? Can it be supported long term? What internal and external expertise will be required? Do you want this database link into any other databases such as SAP?

Some of the options include an Access database, a SQL Database with an Access front end, and an Oracle database with a web front end to name a few. The database scheme can be broken down into two parts; the database and the user front end. The user front end gives the user access to the database and is the most complicated part of the system since in many cases the front end includes security, reporting, viewing, editing, and importing of the information. The selection of the database will depend on company preferences, size of the application, and data access speed requirements.

At Solutia, we wanted the application to be corporate wide software so we had a standardized way of entering the PRS. We identified engineers, operations, and maintenance as users of the database. A prime area of discussion is how much access is required for the operators and operations? To give operators all the technical information from the PRS would tend to just confuse the operators and not be a useful tool for them. We looked at two approaches; 1) develop a simplified report for the operators and just print out the information for the operators, 2) develop some screens specific to the needs of the operator and give them the tools to search for the information they need. In processes where there are not many IPFs, making a printed copy of the report or even creating a PDF of the report can be an easy approach to this issue. But where there are many IPFs, generating a printed or electronic file is not very practical and wastes all the work done to create the relational database.

This paper provides examples from a centralized platform chosen for its long term supportability, web-based data interface, and integrated security with existing company network policies. The platform decision was based on underlying goals of consistent documentation across sites and the cost of developing the application.

3. ORGANIZING THE IPS RELATIONSHIPS

Let's start out by defining a couple of types of relationships between tables used in databases. In the first case, we will cover a parent/child relationship and then the associative relationship. With the parent/child relationship there is one record in the parent table, but there can be many records in the child table. The one restriction is that the child record can not be associated with any other parent table. For instance with an IPF, each proof test can be unique to the specific IPF. In addition there can be more than one type of proof test for that IPF such as an end to end test, a partial test, and an inspection. So the proof testing table would be a child of the IPF table.

With the associative relationship there are two tables with records in them. Any one record in the parent table can be associated with many records in the child table, but unlike the parent/child relationship, the records in the child table can be associated with more than one parent. Generally if there is a one to one relationship between records, then it is better to just put that record in the parent table and eliminate the child. Using these relationships you can eliminate duplication of information and save work.

Each item in the PRS list is related to each other in reality similar to a family tree or taxonomy. Understanding each of these relationships helps organize the required information. One of the powerful features of a relational database is that you only need to enter data once and use it in many places. A good way to get started is to pick out several PRS within a process. Can you see any duplicate information from one PRS to another? If so, you have identified your first relationship. Now, look at the relationship. Will there only be one parent of the information or will there be many parents? This will tell you is the relationship will be a parent/child relationship or an associative relationship. Continue going through the information till all the relationships have been identified. Make sure to group the records which have relationships into as few tables as possible. The team reviewed all the data and categorized them into seven main tables; Function, Input, Action, Components, Hazards, Drawings, and Proof tests as shown in figure 1.

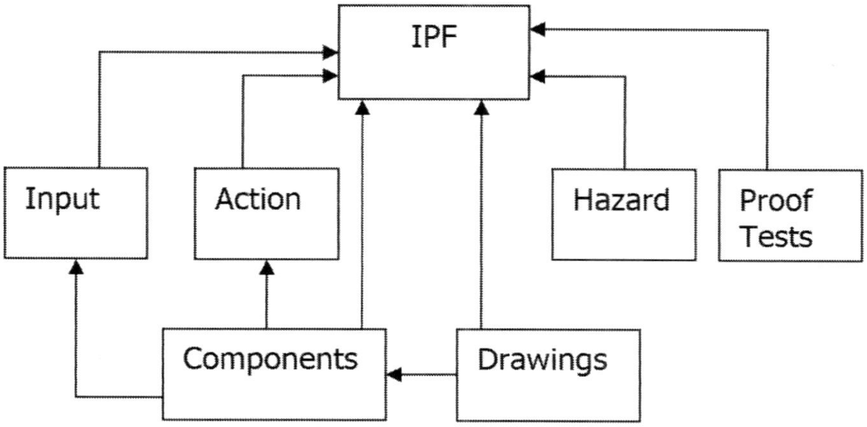

Figure 1

At Solutia, the function is typically the focus, so the IPF table becomes the main table of the database. The input, action, hazards and proof tests were identified as child tables of the function. The hazards table could also have had an associative relationship, since in some cases there could be several IPF which are needed to reduce the risk to an acceptable level. Since the duplication in this area was not considered significant, we decided that it was acceptable to just duplicate these records and make them child relationships to the function.

The components table created a special relationship with the function table. In some cases such as the logic solver and the support systems, the relationship was direct with the function, but in the case of sensors and final elements, the relationship was not straight forward. In some cases, more than one process variable will need to be monitored to identify the hazard, so an additional table was needed to show that relationship and was called the input table. Likewise, there may be several actions the IPF needs to take, so an action table was created. Since the relationships of the action and input tables to the function and components were similar, we ended up combining the tables into one action/input table and creating a variable to identify whether the record was for an input or for an action. The component table then had an associative relationship with the action, input, and function tables.

The drawing table also created a special challenge since there are drawings associated with the functions, hazards, components, and proof tests. So an associative relationship was created with all of those tables and the drawing table.

Since this database will be used for a whole corporation, it was necessary to provide some additional ways to sort the data by division, plant, business, and area. Additional tables were added to provide this capability.

For each of the associative relationships an intermediate table needed to be created to store the relationships between the two tables. Attachment 1 shows the final database structure.

Once the structure of the database was identified, each field in the table was assigned a data type. This is the time to identify constraints on data entry. Boundaries on data types and valid entry lists can be used to help drive consistency across multiple users and sites.

Tables/Fields		Data Type	Number of Characters	Description
prs_function				
PK	**function_id**	NUMBER		Unique index number for the IPF
FK2	**Area_id**	NUMBER		*Look up from Area table. Filter by assigned plant. Display columns (plant_name, business_unit_name, area_name).*
	function_title	VARCHAR2	255	Title for the Instrumented Protective Function
	function_type	VARCHAR2	25	Enter the type of Function
	description	VARCHAR2	1000	Statement of the Function
	identification_num	VARCHAR2	50	The IPF identification number
	revision_num	NUMBER		The revision number for the IPF in the format of X.Y where a major revision will increase the X and minor revisions will update the Y.
	revision_comment	VARCHAR2	4000	Any comments on the reason for the revision including the changes made.
	moc_reference_num	VARCHAR2	255	If Management of change was used the MOC document identification.
	safety_flag	VARCHAR2	1	Identifies that this function is required for Safety

Table 1 – Sample of Data Type Spreadsheet

4. INTERFACING WITH OTHER DATABASES

There may be other databases that have information needed for the PRS, specifically the maintenance system. Providing the capability to access this information reduces the need to duplicate the information. At Solutia we use the SAP PM module to manage the process equipment. Rather than providing just a reference, the PRS database was designed to provide a link to download the SAP data directly to the PRS component as long as the user provides an SAP equipment number. The user then can press a button and pull in the data, thus eliminating duplication of the information.

5. DEFINING THE INTERFACE

With the table structures identified, the viewing/editing interface can be developed. Not all users will be well versed in the use of relational databases, so it makes sense to develop an interface that is intuitive for the structure of protective functions. A specification was developed by the team to identify layouts of screens presented in the interface. This up front specification was key to reaching agreement between sites and keeping development costs minimized.

The team decided on two main entry points to the data: the IPF level and the component level. By generating common search fields for each of these interface points, users across the company would view data in the same format by simply selecting search criteria such as business division, plant, area, and unit. The team decided that filtering down to the individual component or IPF level would provide the flexibility to cover any user's search requirements.

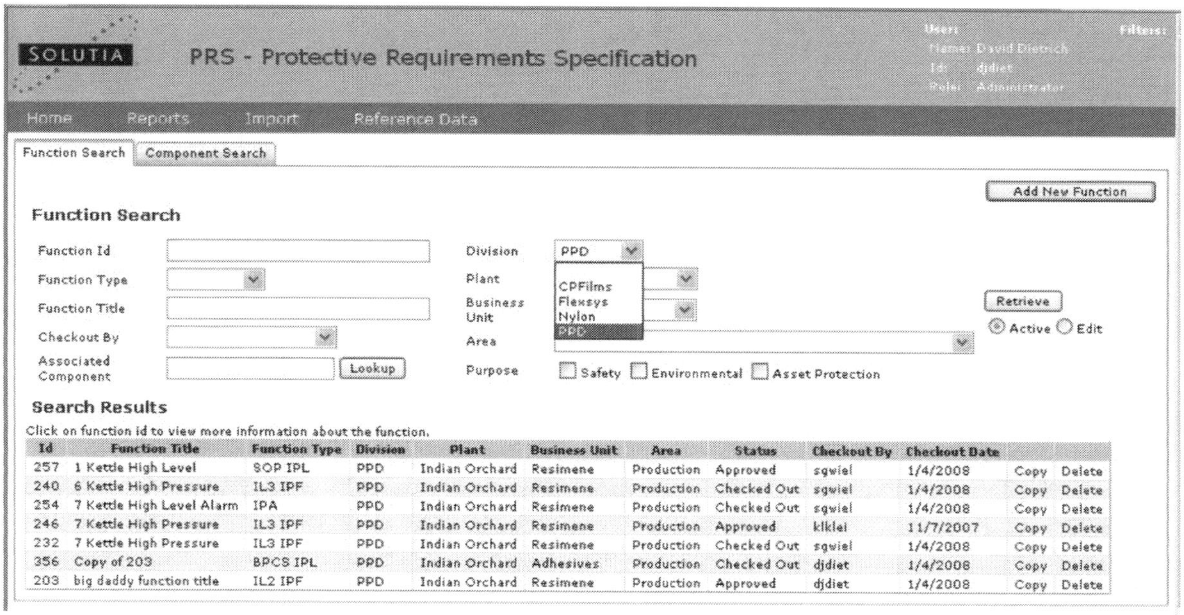

Figure 2
Web page layout showing drop down selections for IPF searching.

Along with data presentation, the specification defined full functionality of the interface. This included aspects such as reference data entry and presentation, security capabilities, reporting, and editing features.

Figure 3
Key interface touch points defined by specification

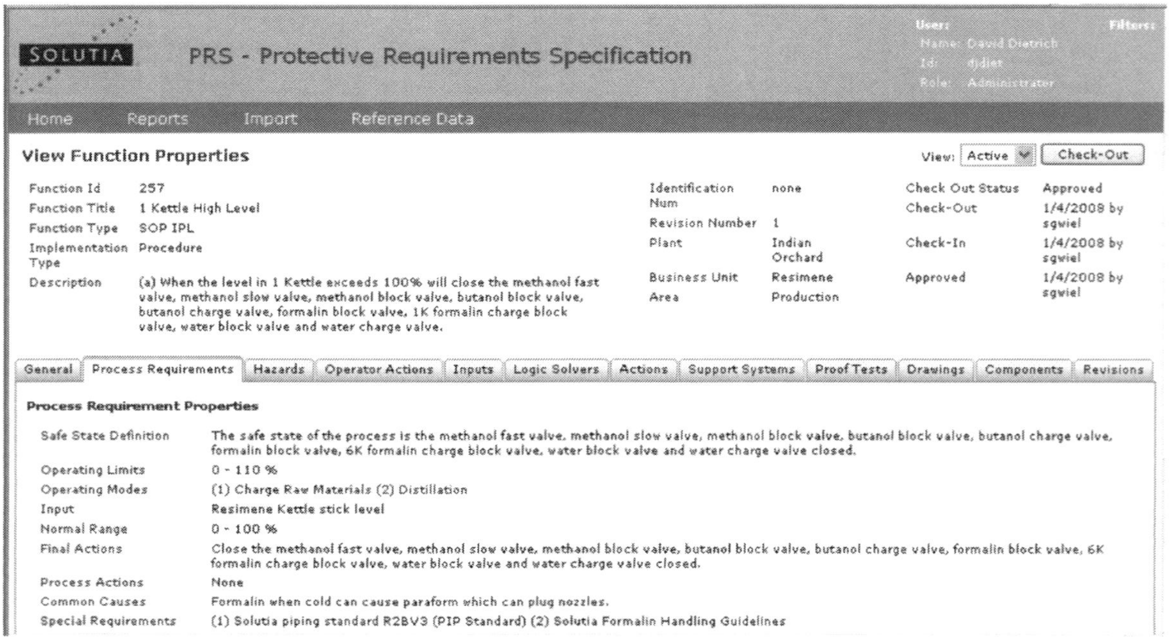

Figure 4
Example component viewing interface

6. DATABASE SECURITY

The other part of change management is access security which is required to prevent unauthorized changes to the database and to track who is making the changes electronically. As a web application, the software could detect who was logged on and automatically set the privileges of the user. There were five roles identified viewer, editor, approver, super approver, and administrator. Most of the roles are self evident. The viewer can view the data, but not change the data. The editor can make changes based on the division, plant, business, and area they are assigned to, but they do not have the authorization to approve the changes. The approver has the authorization to approve the changes to the PRS as well as the privileges to view and edit. The administrator has total control to assign roles, view, edit, and delete functions and components, and modify drop down lists. The Super approver role can assign roles to anyone at the specific site, view, edit and delete functions and components.

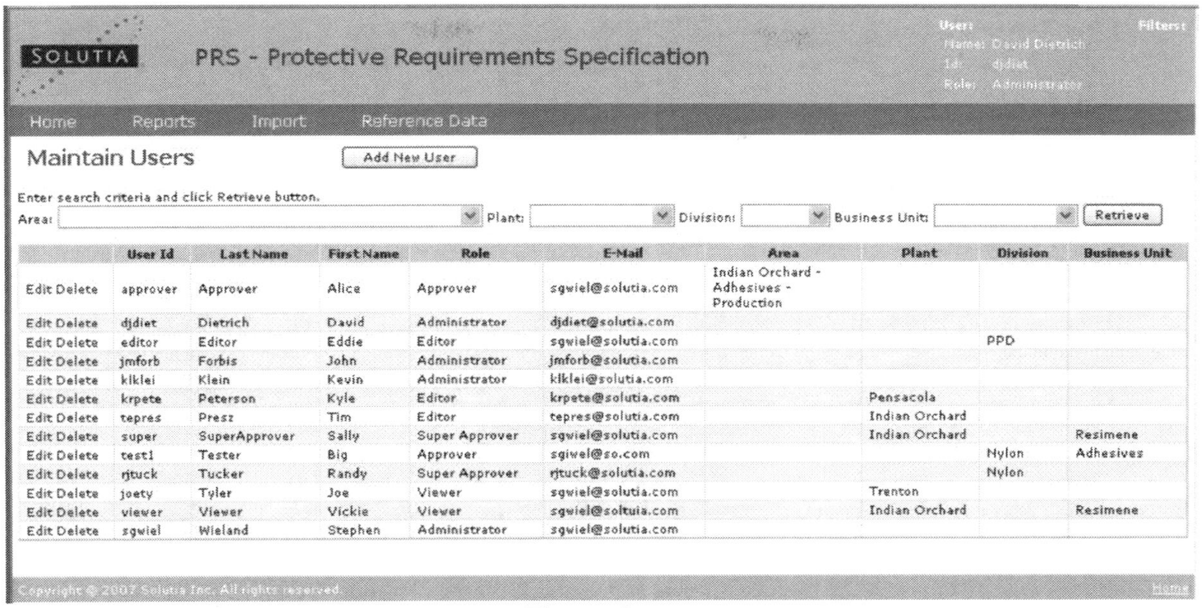

Figure 4
User properties interface

With all companies the information contained in the PRS is mostly confidential information and some businesses within Solutia needed to have additional security due to the intellectual property concerns. As part of the security, each function and component was assigned a division, plant, business unit, and area. Each user outside of the administrator was required to have at least a division level assigned to their user account. For approvers, editors, and viewers, they also needed to have at a minimum a plant assigned and would only be able to view records from their plant.

7. CHANGE MANAGEMENT

The PRS is part of the process safety information files and as such revisions and changes need to be managed. So an important part of this project was to design into the database a way to track changes and provide an approval path. This presented a unique challenge for the team. When a component is checked out for revision do you also check out the function? In the same way when you checkout a function to make a revision do you then also check out all the components? If a component is used in many places, do you need to also check out all the functions associated with that component? You can see at this point things were getting out of hand. If a person were to check out a PLC Logic Solver for revision, you could have most of the IPF in the business unit checked out at one time. This was not practical, so instead we chose to track changes to the IPF separate from the component changes. This way you would limit the impact of checking a record. The only caveat to this in that now you need to be sure to train the editors of the database to perform the due diligence of identifying the impacts of making a change to a component with multiple IPF.

How you track editing changes can have a major impact on your database complexity. For instance, when we started development of the database, we chose to have three different databases with identical structures, one for active records, one for edit records and one for archive. The difficulty of this and eventually the decision to scrap this idea was how to maintain the referential integrity of the database. Because when you check out a function, you then need to record all the associations for that record. Now if I make a change to one of the associated records, such as a component which is used in multiple IPF, is that allowed? In addition when you have completed the changes what do you archive? Just the IPF, or do you include all the components? Will you be able to go back in time and accurately identify what the specific IPS looked like at any one time? After several attempts at making this work, we decided this was not practical and would be complicated to implement. Solutia chose to use a single set of tables for edit and active records and use a flag to differentiate the two. When an IPF was checked back in, a complete IPF report was generated and stored in the database as a PDF. This way there was a snapshot in time made for the IPF. This eliminated the need for an archive table. Using this strategy only one database could be used.

Now with the structure simplified, when you check out a function or component, the program can just make a copy of the function or component along with the child records and relationships. A flag was created to segregate the edit version verses the active version. When the record is checked out for editing the record becomes locked and only the editor checking out the record can edit it. This will provide a way to track who made the changes. Once the record is changed, the editor checks the record back in and the database sends the approver for that location an e-mail that the function or component is ready for approval.

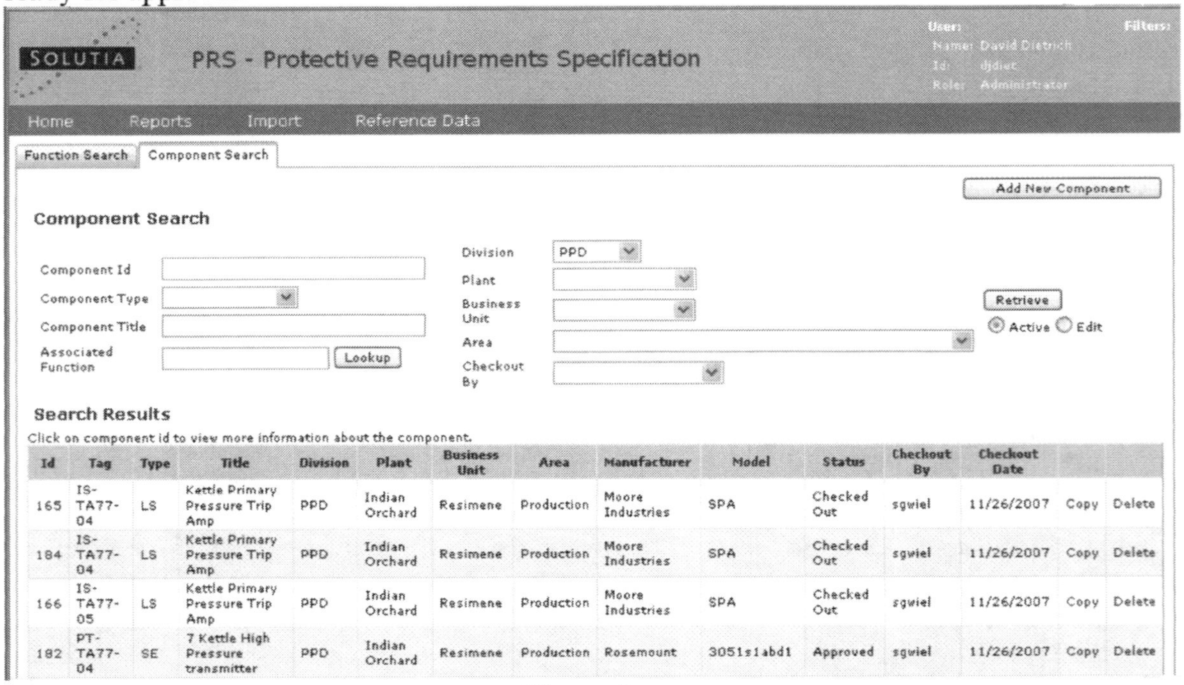

Figure 5
Edit status display provided with component record list

Since only the person checking out the record has access to make changes, if someone else needs to make a change during that time, they would need to look up who has the record checked out and either wait till the record is checked back in and approved or could work with the editor to make the changes required. But in the case of projects, it is important to develop the PRS early in the project life which can last for many months or years depending on the size of the project. You need to consider how you want to handle these situations. Do you allow the project engineer to go in and take out all the records for the project and tie them up for the duration or do they wait until the equipment is installed to make the changes? Are there other ways to handle these changes? We identified two ways to handle project changes; 1) for new functions and equipment, the records could be managed locally using a spreadsheet and then the records can be imported, 2) the project engineer goes ahead and checks out the records making the changes as needed and then if a change needs to be made outside of the project, then the project engineer could delete the changes, and allow the other changes to be made and check out the record again. The team agreed that generally changes to existing IPS should not be frequent and this would be an acceptable approach.

Reference data accessible to the database administrators coupled with data entry fields round out the editing forms provided by the interface. Depending on user rights, records can be modified and returned for review and approval by appropriate users.

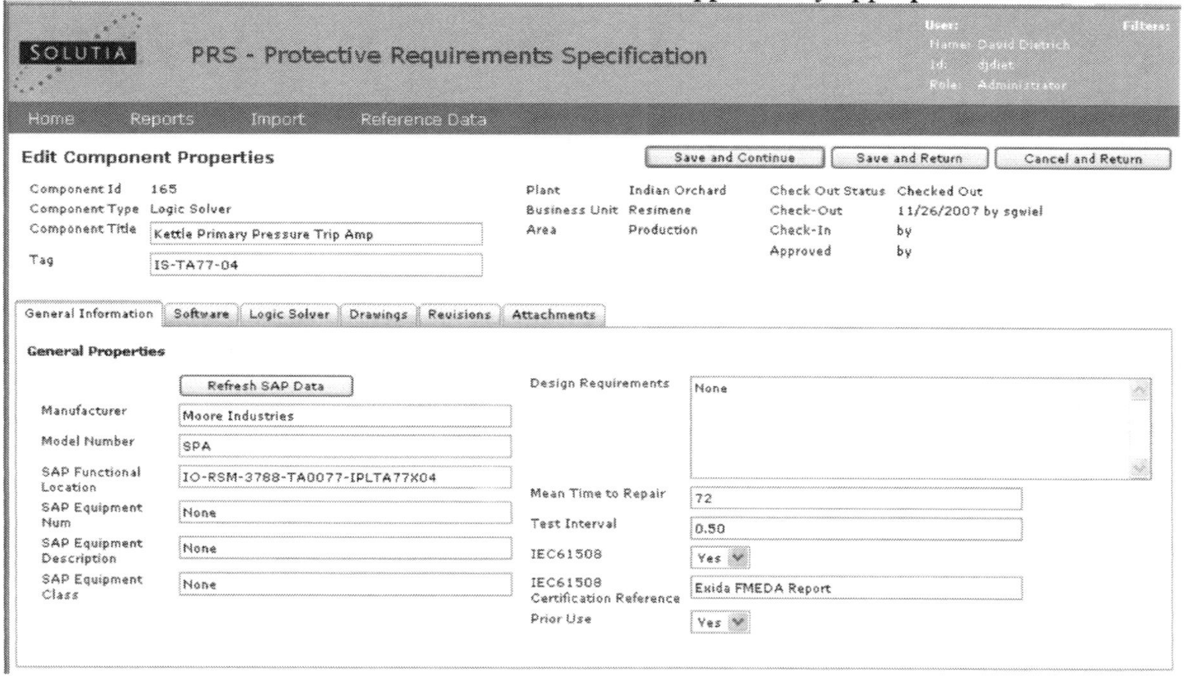

Figure 6
An edit display showing accessible entry fields

8. IMPORT

Across Solutia there were many different ways users were documenting their IPS from word documents to access databases. The data already collected took many hours to collect and was still of value. A method of importing the records needed to be developed. But how extensive will the import capability need to be? Is it necessary to import a complete IPF with all the child records and the components already attached? If so how do you manage the referential integrity between the records? How do you identify which components go with what functions and how much will this cost? Will your IT programmer put out a hit on you?

Ultimately, in order to keep the design simple, we decided to design the import program to be able to upload information to three of the tables; the IPF, the components and drawings. The IPF table also included the hazards child table. Once the data is uploaded, the user manually established the links between tables.

When importing the data, you want to make sure that the information is going into the correct fields in the table and that the data meets the required data type. This should be done before importing the data, as most database applications will reject entry of an improper format. Import templates were developed in a spreadsheet format to provide users with a bulk data entry tool. We included an intermediate interface in the application where users could see their data before it is incorporated in the PRS records. This provides the opportunity for users to cross check data entry. It also can be used as an application error check to indicate missing fields and format violations before final data import.

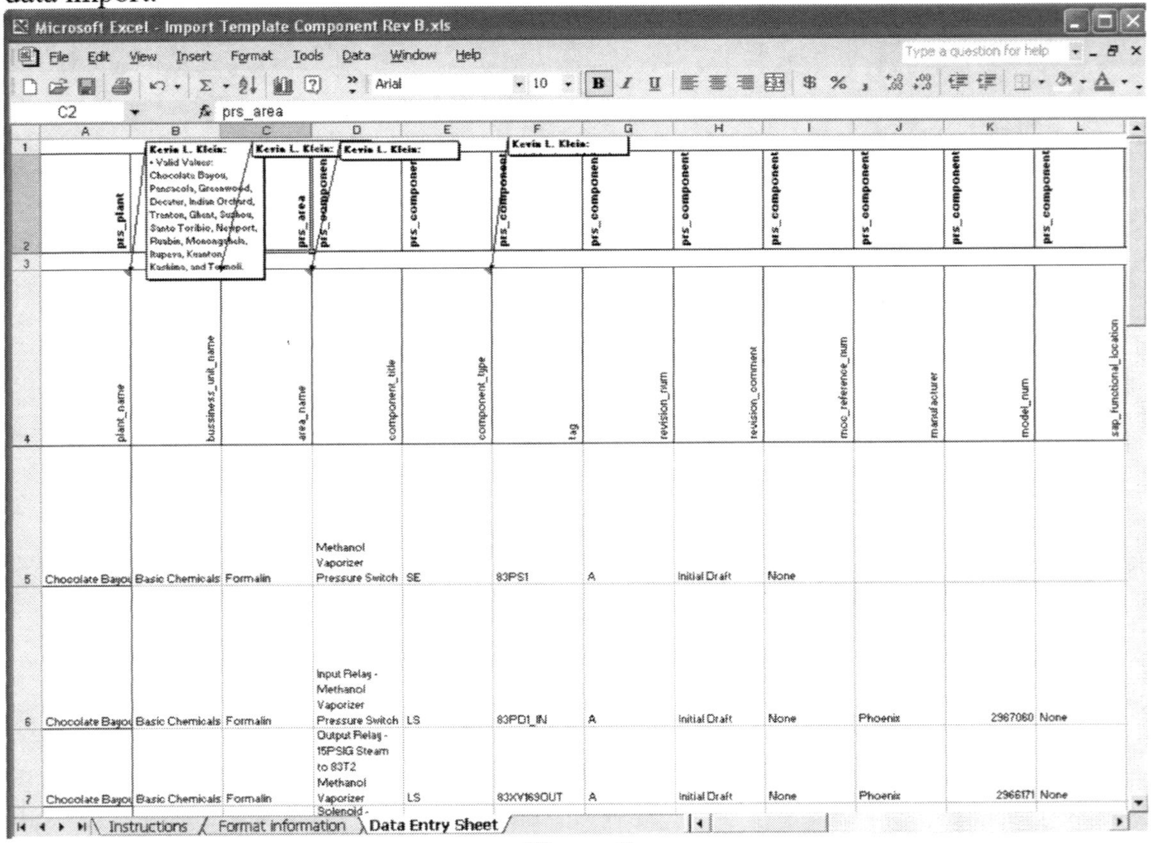

Figure 7

Component import template in Microsoft Excel

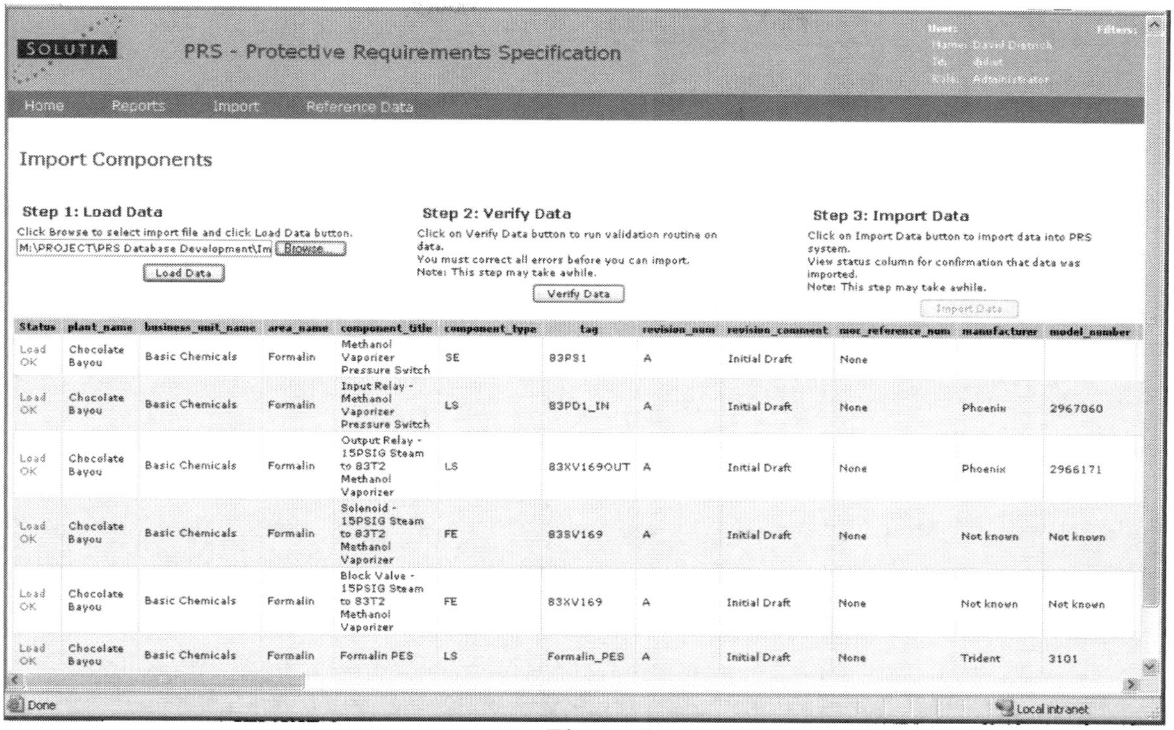

Figure 8
Loaded data ready for verification and import.

9. REPORTING

Since the PRS will be used by many different types of people through out the lifecycle of the IPS, it was important to have a system which would allow the information to be displayed differently as needed. We identified several specific groups which need to print out and use reports from the PRS database; the engineers, maintenance and operations. When looking at the required functionality for the operators, you need to decide whether to document the IPF directly in the operating procedures and just reference the operating procedures in the PRS, or do you reference the PRS document from the operating instruction and provide a report for the operators? In our design we provided both options, a report for the operators as well as the ability to attach the operating instructions to the database. Since the operators do not need all of the details provided in the PRS, the operator report needed to provide a summary of the important features for the operator to use. These include a basic description of the function, the classification, the hazard, revision number, a listing of the components and the trip points.

For the maintenance personnel, they need to be concerned with what components are being used and how to maintain them. The maintenance reporting focused on the component portion of the database by providing a listing of the components and the locations used and a way to search for similar components being used elsewhere in the company.

10. DEVELOPMENT OF SOFTWARE

As with the development of an IPS, you will need to document all the choices you have made so the design can be reviewed by the users. The design basis for the software includes all the details on the security, database architecture, models of user screens, editing, import, and reporting. Once a consensus is arrived at, the design basis is delivered to the programmers to start the programming. As the programming is developed, reviews are performed to ensure the programming meets the expectations of the users defined the program specification.

11. TESTING AND ROLL OUT

Once the program is up and running, this is the time to see if we can break it. This is the objective of testing to verify that the programming is correct and robust enough to operate without failure. When selecting the users to test the program consider identifying several users from each of the user types such as approvers, editors, viewers, operations, and maintenance. This will ensure that all parts of the program are working properly and that the users understand what to do. During this time period the user manual can be developed by the test users. The user manual should provide clear instructions on how to navigate through the program and step by step procedures for entering, editing, approving, reporting, and deleting a PRS.
With testing completed and user manual in hand, it is time to start training the users on the database.

12. CONCLUSION

Providing effective tools such as relational databases can greatly enhance the productivity and usefulness of the PRS. The PRS family of information is complex, but can be effectively managed with a relational database. There continue to be many options for documenting the IPF. This paper has showed you what one company has chosen to do, but the options are endless with relational databases. You just need to get into the family.

13. REFERENCES

1. CCPS/AIChE, *Guidelines for Safe and Reliable Instrumented Protective Systems*, American Institute of Chemical Engineers, New York, NY (2007)

2. ISA, ANSI/ISA84.00.01-2004 Part 1– *Functional Safety: Safety Instrumented System for the Process Industry Sector*, The Instrumentation, Systems, and Automation Society, Research Park, NC (2004).

14. ATTACHMENT

Protective Requirements Specification (2.1)

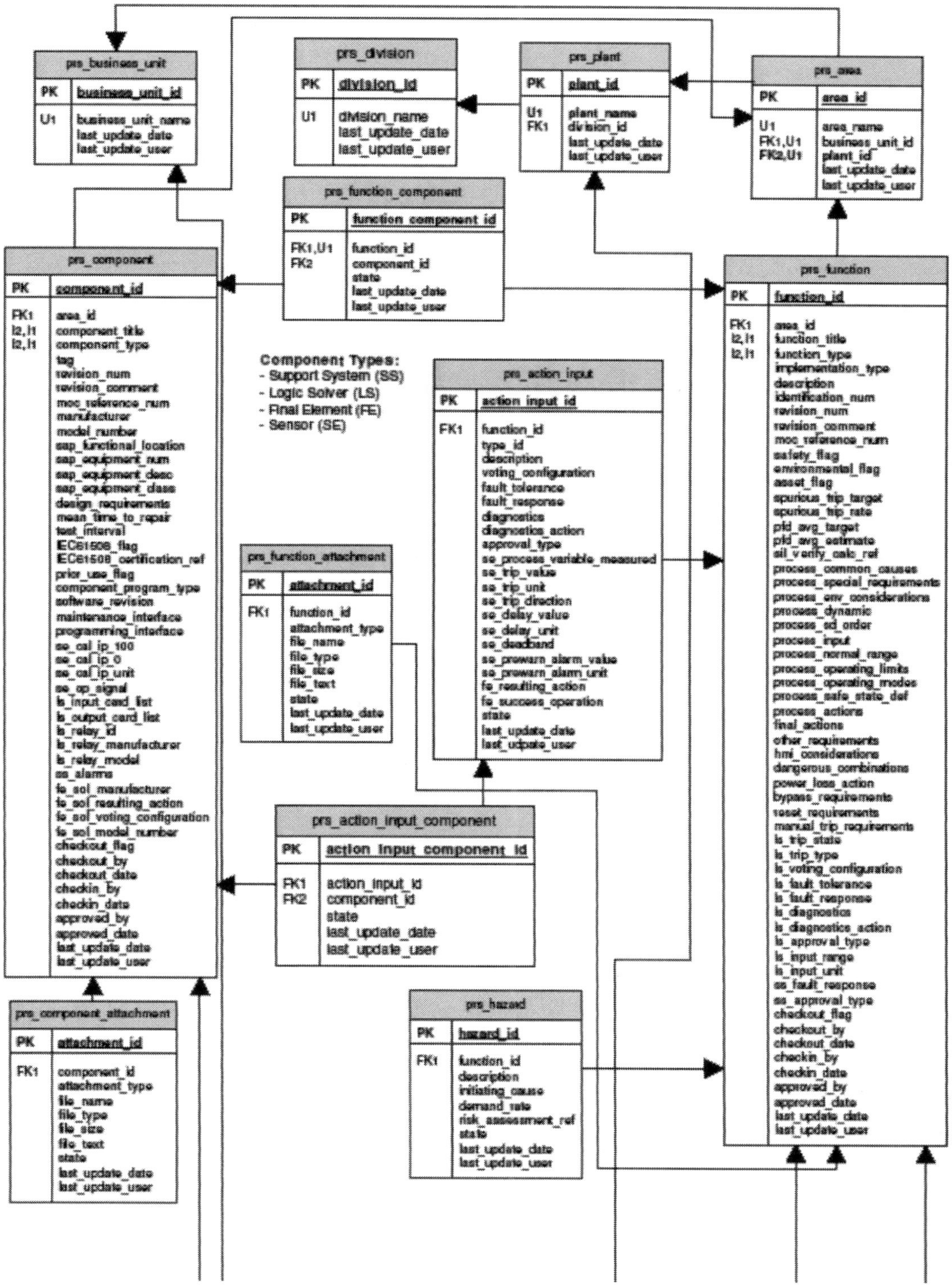

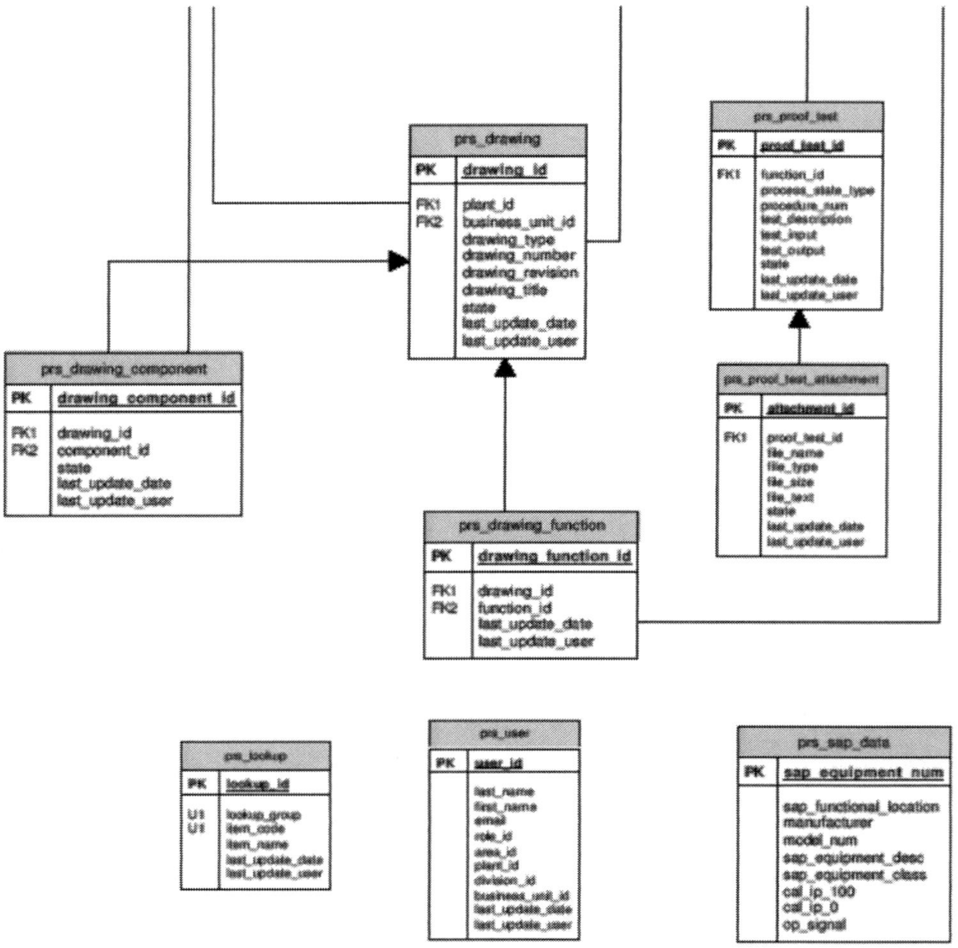
prs_drawing
PK drawing_id
FK1 plant_id
FK2 business_unit_id
drawing_type
drawing_number
drawing_revision
drawing_title
state
last_update_date
last_update_user
prs_proof_test
PK proof_test_id
FK1 function_id
process_state_type
procedure_num
test_description
test_input
test_output
state
last_update_date
last_update_user
prs_drawing_component
PK drawing_component_id
FK1 drawing_id
FK2 component_id
state
last_update_date
last_update_user
prs_proof_test_attachment
PK attachment_id
FK1 proof_test_id
file_name
file_type
file_size
file_text
state
last_update_date
last_update_user
prs_drawing_function
PK drawing_function_id
FK1 drawing_id
FK2 function_id
last_update_date
last_update_user
prs_lookup
PK lookup_id
U1 lookup_group
U1 item_code
item_name
last_update_date
last_update_user
prs_user
PK user_id
last_name
first_name
email
role_id
area_id
plant_id
division_id
business_unit_id
last_update_date
last_update_user
prs_sap_data
PK sap_equipment_num
sap_functional_location
manufacturer
model_num
sap_equipment_desc
sap_equipment_class
cal_ip_100
cal_ip_0
op_signal

Risk Analysis for Operation of Aluminum Heat Exchangers Contaminated by Mercury

S. Mark Wilhelm, Ph.D.
Mercury Technology Services
Tomball, TX 77377
smw@hgtech.com

A 10th PPSS Anniversary Paper

ABSTRACT

Brazed aluminum plate-fin heat exchangers are extensively used in gas separation processes including LNG, LPG, NGL, nitrogen rejection and olefins manufacture. In situations where mercury is a trace component of feed gas or liquid feeds to crackers, condensation of liquid or precipitation of solid mercury can occur in heat exchanger passes, even with functional mercury removal systems in place. Mercury in liquid phase causes, under certain well-defined conditions, liquid metal embrittlement of susceptible metallurgy or amalgam corrosion of core fins, both of which can lead to sudden loss of pressure containment. Mercury-contaminated aluminum heat exchangers require close scrutiny and quantitative risk assessment to allow safe operation, remediation or to justify replacement. The risk analysis procedure involves computational prediction of mercury deposition, inspection of critical areas, detailed assessment of metallurgy and fabrication, strain analysis of temperature changes during trips and shutdowns and oxide fatigue analysis. Differentiation of leak and rupture failure modes can be accomplished based on calculated amount of deposition and on location of mercury deposits as determined from focused inspection. Assigning probability of failure requires an intimate understanding of the mechanistic influences to Liquid Metal Embrittlement (LME) crack initiation and propagation. Failure statistics are essential to assignation of probability-based risk factors.

1. INTRODUCTION

Deposition of liquid elemental mercury in aluminum heat exchangers (AHXs) can compromise their structural integrity. One mechanism by which this can occur is referred to as liquid metal embrittlement (LME), which has been responsible for a number of failures of aluminum equipment in the gas processing industry over the past 30 years. A second mechanism by which mercury can attack aluminum is amalgam corrosion (AMC), which requires both mercury and water in order to oxidize and thus degrade aluminum equipment.

In most cases, the LME failure mode is a leak in proximity to a weld but occasionally LME cracks propagate to greater distances to produce a rupture that allows sudden discharge of large quantities of gas and/or liquids. The AMC failure mode varies in location but fin corrosion can allow plate separation, again resulting in a sudden gas release. The consequences of sudden, unexpected loss of pressure containment by gas processing equipment are well known to the industry and necessitate stringent safety engineering, risk analysis and risk minimization efforts. If mercury is discovered in aluminum equipment, or if estimated to have deposited due to a plant upset, operators must make decisions on plant operation, remediation and possible replacement of critical equipment. Risk analysis then becomes a key tool to assist educated decisions thus to ensure safety and plant integrity.

Several steps are required to analyze and quantify risks associated with operation of aluminum plate-fin heat exchangers potentially contaminated by mercury. The major risk analysis steps are:

- Acquisition of analytical data for mercury in feed to the cryogenic equipment both current and historical
- Computational analysis to predict amount and location of mercury deposition
- Heat exchanger inspection to quantify degree of contamination, location and size of deposits and to confirm predictions
- Compilation of circumstantial evidence including historical operating conditions and practices
- Scrutiny of metallurgy, welding procedures and heat exchanger design

Information from these steps is then compared to the statistical compilation of circumstantial causes of prior AHX failures so as to assign probability of leak or rupture. The consequences of failure are assessed as a separate exercise to complete the risk assessment in a quantitative manner. The procedures discussed herein comply with those specified in API Recommended Practice (RP) 580 (1). LME or AMC failures of aluminum equipment reside in a special category of failure mechanisms, however, and thus require an extraordinary understanding of mechanism to assign risk probability. Failure consequences can be quantified using the prescriptions of API (RP) 580 but assignment of probability of failure is the more difficult task, especially in the case of LME or AMC.

2. MERCURY in ALUMINUM HEAT EXCHANGERS

Mercury is a naturally-occurring trace component of natural gas, liquids associated with gas, and crude oil. Of the variety of mercury compounds that may exist in gas reservoirs, elemental mercury is the dominant mercury species in gas after separation of produced water and condensate. In crude oil and condensate, several other mercury species are typically present in addition to the elemental form (2, 3). Elemental mercury is the primary mercury species responsible for damage to equipment. Throughout the following discussion, when the term "mercury" is used, elemental mercury is meant unless otherwise stated.

Aluminum heat exchangers used in cryogenic natural gas separations receive mercury in the gas being liquefied and fractionated, in which case elemental mercury is the vastly dominant species. For processes with liquid feeds such as naphtha or gas condensate, the various mercury species dissolved or suspended in the feed liquid typically are converted to the elemental form by conditions (heat + hydrogen) in the furnace or cracker. Thus for olefins and syngas processes also, the elemental form of mercury is the mercury species that enters the AHX.

Details of construction of AHXs can be found in the specification document from the Aluminum Plate-Fin Heat Exchanger Manufacturers' Association (4). For most equipment manufactured before year 2000, and many since then, high strength aluminum alloys (5083 and 6061) were used to construct headers, nozzles, flanges and manifold piping (see Figure 1). The high strength aluminum alloys that contain magnesium as the strengthening agent give rise to metallurgical features potentially susceptible to LME (5). AMC is not alloy dependent.

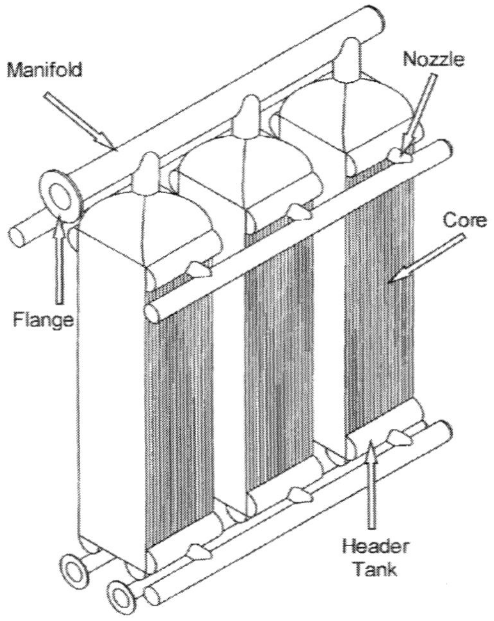

Figure 1 – Aluminum Heat Exchanger

3. MECHANISM of LME

Mercury deposits in cryogenic heat exchangers by condensation (to liquid Hg) of mercury vapor in gas or by precipitation (to solid Hg) of mercury dissolved in liquid hydrocarbon. The phase transitions are caused by cooling in the heat exchanger itself or by external cooling from a Joule-Thomson (J-T) valve or turbo-expander. Mercury can accumulate in the aluminum equipment as either liquid or solid deposits. In cold liquid streams, solid (frozen) mercury particles can carry long distances and thus mercury deposits are often found in areas of limited flow far from the point of actual condensation or precipitation. Solid deposits become liquid when heat exchangers warm during a shutdown or trip.

LME requires mercury to be in liquid phase and in physical contact with metal, not just the oxide on the surface. LME cracks in high strength aluminum alloys propagate in an intergranular manner. LME is distinct from stress corrosion cracking in that no purely electrochemical processes are involved. The LME mechanism is illustrated simplistically in Figure 2.

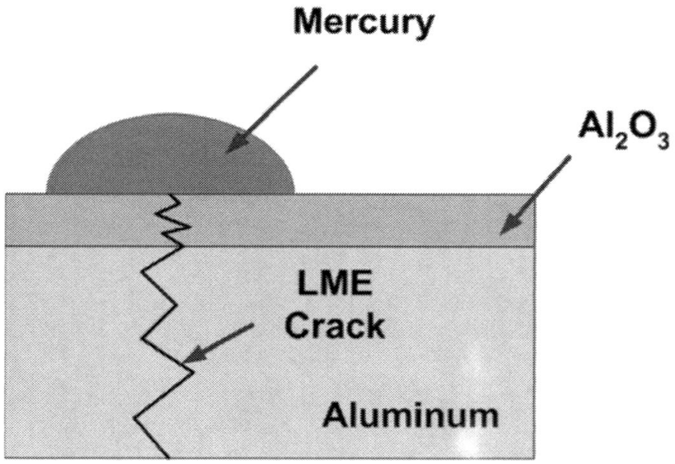

Figure 2 – Liquid Metal Embrittlement of Aluminum

The key ingredients for initiation and propagation LME cracks in aluminum are as follows:

- Temperature above the freezing point of mercury (-38.9 C). *Necessary for both initiation and propagation.*
- Mechanical, thermal or residual stress. *Necessary for propagation.*
- A breach in the aluminum oxide layer that covers the surface to allow mercury to come in direct contact with aluminum metal. *Necessary for initiation.*
- The contacted metal must have a susceptible metallurgy and microstructural condition. *Necessary for initiation and propagation.*

3.1 Metallurgy

Initiation of LME by mercury is more likely in aluminum alloys that contain magnesium as an alloying element. These include the 5000 and 6000 series aluminum alloys that have magnesium additions from approximately 0.2 to 6.0% to provide an ultimate tensile strength of approximately 20 to 50 ksi in the annealed condition. Al-Mg alloys are welded using filler metal alloys that also contain magnesium such that the weld achieves the strength of the base metal. Aluminum alloys with more than 2 - 3% magnesium have the potential for sensitization.

Sensitization is the partition of the non-metallic or inter-metallic compounds to grain boundaries primarily during welding, heat treatment or high temperature service. In the case of Al-Mg alloys, the inter-metallic compound Al_3Mg_2 segregates to grain boundaries during welding producing a microstructure especially prone to mercury LME. Mercury reacts with the Al_3Mg_2 in grain boundaries to form a Hg-Mg inter-metallic compound that is insoluble in the mercury in the propagating crack (5, 6).

$$Hg + Al_3Mg_2 \rightarrow Hg_2Mg \qquad [1]$$

Because mercury attacks the Al_3Mg_2 in grain boundaries, mercury LME of Al-Mg alloys is strictly intergranular, i.e. cracks follow the metal grain boundaries and do not cross grains (Figure 3).

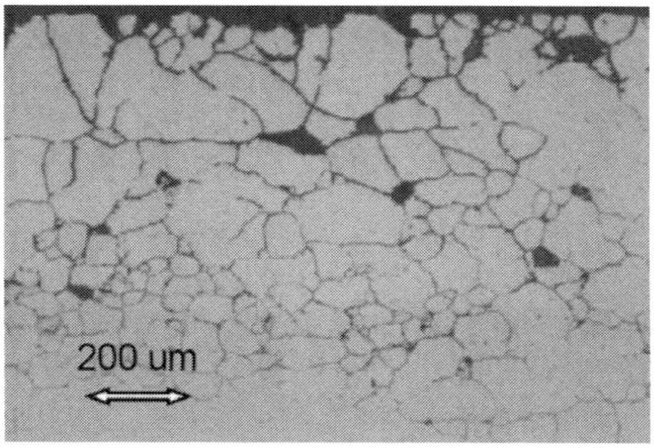

Figure 3 – Intergranular Attack of Al 5083

3.1 Oxide Fatigue

Mercury deposition on the surface of a susceptible aluminum alloy is a necessary, but insufficient condition for LME to initiate. Crack initiation requires some mechanism to allow the surface mercury to reach the underlying metal. The surface of fabricated aluminum is covered by an oxide (Al_2O_3) having variable thickness and mechanical properties that depend on fabrication method. Under normal conditions (isothermal, isobaric, benign chemical environment), the oxide on aluminum is resilient and impervious to mercury that may reside on its process-side surface. Mercury can breach the oxide on aluminum in several ways. Abrasion by particulate matter can damage the oxide sufficiently to allow contact. Corrosion from chemicals introduced upstream can also disrupt the normally protective alumina layer. Most commonly in cryogenic service, the aluminum oxide is damaged by thermal and mechanical strains that produce oxide fractures caused by what is termed "oxide fatigue".

Oxide fatigue is the cumulative disruption of the aluminum oxide crystalline structure by strain induced micro-fracture events. Fractures occur when there is disproportionate strain in the aluminum metal substrate and the aluminum oxide surface layer. In actual practice, AHXs that have failed typically have experienced numerous cycles in temperature and pressure before the mercury that may be present is able to penetrate the oxide completely. This time and strain event dependency is attributed to oxide fatigue (7).

The difference in the coefficient of thermal expansion of aluminum and aluminum oxide is such that aluminum metal expands in physical dimension about 3 times as much as its oxide with increasing temperature. When a heat exchanger is operating cold, the oxide is in compression. When an AHX is allowed to warm, the oxide on the surface will stretch and may fracture to some extent due to the greater expansion of the aluminum metal to which it is attached. In practice, AHXs contaminated with mercury resist some number of temperature excursions before mercury attack (LME) initiates, if all other necessary conditions for LME are present.

The thermal or mechanical strain rate is also important to the amount of damage done to the oxide and hence the time to initiate cracks. Although not precisely quantified, rapid changes in pressure or temperature are especially detrimental to the oxide's ability to resist penetration by mercury (7, 8). Sudden pressure changes used to dislodge particulate material in clogged heat exchanger passes are extremely detrimental. If mercury contamination is present in the clogged heat exchanger pass, efforts to blow out the solids by sudden pressure release upstream can cause LME failure in only a few attempts.

Oxide fatigue can be quantified by measuring cycles to failure in the laboratory or from correlations of strain events in service to actual failure time (7). Time to failure depends on the temperature (or pressure) difference in the cycle, the rate of temperature (or pressure) change and on the thickness and mechanical properties of the oxide. Oxide fatigue curves are used to estimate time in service to crack initiation by examination of the service history of the heat exchanger and comparison to experimentally measured cycles to failure under conditions similar to those seen in service. The thorough examination of service history typically reveals number of trips and shutdowns and temperature change associated with each.

Other mechanisms, such as abrasion and chemical attack, also exist whereby the normally protective surface oxide on aluminum can lose its ability to prevent contact of surface mercury with the underlying metal. Statistically, oxide fatigue is more common in that probability of failure is strongly correlated with AHX length of service.

3.3 Fracture Mechanics

One unusual aspect of LME, as opposed to other fracture processes, is that the crack propagation rates can be exceedingly rapid and the stress intensity required for crack propagation can be very low (9, 10). Crack velocities on the order of cm/s have been measured in the laboratory. Theoretical and empirical studies have suggested that LME crack propagation is influenced by numerous factors that include rate of liquid mercury surface and bulk diffusion in cracks, by the concentration of Al_3Mg_2 in grain boundaries (degree of sensitization) and by the dissolution rate of aluminum into mercury at the crack tip and by solid-phase fracture mechanics parameters (9-14).

Low temperature crack propagation rates have not been measured. Thus it is not certain as to whether cracks that may initiate at ambient temperature will propagate at cryogenic temperatures. If crack propagation rates at low temperatures are proportional to the rate of liquid mercury diffusion and aluminum dissolution into liquid mercury, then propagation rates should be essentially zero below the freezing point of mercury. The aspects of fracture mechanics studies that impact risk assessment are:

- Propagation requires very low stress hence efforts to eliminate stress are typically futile. In assessing the likelihood of crack propagation in equipment, the stress criterion is always satisfied.
- Propagation rates are fast, especially at ambient temperatures. This means that methods used to detect slow-growing cracks (acoustic emission, ultrasonic techniques) typically are ineffective to locate LME cracks.
- Metallurgical defects that impact crack initiation in other mechanisms likely have little influence on LME. For example inclusions such as tungsten in welds do not encourage crack initiation.

4. MECHANISM of AMALGAM CORROSION

Amalgam corrosion of aluminum is similar to LME in the sense that the oxide on the surface must be breached to allow physical contact of mercury and aluminum. Contact of mercury with any aluminum alloy will produce amalgamation (simple dissolution). Amalgam corrosion occurs spontaneously if water vapor or liquid water contacts the amalgam. The rate of amalgam corrosion penetration is proportional to humidity. The reaction scheme is:

$$Hg + Al \rightarrow Hg(Al) \text{ amalgamation} \quad [2]$$

$$Hg(Al) + H_2O \rightarrow Al_2O_3 \cdot 3H_2O + H_2 + Hg \quad [3]$$

$$Hg + Al \rightarrow Hg(Al) \quad [4]$$

Amalgam corrosion (Figure 4) does not consume the mercury and hence is self-propagating so long as mercury is in contact with aluminum metal and water is available. If sufficient moisture and mercury are present, aluminum structural components can be penetrated rapidly. The rate of attack is mass transfer limited and the aluminum oxide corrosion products inhibit the access of moisture to the amalgam surface eventually. The amalgam corrosion reaction is not alloy-specific. All aluminum alloys can be affected.

Figure 4 – Core Separation due to Amalgam Corrosion

If mercury deposits are discreet and not continuous and liquid water is present, amalgam corrosion can be localized in the form of pits. The pits propagate when the exchanger is warm and water collects in low areas that have surface mercury contamination. In borescope inspections mercury pits are rarely but occasionally seen in manifold piping and appear as white spots on the shoreline of water deposits.

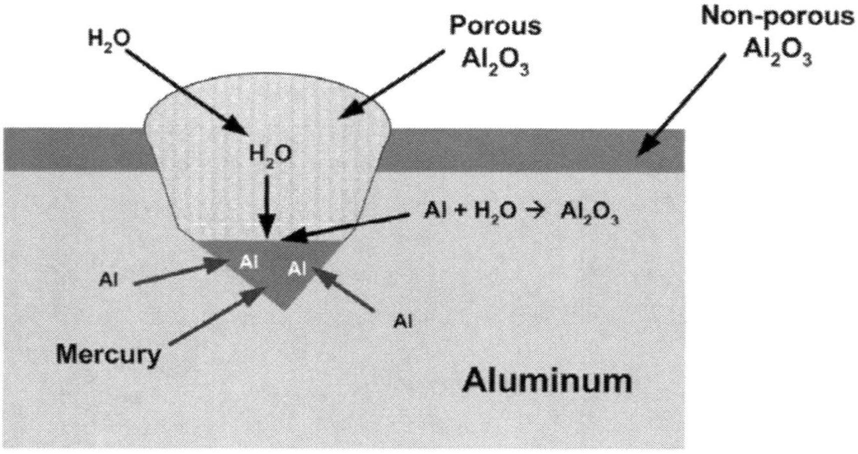

Figure 5 – Mercury Pits on Aluminum

5. PREDICTION of MERCURY DEPOSITION

A computational analysis can be performed to determine the conditions under which mercury will condense in the AHX or across J-T valves and turbo-expanders (any cooling process). The following assumptions typically are employed in the modeling of mercury in the process:

- All mercury is in the elemental form.
- Separations are at equilibrium.
- The solubility of mercury in liquids depends on liquid composition but constant with pressure.
- The vapor pressures of mercury and hydrocarbons can be estimated using a thermodynamic equation of state.

Equilibrium concentrations of Hg in gas and liquid phases are calculated using partition factors (solubility/vapor pressure) which depend upon non-ideality in the gas phase and compressibility of the liquid phase. The influence of hydrocarbon composition on mercury vapor pressure and solubility is empirically verified in the range 0 - 100 C but becomes increasingly uncertain below the freezing point of mercury and at very high pressure. Empirical data are used to augment model predictions.

Predictions of actual accumulation of mercury deposits (Figure 6) depend, in part, on piping and equipment orientation since gravity influences deposition. Fluid dynamics must also be considered because precipitated solid mercury can travel long distances suspended in liquids. Predictions of the locations of deposition rely heavily upon experience gained from inspections.

As computational analysis techniques have improved, so has understanding of mercury concentration limits for precipitation of mercury in cryogenic systems. Likewise analytical measurements of mercury concentrations in gas streams have become much more sensitive, to where detection limits now have approached 0.01 $\mu g\ Hg/Sm^3$ of gas. For some cold gas separations, functional and well-designed mercury removal systems positioned upstream of the AHX are now seen as marginally able to remove mercury to level sufficiently low that some mercury precipitation or condensation does not occur. Even though rates of deposition may be small, over time significant deposits may develop.

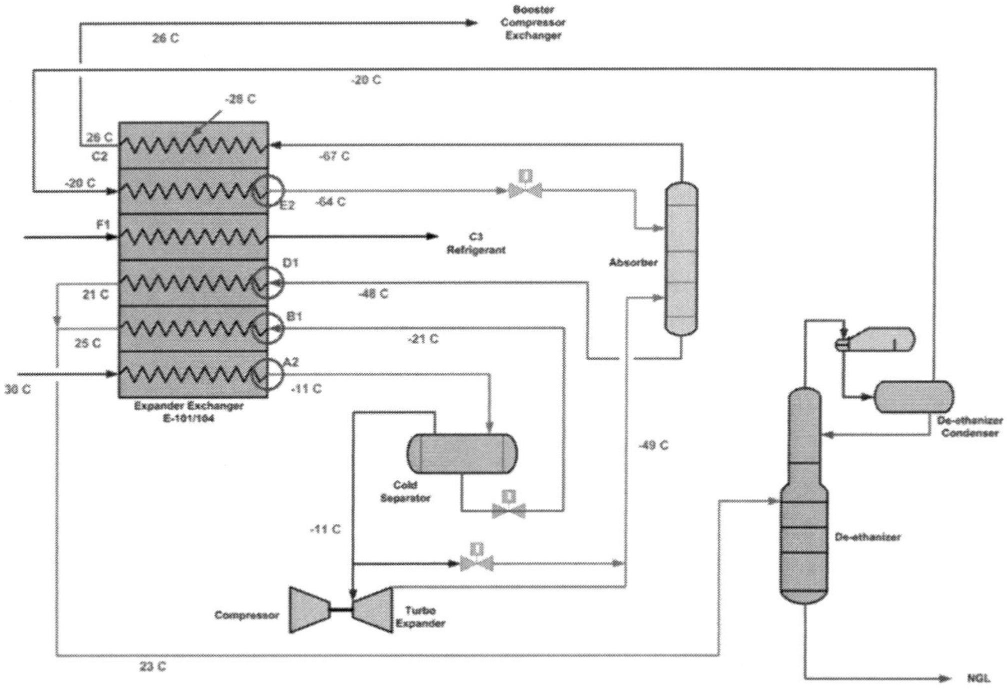

Figure 6 – Locations of Mercury Accumulation from Computational Analysis

6. INSPECTION

If mercury deposition is predicted by computational analysis or circumstantially suspected, an inspection is warranted. Inspections are normally performed during a scheduled turn-around. Since warm conditions are potentially harmful to aluminum heat exchangers, a shutdown for the sole purpose of inspection adds to the risk of oxide fatigue. The primary objective of the inspection is to document the location and quantity of mercury in the AHX so as to verify or refute predictions obtained from the computational analysis and to gain additional information not available from other sources (water, corrosion, etc.). The size and complexity of most AHXs preclude inspection of all piping welds, headers, nozzles and manifolds. Locations for inspection are selected based on the computational analysis and based on statistical probability.

Heat exchanger inspection is essential to assignment of failure probability. Non-intrusive radiographic inspections and invasive video-borescope inspections are used as well as ultrasonic techniques in some instances. Mercury deposits are easily detected using conventional (film) or computed χ-ray or γ-ray techniques. The trade off is better resolution with χ-ray and easier source orientation with γ-ray. Figure 7 shows a conventional radiograph of mercury deposits adjacent to a circumferential weld and backing ring.

The utility of borescope inspection is that locations inaccessible to radiography can be examined. In addition, corrosion processes and the precursors of corrosion processes can be identified by borescope inspection but not by radiography. Borescope inspection should be conducted when the equipment is under nitrogen purge so that moisture cannot enter the equipment. Figure 8 is a photograph extracted from a borescope video showing a nozzle and core distributor plate.

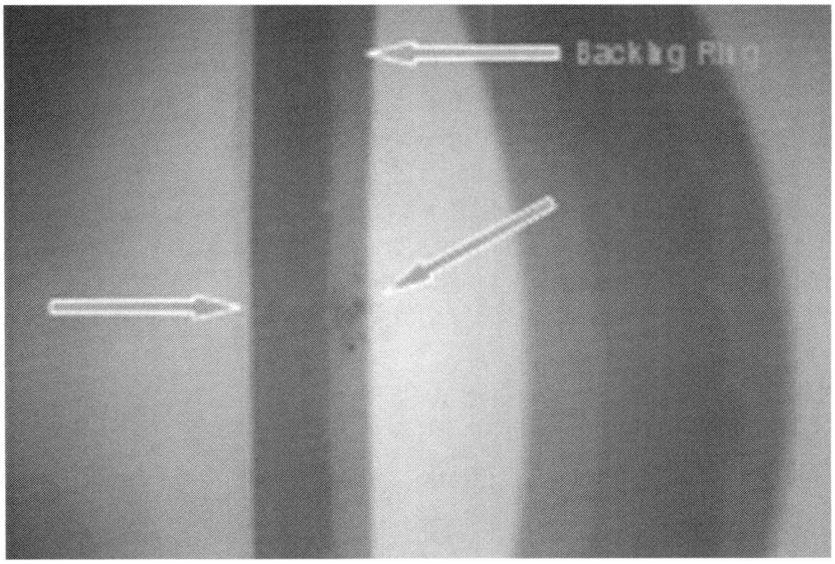

Figure 7 – Radiograph Showing Mercury behind Backing Ring

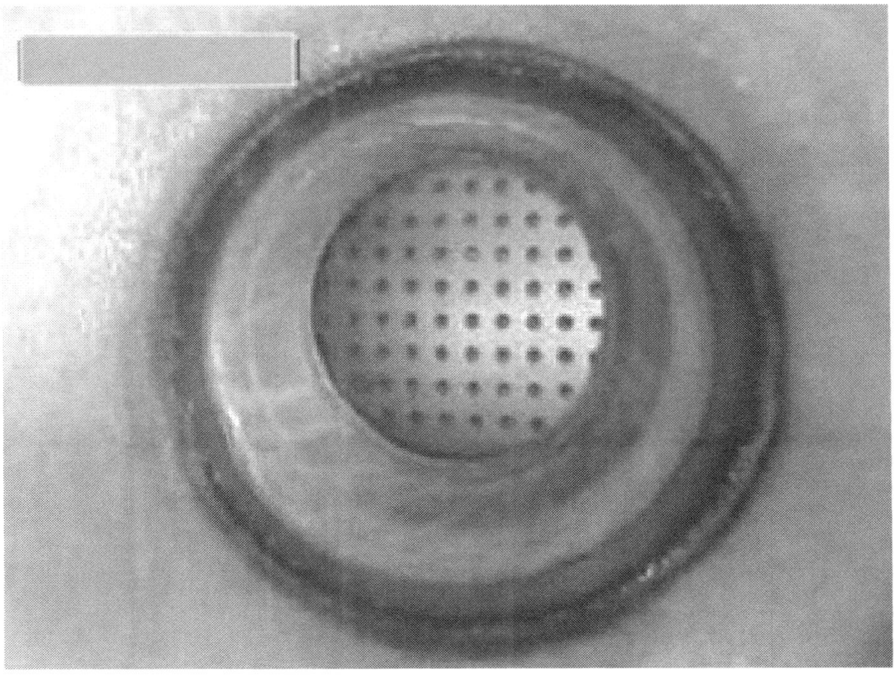

Figure 8 – Borescope Picture of AHX Nozzle

7. QUANTIFICATION of RISK

7.1 Circumstantial Risk Factors

An AHX that becomes contaminated by mercury is at some risk of failure but risks often can be managed and minimized to allow safe operation if the amount and location of the mercury in the equipment can be assessed accurately and circumstantial risk factors can be eliminated or minimized. The initial phase of the risk analysis attempts to determine the likelihood of failure of the aluminum equipment due to contact with condensed liquid mercury. The likelihood of failure, defined as loss of pressure containment due to LME or amalgam corrosion, depends on circumstances, both historical and those existing at the time of analysis. In the analysis of circumstances that impact risk of LME or AMC, answers to the following questions are typically sought:

- Is (or was) mercury present in the feed to the plant?
- What is (or was) the concentration of mercury in gas entering the aluminum equipment?
- Is there an upstream mercury removal unit (MRU) and does it perform to specification?
- Can mercury condense to a liquid or solid phase in the heat exchanger passes due to cooling?
- Are mercury deposits known to be present the aluminum heat exchanger (AHX)?
- What is the amount and distribution of the contamination?
- Is liquid mercury in contact with aluminum having a susceptible metallurgical microstructure?
- At the locations of susceptible metallurgy, are stresses present that have sufficient magnitude to allow LME?
- What is the frequency of strain events (trips, shutdowns, upsets)? Are strains dynamic?
- How old is the equipment? Who made it and what are the construction details?
- Do welds have backing rings?
- Is the pressure drop across the equipment constant or increasing?
- Is the upstream dehydration system working properly?
- What other circumstantial factors could affect AHX integrity as affected by mercury (methanol injection, derime procedures, particulate intrusion, etc.)?

7.2 Probability

The probabilities of failure by leak or rupture are estimated by comparison of AHX circumstances and inspection data to those of heat exchangers known to have failed. Failure statistics, although limited in some categories, are available for both leaks and ruptures and include the following selected examples:

- Failure frequency correlates with equipment age.
- Failure frequency correlates with frequency of trips and shutdowns.
- Mercury deposits behind backing rings (Figures 7, 9) have a high probability to allow crack initiation but the amount of mercury in the pipe/ring crevice is seldom sufficient to allow extended circumferential cracks. Thus leaks at backing rings are much more likely than rupture. Other locations of high probability are revealed from failure statistics and include circumferential welds in manifolds, nozzle to header tank welds and reboiler piping welds.

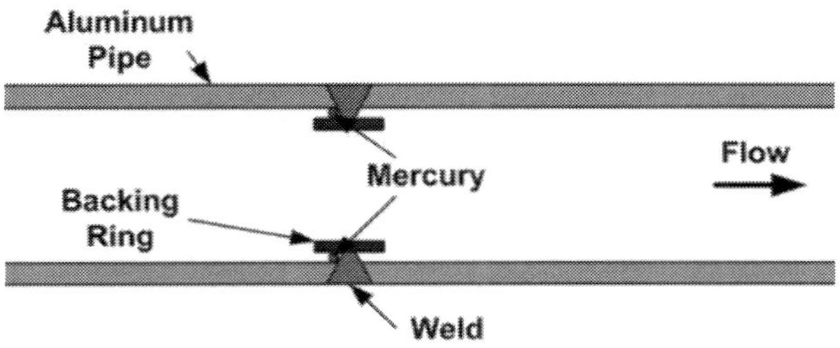

Figure 9 – Mercury in Backing Ring Crevice

- Leaks are more likely when equipment is pressurized after a scheduled shutdown of prolonged duration (days). Leaks in service associated with trips and unscheduled shutdown events of short duration are statistically less likely than leaks after long plant shutdowns.
- Ruptures are rare but they do occur (15). Ruptures require sufficient mercury on the surface to feed dissolution inside the crack (Figure 10). The mercury inside a crack is saturated with aluminum (high Al surface area, low Hg volume). Dissolution at the crack tip can only occur if there is unsaturated mercury on the surface to feed mercury to the crack and to accept aluminum dissolving into mercury in the crack and diffusing according to the existing concentration gradient. Substantial accumulation giving rise to large deposits of mercury is therefore required to provide conditions conducive to rupture.

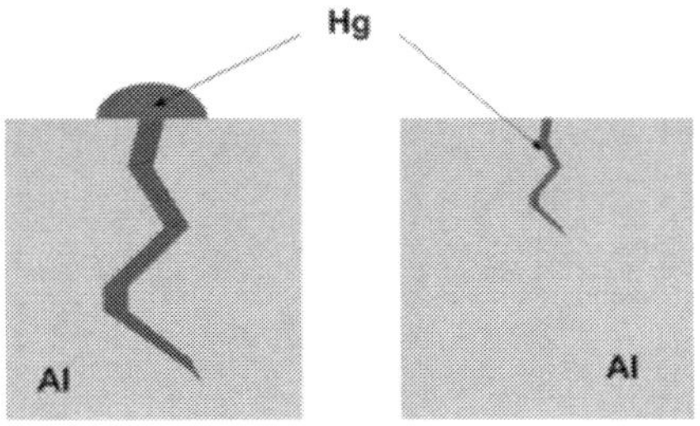

Figure 10 – Mercury Deposits and Crack Length

- AMC failures correlate to the same characteristics as LME crack initiation but also to the performance of upstream dehydration equipment. Upsets in molecular sieve or glycol dehydration systems, when coupled to mercury deposition in the AHS, give rise to increased probability of AMC.

8. FAILURE CONSEQUENCE and RISK MINIMIZATION

The estimation of failure consequence is a separate exercise to the assignment of failure probability. Consequences for leak or rupture scenarios are estimated using conventional methods. The recommendations of API RP 580 serve as a good guide to the methods typically utilized. Incidents of AHX failure due to LME have been reported that have led to major fires (15) and personnel fatalities in some instances.

A major potential consequence of an LME or AMC failure is plant down time. AHXs are in high demand now for LNG and LPG manufacture and there is a limited number of manufacturers. Orders for new equipment can have long lead times. When assessing failure scenarios down time becomes a major cost factor, especially when contrasted with repair.

Repairs and/or partial replacements to an AHX or associated piping can be accomplished if mercury deposits are small and localized. For example contaminated header tanks, nozzles and piping sometimes can be replaced in the field. Such exercises require a separate risk analysis that examines LME resistance of field repair welds, the procedures for pressure testing and the risk involved with prolonged down time in warm condition. Pressure and leak tests may be necessary to meet pressure vessel code but they carry risk to initiate and propagate LME cracks.

The product of LME or AMC failure probability and consequence can be summarized on a numerical scale or using a matrix approach (1). From the calculated risk for leak and rupture scenarios, one can then address remediation and risk reduction. The process to manage risk and thus remain in operation is summarized in Figure 11. The decision on operation depends mainly on the amount of contamination discovered by inspection or, is some cases, predicted by computational modeling.

Risk minimization plans address at least 4 major areas:

- Improving the accuracy and frequency of analytical measurements such that better predictions of accumulation are possible
- Preventing mercury deposition in the equipment
- Using aluminum equipment tolerant of some contamination
- Minimizing conditions that allow oxide fatigue or damage

Mercury removal units (MRUs) employ sorbents to capture elemental mercury in either gas or hydrocarbon liquid streams and are the most often utilized AHX risk minimization method (16). A properly functioning and well maintained MRU, coupled with an analytical monitoring plan that targets MRU performance will prevent mercury accumulation in most processes. Over the course of time, however, even well designed mercury removal systems may allow some mercury deposition. These situations develop when MRU sorbent beds for are damaged by liquid or contaminant intrusion, when beds spend prematurely or when feed composition changes abruptly.

If an inspection finds moderate to high levels of contamination, the options for remediation are limited. Possibilities to remove mercury deposits from aluminum equipment include thermal desorption, solvent cleaning and chemical cleaning. All approaches have limited chance of complete success if mercury deposits are substantial. Chemical cleaning is possible but seldom attempted due to the risk of damaging the oxide and spreading contamination within the equipment.

Remediation or replacement of aluminum equipment also must consider upstream mercury accumulation in steel piping and vessels. Steel absorbs mercury reversibly such that mercury absorbed in warm upstream piping and equipment can be reintroduced into the gas stream entering the cold AHXs. Thus upstream remediation may be required to prevent mercury deposition into new or otherwise clean equipment.

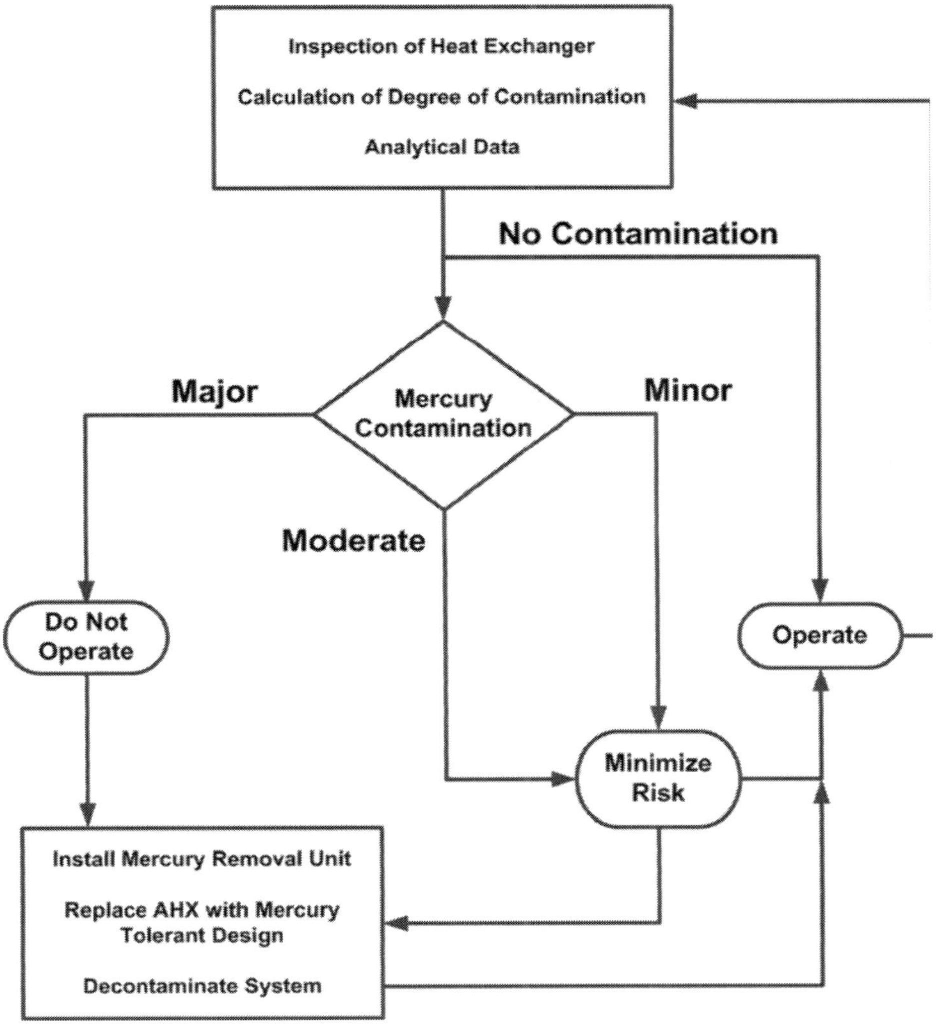

Figure 11 – Decision Tree for Risk Minimization

9. CONCLUSIONS

Mercury's adverse interaction with aluminum requires careful scrutiny of processes that employ aluminum heat exchangers. Most feeds to natural gas and syngas separation processes contain some mercury and many have concentrations that, although low, are significant for aluminum cryogenic equipment. The need for concern is coupled to the fact that many AHXs have entered, or will enter soon, a period of service considered old (10 – 20 years or more of operation) from the standpoint of metal fatigue and oxide fatigue. Periodic inspection and risk analysis for operation are prudent exercises that are needed to safeguard personnel safety and plant integrity. The probability of LME or AMC failure can be quantified by examination of failure statistics in comparison to the circumstantial risk factors revealed by an understanding of the LME and AMC aluminum degradation mechanisms.

10. REFERENCES

1. API Recommended Practice 580, *Risk-based Inspection,* American Petroleum Institute, 2002.
2. Wilhelm, S.M. and Bloom, N.S., Mercury in Petroleum, *Fuel Processing Technology*, 63, 1, 2000.
3. Bloom, N.S., Analysis and Stability of Mercury Speciation in Petroleum Hydrocarbons, *Fresenius' J. Anal. Chem.*, 366, 5, 2000.
4. Aluminum Plate-Fin Heat Exchanger Manufacturers' Association (ALPEMA), *Standards for Brazed Aluminum Plate-Fin Heat Exchangers,* 1994.
5. Nelson, D.R., Mercury Attack of Brazed Aluminum Heat Exchangers in Cryogenic Gas Service, *Proceedings 73rd Annual GPA Convention*, Gas Processors Association, 1994.
6. Bell, R. N., Understanding and Preventing Failure of Aluminum Equipment in the Presence of Liquid Mercury, *Proceedings - American Institute of Chemical Engineers Spring Meeting*, Atlanta, Georgia, 2005.
7. Wilhelm, S.M., Methods to Combat Liquid Metal Embrittlement in Cryogenic Aluminum Heat Exchangers, *Proceedings 73rd GPA Convention*, Gas Processors Association, 1994.
8. Wilhelm, S.M. and Kane R.D., Use of Slow Strain Rate Tests to Evaluate the Embrittlement of Aluminum and Stainless Alloys in Process Environments Containing Mercury, *ASTM STP 1210*, Philadelphia, Pennsylvania, 1993.
9. McIntyre, D., English, J., and Kobrin, G., Mercury Attack of Ethylene Plant Alloys, Paper 106, *Proceedings Corrosion 89*, National Association of Corrosion Engineers (NACE International), 1989.
10. Kapp J.A., Duquette, D., Kamdar, M.H., Crack growth behavior of aluminum alloys tested in liquid mercury, *J. Eng. Mater. Technol.*, 108, 37, 1996.
11. Liu, H.W., Fang, L., Effects of surface diffusion and resolved shear stress intensity factor on environmentally assisted cracking, *Theoretical and Applied Fracture Mechanics*, 25, 31, 1996.
12. Joseph, B., Picat, M., Barbiera, F., Liquid metal embrittlement: A state-of-the-art appraisal, *Eur. Phys. J. AP*, 5, 19, 1999.
13. Lynch S.P., Metal-induced embrittlement of materials. *Materials Characterization*, 28, 279 1992.
14. Coade, R., Coldham, D., The interaction of mercury and aluminium in heat exchangers in a natural gas plants, *International Journal of Pressure Vessels and Piping*, 83, 336, 2006.
15. Aon Advanced Risk - Finance Conference 2007, How does an organisation treat and manage a large complex material damage and business interruption loss? Claims case study – Moomba explosion; http://www.aon.com.au/pdf/speakers_corner/2007/session4_claims_case_study_moomba_explosion.pdf, 2007.
16. Wilhelm, S., Conceptual Design of Mercury Removal Systems for Hydrocarbon Liquids, *Hydrocarbon Processing*, April 1999.

Improving Process Safety by Addressing the Human Element

John Haesle
Celanese Chemical Corporation
Pampa Texas
John.haesle@celanese.com

Chris Devlin
Celanese Chemical Corporation
Bishop, Texas
CPDevlin@Celanese.com

Jack L. McCavit
JL McCavit Consulting, LLC
Pampa, Texas
jlmconsulting@sbcglobal.net

ABSTRACT

Celanese had several process safety incidents and near misses in 1996. Although there were several root causes, this paper will address only one of them; human error. The system which was developed to improve this human element is the Conduct of Operations system[1]. The Conduct of Operations system has improved process safety performance by focusing on three operational areas: training, procedures and general operations.

This paper will describe the need for and the content of Conduct of Operations system as implemented by Celanese.

1. INTRODUCTION/BACKGROUND

In 1996, Celanese experienced an increase in the number of process safety related incidents and significant near misses, even though all the elements of OSHA 1910.119 had been implemented. An investigation across multiple sites revealed a number of root causes, one of those being human error. Further studies of the human error root cause revealed a number of interdependent human elements issues.

The human element has a significant impact on process safety. Celanese began to address the human element in 1996 and approved a corporate process in 1998. This process is called Conduct of Operations.

2. WHAT IS CONDUCT OF OPERATIONS?

Conduct of operations (*operations*) is the execution of operational and management tasks in a deliberate and structured manner. [2]

Since most of us travel by air, let us think about the airline industry's "process safety" program which ensures a safe and uneventful trip. Have you seen the pilot doing a walk around the plane checking different areas? They are not out there for their exercise. They are making sure the plane is fit for duty. Have you looked into the cockpit as you board the plane and seen either the pilot or copilot with a clipboard and checking something? Why do they have a preflight checklist that both pilot and copilot complete? They are qualified pilots and most have thousands of flying hours yet they still use a checklist to ensure all functions are operating and switches are in the proper position. Why are pilots required to state their airplane call sign, repeat the flight instructions back to the air traffic controller and then the controller must acknowledge the information? Sounds like a redundant time consuming effort especially when this occurs hundreds of time a day throughout the year, however; this ensures clear communications. Why maintain a "sterile cockpit", no non-flight discussions, until a specified altitude? They are experienced pilots yet they will not allow themselves to be distracted. Why does every major airline have a specific recertification time period and process? This is expensive and does not make sense because they fly hundreds of hours a year and often the same model of plane; however, the reason is for operational discipline. The airline industry has instituted a number of standard practices to improve their safety record and more importantly the safety of the people that fly…you and I. Most people take it for granted when they board an airplane that they will arrive safely at their destination but very seldom think about the operational discipline of both the pilot and support crews.

The Celanese Conduct of Operations process gives written requirements for the execution and control of unit operations activities. The objective of Conduct of Operations is to:

- Define the requirements and expectations for conducting operations at a high standard of total safety and effectiveness
- Reduce variability of operational practices in order to reduce human error, increase safety, and enhance effectiveness
- Perform tasks according to expectations, policies and procedures in a safe, businesslike, professional manner

In Celanese, this process applies to all unit operation and operations-maintenance interface activities. Unit operations include manufacturing units, tank farms, loading/shipping activities, operating units, utilities, waste treatment, etc.

3. WHAT ARE THE BENEFITS?

A plane crash makes sensational news whether it's on the evening news or in the morning newspaper. It captures our attention because of the number of deaths and the visual images of the aftermath or in some case the actual video footage of the sequence leading to the crash.

Celanese experienced a major process safety related incident in their Pampa Plant in 1987 which resulted in 3 deaths, multiple injuries and the destruction of the facilities. This made national news just like an airplane crash. The chemical and refining industries should have processes similar to the airline industry to improve the human element.

Within the chemical and refining industry, most companies have policies for, operator certification, effective communication, and safe behavior leading to the expectation that all safety related near misses and incidents are eliminated. Nevertheless, the industry continues to experience incidents. The technical staff continuously improves mechanical systems, designs new and improved controls but often overlooks the human element…the operators. Risks include bypassing the hardware systems we have in place to prevent accidents. Operators can skip a step in a procedure because of a lack of understanding and because they have skipped the step in the past without incident. It becomes "their way" of performing the task. Initial job certification can be completed by an on-shift buddy that just "had that gut feeling" that the trainee knew the job and signed off without verification of knowledge or performance of the tasks. These are just a few examples of risks that are self managed by operators and as long as nothing happens they will become normalized. This phenomenon has many other names such as normalization of deviance and drift. As a result of this drift, management often uncovers lack of operational discipline following an incident. Management is surprised, shocked and then angered because of the drift in operator behavior and that their systems allowed such behavior.

Conduct of Operations is a proactive approach to improve the current practices, improve the support systems and the behavior of the operator. The benefits include:

- Consistent application of training and procedures
- Reduction in human error

> As the complexity of operational activities increases, a commensurate increase in the formality of operations must also occur to ensure safe, reliable, and consistent performance of critical tasks. [3]

4. WHAT ARE THE ELEMENTS

Celanese corporate Conduct of Operations policy contains elements that address all aspects of unit operation which impact normal job duties of the operator. These elements contain the appropriate level of detail and information so consistent application by all operators company wide is possible.

These elements are listed below.

4.1 TRAINING

The training process is based on proven methodology that incorporates a formal structured process to achieve a desired job performance. This structured process includes a systematic approach to ensure:

- all knowledge and skills are identified
- a consistent application of the training process independent of the trainer such that all operators' knowledge and skills are consistent
- operator testing to an established standard
- clear and traceable documentation

The training process reduces human error associated with operating processes.

4.1.1 TRAINING PROGRAM

The training program is performance based, objective centered and evaluated to a standard. This ensures consistency by requiring that all employees receive the same training regardless of their trainer or evaluator. A formal training program reduces the variability of the knowledge and skills for new employees, it ensures all trainees are held to the same standards for performance and evaluation of tasks or functions, and it reduces the amount of training time for operator qualification. For existing operators, the requalification process ensures knowledge retention and standardized task performance.

4.1.2 ON THE JOB TRAINING TRAINER QUALIFICATIONS

The training and subsequent qualification of operators ultimately takes place on shift and in the unit. The trainees are normally trained on the equipment for a given task with a qualified operator assigned as trainer. An on-the-job training (OJT) program includes the selection of trainers and training to teach the OJT trainers how to train. The training and evaluation are consistent and the variability for the performance of the task and job duties is reduced.

4.1.3 TRAINEE CONTROL

New employees enter units with enthusiasm and the willingness to learn their job duties. Training systems must control this energy to prevent operating equipment or processes until the necessary knowledge and skills to safely perform the tasks are obtained. Requirements must be in place to control the on shift training and trainee. These requirements include the responsibilities of the OJT trainer to closely monitor the trainee and to remain in a position to intervene, or prevent an unsafe act.

4.1.4 TESTING

A process that provides the requirements for the testing of trainees is developed and applied for assessment of all operation training activities. The requirements include written and oral tests for knowledge and performance tests for tasks. The process includes the following:

- testing immediately after training
- testing again at a later time to verify retention of knowledge and skills,
- establishment of a pass/fail criteria for both knowledge and performance
- evaluation system that separates the trainer and the evaluator
- requalification process.

4.1.5 REQUIRED READING

Often there is confusion between items that require training and items for information only. The difference is clarified to distinguish between the two. Required reading is used to document communication and is not training. Examples of required reading include:

- notes in a shift log,
- minor changes in policies or procedures
- incident sharing information.

The intention of required reading is to efficiently convey information, however; it does not replace formal training when formal training is required.

4.2 PROCEDURES

Celanese developed a global written policy that established administrative requirements for written procedures commonly referred as Standard Operating Procedures (SOP).

4.2.1 PROCEDURE USAGE

Operators are required to follow procedures. Where procedures are found to be incorrect or incomplete, a process was developed to allow an operator to deviate from the procedural step and to initiate a process to change the SOP if required. This process is to ensure a proper review is completed prior to deviation from the step and for the modification of the SOP, if required. This review process is to generate a conversation between appropriate operation personnel to ensure deviation from the procedure will not place the personnel, process or equipment in an unsafe condition. SOPs are continuously reviewed to ensure they are consistent with the current process and as a result they are followed.

Each SOP is classified based on frequency, complexity and consequence of deviation. Based on the results, the SOPs are then classified as Critical or Reference using the point system described below.

SOP Classification Point System:

Frequency	Pts	Complexity	Pts	Consequence	Pts
Daily	1	Information Access	1-3	Low	1-3
Weekly	2	Mental Loading	1-3	Medium	4-6
Monthly	3	Action Loading	1-3	High	7-9
Semiannual	4	Communications	1-3		
Annual	5	Stress	1-3		

- Reference SOPs have a total point value of 10-20
- Critical SOPs have a total point value of 21-30

Example:

Procedure	Frequency	Complexity	Consequence	Total	Type
R-137A S/U	2	9	2	13	Reference
R-35 S/U	4	13	8	25	Critical

Critical SOPs require initials of the operator after completion of each step and a signature after completion of the SOP. Critical SOPs are required to be "in hand" as each step is performed. An example of a critical SOP is a startup of an exothermic reactor. Within the industry there have been multiple incidents involving exothermic reactors. The critical SOP system adds another layer of protection.

Operators are still required to follow Reference SOP as written, however; they do not require an initial after each step or a signature of completion. An example of a reference SOP is a startup of an R-134A refrigeration system. The Reference SOP system is efficient and yields consistent results.

4.2.2 PROCEDURE CONTROL

New or modified SOPs are required to be validated and verified prior to being issued. Validation requires the operators to ensure the task can be accomplished safety if the steps are followed as written. Verification requires a technical review by the appropriate engineers and specialists to assure all process safety and EHS concerns are appropriately addressed. Other requirements include:

- document control process
- assigned responsibility for procedure approval
- system to ensure annual certification of SOP

4.2.3 PROCEDURE DEVELOPMENT

Procedures are written to give the operator clear and concise instructions for the safe completion of a task. Procedure development methodology is defined and procedure writers trained to ensure SOPs are written consistently thus reducing operator confusion. Some elements of the procedure development methodology are:

4.2.3.1 Page Layout

The presentation of the steps and words on the page is important as the user sees the overall layout before reading the individual steps or words. The user may decide against reading the procedure if the text is packed too densely on the page. An easy-to-read page is more important then the desire to shorten the number of procedure pages. Spacing and font is a subjective human factor. As a rule of thumb, 12 point fonts are easily read under most lighting conditions and anything less than 8 point fonts hard for most users to read. Select a font that is simple and utilizes straight lines with little or no extra "flags" or "curls".

4.2.3.2 Writing Style, Format and Terminology

Some of the elements within this methodology include: keep sections in the same order within and between procedures, use the same page and step numbering schemes, always identify Notes, Warnings and Cautions consistently, write in present tense, use active voice, begin with an imperative, keep superfluous, "nice-to-know" information to a minimum.

4.2.3.3 Word Usage

The verb is the most important word in the step; therefore, attention must be given to the selection that best describes the desired action and then use it consistently in that context. Use vocabulary that is easy for expected users to read and understand. Use short, common words except where standard industry terms or technical words are required. Avoid the use of synonyms and do not use contractions.

4.2.3.4 Action Step Construction

The structure of a step affects the rate at which it is read and understood. Short steps enhance the user accuracy and efficiency and reduce the chance an action is missed. Keep steps as simple as possible with a maximum of two related actions for each step. Begin a step with an action verb followed by the direct object of the action.

4.2.3.5 Mechanics of Style

This element includes spelling, punctuation, use of colon, commas and hyphen, parentheses, periods, quotation marks, capitalization, tolerances, acceptable ranges and nominal value.

Examples:

Incorrect: The boiler pump must be started
Correct: **START** "C" Boiler pump

Incorrect: DO NOT proceed to step 23 unless inlet valves are open
Correct: **IF** inlet valves are OPEN,
THEN GO TO Step 23.

4.3 OPERATION

Included in the operation element are the following:

4.3.1 PROCESS SAFETY OFFICER (PSO)

The PSO is an added headcount during critical operations whose sole purpose is to observe and intervene to ensure safe operations. The primary responsibility of a PSO is to prevent process safety incidents during critical startups, shutdowns and other operating modes by providing an objective, independent safety-oriented perspective.

The PSO is required to have an in-depth knowledge of the process. Normally a PSO is an experienced operator, unit specialist, or Production engineer or unit leader.

The determination of those tasks requiring a PSO is the responsibility of unit supervision with input from appropriate resources. By default, SOP's defined as critical require PSO involvement. There may be other unit operations designated by unit supervision needing a PSO.

4.3.2 OPERATIONS SHIFT CHANGE

A formal operations shift change procedure was implemented to ensure communication of necessary process related information, maintenance activities and other work performed during the shift and crossing into the next shift. Each operating unit is required to have a documented shift change protocol which includes a shift change checklist and logbook.

Incidents and near misses have occurred when operators fail to communicate the status of the process or maintenance activities to the next shift.

4.3.3 MAINTENANCE OPERATIONS TURNOVER

The purpose of a formal maintenance operations turnover is to ensure equipment is ready for maintenance. Operations and maintenance personnel perform a field verification to ensure the equipment is ready for maintenance. Maintenance activities cannot begin until approved by operations. After handover from operations, maintenance assumes ownership until the equipment is ready to be returned to operations.

4.3.4 VERIFICATION AFTER MAINTENANCE

Provides requirements for the field verification of equipment readiness and proper line up prior to putting the equipment safely back in service after maintenance. Verification is usually performed by a single qualified operations person to ensure that equipment status is as expected; however, in those cases where a reasonable potential exists for a component not being in their correct position or the consequence of error is significant an independent verification by a second qualified operations person is required.

4.3.5 COMMUNICATIONS

The communications element describes a formal communication process between operators, between operators and maintenance, or between operators and other support personnel to ensure the effective, reliable, and accurate transmission of verbal information. All appropriate personnel are trained in the techniques in "closed loop" communication. This technique requires all operating directions given verbally to be acknowledged by the recipient. The acknowledgement can be a verbatim repeat back or paraphrasing the direction to ensure the instructions are understood by both. This results in reduced confusion.

Also a verbal verification process was established to ensure operators are safe by requiring periodic contact with those operators working in the field alone.

4,3,6 INHIBITING SAFETY SYSTEMS

The objective is to eliminate the unauthorized bypassing or disabling of safety systems such as:

- Safety Instrumented Systems (SIS)
- Process and other critical interlocks
- Pressure relief devices (PRD's)
- Fire detection and suppression equipment/systems
- Flammable or toxic area monitors
- Alarms

The process includes:

- Approval requirements,
- identification of status (labels, signs, etc.),
- communication to all effected operation personnel. .

4.3.7 LABELING

Proper labeling of equipment and piping is essential to minimize human error and reduce confusion during plant operations and maintenance activities. Each site has a policy defining equipment and piping identification requirements. Minimum corporate labeling requirements include:

- rotating equipment (pumps, compressors, etc)
- fabricated equipment (vessels, tower, etc.),
- control valves
- instrumentation (transmitters, analyzers, etc.)
- Shutdown Instrumented System (SIS) components are uniquely identified
- piping that enters and exits an underground roadway crossing.
- utility drops
- safety critical bleeders (double block and bleeds) intended to remain open

4.3.8 HOUSEKEEPING

Operation areas shall be kept clean and organized for safety and reliability. This improves morale.

4.3.9 MANAGEMENT OF CHANGE (MOC)

Celanese had a corporate MOC policy in place prior to the development of the Conduct of Operations but because of the importance of this system it was included in this policy. Operations and Maintenance personnel are the last line of defense in preventing unauthorized work or changes to be performed in the process unit. Operators are trained in changes prior to operating the changed process. When issuing work permits in the unit, operations personnel ensure all maintenance and construction activities are properly authorized. When asked to perform work in a unit, maintenance personnel ensure that work is properly authorized. Other unit support personnel should not ask for work to be performed in a process unit without proper authorization.

4.3.10 LIGHTING

Operations personnel must have a well-lit work area to identify equipment, read instruments and see developing problems. The operating staff is directed through this policy to ensure that light bulbs are replaced as necessary.

4.3.11 TOOL MANAGEMENT

As operators are asked to do additional maintenance work, they must be provided with the proper tools. Operating personnel are responsible for the selection, use, and upkeep of their tools.

4.3.12 CONTROL OF ACCESS TO THE UNIT

Celanese instituted a process to ensure control over the entrance into a unit by maintenance, contractor, laboratory, or other support personnel and to provide information on the hazards of the unit. In addition, when an emergency occurs control of access to the unit aids in the accountability of people in the unit.

4.3.13MANAGEMENT REVIEW

Celanese has implemented a system called Management Review in which the site periodically checks the health of its various PSM systems in monthly meetings with the Plant Manager and the leadership team. Conduct of Operations is one of those systems that are checked annually to prevent drift.

4.3.14 OTHER POSSIBLE CONDUCT OF OPERATIONS ELEMENTS

The Risk Based Process Safety CCPS book released in 2007 has a chapter on Conduct of Operations that includes additional elements listed below [1]:

- Assign adequate resources
- Validate program effectiveness
- Adhere to safe operating limits and limiting conditions for operation
- Emphasize observation and attention to detail
- Promote a questioning/learning attitude
- Train workers to recognize hazards
- Establish standards of conduct
- Maintain fitness for duty

5. CONCLUSION

Celanese addressed the human element by implementing a Conduct of Operations system focusing on training, procedures and general operations.

The Celanese Conduct of Operations process gives written requirements for the execution and control of unit operations activities. The objective of Conduct of Operations is to:

- Define the requirements and expectations for conducting operations at a high standard of total safety and effectiveness.
- Reduce variability of operational practices in order to reduce human error, increase safety, and enhance effectiveness.
- Perform tasks according to expectations, policies and procedures in a safe, businesslike, professional manner

Conduct of Operations is a proactive approach to improve the current practices, improve the support systems and the behavior of the operator. The benefits include:

- Consistent application of training and procedures
- Reduction in human error
- Less process safety incidents

6. REFERENCES

1. Kieffer, H.E., "Conduct of Operations", internal Celanese document, 1998
2. "Risk Based Process Safety", Center for Chemical Process Safety, American Institute of Chemical Engineers, New York, 2007
3. Lorenzo, Donald K., "Conduct of Operations – the Glue Between Training and Operational Performance", Ninth Process Plant Safety Symposium, American Institute of Chemical Engineers, Houston, 2007

Evaluating Operational Discipline in Psm Audits

James Klein, E. I. du Pont de Nemours and Company, 1007 Market Street, N5448, Wilmington, DE 19898

Audits of site PSM systems are commonly conducted to help ensure compliance with regulatory and corporate requirements, to help identify continuous improvement opportunities for site PSM programs, and to serve as leading indicators of PSM performance. Operational Discipline is the ability of an organization to rigorously execute PSM system requirements correctly every time in order to avoid injuries and incidents and to achieve high levels of PSM performance. Operational Discipline should therefore be an important focus in PSM audits, although this is often difficult and there is only limited guidance on effective auditing approaches. The purpose of this paper is to discuss how Operational Discipline can be evaluated to help improve the quality and benefits of PSM audits.

A Site-Based Workshop for Evaluating and Improving Operational Discipline

James Klein, E. I. du Pont de Nemours and Company, 1007 Market Street, N5448, Wilmington, DE 19898

No abstract available for this paper at the time of publication.

PSM Knowledge Management and Applications across Common or Sister Plant Technologies

David E. Cummings, MS, CIH

Global PSM Competency Leader
DuPont SHE and Sustainable Growth Center
1007 Market Street
Wilmington, DE 19898
(302) 774-9558
david.e.cummings-1@usa.dupont.com

A 10th PPSS Anniversary Paper

ABSTRACT

Systems to share and apply process safety management knowledge and experiences across common or "sister plant" technologies (SPT's) within a company or operating division are important to leverage and apply risk management information in a consistent and sustainable manner at all applicable plant locations. This typically may be an area of weakness in a corporate PSM program since it can involve multiple sites and geographical regions on a global basis, may be managed in a decentralized fashion, and is not easily included in traditional site or unit based PSM auditing systems or performance reviews. Many opportunities exist to leverage the routine PSM knowledge and learnings generated from plant process operations involving "sister plant" technologies, including elements such as process technology, mechanical integrity, hazards analysis, MOC, incident investigation, and operating procedures. Leveraging this knowledge in a focused manner is particularly important in a business and operating environment when significant changes may occur as part of new products, six sigma efficiencies, and enhanced risk reduction measures. This paper will review a global study conducted by DuPont to analyze the current state of PSM knowledge leveraging across "sister plant" technologies, implement improved and sustainable systems to ensure that knowledge is applied consistently, and identify the key success factors necessary to improve and maintain this sharing both at a plant and business level.

1. BACKGROUND ON DUPONT

DuPont is a global science based product and services company that operationally is organized around five main business platforms, each with several groupings of businesses which serve over 200,000 customers worldwide with annual revenues of over $30 billion. Founded in 1802, DuPont operates at more that 175 manufacturing sites in 70 countries involving 64, 000 employees

The company operates many different manufacturing technologies involving both higher hazard processes and lower hazard (typically involving mechanical operations). These technologies vary widely including the following examples:

Chemical Solutions Enterprise (e.g., oleum, aniline, sodium, cyanides, methylamines,)
Engineering polymers (polyester and acetyl resins, glass laminating products)
Agricultural products (e.g., herbicides, fungicides, insecticides)
Fluoroproducts (e.g., refrigerants, electronic gases, fluoropolymer solutions)
Performance Coatings and Color technology (e.g.; automotive paints and titanium dioxide)
Advanced Fiber Systems (e.g. Tyvek®, Sontara®) and Non-Wovens (e.g., Nomex®, Kevlar®)
Electronics, Displays, Imaging technologies (electronic polymers, circuit material, LCD's)
Pioneer Seed (corn, soybeans, alfalfa) and Solae (soy protein technology food products)
Surfaces (Corian® and solid surface materials)

In addition to the five business platform structure, Global Operations is also segmented into four worldwide regions including North America, Latin America, Europe, and Asia Pacific. The company's internal systems and standards for process safety are applied to all sites within all regions and manufacturing technologies through one set of global policies and procedures.

The DuPont corporate structure for PSM relies on a small centralized, matrixed global PSM competency team who has primary ownership and responsibility for internal standards, technical training, auditing, and internal initiatives. There is also a relatively small group of highly skilled PSM consultants within the corporate Engineering group who can provide complex PSM risk management support to sites and businesses including QRA's, facility siting, project support, and complex PHA facilitation, however; the basic PSM model and philosophy is intended to develop PSM related knowledge and skills within the site line and support organizations for operations, maintenance, technical, and R&D personnel.

A broad set of PSM performance indicators are used both at the corporate, business, region, and site levels to measure and report periodic PSM results.

2. SISTER PLANT TECHNOLOGIES WITHIN DUPONT

Each of DuPont's main businesses and products vary considerably in terms of materials, technologies, and physical assets, however, there are more than ten distinct business units across approximately fifty (50) total sites that use common manufacturing technologies. DuPont has established a definition of common or "sister plant" technologies (SPT) within our corporate process safety standard as follows:

Product manufacturing technologies at multiple site locations within a DuPont business platform that involve similar or identical process design basis (i.e., chemistry or process conditions), hazardous substances, equipment design, and process hazards unique to the specific business

Dependant on the business structure, multiple locations of sister plant technologies may exist solely in North America, or exist across one or all of the other remaining three regions. Examples of these include but are not limited to:

Performance Coatings – twelve (12) plants worldwide that manufacture paints resins and coatings with an extensive set of reactor safety risk management controls

Sulfur Products – four (4) unit plants manufacturing fuming sulfuric acid which handle oleum, sulfur dioxide, sulfur trioxide and three (3) acid refinery recovery operations which handle strong acids

Performance Materials – multiple plants that manufacture specialized engineering polymer products and resins such as Hytrel®, Zytel®, and Delrin®; multiple plants that operate ethylene co-polymer processes using high pressure autoclave technologies and flammable gases.

Titanium Technologies – five (5) plants worldwide that manufacture titanium dioxide and associated products which handle chlorine, titanium tetrachloride, flammable liquids, and other higher hazard materials

Tyvek ® - two plants (2) worldwide involved with the manufacturing of Tyvek® products which handle a variety of hazardous substances such as solvents.

Solae Soy Protein – eight (8) plants worldwide which involve production of soy based food components which may handle acids, reactive solids, and flammable liquids

DuPont Performance Elastomers – multiple plants that manufacture elastomers using a neoprene monomer process

Nitration Reaction Technologies – multiple plants that operate nitration processes with significant chemical reactivity hazards

Highly Toxic Material Guardian Committees – a series of eighteen (18) individual committees comprised of global plant sites that handle substances which DuPont has classified as "Highly Toxic Materials" (HTM's) such as chlorine, anhydrous ammonia, phosgene, fluorine, hydrogen fluoride, anhydrous hydrogen chloride, etc.

An effective and consistent SPT network across all applicable businesses and locations is important for global risk management programs to supplement the traditional site based PSM programs to help ensure that the following:

1) Learning and applying the experiences of all sites operating a specific technology with the specific goal of preventing incidents
2) Ensuring a high level of consistency involving layers of protection across similar sites and hazards
3) Achieving efficiencies in the PSM process through routine collaboration
4) Reducing liabilities and potential regulatory issues due to inconsistent approaches
5) Maintaining alignment and focus over time to understand and prevent divergence in design/operations as they relate to risk management
6) Understanding where differences across common technologies exist and verifying that these inconsistencies are either acceptable or require remediation.

3. RECENT FOCUS ON IMPROVEMENTS FOR KNOWLEDGE MANAGEMENT - SISTER PLANT TECHNOLOGIES

In 2006, DuPont designed and conducted a broad based internal PSM self-assessment (i.e., DuPont PSM Discovery Process) to evaluate systemic opportunities to further strengthen our global systems for risk management. In part, this analysis was conducted due to a significant company reorganization in late 2004 which consolidated twenty-two strategic business units into the five platforms while also adopting a regional based operations approach. One of the four key work streams with this internal assessment focused on overall effectiveness of PSM knowledge sharing and intensification with specific emphasis on leveraging across similar plants and operations. A general problem statement was subsequently developed as part of the initial analysis:

PSM hazard data, technology, and general risk management information may not be effectively identified, shared/leveraged, and available within and across multiple sites, platforms and regions with common technologies in order to ensure maximum knowledge intensification and application.

The terminology focusing on "intensification" of knowledge in this context was to broadly strengthen the management systems related to PSM by identifying improved methods to leverage and apply PSM knowledge and experience across similar sites.

Intensification – tending to strengthen or increase and concentrate the review, sharing, application and retention of information, methods, experiences, and new learnings in a consistent and sustainable manner

A series of detailed questions and data collection sources were identified for assessment as well as stakeholder interviews with over 120 employees in operations, PSM, regional, business and site roles. These key questions and data collection leveraging PSM knowledge across sister plant technologies involved the following:

- Where do sister plant technologies exist and at how many locations across each business and region?
- What is the current state for sharing and leveraging PSM knowledge with these sites?
- Who is responsible for leading and managing each SPT process where they exist?
- What PSM elements and knowledge are currently shared and how effectively is this conducted on an ongoing basis?
- How do businesses and regions confirm that SPT network leveraging is active and effective?
- What tools and approaches are in place and are effective to facilitate leveraging of this knowledge?
- What additional elements are needed to strengthen current SPT approaches?

4. CRITICAL SUCCESS FACTORS – RESULTS

As part of the detailed evaluation and results, DuPont identified nine critical success and performance factors that should be in place and maintained in order to achieve long term and successful PSM knowledge sharing and implementation across each SPT network. These critical success factors include the following (general summaries):

4.1 A formal designated leader for each applicable SPT network

Sister plant technologies are typically aligned with one business group but physically exist in varying geographical locations and in multiple regions. The designation of one key person who has overarching responsibility and authority to lead and maintain a sister plant network to apply PSM knowledge and experiences is key to a sustainable and effective leveraging process. While this is not typically a full-time role, this individual's responsibilities should be addressed by his/her line management in annual goal setting as well as the year end performance appraisal process. The individual should possess a thorough technical understanding of the common technology, process safety requirements, and strong team building and organizational skills. Team members on the SPT network also should be supported by their line management to ensure active participation and recognition.

4.2 A senior management sponsor within the Platform

Identification of a senior operations or technology sponsor within the business who can advise and support the sister plant technology leader, serve as counsel, set management expectations, and remove potential barriers is also important to support this effort. The "care and feeding" of these teams over time needs to be considered in terms of senior management support, progress reviews, interventions where necessary, and positive recognition for accomplishments.

4.3 A written procedure or program describing the PSM SPT Knowledge Intensification process

As with other key PSM risk management related program elements, developing and maintaining a set of written procedures or internal policies for a sister plant technology network is important to clearly document responsibilities, team activities, and the overall structure. This procedure can also clearly define expectations and goals for individual team member performance. A written program also can be instrumental in sustaining the team's work, knowledge, and membership over time as people change. (refer to Section 5 of this paper for specific technical program elements which should be described in the written program for each SPT in terms of the "what's, how's, and who's" of leveraging knowledge)

4.4 A formal planning process - annual goals and objectives with budget for continuous improvement

Each sister plant technology team operates independently within the broader context of its PSM and manufacturing role within DuPont. The team should establish a formal set of annual goals and objectives and identify any financial support needs though a budget which can be reviewed with the sponsor and supported by the business as needed. This might include funding and support for new tools to improve team effectiveness, travel to internal or external meetings, or engaging a subject matter expert for technical assistance to benefit all common technology sites on a specific process safety issue or need. These SPT budgets typically are very small, but need to be considered in a proactive manner as part of the overall business planning process.

4.5 A team culture of openness and motivation to share across all sites versus competition

It is not usual for plants that operate common technologies and manufacture similar or identical products to maintain a spirit of independence, competitiveness, and "not invented here syndrome" even within the same business and company. This cultural aspect can be a barrier to effective sharing and leveraging of PSM knowledge and experiences. A spirit of teamwork, openness and an obsession for sharing and applying information across sister plants is a fundamental need in this area

A second cultural and behavioral aspect of leveraging knowledge broadly simply involves taking time to share the information. Most plant site people interviewed have a very full plate of existing work to be done to support site operations. When a conflict develops, the site needs typically take precedence which is understandable as a short term intervention, however this can negatively impact the ability and quality of leveraging if a site develops a chronic pattern of non-participation, Individuals with a specific discipline and motivation to share information and fully participate in the SPT process are essential.

4.6 Leveraged electronic database for sharing and retaining SPT information

One of the most critical elements to ensure effective sharing and leveraging in a sustainable fashion for each SPT network over time is to develop and maintain a database or team room where all sister plant technologies can retain information, knowledge. and learnings. There is a myriad of documentation that can be generated by each site and embedded in a common location for both active and passive review by other applicable sites including new hazard information, completed PHA reports, incident reports and RCFA's, audit reports, MOC's, common procedures, etc. Attempting to share these documents via email without a means to retain and provide access over time has proven difficult based on reported site experiences. In some cases where a mature SPT network exists, data and knowledge is centrally available and has been retained for over twenty years and several working generations of employees.

Most team rooms in current use within DuPont SPT networks also have tools to assist teams in their routine work such as meeting agendas, follow-up and tracking of team action items, and assisting with collaboration (e.g., drafting a new document or technical specification with comments from each of ten sister plant technology sites). These leveraged databases may be an intranet web based system, a "Lotus Notes" collaborative team room, or another electronic means to update and retain key information.

Database security needs to be considered prior to adopting an SPT team room in terms of who has access and how the documents are protected and maintained since a significant amount of proprietary or confidential information related to the manufacturing technology may be retained for process safety purposes.

4.7 Routine, scheduled SPT meetings with active participation (including a periodic face to face meeting)

An effective sister plant technology team must schedule and conduct routine team interactions in order to improve and maintain knowledge and sharing activities. Periodic meetings (conference calls and/or web meetings) at a minimum once per quarter have been shown to be effective. Teams typically struggled to maintain continuity of work and purpose when meetings are held at lesser frequencies. Some SPT teams have chosen to vary the frequency based on current work and may conduct monthly teleconferences during peak activity times. Conducting a less frequent face to face (interactive) multi-day meeting is also desirable to improve team building, sharing and application of recent common knowledge, or work collaboratively on a particularly difficult problem or new strategy.

4.8 A periodic review or report of performance and results across sites and within team (e.g., annual)

By definition, sister plant technology teams involve activities across two or more sites and may be managed beyond the typical plant "fence line" within an operating business platform. This structure is not as conducive to periodic review and auditing as that of a traditional PSM audit which focuses on a specific unit or operation at one location. During traditional site based PSM audits, DuPont has found it difficult to measure the effectiveness of leveraging knowledge at one site whose origin may have come from one of several other locations. Due to these factors, conducting an annual evaluation of the SPT network's results (typically with the sponsor and business) is currently required. This evaluation is typically completed in the first quarter of the proceeding year and will review team activities, accomplishments, and future focus areas. The SPT network review is not intended to replace "ground level" first and second party process safety audits conducted every three years at each manufacturing plant site which are typically administered by each of the four global regions.

4.9 MOC – Personnel process with specific focus on team participation

As sister plant technology team members and organizational changes occur, there needs to be an overarching MOC – Personnel process to integrate new members and ensure ongoing effectiveness. This should involve key roles such as the sister plant technology team leader, individual members at sites, the team sponsor, or other key functional roles as part of or support the team on a frequent basis.

5. SPECIFIC INFORMATION (ACCESS, SHARING AND RETENTION WITHIN COMMON TECHNOLOGY DATABASE)

The following sections describe a set of PSM related information and data that are now included in DuPont internal policies as they apply to existing SPT networks. Figure 1 below represents one "typical" model of an SPT network within the Performance Coatings business.

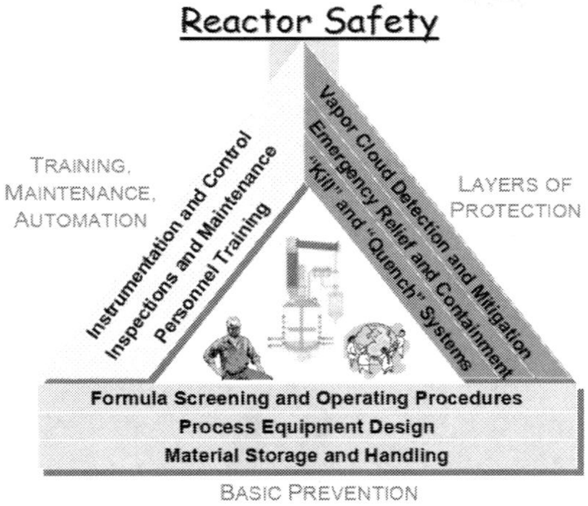

Figure 1: Example of DuPont Performance Coatings Reactor Safety Network

5.1 Sharing Incidents with self evaluation and feedback/closure across common applications

Each site within a SPT typically reports, conducts and publishes an incident report where a formal investigation was conducted. These may be shared broadly within the company on email; however, little feedback or confirmation of evaluation and implementation is typically conducted. SPT's should broadly share their own incident reports and learnings, and use the team approach to document and verify the implementation or correction of incident learnings at all sites in order to achieve and sustain maximum prevention.

5.2 Process technology information (hazards of materials, process and equipment design)

By definition, SPT networks involve sites with a common technology basis which involves a clear identification of all process hazards, process design basis, and equipment design basis. Where one site may have conducted local reactivity testing and evaluation of a raw material or intermediate which had not been previously completed, this new data should be shared with all other SPT sites where applicable. Leveraging information on equipment design to identify similarities and differences should also be periodically conducted

5.3 Hazard information and layers of protection (PHA reports, recommendations, interlock event classifications and engineering/administrative controls)

Cyclic PHA reports for existing processes and baseline PHA reports for projects and changes should be shared across sister plant technologies in order to share a joint understanding of each phase of a PHA such as hazard identification, evaluation , human factors, facility siting, inherent safety, and final recommendations to further reduce risk. Where similar or identical hazards and consequences exist, the layers of protection for both engineering and administrative controls should be equivalent. This should include safety instrumented systems, detectors, containment systems, and similar passive and active controls. Where differences in protective layers exist due to age of the plant or other local variations, the business and sites should have a clear understanding on what and why they are different, and determine if this variability is acceptable.

5.4 Management of Change – Technology and Facilities

Each site typically determines what changes to technology, facilities and subtle changes may be appropriate based on inputs from a variety of change agent sources, including six sigma projects, new products, local improvements, regulatory requirements, etc. Establishing a means to share local MOC initiatives and results across the SPT network is valuable to leverage information and results

5.5 Mechanical integrity knowledge for common equipment

Where similar or the same physical equipment and process conditions may exist across SPT's, collaboration across the broad spectrum of mechanical integrity elements is important to help ensure consistency and efficiency. This may include

- Leveraging common equipment specific maintenance procedures
- Leveraging test and inspection (T&I) methodologies
- Sharing the results of T&I and reliability engineering analyses results in order to compare and manage similar equipment across multiple sites consistently

DuPont's recent evaluation found this to be the most common and significant SPT technical knowledge area that is not typically leveraged as broadly as possible. This is attributed in part due to past practices to focus more information sharing on technology aspects of the PSM system and not as much emphasis on leveraging facilities information such as mechanical integrity. This also highlights the importance of SPT network participation and sharing by plant reliability engineers and use of common material management and MI software packages such as Ultrapipe and the SAP maintenance module as much as practicable.

5.6 Coordination on Operating procedures and SOC's

Operating procedures, Standard Operating Conditions (SOC's), and safety limits at all SPT locations worldwide should be substantially similar in terms of technical content and requirements. Formats for procedures may differ depending on the county and historical perspective, but the content should be aligned within a common framework in terms of inputs for the process control system.

5.7 PSM audit results and unique findings/issues

DuPont currently maintains internal requirements and procedures to conduct first and second party PSM audits at all worldwide manufacturing operations. These audits are conducted and scored using a standard set of 300 questions across fifteen elements. The results of these performance audits should be shared across all SPT's in order to leverage any findings that may have broad application to all sites for technology, facilities and personnel.

6. ADDITIONAL CONSIDERATIONS

In addition to the nine critical success factors in section 4 and the specific PSM related technical elements in section 5, additional (albeit limited) internal practices were identified during DuPont's evaluation that can also add value and may be effective in order to apply PSM broadly across worldwide common technologies. These include

6.1 Multi-site participation/coordination within cyclic PHA's

In specific cases and in addition to sharing PHA's conducted independently at each site, some SPT's have conducted cyclic PHA's across multiple sites where team membership is comprised of various technical resources and subject matter experts. This requires substantial planning and coordination, and additional time commitments and organizational resources. The principal benefits include additional synergies across multiple sites involving detailed knowledge, experience and collaboration. There are also potential negative impacts in terms of maintaining the effectiveness of team interactions, broadness of scope, failure to recognize subtle differences in design and operations, and logistical issues associated with multi-site interfaces over extended periods of times.

6.2 Independent first and second party auditing criteria and systems for unique technologies

In specific cases, SPT's have established additional first and second party PSM audits beyond the generic PSM audits conducted at all sites to evaluate specific technologies and controls unique to that technology. These audits involve operations personnel who are "experts" in the technology involved and usually involve a SPT site conducting an independent audit at another site.

6.3 Common PSM procedures and checklists

Development and use of common or unique PSM procedures and checklists specific to the applicable sister plant technology may be developed in specific cases such as a common PSSR checklist for equipment sets, a leveraged set of operator training materials for periodic refresher training and requalification on common equipment or process segments, common quality assurance checklists for purchase of replacement equipment, etc.

7. SUMMARY

The need for sister plant technology networks may exist in many moderate and larger sized companies where similar or identical manufacturing processes exist. These networks can provide a very effective, efficient and important risk management process for a given enterprise across multiple locations in order to prevent incidents, leverage and apply knowledge, deliver consistency in PSM to greatest extent possible, and gain the most benefit of knowledge and experience involving the same type of equipment, products and associated hazards. In this information technology age of email, internet and other means of sharing and distributing communications and "data overload" from both internal and external sources, a formal SPT network can also serve as a reliable focus and rigorous mechanism to filter, exchange, and "intensify" key risk management information and help leverage the application of knowledge where it can make the most difference.

PSM knowledge sharing and intensification across all sister plant technologies worldwide can be a labor intensive and challenging process which does not maintain and sustain itself naturally unless there is visible and active senior leadership support and a formal SPT management structure. The critical success factors described within this paper, as well as the technical elements of PSM information for sharing, provide one means of recommended practices to achieve these multiple benefits with SPT networks for longer term sustainable gains in risk management.

“Yellow Lining” for Re-Commissioning and Start-up

Doug Burroughs, Bishop Facility, 000, New York, NY 10016

To prevent exposures to personnel, or releases to the environment, procedures are used to safeguard the status of equipment during upsets, start-ups, or shutdowns.

After a production outage, a unit shut-down, or re-commissioning, we use a process called “Yellow Lining” of P&IDs to verify all equipment, associated piping, valves, and instrumentation have been returned to “Normal Operating Status” before the introduction of “Acute Chemicals” back into major pieces of equipment.

“Yellow Lining” of P&IDs uses verification by two or more individuals to ensure all equipment is back in “Normal Operating Status”. Discrepancies are noted directly and accountability is provided by placing names, dates, and time on the actual P&ID. Hazard assessments are reviewed with the field operator and supervision, allowing actions to be taken prior to start-up. P&IDs are retained for PHAs and P&ID corrections.

Done correctly with start-up procedures, “Yellow Lining” will mitigate exposures and releases while keeping P&IDs and unit knowledge up to date.

LOPA Misapplied: Common Errors Can Lead To Incorrect Conclusions

Karen A. Study
Rohm and Haas Company
Deer Park, TX
kstudy@rohmhaas.com

John W. Champion
Rohm and Haas Company
Deer Park, TX
jchampion@rohmhaas.com

A 10th PPSS Anniversary Paper

ABSTRACT

Layer of Protection Analysis (LOPA) is a powerful tool for quantitative risk assessments. If applied correctly, it can provide quick and efficient guidance on what additional safeguards are needed, if any, to protect against a given scenario. If misapplied, an overly conservative calculation of risk may result in over-instrumentation, additional life-cycle costs and spurious trips. Even worse, a non-conservative calculation of risk could result in an under-protected system and unacceptable risk of an undesired consequence occurring.

This paper describes several categories of common errors, some overly conservative and some non-conservative. Case studies of actual plant scenarios are used to illustrate.

Keywords: Layer of Protection Analysis, Risk, Quantitative, LOPA, Case Studies

Introduction

Layer of Protection Analysis (LOPA) has been in widespread use since an early methodology was published in the 1993 CCPS publication, *Guidelines for Safe Automation of Chemical Processes*[i]. As the early practitioners have gained experience in using the tools, they have learned many of the pitfalls and limitations of the technique. LOPA can provide quick and efficient guidance on what safeguards are needed in a system to protect against a given scenario. If misapplied, however, an overly conservative calculation of risk may result in over-instrumentation, additional life-cycle costs and spurious trips. Even worse, a non-conservative calculation of risk could result in an under-protected system and unacceptable risk of an undesired consequence occurring.

With publications such as CCPS 2001[ii] and the requirements of standards such as IEC 61508/61511[iii,iv] and ANSI/ISA-84.00.01-2004[v], more practitioners are learning and applying LOPA every day. The intent of this paper is to illustrate seven common errors that can be made when conducting a LOPA, using real plant examples.

Common Error #1 – Not considering the operating culture

How a system is operated needs to be understood fully to establish initiating cause frequencies and the probability of failure on demand (PFD) of protection layers. Assumptions made by the analyst to calculate initiating cause frequencies and protection layer PFDs need to be verified by operator interviews or observations.

Example 1

Overflow protection that uses an automated response such as a high level - SIF – to shut off flow to a tank can lead to complacency and a more frequent challenge of protection layers. In this case, how the system operates directly affects the initiating cause frequency. A typical design is illustrated below in Figure 1.

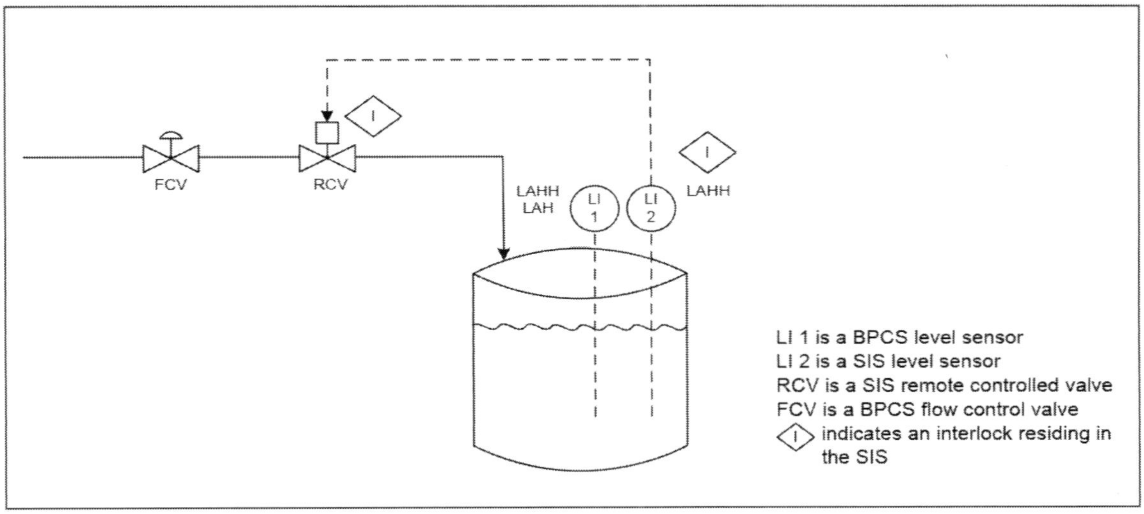

Figure 1: Common Overflow Protection Design

If the initiating cause chosen for this design is failure of the flow control valve, the calculated frequency of overflow depends on how the transfer is done. If the analyst incorrectly assumes that the operator monitors the transfer using LI 1 and stops flow once the desired level is reached, the initiating cause frequency estimate will be too low. Conversely, if the analyst incorrectly assumes that the operator doesn't monitor the transfer, the initiating cause frequency will be too high. This is illustrated in LOPA Table 1.

In the LOPA table, the frequency of tank overflow is calculated for two cases. In Case 1, flow into the tank is monitored by an operator via the basic process control system (BPCS) level indicator, LI1. The flow is stopped by the operator once the desired level is achieved. The safety instrumented function (SIF) protection layer is only activated if the operator fails to stop the flow. In Case 2, the operator relies on the SIF to stop the transfer into the tank. The level is not monitored and the BPCS alarm is ignored since it would always alarm when a transfer occurs. So, the SIF is needed for every transfer. If the SIF fails to stop a transfer, the tank will overflow.

LOPA TABLE 1: Overflow Protection using SIF only versus SIF and BPCS

Consequences	Initiating Cause Frequency	/year	BPCS PFD		Operator Response to Alarms PFD		SIF PFD		Mitigated Consequence Frequency /yr
Case 1: *Operator monitors level during transfer and stops transfer when desired level reached*									
Case 1 Tank Overflow; Potential spill of flammable and/or toxic material	BPCS (LI1 or FCV) Loop Failure During Transfer	1E-01	n/a, initiating cause is BPCS failure	1	n/a, not independent of BPCS	1	Independent high level signal will close inlet RCV (LI2 and RCV)	2E-02	2E-03
Case 1 Tank Overflow; Potential spill of flammable and/or toxic material	Operator Error Frequency = 20 transfers/yr x human error rate of 1/100 transfers	2E-01	n/a, BPCS does not automatically stop transfer	1	High level alarm. Same operator responds that makes error. No credit	1	Independent high level signal will close inlet RCV (LI2 and RCV)	2E-02	4E-03
Total Overflow Frequency for Case 1									**6E-03**
Case 2: *Operator depends on SIF to stop transfer and does not monitor the level*									
Case 2 Tank Overflow; Potential spill of flammable and/or toxic material	SIF Loop Failure During Transfer (LI2 or RCV)	4E-02	n/a, BPCS does not automatically stop transfer	1	n/a, alarm ignored by operator, normally sounds during transfer. No credit	1	Initiating cause is failure of SIF. No credit	1	2.E-02
Total Overflow Frequency for Case 2									**2.E-02**

Common Error #1 – Not considering the operating culture

Example 2

Figure 2 shows a distillation column that is fed an additive that vaporizes in the presence of heat. During normal operation, a small additive flow is fed to the column and the resulting vapor leaves via the normal vent system. However, if the additive is not stopped when the column is shut down and cooled, it can accumulate in the column sump. Upon subsequent heat-up, rapid vapor generation can overwhelm the normal and emergency venting systems, resulting in over pressurization of the column.

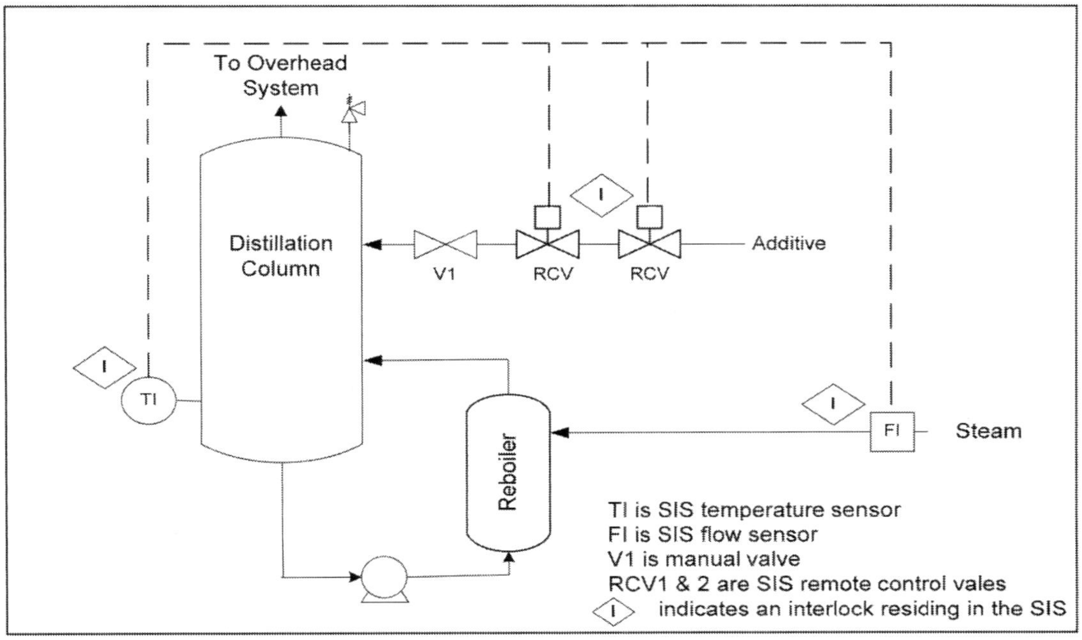

Figure 2: Additive feed to distillation column

Originally, the only protection against this scenario was the operating procedure for stopping additive flow as part of the column shutdown. The HAZOP team recognized that this was inadequate protection for such serious consequences, so a LOPA was performed. Based on the LOPA, a new protection layer was required to ensure that the column did not overpressure due to accumulated additive in the column during a startup. The required PFD for the new independent protection layer was equal to or less than 1E-02 to meet the risk criteria. The LOPA team decided that the new protection layer would be a SIF that closed the additive flow on low steam flow to the distillation column. However, the LOPA team recognized that, once installed, the operators would likely rely on the automated shutdown action and would not manually block in the system anymore. So while the LOPA added one protection layer, human nature removed another. As shown in LOPA Table 2, a net 0.1 risk reduction is achieved when the 1E-02 PFD SIF is installed when comparing the Base Case to Case 1.

So the conclusion was to add a SIF with 1E-03 PFD to achieve the desired risk reduction, as shown in Case 2 of LOPA Table 2. The new SIF would close redundant RCVs on the additive to the column if the distillation column steam flow or temperature was low. One out of two voting would be used for the sensors and the final elements of

LOPA TABLE 2: Additive Feed to Distillation Column

Consequences	Initiating Cause Frequency	/year	BPCS PFD		Operator Response to Alarms PFD		SIF PFD		Mitigated Consequence Frequency /yr
Base Case: *Operator stops additive flow when column shut down*									
Overpressure of distillation column due to excess additive during s/d	Operator fails to manually block additive flow to column (1 sd/yr with 0.1 error rate)	1E-01	n/a	1	no relevant alarms	1	n/a	1E+00	1E-01
SIF Case 1: *Operator depends on SIL 1 SIF to stop additive flow when column shut down*									
Overpressure of distillation column due to excess additive during s/d	Operator fails to manually block additive flow to column (1 sd/yr with 1.0 error rate - operator depends on automated shutoff)	1E+00	n/a	1	no relevant alarms	1	low steam flow closes additive RCV	1E-02	1E-02
SIF Case 2: *Operator depends on SIL 2 SIF to stop additive flow when column shut down*									
Overpressure of distillation column due to excess additive during s/d	Operator fails to block additive flow to column (1 sd/yr with 1.0 error rate - operator depends on automated shutoff)	1E+00	n/a	1	no relevant alarms	1	low steam flow or low temperature closes 2 additive RCVs	1E-03	1E-03
								Target Frequency	**1E-03**

the SIF. If the analyst had just added the original SIF, without regard for the cultural effects of the new automation, it would have resulted in a less protected system than intended.

Common Error #2 – Not considering common failure modes

When estimating the probability of failure on demand for protection layers, common causes must be accounted for to obtain an accurate frequency estimate for a given scenario. Common failure modes can exist between the initiating cause and the protection layer as well as between multiple protection layers.

Example 1

One example of a single common failure involves operating a column in a fouling service. If pluggage can result in failure of pressure control, any pressure relief device may be susceptible to the same pluggage issues. Hence, pluggage can initiate a high pressure consequence AND disable the protective device (the pressure relief valve).

Example 2

Another example involves implementing control and protection layers in a common system. In Figure 3, a distillation column is designed such that overhead vapors are condensed in a partial condenser. The condensed liquid is collected in a receiver and is refluxed back to the column via a flow control valve (FCV) and/or removed via a level control valve (LCV). Failure of the level control can result in high level in the receiver. If this high level blocks the vapor outlet path, the column can overpressure. To prevent this scenario, a secondary level instrument (LI2) is used to shut off feed and steam to the distillation column on high receiver level. In this example, the receiver level control and the protective layer are both implemented in the BPCS.

Several common failure modes exist. If LI and LI2 both go to the same card in the BPCS, a single card failure can result in high level AND disable the process logic function that protects against this scenario. Using separate cards for each level instrument provides more protection, however there are still other common failure modes that could disable both. See Figure 4 for examples of some common elements in the BPCS.

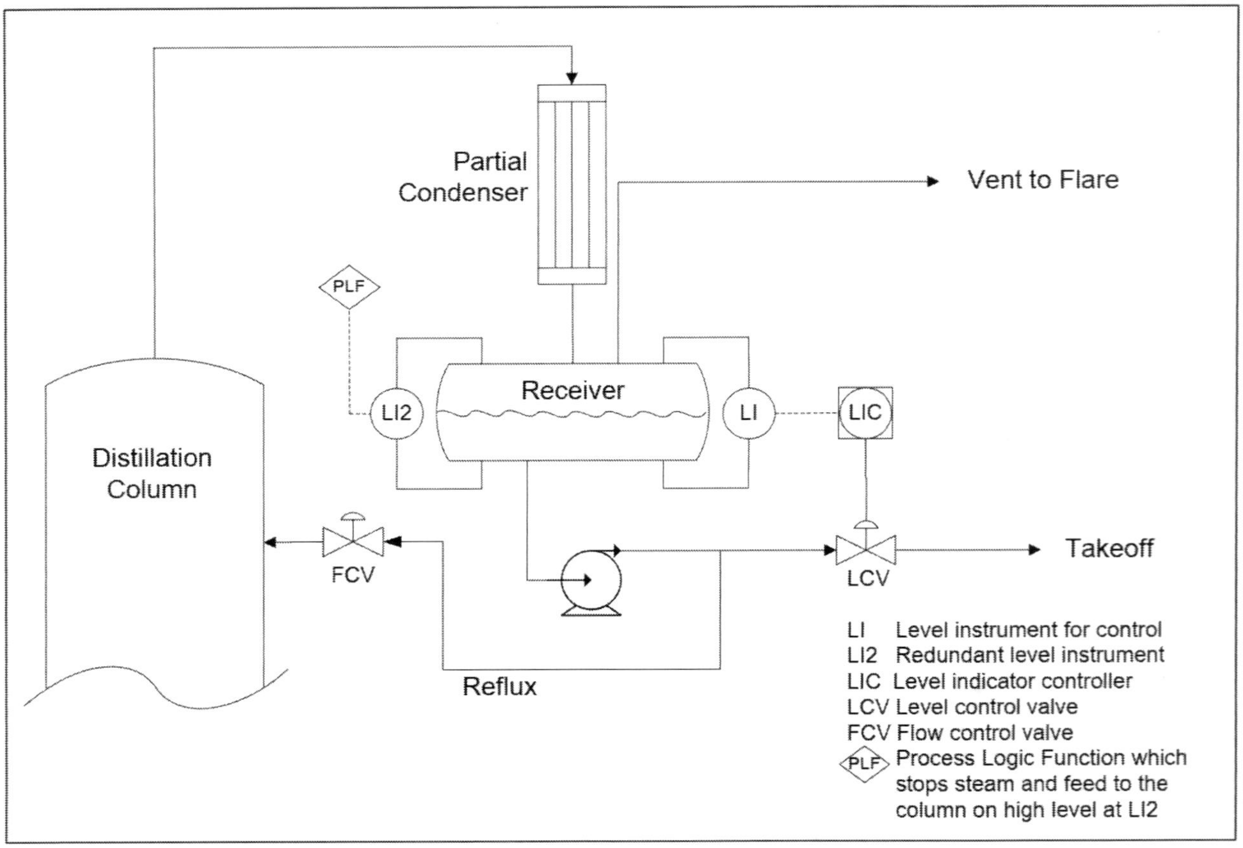

Figure 3: Distillation Column and Overheads System

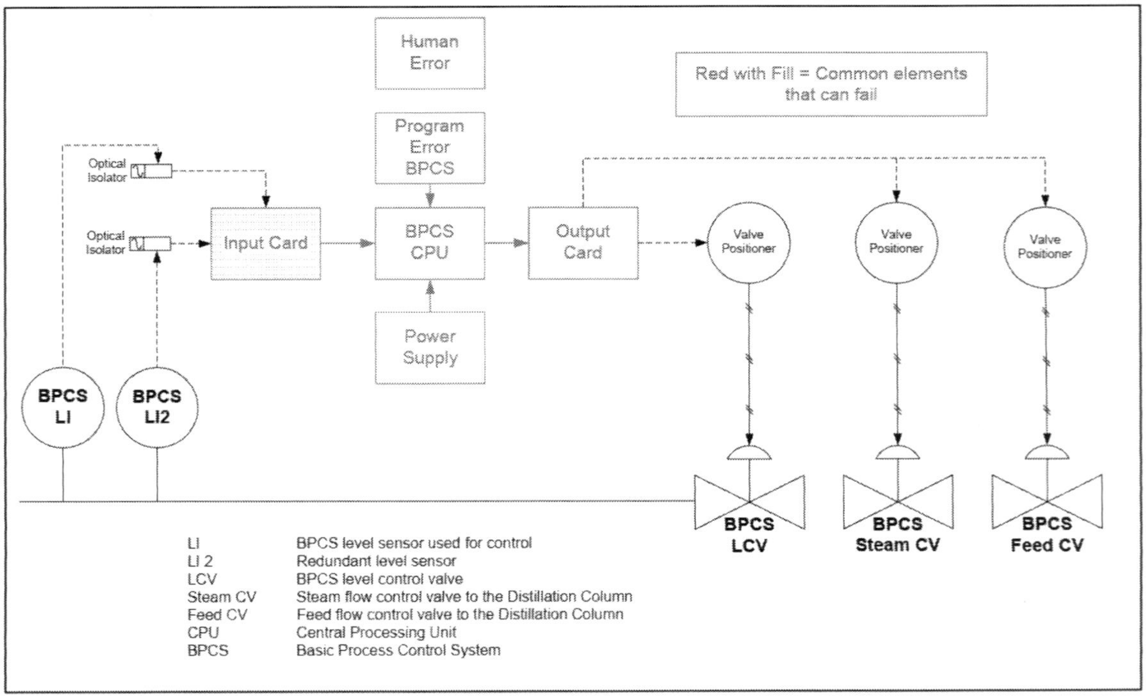

Figure 4: Typical BPCS Common Elements

ANSI/ISA-84.00.01-2004 provides guidance for the maximum credit that can be taken for the BPCS: "The risk reduction factor for a BPCS used as a protection layer shall be below 10". This equates to a PFD of less than 0.1. For the purposes of this article, a 0.1 PFD will be used for simplicity. So, if there are multiple alarms or process logic functions residing in the BPCS that protect against a singe scenario, a PFD of 0.1 applies to the combination of all the functions in the BPCS protection layer. Furthermore, if the initiating cause is failure of a component within the BPCS, protection layers implemented in the BPCS may be compromised.

Example 3

Another example of common failure modes can be illustrated using human errors. If the initiating cause for a flammable spill is an operator forgetting to close a manual bleed, use caution when giving any credit for the same operator later preventing or mitigating the spill. There may be common causes involved that would reduce the likelihood (probability) of the operator detecting an initial error. Suppose an operator is busy elsewhere and forgets to close the bleed. The same operator may still be too busy to double check the line-up prior to introducing chemicals to the system. Similarly, if the operator is distracted due to personal issues, his performance may be hampered during the entire scenario. Also, if the operator is injured at the beginning of the release, he may not be available to take corrective action.

Common Error #3 – Not considering the impact of an IPL on the entire system

This example illustrates the importance of looking at the entire system, not just the specific unit operation of initial concern. Figure 5 shows a flammable feed to a distillation column coming from a small feed tank. This column had a history of tray damage due to high liquid level excursions and in one instance had lost containment due to tray parts damaging the reboiler circulation pump bowl. The HAZOP and subsequent LOPA had identified the need for a SIF to protect against high level in the distillation column. The SIF action would be to shut off steam and feed to the column on detection of high level.

If the implementation of this SIF was done without considering the impact on the feed tank, the result could have been more frequent overflows from the feed tank. In this particular case, the implementation of the distillation column SIF would increase the challenge rate to the feed tank by tenfold. As a result, the feed tank also needed an SIF to trip its feed on high level detected in the feed tank.

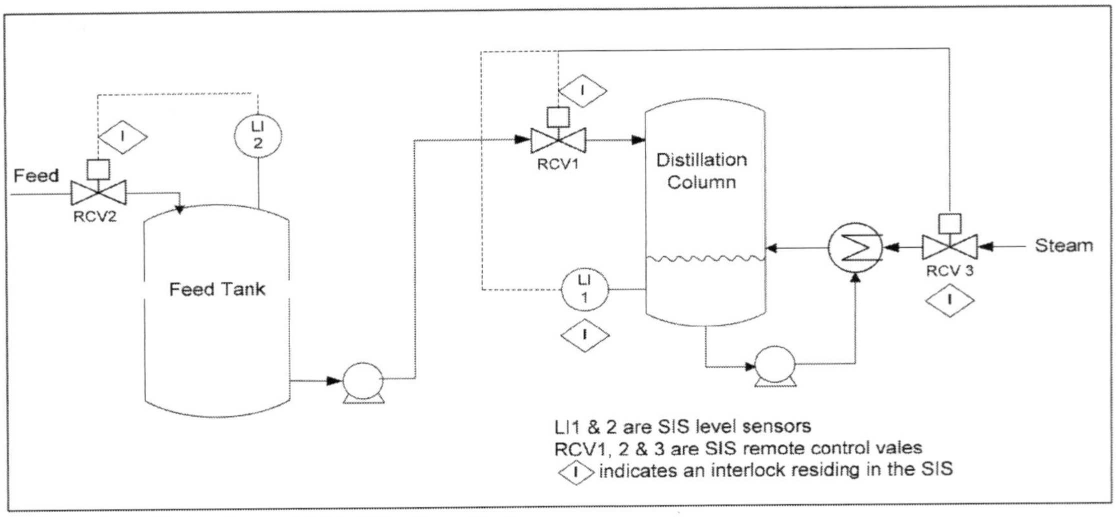

Figure 5: Feed Tank and Distillation Column

Common Error #4 – Using industrial databases incorrectly

When using failure rates and PFD estimates from industrial databases, the analyst must be aware that these databases do not typically include failures due to loss of utilities such as power, cooling systems, instrument air or nitrogen. For example, if loss of a feed pump can initiate a runaway reaction, the pump failure frequency from an industrial database may be used to estimate one initiating cause frequency. However, a separate pump failure cause due to loss of power may also need to be added to the list of LOPA scenarios that result in a runaway reaction. Loss of utilities constitutes a large category of potential initiating causes. Table 1 summarizes some typical initiating causes caused by loss of utilities.

Table 1: Utility Failures That Can Initiate Hazardous Scenarios

Scenario	Utility	Potential Initiating Cause
Tank Overflow	Power	Pump Failure
	Power	Loss of BPCS
	Power	Loss of Radar Level Indication
	Instrument Air or Nitrogen	Control Valve Failure
	Instrument Air or Nitrogen	Loss of Pneumatic Level Indication
Runaway Reaction	Power	Pump Failure
	Power	Loss of BPCS
	Cooling Tower Water	No/Low Cooling on Reactor
	Refrigerated Water	No/Low Cooling on Reactor
	Instrument Air or Nitrogen	Control Valve Failure
High Pressure in a Column	Power	Loss of BPCS
	Instrument Air	Control Valve Failure
	Cooling Tower Water	No/Low Cooling on Overhead Condensers
	Refrigerated Water	No/Low Cooling on Overhead Condensers

Common Error #5 – Not understanding high demand initiating causes

The CCPS book, Layer of Protection Analysis,[ii] defines high demand mode as "when the challenge frequency to an IPL is higher than twice the test frequency for the IPL." When in high demand mode, the initiating cause frequency is the failure frequency of the first IPL. The following example demonstrates how missing this concept can result in an overly conservative result.

Process vent gas containing mostly air with a small amount of waste organics are fed to a gas fired thermal oxidizer to both abate the emissions and as supplemental combustion air. See Figure 6 for details of the system. Under certain process upset conditions, the organic concentration in the vent gas stream can spike up to the flammable range. So the undesired consequence in the LOPA is high organic concentration in the vent gas causing a deflagration in the thermal oxidizer and associated vent piping. The independent protection layers (IPLs) include a SIF that diverts the vent gas stream away from the thermal oxidizer if the vent gas becomes flammable.

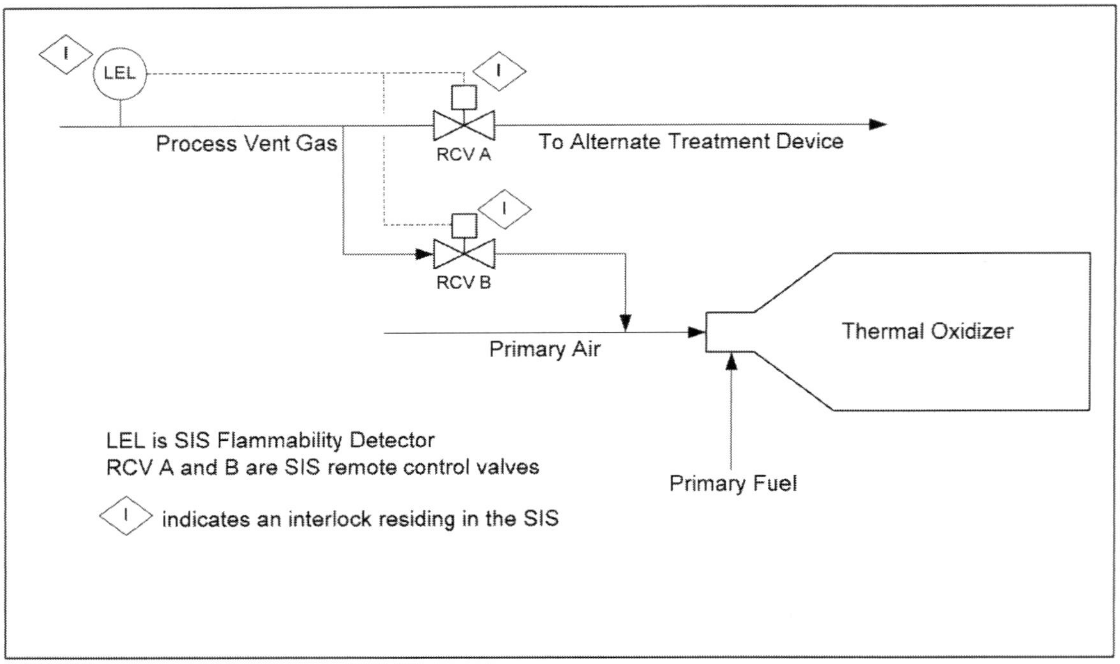

Figure 6: Process vent gas to Thermal Oxidizer

A high organic concentration occurs approximately 20 times per year, so following normal low-demand LOPA practices would result in Case 1 in LOPA Table 3. The mitigated consequence frequency is estimated at 1 per 5 years.

Following the CCPS methodology for high demand mode in Case 2 gives a different result. The mitigated consequence frequency is 1 per 50 years, or one order of magnitude lower. So, not accounting for high demand mode could mislead the team into applying a higher integrity SIF than is actually needed. See CCPS (2007)[vi] and ISA (2005)[vii] for additional discussion of high demand mode.

LOPA Table 3 – The effects of high demand methodology on the undesired consequence frequency

Consequences	Initiating Cause Frequency	/year	BPCS PFD		Operator Response to Alarms PFD		SIF PFD		Mitigated Consequence Frequency /yr
Case 1: *High demand methodology not applied*									
Deflagration in thermal oxidizer and vent piping	High LEL in vent gas stream to thermal oxidizer; 20 times/yr	2E+01	n/a	1	Operator cannot respond fast enough to mitigate	1	Diversion of vent gas stream to safe location	1E-02	2E-01
Case 2: *High demand methodology applied*									
Deflagration in thermal oxidizer and vent piping	High LEL in vent gas stream to thermal oxidizer - high demand, so use 2 * test frequency * IPL PFD	2E-02	n/a	1	Operator cannot respond fast enough to mitigate	1	included in initiating cause frequency	1E+00	2E-02

Common Error #6 – Not considering time at risk

Time at risk can significantly change the frequency of certain scenarios. For example, if an agitator fails during the reaction step of a batch, there could be a runaway exotherm. For this example, the initiating cause frequency is calculated by

Agitator failure frequency X fraction of time the reaction step is in progress.

Assuming that two batches are processed weekly and the reaction step for a batch takes 5 hours to complete, the fraction of time at risk would be

104 batches/year x 5 hours per batch = 520 hours per year
(~6% of the year).

If the agitator failure rate is determined to be 1E-02 per year, the initiating cause frequency for this scenario is

1E-02 per year x 6% = 6E-4 per year.

This frequency is significantly less than the frequency using only failure of the agitator, 1E-02 per year (once every 100 years).

For this particular example, it is assumed that a batch would not be processed if the agitator was not functioning. If batches are processed without first confirming that the agitator is functioning, the initiating cause frequency will not be reduced by the fraction of time the reaction takes place. In this case, failure of the agitator can occur at any time but will not be detected until problems arise during processing the next batch.

Common Error #7 – Not considering the consequences of spurious trips

A LOPA is typically focused on calculating the frequency of an undesirable consequence and the PFD provided for IPLs is based on the dangerous failure rate. However, the consequences of spurious trips should also be considered anytime a SIF is designed. In the following example, a SIF was being designed to protect against an undesired consequence, however it was recognized that a spurious trip would result in a different undesired consequence.

A small tank of a self-reactive catalytic material was being installed (see Figure 7). The HAZOP team recognized that a SIF was needed to protect against runaway reaction due to potential catalyst contamination. The recommended protective measure was to add quench water to the tank if high temperature was detected. The quench water would suppress the runaway reaction of the catalyst. Unfortunately, the quench water would also dilute the catalyst, causing a stalled reaction in the downstream process reactor, resulting in a hazardous condition. LOPA was used to examine both the original undesired consequence and the spurious trip condition. The results of this comparison are shown in LOPA Table 4. Because of the much higher likelihood (frequency) of a spurious trip than a real trip (driven by frequent power failures), the overall system risk is actually higher with the SIF, as it was originally designed.

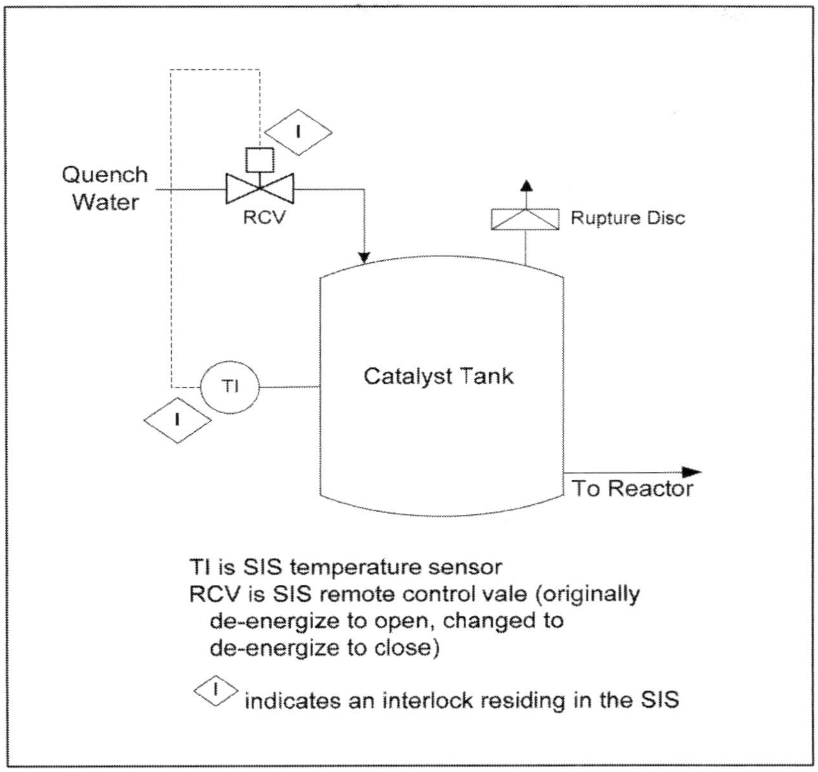

Figure 7: Self-reactive Catalyst Tank

Normally, the remote control valve for an SIF would be designed to "fail safe" in the trip position (that is, the valve would be specified for its "shelf" (deenergized) position to be the tripped position). This is known as de-energize to trip. For this particular application, it was advantageous to install an energize to trip valve on the quench water. While the energize to trip valve increased the PFD of the SIF somewhat, the risk reduction achieved was still acceptable. And, it also greatly reduced the frequency of water contamination due to a spurious trip.

LOPA Table 4 – The consequence of spurious trips

Consequences	Initiating Cause Frequency	/year	Existing Op Response to Alarms PFD		Existing SIF PFD		Other Mitigation PFD		Mitigated Consequence Frequency /yr
Case 1: *Contamination of day tank*									
Overpressure and catastrophic rupture of day tank	Contamination of tank with reactive material	1E-01	Operator cannot respond in a manner to mitigate	1	High day tank temperature opens quench water RCV	1E-02	Day tank rupture disc	1E-02	1E-05
Case 2: *Spurious trip of quench SIF resulting in downstream runaway reaction*									
Overpressure of reactor due to stalled reaction and accumulation of reactants	Spurious trip of high temperature SIF in raw material day tank, resulting in dilution of raw material. Dominated by annual power failure	1E+00	Reactor temperature monitoring and abnormal response procedure	1E-01	N/A	1E+00	Reactor rupture disc	1E-02	1.E-03

Conclusion

Layer of Protection Analysis is an excellent semi-quantitative tool that can be effectively used to estimate risk. However, a thorough understanding of common pitfalls is necessary to apply the methodology accurately. The 10 real-life examples described in this paper illustrate 7 of the more common pitfalls that can be encountered when performing a Layer of Protection Analysis. As demonstrated in the examples, the misapplication of LOPA can cause errors of an order of magnitude or more. An overestimation of the risk, as seen with the high demand case, can lead to greater capital expenditure and life-cycle costs. An underestimation of the risk, as seen with the industry database error, can lead to a system that isn't adequately protected. Fortunately, each of these errors was discovered during the LOPA or during the LOPA peer review process, prior to actual implementation in the plant. The keys to catching errors at this stage are solid LOPA training, well developed LOPA guidelines and a strong peer review process.

Disclaimer

---

[i] CCPS (1993), *Guidelines for Safe Automation of Chemical Processes*, New York: American Institute of Chemical Engineers, Center for Chemical Process Safety.

[ii] CCPS (2001), *Layer of Protection Analysis: Simplified Process Risk Assessment.* New York: American Institute of Chemical Engineers, Center for Chemical Process Safety.

[iii] IEC (1998), IEC 61508, *Functional Safety of Electrical / Electronic / Programmable Electronic Safety-Related Systems*, Geneva: International Electrotechnical Commission.

[iv] IEC (2001), IEC 61511, *Functional Safety of Instrumented Systems for the Process Industry Sector*, Geneva: International Electrotechnical Commission.

[v] ISA (2004), ANSI/ISA-84.00.01-2004: *Functional Safety: Safety Instrumented Systems for the Process Industry Sector*, Research Triangle Park, NC: Instrument Society of America.

[vi] CCPS (2007), *Guidelines for Safe and Reliable Instrumented Protective Systems*, Center for Chemical Process Safety, American Institute of Chemical Engineers, New York

[vii] ISA (2005), ISA-TR84.00.04-2005 Part 1, *Guidelines for the Implementation of ANSI/ISA-84.00.01-2004 (IEC 61511 Mod)*, ISA (International Systems and Automation Society), Research Triangle Park, NC

Process Safety Walkthrough Reviews

I Made Sukrajaya
BP West Java
Jakarta, Indonesia
sukrajim@bp.com

Margaretha Thaliharjanti
BP West Java
Jakarta, Indonesia
margarett@bp.com

A 10th PPSS Anniversary Paper

ABSTRACT

Safe design for process safety will not remain the same if process safety is not sustained for safe operation and maintenance. One key element of Process Safety Management is to conduct routine process safety audits. Many such audits rely on documentation, but this may not be enough to observe any improvements on process safety results. In fact, sometimes, the results may be surprising when a formal audits are completed. For this reason, conducting "Walk Through Reviews" of Process Safety will ensure that the actual conditions meet the process safety expectation. The additional important impact on the result is we know how they operate for Process Safety in the field.

What is the focus on the Walkthrough Review? As part of "sustaining process safety"; the focus will be on:

- *prevention of hazards and failures;*
- *control of process safety parameters;*
- *sustained safe operations and maintenance;*
- *operating and design windows;*
- *management of change;*
- *safe work procedures;*
- *understanding the facilities and other process safety measurements.*

Who should conduct the Walkthrough Review? These reviews can be conducted by supervisors and managers at every level; including of course; senior Process Safety Engineer. They can also be done individually or in groups.

When can they be done? They can be done anytime, in accordance with routine schedules or management visits.

How can they be conducted? A Walkthrough review check list and the documents to be reviewed are required. It is recommended that the review modules be done based on Process Equipment modules rather than all the equipment to ensure the effectiveness of the review.

The findings from the review must be recorded. In addition the recommendations must be addressed, tracked and have individually responsibility parties to close out the findings and finally, must be documented.

Futher detailed requirements of Process Safety: "Walkthrough Reviews" will be described in this paper.

1. INTRODUCTION

Facility design, construction, installation and operation needs to protect the health and safety of individuals, the safety of the Company's assets and operations, and the environment within which its' operations are carried out to standards that afford an acceptable level of protection.

However, the safe design for process safety will not remain the same if process safety is not sustained for safe operation and maintenance. One key element of Process Safety Management is to conduct routine process safety audits. Most such audits rely on documentation, but this may not be enough to observe the improvement on process safety results. In fact, sometimes, the results may even come as a surprise when a formal audit is completed. For this reason, conducting "Walkthrough Reviews" of Process Safety will ensure that the actual conditions meet the process safety expectation. The additional important impact on the results is we know how they operate for Process Safety in the field.

The Process Safety Walkthrough Reviews are a method of assessing process safety concerns. They can be used by people who want to take positive action to make process safety in operation safer. Process Safety Walkthrough Reviews can lead to an improvement in the physical environment in ways that will reduce the opportunities for fatality, and damage to facilities and the environmental.

The Process Safety "Walkthrough Review" is intended to identify, evaluate and control Health, safety and Environmental Hazards associated with all Facilities, Equipment, Procedures and Activities, especially for mature and aging facilities. This method is also implemented to ensure that the necessary equipment, safeguards and procedures are in place to eliminate or reduce the risks to an acceptable level.

The greatest advantage of The Process Safety "Walkthrough Review" is that it is easy to apply. The principle behind the technique is simple - comparing a list of predefined questions to the process to aid in the identification of hazards. This method is a simple assessment which can be performed by inexperienced practitioners. Little experience is required to perform this straightforward assessment which involves of checking the list against existing conditions.

This paper addresses the following five critical areas:

- Focus and Objectives of the review
- Selection of the review team
- Timing for the review
- Methodology or Review Process
- Case Study

2. FOCUS AND OBJECTIVE OF THE REVIEW

As part of "sustaining process safety"; the focus of this Process Safety "Walkthrough Review" will be on Prevention and Control of hazards and failures; process safety parameters, sustaining safe operation and maintenance, operating and design windows, management of change; safe work procedure, understanding the facilities and other process safety measurements.

The hazard management illustrated below in the "swiss cheese' diagram shows how Process Safety should be managed in the existing facilities. The Process Safety - Walkthrough Reviews" are part of the hazard management system, and applying this can continually reduce Major Hazards Incident.

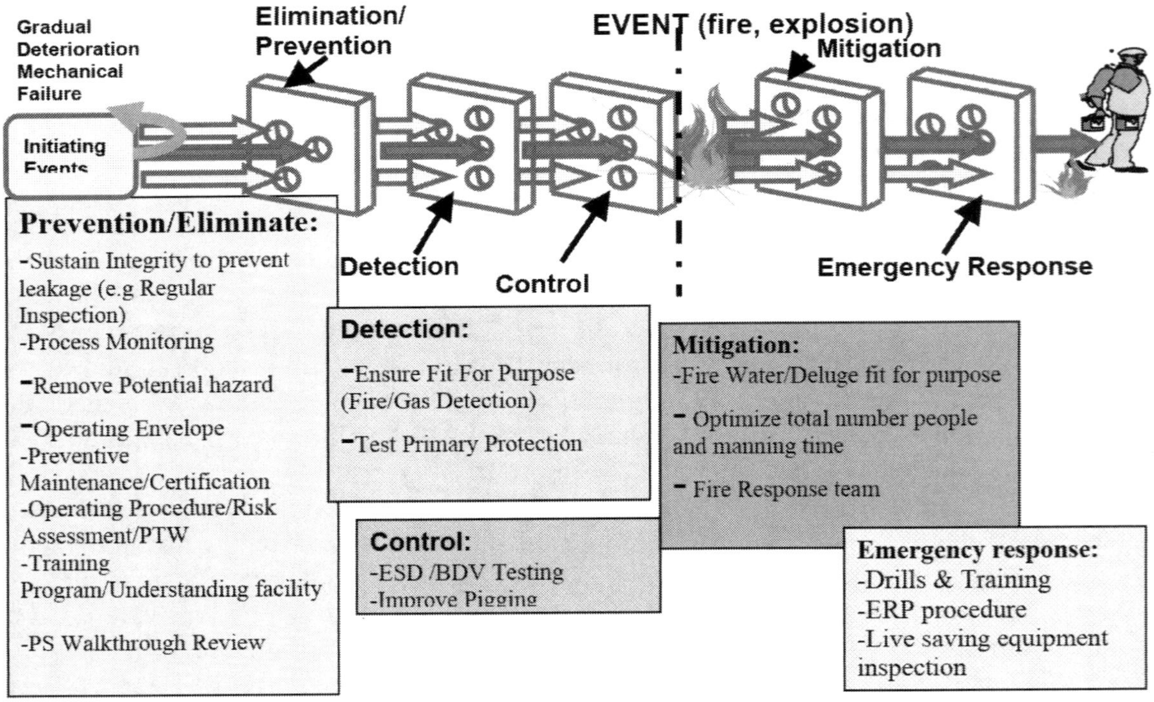

The purpose of the Process Safety "Walkthrough Review" is to provide a systematic approach to prioritizing and evaluating production facilities so as to reduce the risk of:

- Personnel injury
- Environmental damage, and
- Loss or damage to company assets
- Operability and Maintenance problems

The objective of the Process Safety "Walkthrough Review" is ensure that the necessary equipment, safeguards and procedures are in place to eliminate or reduce the risks to an acceptable level for the facility. This method is also intended to provide assurance to Business management that a design; whether for a new facility, a modification to an existing facility or an operating plant; is acceptably safe.

The review results can be used as a day to day reference for the operation site, during planning a work or carrying out risk assessment, evaluating action when critical plant items are found to be deficient. The review can also describe the key risk or major hazard and initiate improvement of a safety critical element of the facility.

3. WHO CAN CONDUCT THE REVIEW?

Anyone can take a part in the review, from supervisor through management level. The review also can be conducted individually or by a group. Small and simple facilities can be reviewed by one person, but more complex facilities should be reviewed by a group.

As a minimum requirement, a review team should have experience in using the selected hazards review technique, good interpersonal and leadership skills and have sufficient expertise to understand the process or operation under review. The size of the team varies according to the needs of the review. The important issue for selecting the team is an optimum number of team members to get good quality review results. If the team is too small, the team will not have the depth of knowledge or experience required to conduct an effective review. On the other hand, if it is too large, it may become too unmanageable for the team leader to handler.

The team composition may consist of Process and Process Safety Engineers, Field Engineer, Operation Engineer and Maintenance Engineer. A team leader should be appointed. As the team leader, he should have already completed formal hazards analysis review training and participated in other hazards analysis review. The team leader responsibilities are:

- Provide direction for the review
- Organize and executes review activities
- Facilitate team meeting
- Usually the most experienced member in the application of the selected hazards analysis technique
- Together with the review sponsor, establish the team composition.

The remainder of the team is composed of experts in various aspects of the facility design and operation. They are selected based on their specific knowledge in equipment design, operating procedures, instrumentation or electrical and maintenance practices.

The team members should also:

- Be involved as frontline personnel who have the most accurate knowledge of the process equipment and procedures.
- Be highly motivated to identify and eliminate hazards.

4. WHEN CAN PROCESS SAFETY BE DONE?

The process safety audit can be done anytime and should not depend on the readiness of the operation staff. The duration of the review also can also vary, depending on the complexity of the facility. The ideal time to allow for a "first audit" is approximately one or two full days for the review of each module. Therefore, it is necessary to define and break down the facility into several modules, rather focus on entire facilities, except for a small facility. Typical module setups can be applied to other facilities.

5. WHAT IS THE METHODOLOGY OR REVIEW PROCESS?

The Process Safety "Walkthrough Review" is real time, and facilities (plant) review based. This method is best applied to existing facilities, especially in mature facilities that have been operating for several years. This review is not suitable during either the design or construction phases, as this review is more focused on the operational aspects and application of the process safety design.

If, in unusual circumstances, it is considered that a Process Safety "Walkthrough Review" is worthwhile based on incomplete or informal information, the full nature of the documentation should be recorded together with appropriate qualifying comments.

The Process Safety "Walkthrough Review" can be divided into three stages:

- Pre- review stage or Preparation
- Review Stage
- Follow-up Stage

5.1 Preparation

In this stage the team will divide, prioritize and select facilities based on risks exposure. The steps include:

1. Dividing the facility into modules such as :
    - Gathering
    - Well Operation
    - Separation
    - Compression
    - Fuel Gas System
    - Etc
2. Completing the module datasheet
3. Calculating the Relative Risk Ranking of facilities
4. Ranking the modules
5. Prioritizing and Select the modules for review
6. Selecting the modules with the highest risk exposure level to perform a walkthrough review

Collection of some important documents should also be done to support the walkthrough review. PIDs, SAFE Charts, Equipment Lay Out and Standard Operating Procedures are some example documents which will be needed during the walkthrough review process. All documentation should be signed off to "as-built" and the latest standard operating procedure before the review commences. This is to ensure an effective review and cut any unnecessary discussion regarding incorrect documentation.

The Relative Ranking of the facilities can be done qualitatively or semi-quantitatively. Simple facilities don't need to quantify the risk number, unless we are dealing with a complex facility which will need comprehensive review to determine which facilities have higher risk than the others. The qualitative risk ranking can be determined by the team and also operation members involved during the review. The important thing for this ranking is agreement between reviewers and all of those

justifications should be written to avoid any doubt in the future. The semi-quantitative risk ranking can be determined by using a table which includes as the minimum:
1. Process Equipment/Complexity of Asset
2. Environmental Impact
3. Personnel Exposure
4. Business Exposure

In the pre-review stage, we also determine the checklist to be used for the walkthrough review. It usually comprises a safe design review as per API 14C or API 14J. If the company has it own standard, the checklist can also follow that standard. The main checklist can consist of reviewing operating pressures, temperatures, and flow rates.

A summary of the pre-review stage should be documented as per the protocol. The protocol also includes team member, schedule review, facility description and ranking, review check list and form for a review.

For those companies that have already addressing the Major Hazard Register; the risk ranking for the facilities may not need to follow the above guidelines; but they can be used as reference to develop the Process Safety Walkthrough Review Module.

In principle, it is recommended that the prioritizing of the facilities start first with the Highest Risk Module.

5.2. Review Stage

In this stage, the steps you need to follow are:
1. Tour the plant
2. Fill out a walkthrough checklist
3. Identify findings and comments
4. Create a report

After identifying the modules that will be reviewed, a walkthrough review can be started. The walkthrough review is started with a tour of the plant to gain familiarity. The team will verify that the PID drawings reflect what is actually in the field. If there are any differences, the team will modify the drawings and note these as findings. To help with the comprehensive review, the team can use a checklist that has been developed before the review has commenced. During the tour, the team can take any pictures that reflect the findings or deviations on the PIDs.

To ensure the comprehensiveness of the review, the team should be accompanied by the Offshore Installation Manager or at least duties supervisor/superintendent.

In addition to the PID review, the walkthrough review can also check the Operating Procedures, Equipment Lay Out and SAFE Chart. By reviewing the Operating Procedure, we can identify whether the facility still operates within the design operating condition or not.

Once the review stage has been completed, the team summarizes all findings from the walkthrough review onto finding forms. Some recommendations can be raised up for corrective action. To determine the urgency of the recommendation, a qualitative risk assessment can be carried out.

All of the results, including findings, concerns and recommendations should be documented and reported. It is essential that the findings, concerns and recommendations be clearly documented so that the operation team can take corrective actions appropriately.

The report should include a cover letter and working documents. All document details should include the following:

- Date of review
- Team members
- Scope of review and boundaries
- Comments on the team's overall findings in general terms
- Short narrative on the principal Hazards
- Discussion of critical safety equipment, instruments and procedures
- Future team meetings and issues to be resolved, including an understanding of the follow up responsibility

The working documents should include the following:

- Marked-Up Drawings (P&ID's), SAFE chart and Equipment Lay Out
- "Walk-Through" check list
- "Walk-Through" review
- List of critical safety items
- Human Factors Findings
- Critical Operating Procedures
- Area Classification Drawings
- Photograph (slides) that can be used for future safety meetings.
- Summary of finding form

Once the review documentation has been completed and issued, the review team's job is finished.

5.3. Follow Up Stage

The results of the reviews must be documented and forwarded to the management responsible for the follow up. The Process Safety "Walkthrough Review" results are normally not the responsibility of the review team. However, it is not unusual for team members to be consulted and even asked to participate in this phase. Recommendations may be accepted, modified, referred for further consideration, or rejected. The reasons for modification, referral, or rejection should be clearly stated in writing. Formal notes should be kept of all such decisions, which can be accessed in the future if required.

Line management should also ensure that actions accepted, referred, or modified are followed through to a successful conclusion. Someone should be nominated to do this and instructed to report formally at regular intervals while the action remains outstanding.

It is necessary to resolve all concerns that have been raised. Each concern needs to be assigned responsibility and acted upon. Action can take the form of modification to rectify a deficiency and assessment can be carried out with proper justification and documentation, that a risk is acceptable.

6. WEST JAVA FACILITY CASE STUDY

This section describes a Walkthrough Review at the West Java facilities as one example of a Process Safety Walkthrough Review. The location of this review was at MB-2 onshore remote facilities.

BP West Java - Indonesia, uses this walkthrough review as a powerful tool to ensure that the process safety and operational safety are addressed within its operation. Due to its aging facilities, which have been at least 30 years in operation, BP West Java realized that the focus of its operation must be on sustaining the process safety and keeping the facilities operating within the Safe Operating Windows.

An example of the Process Safety Walkthrough Review in West Java is the one that was conducted in the MB onshore remote facility. The modules in these facilities were simple, consisting of the following:

1. Well production facilities and gathering module;
2. Separation module;
3. Gas Lift module;
4. Launchers and Receivers module;
5. Drain module;
6. Vent module

The risk ranking of the facility was done by qualitative risk ranking, started from the vents, separation, gas lift, well production, drain and launcher modules.

Walkthrough checklist content is as follows:

- Check the pressure and temperature integrity of the piping system based on the information contained
- Actual design pressures and temperature should meet or exceed the required design pressure and temperature.
- Piping specification class for each line in and out of the equipment shown on the MFD (P&ID) should be checked and the location of the spec breaks reviewed.
- Assures that pressure buildup in a flare, vent or relief header does not affect the ability of the relief devices to handle design flow rates.
- Assures that flare scrubber design is adequate
- Assure that flare or vent location does not present a thermal radiation hazard under normal or abnormal conditions.
- Assures that the drain system collects and directs spilled, leaked or escaped liquid hydrocarbons to a safe location.
- Assures that liquid seals are installed to prevent gas flashing from liquid entering the drain system in one area from migrating to other areas.
- Assures that each segment of the wellhead and the flowline has the proper safety sensing and shut-in instrumentation.
- Assure that the instrumentation logic is appropriate

7. CONCLUSION

The Process Safety Walkthrough Reviews are a powerful tools for seizing control of operations to sustain the process safety, keeping the hydrocarbon in the pipes and operating within safe operating windows. It is used also to define major hazards in the operations and alert the operation while maintaining them. In addition, the summary of the walkthrough review acts as an effective communication tool from the engineering to the operation team to explain the philosophy of the operations when they were designed and also serves as feedback operation to engineering about what the problems in achieving the standard design. It is important to remember that the findings of the Process Safety Walkthrough review should be clear and specific so that the operation can have clear guidelines when defining the corrective actions.

8. REFERENCES

1. Stewart, Maurice, Dr. P.E. Experience – Based HAZOP at VICO Facilities
2. Guidance and Practice for Hazard and Operability Studies (HAZOP) Study, Engineering Technical Practice GP 48-02
3. Center for Chemical Process Safety (CCPS), Guidelines for Risk Based Process Safety, 2007
4. Guidance on Practice for Major Accident Risk Process, GP48-50

Evaluating Protection Layers That Are Not Independent

R. J. Stack; Dow Chemical; Hillsdale, MI 49242

A 10th Anniversary PPSS Paper

Abstract

The overall performance of a system that shares inputs and outputs between the BPCS and SIS can be better than the performance of a BPCS system and SIS system with dedicated inputs and outputs. Layer of Protection Analysis (LOPA) assumes that independent protection layers (IPLs) are independent of the initiating event (IE) components and any components of any other IPL in the LOPA scenario. Often it is found that protection layers are "almost" independent, that they have all separate components except they may:

- share common utilities such as plant air, electricity or cooling water
- share components in instrumented loops with other instrumented IPLs or the initiating event (IE)
- be in the same location and share vulnerability to a common failure such as a fire, or extreme weather

This paper looks at each of these situations using simple examples and by using a fault tree model of the LOPA scenarios to address the impact of sharing components or sub-systems. Also, the paper shows the difference between analysis of sharing at the order of magnitude LOPA level and using best estimate component reliability data.

The overall performance of a system that shares inputs and outputs between the IPLS or between the BPCS and SIS can be better than the performance of a BPCS system and SIS system with dedicated inputs and outputs. This performance advantage is shown in the examples. The maintenance and operational management issues involved with linking formerly "independent" systems can be managed so that the shared system performance advantage can be maintained.

Finally, some general guidelines are suggested for dealing with sharing issues in LOPA and the potential benefits of selected sharing and costs associated with assuring the complexities are controlled are discussed.

Introduction

In the Chemical and Petrochemical industries, safeguards or safety protection layers are used to reduce the likelihood or severity of a scenario. A protection layer is a device, system or action that can prevent or mitigate a scenario. The scenarios are sequences of events that can have undesirable consequences. When these protection layers are evaluated together to see if the required risk reduction is achieved, they are typically assumed to be "independent" from the scenario initiating event and from each other. This assumption is not always true.

Simplified approaches to evaluating a group of protection layers such as Layers of Protection Analysis (LOPA) require that each safeguard that can be claimed as a protection layer be independent; hence the name Independent Protection Layer or IPL. The most common LOPA approach (referred to as Approach A in the CCPS LOPA book), requires for each scenario that an independent protection layer be independent from:

- the initiating event
- any enabling events
- any other IPL

In most plants and for many scenarios, truly full independence does not exist for the protection layers because they share some common engineering or operational factors. The most frequent are: shared components, shared utilities and shared environment and procedural support.

Using simple methods to evaluate these scenarios can yield erroneous results. The degree of error is not normally recognized and the cost of the error or of the conservatism for using simple rules is not usually estimated. However, when protection gaps are identified, and there is not any easy fix, the cost of better analysis is but a small fraction of the cost of upgrades including on-going maintenance and testing. Also, any upgrades can be better judged with better IPL analysis methods.

It is also possible to evaluate a group of IPLs in a scenario together rather than one at a time to see just how much the loss of a contributor to independence affects the performance of the all the IPLs together. This group IPL analysis might also be applied to multiple scenarios that share the same IPLs.

Independence As Applied To Safety Systems

In a perfect world, it may be possible for protection layers to be completely independent of each other with no common cause or dependent failure concerns. In actual practice, given the work processes and economic constraints of the petrochemical industry this is difficult to do. Common cause failures are at best minimized not eliminated; so can IPLs be truly independent? Using the strictest definition of independence, the answer is no.

Corporations use common manufacturing processes, common raw materials, common component suppliers, and design to meet the same industry standards. Many manufacturers get sub-components from the same vendor. So even getting components from a different manufacturer may not eliminate a common cause failure. Using diverse measurement technologies may still have some common cause failures because of common sub-components such as electrical components.

How much independence is enough especially if even architectural design constraints can not always assure independence? When in doubt, the approach to take is to analyze the system and verify that the desired safety targets are met. You start from the basis that there are some common cause or sharing issues and that you will have to account for this in any evaluation of system performance. Doing this requires more work than just following simple rules (e.g. LOPA) but brings common sense flexibility when evaluating systems that are not independent and have known common cause or sharing issues.

Examples of Protection Layer Evaluation

The following examples are used for illustrative purposes only.

A. Common Utilities Shared By IPLs

Plant utilities are typically shared by all plant systems that need these utilities. Some common shared utilities are: electricity, steam, cooling water, plant air, plant nitrogen, and abatement system like incinerators, flares, and scrubbers, and waste water treatment systems. Note that some of these utility systems can be are part of protection layers for certain scenarios. Both the operating process and the standby protection layers may have a dependent relationship (dependency) with utility systems. Also, utility systems can be dependent on other utility systems.

Scenarios that involve utilities are: loss of utility scenarios for a process operation, loss of utility scenarios that affect multiple IPLs but not the process operation and loss of utility scenarios that subsequently bring down other dependent utilities leading to process upsets and/or IPL failures.

For cases where the loss of utility is an initiating event for a process upset, the impact of this failure on all IPLs can be checked. For cases where the utility loss affects only IPLs, the unavailability of the common utility failure can be used to determine the combined performance of the IPLs with this common dependency. Cases that involve a utility causing another utility to subsequently fail are important because some utility systems are protection layers for certain scenarios. This dependency can be overlooked with the result being that IPLs are given credit in a scenario where they will not be able to provide protection.

A utility loss example: A reactor has cooling water to the reactor jacket and cooling water to the emergency condenser attached to the top of the reactor. The loss of cooling water will keep both cooling systems from working but will not cause a process upset since the reaction is only mildly exothermic. For this scenario, the initiating event was a process upset, an accidental double charge of catalyst, a co-incident event was that the cooling water pumps failed or where failed during the process upset. Rather than assuming only one cooling system is available, the protection layers can be evaluated with the common utility failure.

LOPA Approach:

Initiating Event	Jacket IPL PFD	Conden. IPL PFD	Freq/yr	Total IPL PFD
0.1/yr	0.01	1 (no credit)	1E-03	0.01

More Detailed Approach: *9% Beta Factor for cooling water system

Initiating Event	Jacket IPL PFD	Conden. IPL PFD	Freq/yr	Total IPL PFD
0.1/yr	0.01*	0.01*	1E-04	0.001

Simple fault trees illustrate the logic of this more detailed analysis:

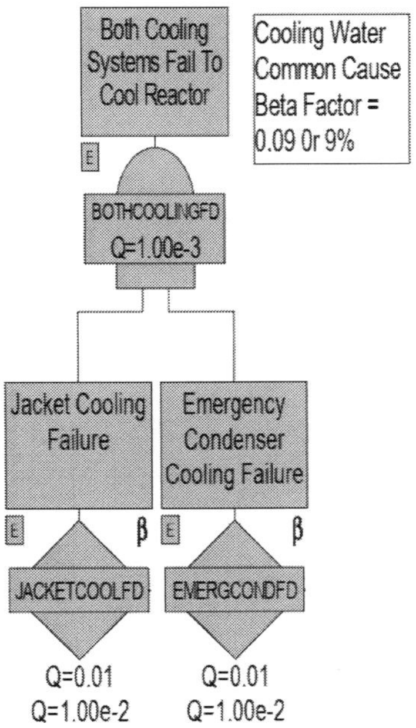

Figure 1: Beta factor used

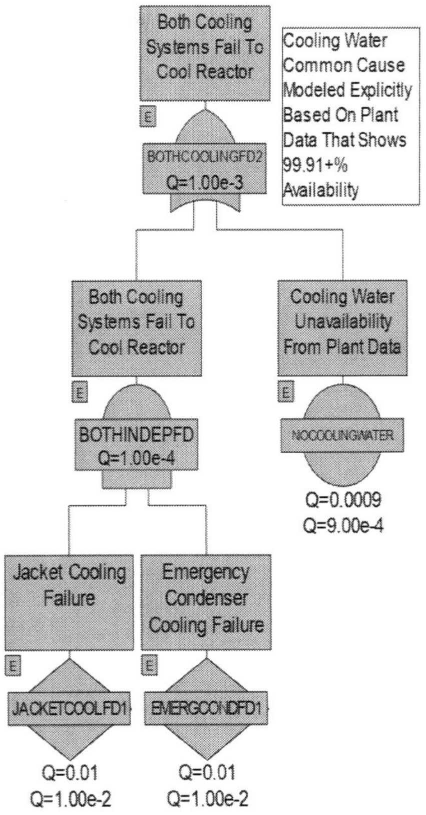

Figure 2: Explicit modeling

The model on the left uses a Beta factor of 0.09, the model on the right explicitly models the cooling water availability based on plant historical data. Both models give the same result and show that the common cause failure limits the overall performance of the protection layers, but also show that the overall protection layer performance is actually an order of magnitude better than the LOPA approach estimated based on the knowledge that the protection layers were not independent. For example, if a target frequency of 1E-04/yr was desired, LOPA would suggest adding an additional layer of protection which more detailed analysis shows is not needed.

B. Shared Components Between Instrumented IPLs

Independence means that there should normally be no sharing of components, sub-systems, or utilities between independent protection layers. However, in practice there are many cases where there is sharing and the strength or probability of failure on demand (PFD) of the shared components are such that this sharing does not result in the need to assume that one of the protection layers is not effective as would be the case when using LOPA (Approach A).

In the previous example, suppose there are two cases: each protection layer has its own temperature sensor(s), logic solver and fail open automatic block valve (no sharing) and another arrangement where there are multiple temperature inputs that are shared.

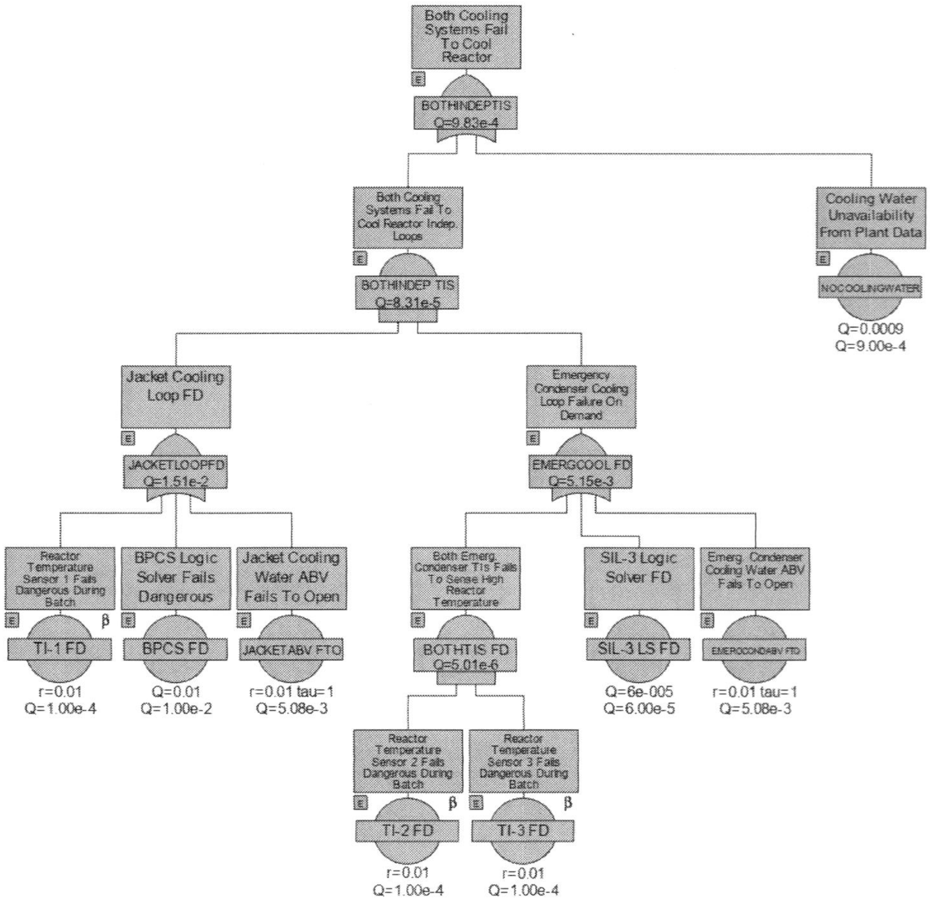

Figure 3: Independent TIs For Each Protection Loop

When all three TIs are shared between both emergency systems, the overall PFD is slightly better and the option of a high select two out of three voting can be used to minimize the chance of false trips.

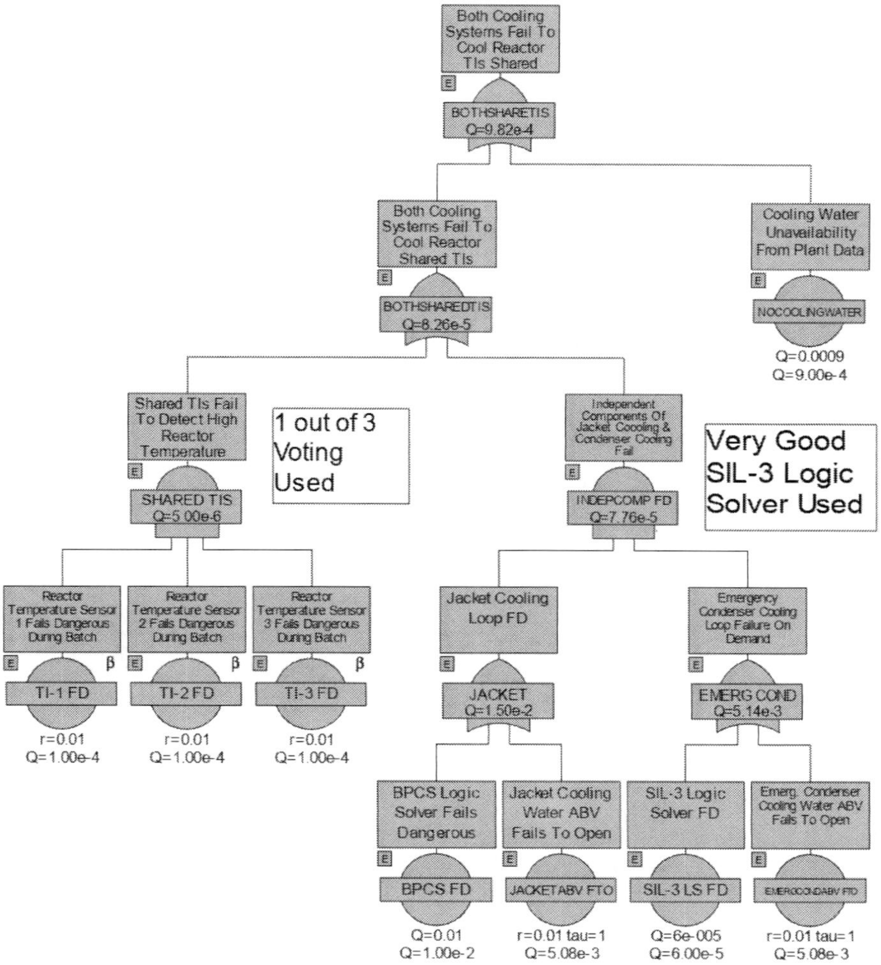

Figure 4: Shared TIs with 1 out of 3 Voting

Using high select two out of three voting will allow one TI to spuriously fail without tripping the emergency cooling. Note that there is almost no change in the calculated performance because the common cause beta factor contribution for multiple TIs failing is much higher than multiple independent TI failures

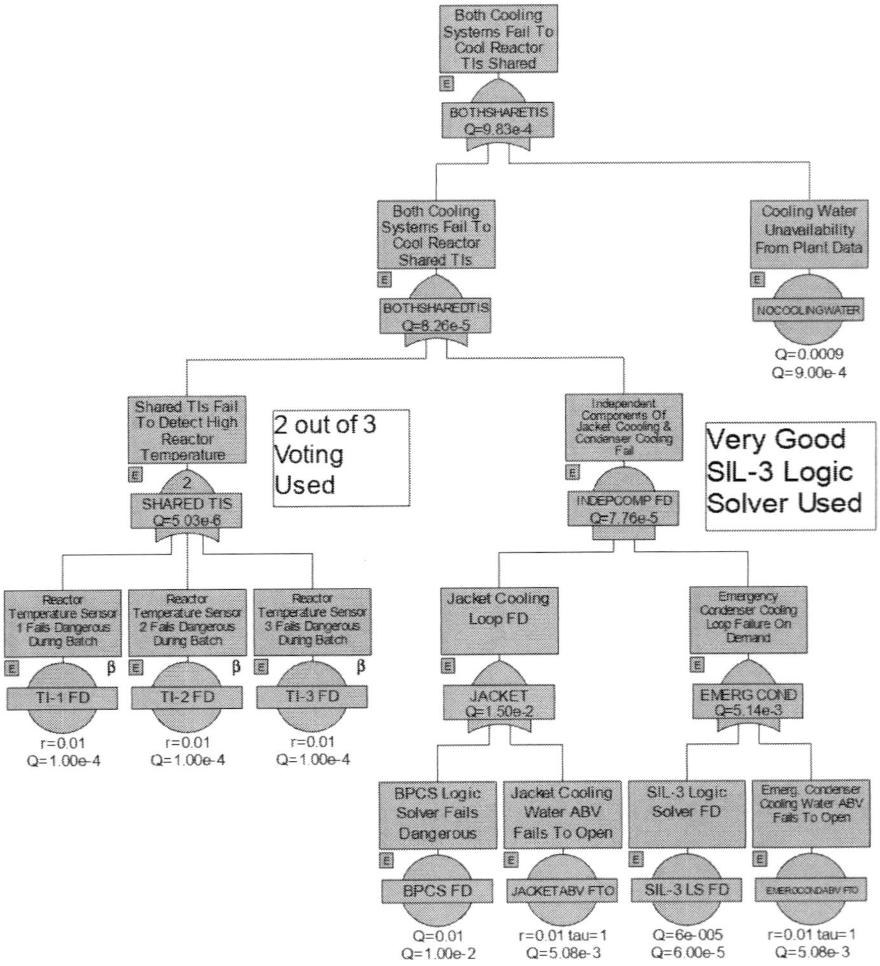

Figure 5: Two of out Three Voting With Reactor TIs

The other component that can be shared if it has a very low PFD is the SIL-3 Logic Solver. For this example, assume that both the Jacket loop and Emergency Condenser loop are using the same SIL-3 logic solver. Also assume that the TIs continue to be shared between the Jacket and Emergency Cooling systems.

The fault tree logic model for the combined systems sharing both TIs and SIL-3 Logic Solver is shown below. Note: using of a common SIL-3 logic solver raises the cooling IPL combined PFD from 2.6E-05 to 8.58E-05.

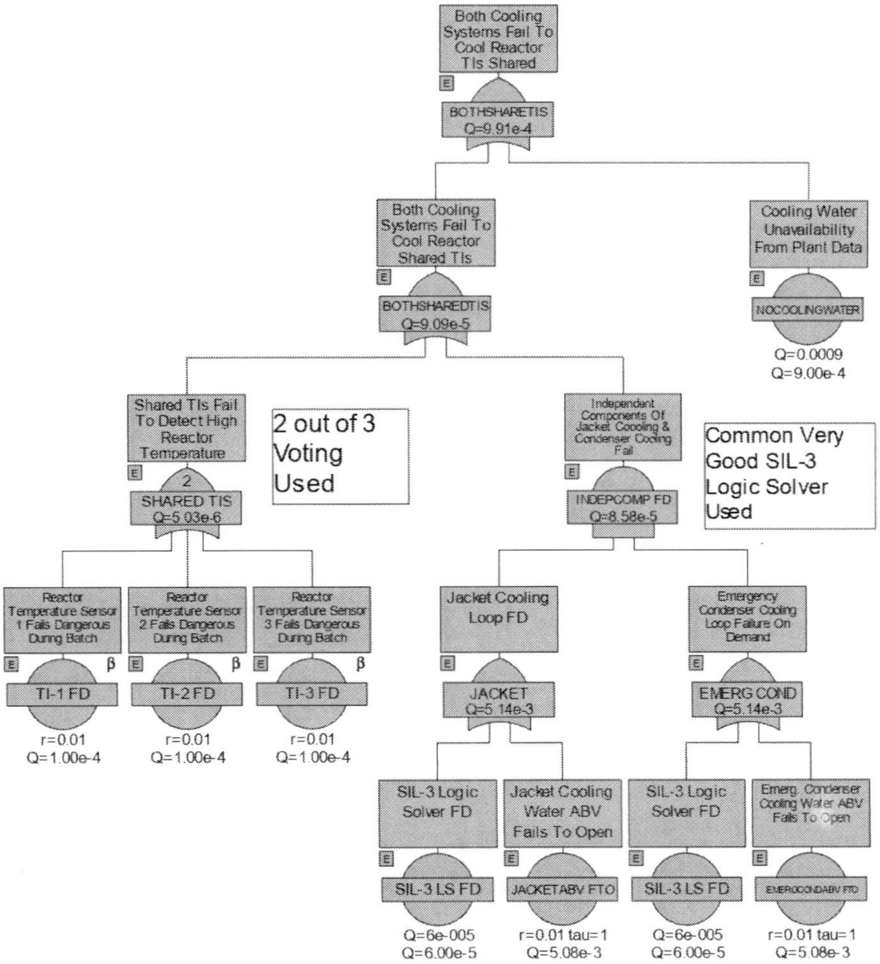

Figure 6: Shared TIs And Shared SIL-3 Logic Solver

If the PFDs of the shared voting TIs and SIL-3 logic solver were not very low, the sharing would have resulted in the combined protective system results failing to meet the 1E-03 PFD target or a risk reduction factor of 1000. Note that for all these cases the PFD target of 1E-03 was achieved.

C. Common location vulnerability

With the cooling water supply unavailability there is the issue of a common vulnerability to external events such as extreme weather. For example, when extremely cold weather causes a plant outage and the utilities can not be promptly restored when this weather exists. A scenario that has extreme weather as an initiating event must look at the performance of every IPL in this situation recognizing that the protection layers all are in this same weather and that they must still be able to function.

In the previous example, it was found that the cooling water availability would not be as good as the historical average data when severely cold weather occurs. This also applies when other types of severe weather exists. The additional protection layer is an administrative procedure to shut down operations when this weather exits or is expected to soon exist.

D. Other Issues: Less Than Perfect Testing

Many IPLs are standby systems, and by design we don't want to use them very often if ever. Thus, these systems must be inspected and tested periodically to assure that they actually will work as designed. Often the non-disruptive test of an IPL will not really allow everything to be checked out to the degree that one can say it is as good as new. For these tests, a test coverage factor can be used to account for the fraction of failures that may not be caught until a thorough inspection and test or even component replacement. The test coverage factor can be used to adjust the model used for calculating IPL PFDs.

Conclusions

In order to efficiently design safety systems and system upgrades, the way forward is to have the LOPA or similar safety team look for and recognize situations where there is instrument or component sharing. For simple and common sharing cases, guidelines can be developed based on a more detailed analysis which the LOPA team can use in their evaluations of IPLs with sharing.

For the unusual cases, additional more detailed analysis, such as fault tree analysis can be done before any recommendations are made to add or modify IPLs based on sharing concerns.

What are the benefits of sharing? Why would you share instruments? If sharing is done properly, the IPLs or safety system's failure probability can be improved or maintained while reducing plant upsets due to false trips and sometimes reducing the total number of instruments that must be maintained.

The payoff can be better safety and higher plant reliability. In summary you can get:

- lower instrumented IPLs (PFDs)
- less plant control failure trips
- less instrumented IPL spurious trips
- better diagnostics due to sensor comparison
- less total installed instrumentation
- less cost of on-going maintenance

The main drawback can be the lack of independence between the protection layers and in some cases, the loss of simplicity in describing and reviewing the shared IPLs. However, with the added guidance for LOPA reviews and additional analysis of selected cases, the complications of sharing can be successfully dealt with. Sharing has advantages but also has dangers. By managing the dangers and complexity, the benefits of sharing can far exceed the drawbacks and complications.

Managing the Intolerable

Fred Henselwood
NOVA Chemicals
Calgary, Alberta, Canada
henselfw@novachem.com

K. Gerry Phillips
GC Phillips Consulting Ltd.
Red Deer, Alberta, Canada
gcphil@telus.net

A 10th PPSS Anniversary Paper

ABSTRACT

Quantitative risk assessments are used within the field of process safety to decide what is and is not appropriate. Typically risk assessments involve the evaluation of probabilistic measures that estimate the average expected value for the situation being considered across a range of potential outcomes. The resulting expected value is then used to determine if a situation represents an acceptable or unacceptable risk based on a threshold value allotted to the scenario. This approach often gives guidance that is at odds with the thoughts and behaviors of some stakeholders as illustrated by the "but what if it does happen?" type of question. This inconsistency results from the inherent limitation associated with expected value approaches in that the methodology is based on whether or not a situation represents an acceptable risk while overlooking the possibility that a situation represents an intolerable event. This paper looks at an adjustment to traditional quantitative risk assessments so as to assess both the acceptability of risk and the tolerability of the associated consequences relative to risk criteria. These adjustments have been found to better represent stakeholder perceptions of risk, more closely relate risk tolerance to corporate values and resources, and to better justify the use of various risk transfer strategies.

1. INTRODUCTION

Quantitative risk assessment (QRA) within the chemical process industry is a technique that has been employed with respect to decision-making processes for a number of years and has been proven to be a valuable tool. The first edition of the CCPS Guidelines for Chemical Process Quantitative Risk Analysis was published in 1989[1]. A second edition of this publication was published in 2000[2] suggesting that ongoing application, development, and refinement of QRA approaches is occurring within the chemical process industry as the related science continues to mature.

The CCPS Guidelines[2] defines Risk Analysis as:

"The development of a quantitative estimate of risk based on engineering evaluation and mathematical techniques for combining estimates of incident consequences and frequencies."

Mathematically risk can be expressed using the following equation:

$$Risk = \sum_i X_i p(X_i)$$

Where X_i is a measure of the consequences associated with a given outcome i and $p(X_i)$ is a measure of the likelihood of outcome *i* occurring, with Risk being defined as the product of these two terms (consequence and frequency) summed over all possible outcomes.

Despite the proven history of QRA approaches and the relatively straightforward mathematical definition of risk, decisions made based on QRA work often fail to generate support from the various stakeholders involved in the process. This inability to gain agreement for decisions based on QRA work most often occurs when catastrophic outcomes are considered, particularly when the catastrophic outcomes are deemed acceptable by the applied QRA methodology.

QRA techniques are based on mathematics related to expected value and decision theory. Mathematicians have for centuries looked at situations such as the St. Petersburg Paradox[3] where this type of decision-making process fails to provide a solution. With respect to process safety the traditional QRA approach is based on the law of large numbers in that as the sample size increases the measured experience will converge with the expected experience. In the case of a scenario that is estimated to occur with a frequency of only once in a million years the opportunity for convergence to occur between experience and theory cannot be reasonably expected and as such the utility of traditional QRA approaches is limited in these situations as clearly the law of large numbers does not apply.

It is this inability of QRA approaches to provide satisfactory answers for low likelihood scenarios that leads to difficulties with respect to gaining the approval of the various stakeholders in the process. Often the question will be raised "but what if it does happen?" Recognizing this limitation with traditional QRA approaches, the focus of this paper will be on methodology adjustments so as to move from a model based on the law of large numbers[4] to a model that accounts for the inability to gain convergence when assessing low frequency events.

2. PROPOSED RISK EQUATIONS

Typically, when conducting a QRA analysis the number of occurrences of an outcome can be modeled using the Poisson distribution function[5]. (There is an assumption that occurrences represent independent events.) The Poisson distribution function can be written as follows:

$$p(y) = \frac{e^{-\mu}\mu^{y}}{y!}$$

Where *p(y)* is the probability of *y* occurrences in a given time period and μ is the expected number of occurrences for the given time period.

Assuming that there is only one outcome of concern the original risk equation can be rewritten as follows:

$$Risk = \mu X$$

Where *X* is a measure of the value associated with the severity of the outcome being considered (assumes all other outcomes have a value of 0) and μ represents the expected number of occurrences for the time period being considered. (Often when looking at risks, frequencies are considered allowing for the assessment of a time period such as one year.)

However, as was discussed above convergence between the expected number of occurrences and the observed number of occurrences may not occur for systems where the likelihood of occurrence for the studied event is low. As such the random nature of the event being examined needs to be considered with the risk measure accounting for the variance related with the system. This adjustment is easily done through accounting for the standard deviation in the expected number of occurrences allowing for the following equation to be generated.

$$Risk = (\mu \pm Z\sigma)X$$

Where *Z* is a measure of the confidence limits[6] desired and σ is the standard deviation associated with the expected number of event occurrences.

Expanding the above equation and focusing on the maximum value gives the following equation:

$$Risk = \mu X + Z\sigma X$$

This approach results in risk being defined based on two terms; μX is the expected value and $Z\sigma X$ represents the variability associated with the expected value.

The standard deviation for the Poisson distribution function is defined as follows[5]:

$$\sigma = \sqrt{\mu}$$

Inserting this relationship into the previous risk equation gives the following function:

$$Risk = \mu X + Z\mu^{1/2} X$$

This arrangement yields risk being related to both the likelihood of a given event and the square root of the likelihood of a given event. It is this second term that is often not considered within traditional quantitative risk assessment approaches.

If over the life of a business or operation a given event (or similar events) is expected to occur often, μ will be significantly greater than $\mu^{1/2}$ resulting in a situation where the above risk equation will reduce to the traditional risk equation:

$$Risk = \mu X$$

However, if over the life of a business or operation a given event (or similar events) is not expected to occur, $\mu^{1/2}$ will be significantly greater than μ resulting in a situation where the risk equation reduces to the following:

$$Risk = Z\mu^{1/2} X$$

As a result the classic risk equation makes sense for common events. However, for infrequent events a new risk equation is proposed based not on the likelihood of occurrence but rather on the square root of the likelihood of occurrence. This resulting dichotomy suggests that organizations require two risk standards so as to efficiently manage the portfolio of risks facing their organization when applying QRA techniques.

3. RISK CRITERIA

Based on the presented equations for risk acceptability, expected events are best managed using the following relationship:

$$Risk \approx \mu X$$

and unexpected events are best managed using the following relationship:

$$Risk \approx \mu^{1/2} X$$

Recognizing that the risk term represents an arbitrary value, as allotted to the situation, the risk relationship for unexpected events can be squared to yield the following relationship:

$$Risk^2 \approx \mu X^2$$

These equations show that for frequent events risk acceptance is linear with both severity and frequency, but for infrequent events the severity relationship now involves a square function, if risk acceptance is made to be linear with frequency.

Working with an assumption that the two criteria should meet at the breakpoint between expected and unexpected events one can envision the following risk criteria as illustrated in Figure 1.

Figure 1.

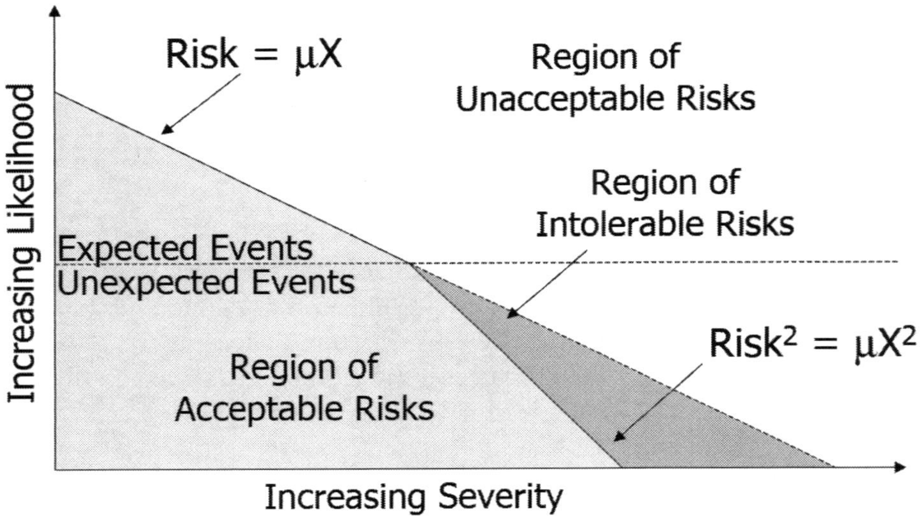

Figure 1 shows the three regions of risk acceptability (unacceptable risk, intolerable risk, and acceptable risk) as defined by the two derived risk criteria for expected and unexpected events.

Relative to traditional definitions for "acceptable" and "unacceptable" the region defined as representing intolerable risks accounts for those scenarios that are positioned between the traditional risk acceptability line and the newly proposed risk acceptability line for unexpected events. Scenarios within this section should be viewed as being unacceptable while recognizing that this unacceptability finding is a result of the magnitude of the consequences and the size of the organization (i.e. larger companies may find certain events to be acceptable while smaller companies may find these same events to be intolerable). As an example a large multinational organization can likely tolerate larger events than a single facility operation.

4. PROCESS SAFETY IMPLICATIONS

When applying the above concepts to process safety related QRA work it is important to consider how the above risk relationships were constructed. In assessing any one situation, particularly as related to catastrophic outcomes, the event by nature will be unexpected. However, when establishing the breakpoint between expected and unexpected events it is important to consider the entire portfolio of risks facing an organization and to define this breakpoint relative to the magnitude of what are and are not expected event severities. If a given outcome is of a severity that would reasonably be expected to be experienced by the organization then that event should be assessed relative to the traditional expected risk acceptance criteria. If however, the potential severity exceeds that which would reasonably be expected by the organization the newly proposed risk criteria for unexpected events likely better represents stakeholder perspectives.

Focusing on financial risks the presented relationships readily explain the use of insurance instruments for risk management (transfer). Working with an assumption that insurance providers exist to be profitable it is reasonable to conclude that the purchasing of insurance represents a situation associated with a negative expected value, which would suggest that the purchasing of insurance cannot be justified. However, if insurance were to be purchased to protect an organization from an unexpected event (based on a high financial severity) then the non-linear relationship would suggest spending beyond the expected value represented by the situation. At the same time from the insurer's perspective, through pooling risks from several sources what was an unexpected risk to a particular organization becomes an expected event for the insurer, explaining why the insurer is willing to allow the risk transfer to occur. That is an intolerable risk is transferred from an organization to another organization for which the event is considered to be tolerable and as such the associated costs and expected values make sense for both organizations. Further, insurance deductibles often give a good indication as to the breakpoint between expected and unexpected financial events to an organization.

Alternatively the determination of the breakpoint between expected and unexpected events can be established based on the reversibility of the assessed event. When assessing the range of threats facing an organization two categories of threats can be identified. The first category represents those impacts that the organization can readily manage and the second category represents those impacts that the organization cannot readily manage. From the perspective of the organization these potential consequences can be considered to be either reversible or irreversible in severity. For reversible consequences time will allow the situation to be repaired and returned to its original state while for irreversible consequences time will not allow the situation to recover.

Examples of reversible events may include the following:

- Injuries
- Environmental damages
- Immaterial financial losses
- Regulatory intervention

Examples of irreversible events may include the following:

- Fatalities
- Environmental change
- Material financial losses
- Regulatory change

What constitutes a reversible or irreversible event will be to some extent a function of the organization making the distinction, with that distinction likely reflecting the size and capabilities of the organization. For example the definition of what defines a material loss will be a function of the financial resources of the organization.

Recognizing that irreversible events will threaten the viability of the organization these events should be considered as unexpected as otherwise one could reach a conclusion that the organization would fail as a result of the occurrence of these irreversible events. Graphically one can go from defining the breakpoint between the two risk criteria based on frequency to defining the breakpoint based on severity as illustrated in Figure 2.

Figure 2.

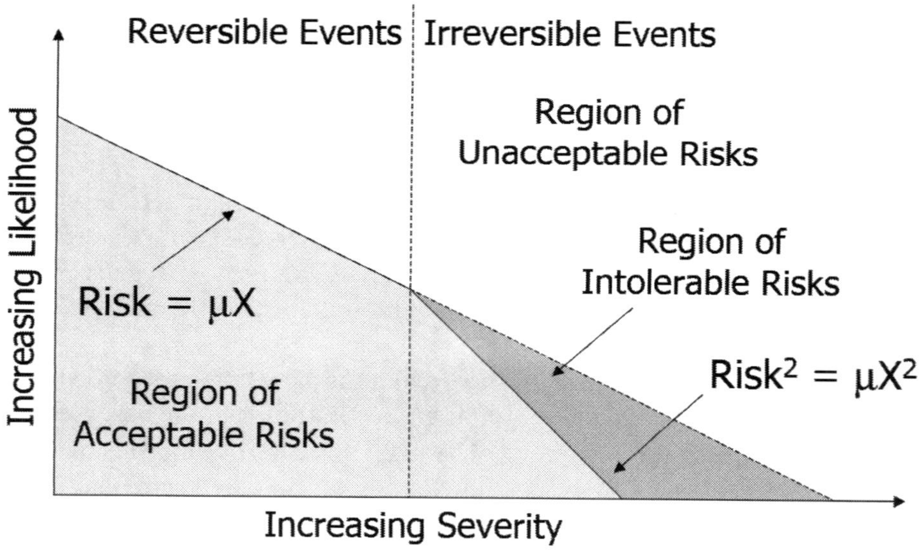

Figure 2 shows the three regions of risk acceptability (unacceptable risk, intolerable risk, and acceptable risk) as defined by the two derived risk criteria similar to that illustrated in Figure 1 with the breakpoint between the two functions defined by the boundary between reversible and irreversible events.

Establishment of the breakpoint between reversible and irreversible events was found to work well for process safety based QRA work as this approach allowed more readily for individual scenarios to be assessed without the need to step back and review the corporate expected risk structures. Further, this approach allowed for risks beyond financial concerns to be accounted for and readily assessed. As an example, if events resulting in multiple fatalities are considered to be irreversible the above approach readily explains the use of non-linear risk acceptance criteria.

5. CONCLUSIONS

Expected value and decision theory, which forms the basis for most QRA work, does not readily apply to low frequency high severity events. As such two camps have developed when assessing these risks, those who go by the traditional QRA approach and those who go by consequences only. Through expanding the derivation of the traditional risk equation to account for the limitations imposed by the underlying assumption in the applicability of the law of large numbers a new risk equation can be generated for these situations. This new risk relationship places a heavier weighting on severity relative to frequency and serves as a midpoint between these two camps. Further, this approach readily explains the use of risk transfer strategies such as insurance and gives a basis for the use of non-linear relationships as often utilized in addressing societal risks[7]. In addition, this approach also forms the basis for logic that suggests that risk tolerability is partially a function of an organization's size and capabilities.

In practice this methodology has allowed for the assessment of various process safety related risks and has given better clarity as to the appropriate course of action relative to the threats being faced. In addition, the determination of the breakpoint between the two risk criteria based on the reversibility of the potential event readily allowed for the proposed approach to be expanded to non-financial impacts facilitating the broader application of this approach.

6. REFERENCES

1. *Guidelines for Chemical Process Quantitative Risk Analysis*, New York: American Institute of Chemical Engineers, Center for Chemical Process Safety, 1989.
2. *Guidelines for Chemical Process Quantitative Risk Analysis, Second Edition*, New York: American Institute of Chemical Engineers, Center for Chemical Process Safety, 2000.
3. Wikipedia contributors, "St. Petersburg paradox," *Wikipedia, The Free Encyclopedia*, http://en.wikipedia.org/w/index.php?title=St._Petersburg_paradox&oldid=186009000 (accessed February 5, 2008).
4. Wikipedia contributors, "Law of large numbers," *Wikipedia, The Free Encyclopedia*, http://en.wikipedia.org/w/index.php?title=Law_of_large_numbers&oldid=187393570 (accessed February 5, 2008).
5. Wikipedia contributors, "Poisson distribution," *Wikipedia, The Free Encyclopedia*, http://en.wikipedia.org/w/index.php?title=Poisson_distribution&oldid=189000264 (accessed February 5, 2008).
6. Wikipedia contributors, "Standard score," *Wikipedia, The Free Encyclopedia*, http://en.wikipedia.org/w/index.php?title=Standard_score&oldid=187737289 (accessed February 5, 2008).
7. *Guidelines for Quantitative Risk Assessment*, The Hague: Committee for the Prevention of Disasters, 1999.

Process Infrastructure PSM - Applying Consequence Analysis and Facility Siting Methodologies to Process Infrastructure Facilities

Douglas J. Ferguson, P.E.
E. I DuPont – DuPont Engineering Technologies – Process Safety & Fire Protection
Houston, Texas
E-mail: Douglas.J.Ferguson@usa.dupont.com

A 10th PPSS Anniversary Paper

ABSTRACT

The process systems for chemical industry operating facilities receive close scrutiny for process safety and consequence analysis concerns because of the potential hazards that may be inherent in the process operations. Process infrastructure facilities and operations, however, may not receive the same degree of analysis because the hazards associated with these facilities are believed to be better understood, to have lower risk, and are accepted as "common or normal" hazards. The safety of infrastructure facilities such as natural gas supply lines, inert gas supplies (such as Nitrogen, Argon or Carbon Dioxide), propane tanks, compressed air, and steam generation can benefit from more in-depth consequence analysis in the design and installation of these facilities.

This paper presents a proposed methodology for conducting consequence analysis and facility siting activities related to the design and installation of process infrastructure facilities.

1. Introduction

Federal and State regulations require that consequence and risk analysis studies be conducted for chemical industry facilities that handle highly hazardous materials as defined in the OSHA 1910.119 and EPA RMP documents. The quantitative consequence analyses conducted for these facilities are used to evaluate the potential impact of a variety of accident scenarios during the Process Hazard Analysis process. Plant personnel use the analysis information to evaluate the adequacy of all current safeguards that are in place to prevent or mitigate the scenarios considered, make facility siting decisions, and develop emergency response plans. They may propose additional safeguards, if needed.

These analyses are usually conducted for a wide variety of release scenarios; and depending on the chemicals involved, the consequences of a release can range from localized flash fires or minor toxic releases, to explosions or toxic cloud releases that may have impact outside the boundaries of the plant facility. Process infrastructure facilities and operations, however, may not receive the same degree of analysis as the hazardous process operations because the hazards associated with these facilities are believed to be better understood and less hazardous, to have lower risk, and are accepted as "common or normal" hazards. Likewise, infrastructure facilities may not receive the same degree of scrutiny or attention when facility siting and emergency response decisions are made; or when changes are made, repairs are needed, or preventative maintenance is required

2. Common Process Infrastructure Operations

Most chemical plant sites in the United States have numerous infrastructure operations that are required to support their process operations and are under site control. Generally the infrastructure operations fall into the following categories:

- Steam generation (Boilers)
- Liquefied Petroleum Gas (LPG)
- Compressed Gas (Air, N_2)
- Cryogenic Gases (Argon, N_2, CO_2)
- Natural Gas (High / Low pressure)
- Brine Systems (Fluorocarbons, Glycols)
- Heat Transfer Systems (DP:DPO, etc.)

The design, construction, operation, inspection, and basic safety requirements of these facilities are generally well established and covered in codes and standards such as the ASME Pressure Vessel and NFPA codes and standards. However, the consequences of a major incident or failure in one of these systems can have a significant impact on adjacent facilities, such as buildings with high personnel occupancy; temporary facilities such as trailers; process equipment; and especially other infrastructure systems, which are often located in close proximity to each other in a "utilities" section of a site layout. Conducting a thorough quantitative consequence analysis, especially in the early stages of a new process facility installation, can provide vital information for facility siting considerations and inherently safer approaches for infrastructure design and installation.

3. Infrastructure Scenarios and Consequence Analyses Considerations

The identification and characterization of process incidents where highly flammable and/or toxic chemicals are involved is required and well defined as part of the process hazard analysis for our chemical industry facilities. Consequence analyses associated with these incidents help us determine the potential on-site and off-site impact should an incident occur. Although there is usually less potential for off-site impact for infrastructure facilities, there are incidents and hazards associated with these operations listed above that can have significant on-site impact. Listed below are types of scenarios that can be associated with the above infrastructure operations and quantitatively analyzed for potential impact:

Steam Generation/Boilers & Heat Transfer Systems

- BLEVE
- Fuel rich – explosion in fire box

- Fuel supply
  - Vapor cloud/ Flash fire (propane, natural gas, DP:DPO vapor)
  - Congested/ Confined space explosion (propane, natural gas, DP:DPO vapor)
  - Fire (propane, natural gas, coal, fuel oil)
  - Dust explosion (coal dust)
- Toxicity/ Environmental (DP:DPO)

<u>LPG Systems</u>

- Fire/ Flash Fire/ Jet Fire
- BLEVE
- Vapor Cloud
  - Unconfined vapor cloud explosion
  - Congested/ Confined space explosion (propane, natural gas)

<u>Cryogenic Gas Systems & Brine Systems</u>

- Asphyxiation (large ground level cloud)
- Cold temperatures – cryogenic burns
- Embrittlement of susceptible materials of construction
- Toxicity/ Environmental (fluorocarbons, glycol)

<u>Compressed Gas Systems</u>

- Overpressure – Vessel burst

<u>Natural Gas Systems</u>

- Fire/ Flash Fire/ Jet Fire
- High pressure (possibly without odorization)
- Flammable Gas Cloud
  - Unconfined vapor cloud explosion
  - Congested/ Confined space explosion (low pressure gas)

4. Consequence Analysis Examples

Quantitative consequence analyses can be developed for the above scenarios for specific installations or in a generic manner so that the information can be used to evaluate various systems and installations in order to determine which facility will provide the required capacity while minimizing the potential for adverse facility siting impact.

An example of using generic data to size a facility to minimize impact would be in determining the size and location of a LPG storage tank. The "worst case" concern for LPG tanks is the potential for a Boiling Liquid Expanding Vapor Explosion (BLEVE). Using the information found in the Center for Chemical Process Safety, <u>Guidelines for Evaluating the Characteristics of Vapor Cloud Explosions, Flash Fires, and BLEVES</u>, American Institute of Chemical Engineers, 1994[1] and calculational tools such as MathCad, a spreadsheet can be developed to determine the overpressures and the thermal radiation generated by a LPG tank BLEVE based on the energy released per a baseline unit volume of the specific gas, such as Propane, for a range of relieving pressure and temperature conditions. These values can be plotted and used to look at the range of potential impacts for different capacity storage tanks as shown below:

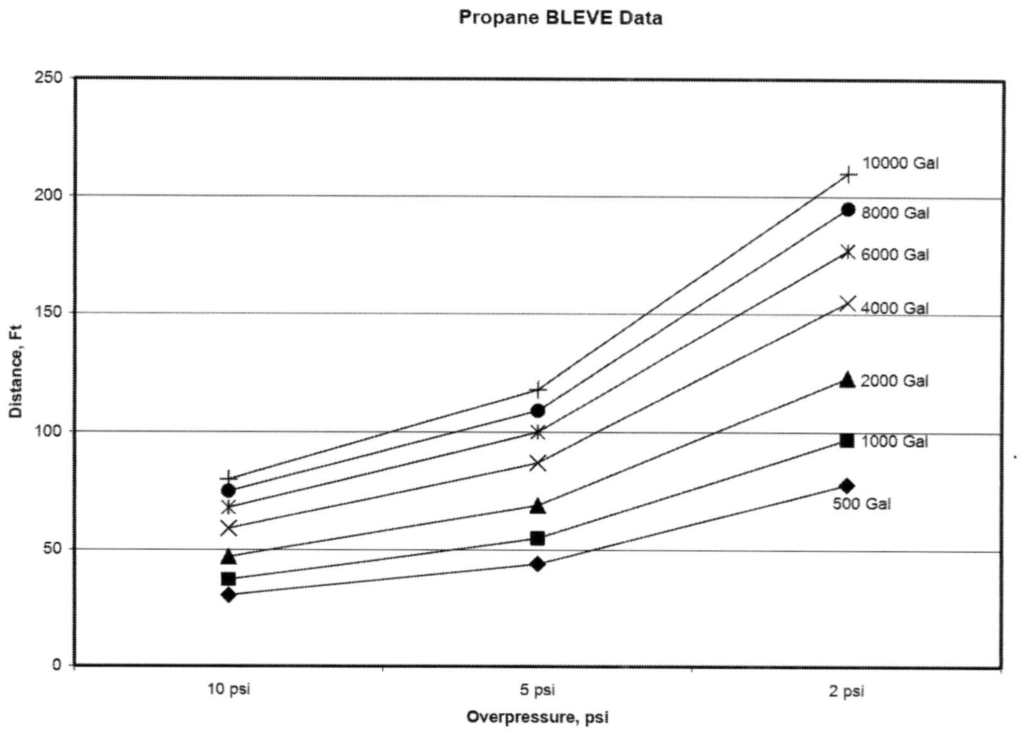

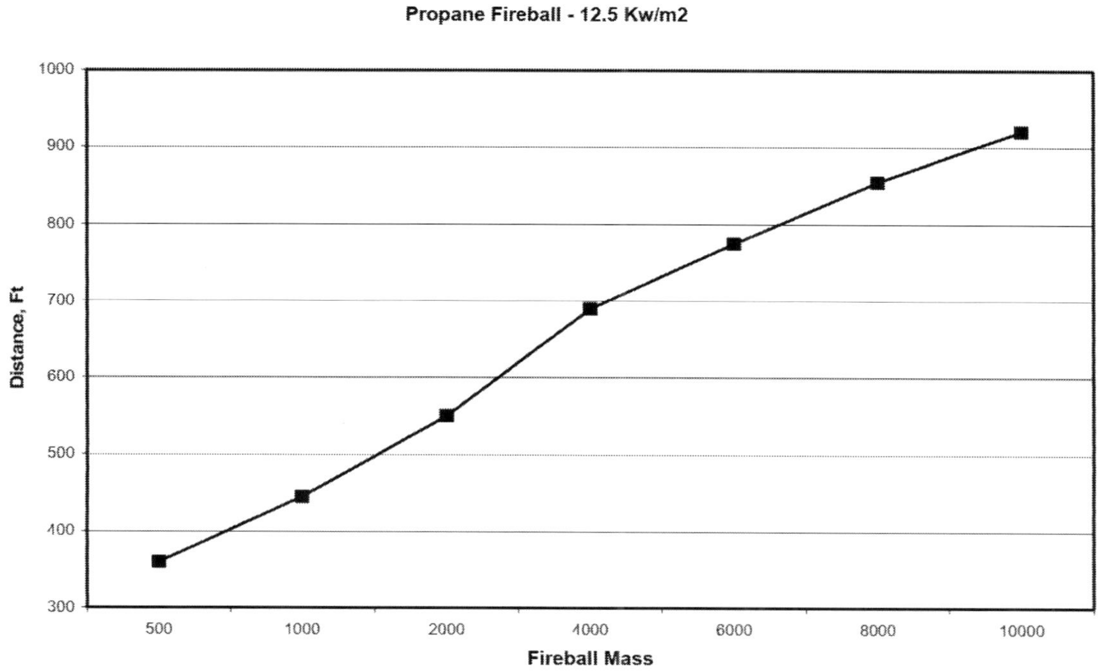

Based on the required demand for LPG by the user and the location of occupied buildings or critical equipment in the area, a storage tank can be selected to provide an adequate LPG supply for the user and then positioned in a manner to minimize facility siting impact. As alternatives, (1) the buildings and equipment could be redesigned or refitted to withstand the potential overpressures and thermal radiation, or (2) the LPG storage tank could be fire protected in some manner – deluge system, fire-proofing insulation, etc. – to greatly reduce the likelihood of reaching a BLEVE condition in the tank.

The same analysis technique can be applied in determining the potential impact of a boiler steam drum BLEVE. Overpressures are the primary concern for a boiler BLEVE since there is no major fireball associated with a boiler BLEVE. The overpressures generated in a boiler explosion will often be greater than those for a flammable material under equivalent conditions because of the greater expansion that occurs as the liquid flashes to a vapor – see graph below for a 185 psi boiler.

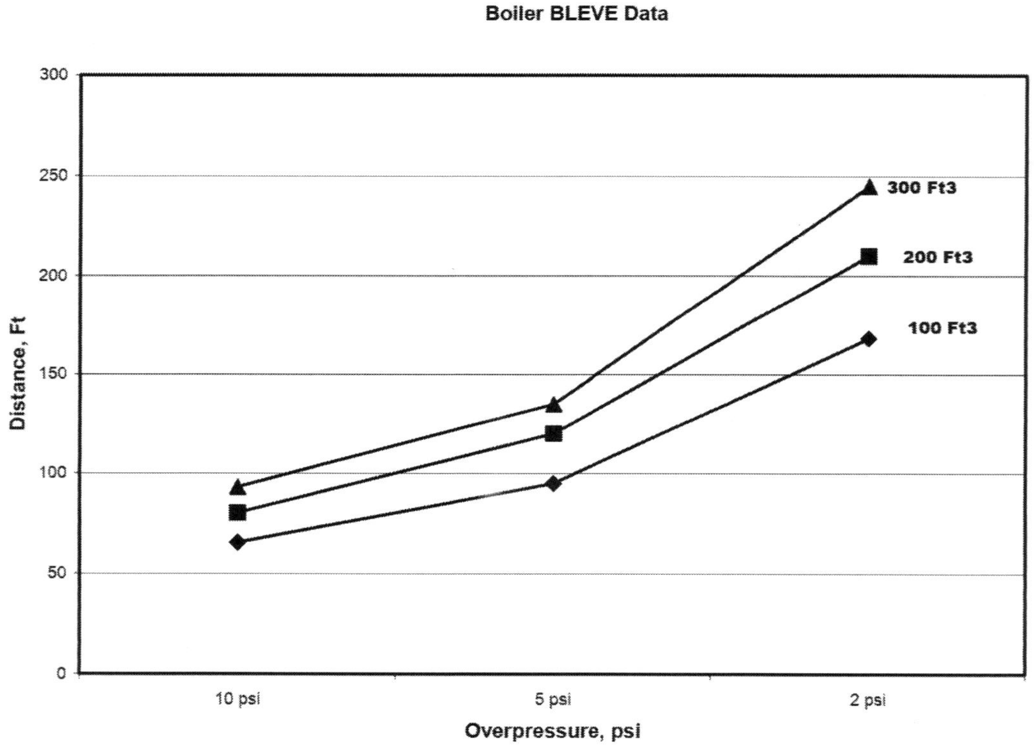

The second greatest concern for several of these type systems is a possible flammable gas release (LPG or natural gas for the boiler). The flammable cloud could result in an unconfined vapor cloud explosion or, more credibly, a large flash fire or a confined/ congested space explosion. Although there are some very sophisticated modeling tools, such as FLACS[2] and CEBAM[3], available for use in modeling flammable cloud behaviors and explosion overpressures in congested areas, the initial analysis can be done using standard dispersion analysis tools such as TRACE® and PHAST®. A variety of scenarios can be modeled just as is done for process flammable or toxic materials. One scenario to model would be the 10 minute release condition for a LPG storage tank or a full line break for a natural gas line to determine the maximum size flammable cloud that could be generated as shown below for Propane:

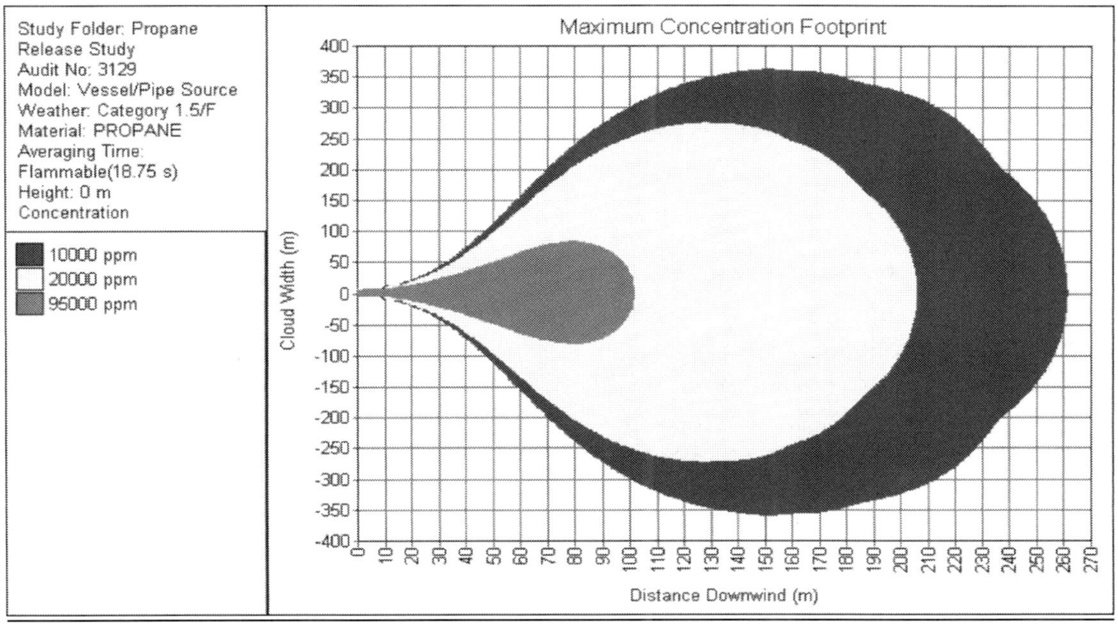

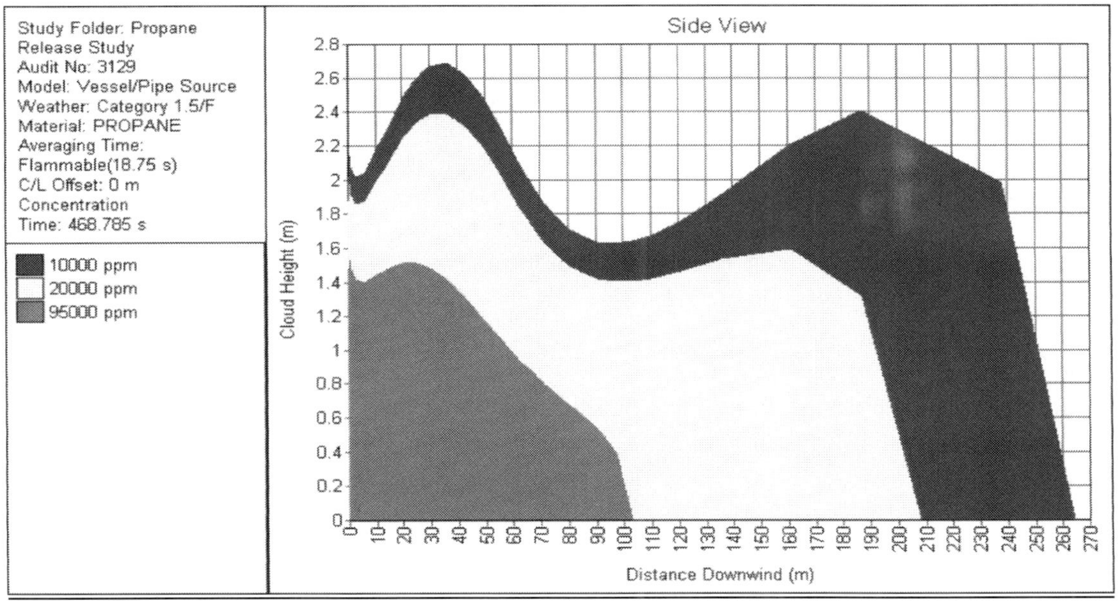

Even though the 10 min release or full line break scenario is very unlikely, it does provide a maximum impact footprint and characterizes the behavior of the gas cloud. For heavier than air LPG (Propane in this case), the cloud is a wide spread, low elevation dispersion. Natural gas, being lighter than air, will be a buoyant cloud that will often tend to "bend" and stay close to the ground for a longer distance downwind as the wind speed increases. More credible scenarios than these cases, such as loading hose failure or pipeline leakage cases, could then be run and their potential for entering congested or confined areas evaluated based on the building or process facilities that the cloud could reach. These gas dispersion analyses can also be used to determine siting issues, mitigation actions, and the need to shut down building intakes in the event of a release.

For high pressure natural gas lines (200 psig and higher), the actual pressure release- even if ignition does not occur during a rupture - can produce dramatic effects itself. The rupture of a 2' diameter, 850 psig gas line in an open field area outside of Union, Michigan in December 2006 created a crater about 60 feet in diameter and 12 feet deep. Ignition did not occur, but the trencher operator was killed due to asphyxiation by the gas cloud.

Another major concern may be releases or leaks in a confined or congested area. This concern may especially be the case for natural gas lines which often enter building areas. The worst case consequence analysis for this scenario is to determine the impact of filling a defined volume to the stoichiometric concentration with the flammable gas and utilizing a proprietary software program, like VCEC[4,5] (Vapor Cloud Explosion Contours), to determine the overpressures and impulses that could be generated upon ignition. The calculated overpressures and impulse curves will depend on characteristics of the affected area, such as degree of confinement and amount of congestion. The second type of analysis that can be done for a release in a fully confined area is to calculate the amount of gas that must be present in a small volume of a room (usually 10%) at stoichiometric concentration which will produce significant room damage should ignition occur. This analysis is based on ideal gas laws and the change in room air temperature following ignition of the vapor cloud. Assuming room volume remains fixed, the number of moles of gas remains essentially unchanged in the combustion process, and the temperature rise during ignition is approximately seven times the initial temperature, a spreadsheet can be created to determine this amount of gas (see below). The room dimensions and temperature, molecular weight of the material being leaked, the stoichiometric concentration in air, and a specified overpressure that will cause damage to the room are the required inputs. The spreadsheet result indicates the total amount of flammable vapor that must be present to generate the overpressure. Knowing the room ventilation rate and the potential leak rate from the line, the need for additional ventilation or measures for reducing the likelihood of leakage can be determined.

Consequence Analysis: Vapor Explosion In a Room

Enter building dimensions on this form.

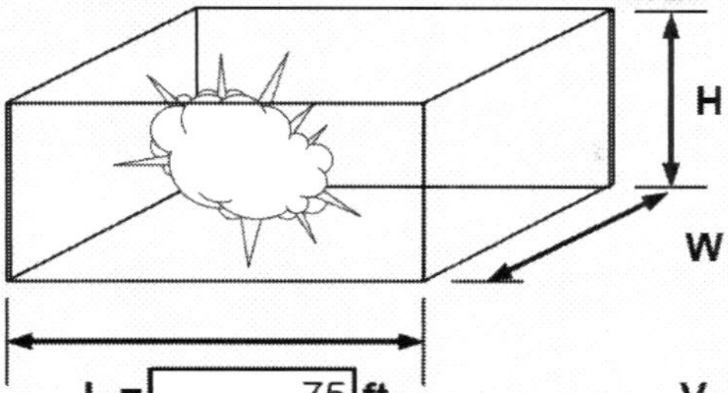

H = 15 **ft.**

W = 50 **ft.**

L = 75 **ft.**

V_b = 56250 **cu ft**

Inputs		
Chemical Evaluated:	Natural Gas	
Molecular Weight of Flammable Vapor =	16.04	**lbs/lb_{mole}**
Stiochiometric Concentration of Chemical (Cst) =	0.10	**volume fraction**
Room Temperature =	75	**°F**
Building Damage Threshold (□P) =	1.3	**psi**

Outputs		
Expanded Volume of Burned Gasses (V_E) =	**61033**	**cu ft**
Volume of Cloud Before Combustion (V_c) =	**683.31**	**cu ft**
Volume of Vapor in Flammable Cloud =	**64.91**	**cu ft**
Minimum Release to Create Damage =	**2.666**	**lbs**

Cryogenic gas systems should also be analyzed using standard dispersion analysis tools such as TRACE® and PHAST®, just as is done for process materials. One scenario to model would be the 10 minute release condition from a Liquid Nitrogen storage tank or a full line break line to determine the maximum size cloud that could be generated and result in possible asphyxiation of personnel. The cloud dispersion should be reviewed for potential to reach building ventilation intakes, just as is done for flammable or toxic releases. Shown below is a 10 minute release for a Liquid N2 tank containing 50,000 lbs of saturated liquid N2 at 140 psig. As noted for the LPG and propane concerns, more credible scenarios such as loading hose failure or pipeline leakage cases could be run and the potential for entering congested or confined areas evaluated based on the building or process facilities that the cloud could reach.

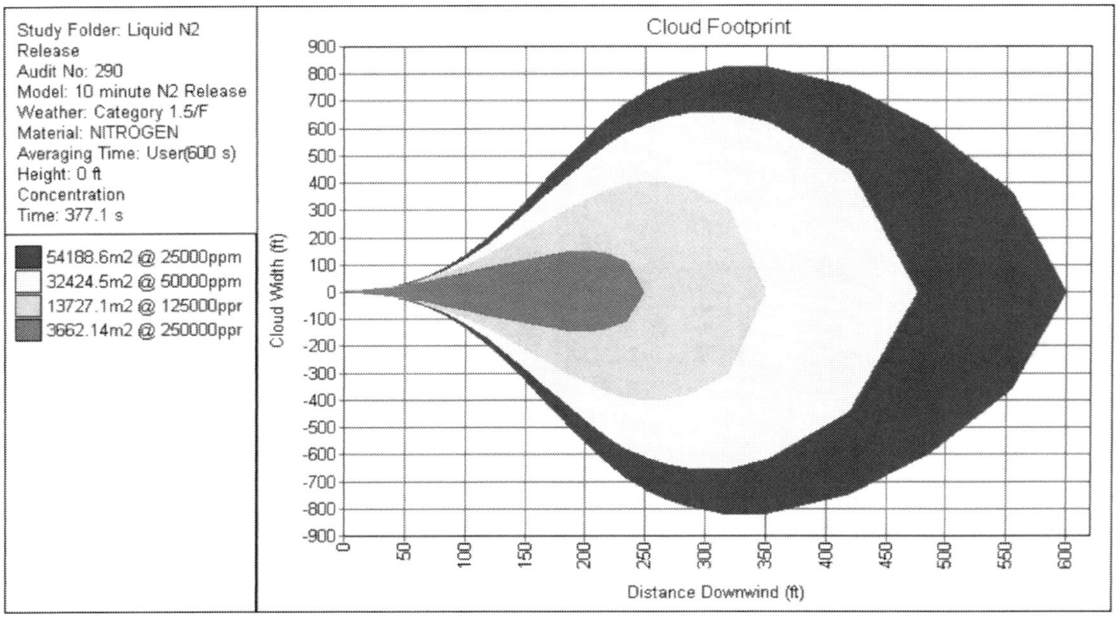

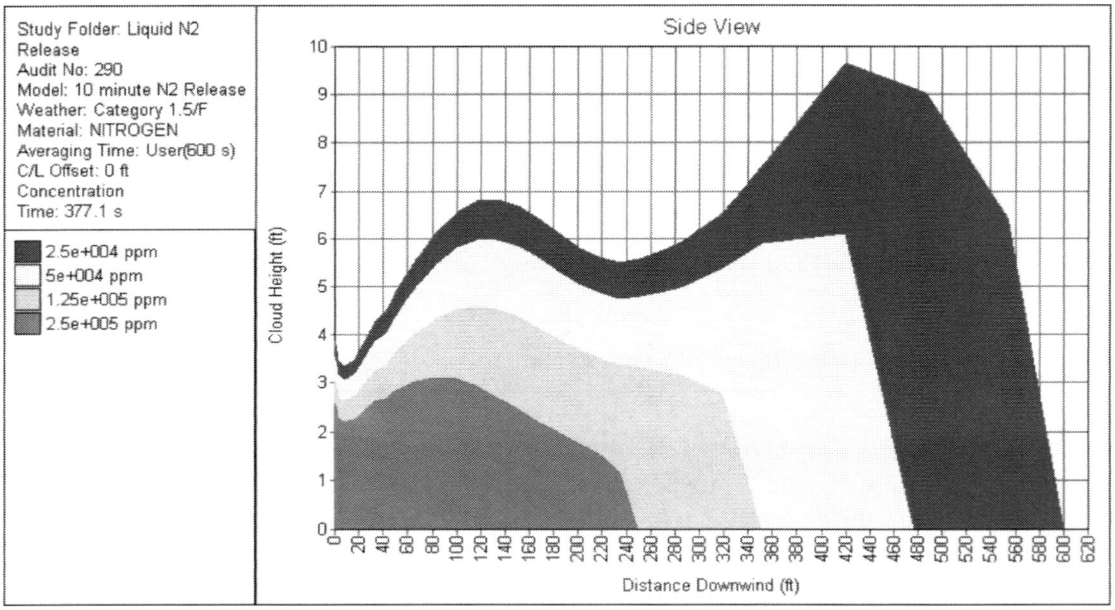

Finally, compressed gas systems need to be analyzed for the impact of overpressure scenarios. These analyses are straightforward vessel burst scenarios and can be made using information from Industrial Explosion Prevention and Protection Flash Fires, and BLEVES by Frank T. Bodurtha, 1980[6]. Using a simple spreadsheet, a series of volumes and pressures can be used to generate overpressure and impulse curves similar to those done in the BLEVE discussion above. Then based on the required gas demand for the site and the location of occupied buildings or critical equipment in the area, a storage tank can be chosen to provide an adequate supply for the users and then positioned in a manner to minimize facility siting impact.

5. Inherently Safer Design Concepts

The use of Consequence Analyses can provide the incentive for investigating a variety of inherently safe design concepts that could be applied in both new installations and existing facilities in a creative manner to mitigate or eliminate the potential for onsite impact for infrastructure systems. For new and existing installations, consideration should be given to issues such as:

1. Minimizing in-process inventory
   - Eliminating the number of or reducing the volume of storage vessels, especially LPG storage
   - Just in time delivery – minimizes inventory for LPG
2. Substitution of less hazardous materials
   - Combustible rather than flammable materials (boiler operations)
   - Use of lower hazard utilities
3. Limitation of effects
   - Storage at lowest pressures possible
   - Design and construct vessels to be strong enough to withstand the largest overpressure that could be generated within the system,
   - Units located to eliminate or minimize onsite impact to personnel or equipment and plant facilities
   - Infrastructure systems located to reduce or eliminate adverse effects from other adjacent infrastructure installations
4. Simplification of equipment design so that it is difficult to create a potentially hazardous situation due to an operating or maintenance error
   - Using passive vs. active controls (e.g., insulation, firewalls, etc.)
   - Using buried or shielded tanks, covered containment pits, etc.
   - Minimizing the number of flanges in flammable gas processes

6. Summary

Although process infrastructure systems are less prone to the type of process upsets and incidents that are characteristically the focus of consequence analysis, facility siting studies, and the application of inherently safer process (ISP) concepts, conducting qualitative consequence analyses for infrastructure systems can contribute vital information for facility siting studies; and safer and more robust infrastructure systems will be the result of applying ISP concepts to infrastructure system design and installation.

7. References

1. Center for Chemical Process Safety, **Guidelines for Evaluating the Characteristics of Vapor Cloud Explosions, Flash Fires, and BLEVEs**, American Institute of Chemical Engineers, New York, NY, 1994.
2. **Kees Van Wingerden, Olaf Roald Hansen, and Pierre Foisselon,** Prediction Blast Overpressures Caused by Vapor Cloud Explosions in the Vicinity of Control Rooms, **Process Safety Progress, Spring 1999 (Volume 18, No. 1)**

3. **Clutter, J. Keith**, et al, Modeling Environmental Effects in the Simulation of Explosion Events, **International Journal of Impact Engineering, March, 2006.**

4. **Baker, Quentin A., et. al,** Recent Developments in the Baker-Strehlow VCE Analysis Methodology, **Process Safety Progress, Winter 1998.**

5. **Pierorazio, Adrian J., et. al,** An Update to the Baker-Strehlow-Tang Vapor Cloud Explosion Prediction Methodology Flame Speed Table, **Process Safety Progress, March, 2005.**

6. **Frank T. Bodurtha,** Industrial Explosion Prevention and Protection, McGraw-Hill Book Company, 1980.

Fire and Explosion in an Explosives Conditioning Bunker

Timothy J. Myers
Exponent, Inc.
21 Strathmore Road
Natick, MA 01760
tmyers@exponent.com

Peter C. Hinze and Harri K. Kytömaa
Exponent, Inc.
21 Strathmore Road
Natick, MA 01760
phinze@exponent.com and hkytomaa@exponent.com

Prepared for Presentation at
American Institute of Chemical Engineers
2008 Spring National Meeting
42nd Annual Loss Prevention Symposium
New Orleans, LA
April 7-9, 2008

UNPUBLISHED

Fire and Explosion in an Explosives Conditioning Bunker

Timothy J. Myers
Exponent, Inc.
21 Strathmore Road
Natick, MA 01760
tmyers@exponent.com

Peter C. Hinze and Harri K. Kytömaa
Exponent, Inc.

Abstract

An explosion occurred in a munitions bunker that conditioned an explosive at an elevated temperature, completely destroying the bunker and damaging nearby bunkers. The explosive was used in the production of missile fuel. Investigators believed that the explosion occurred either due to contamination in the explosives or a fire in the bunker initiated by the heating system. Burn damage to combustible materials and witness statements indicated that a fire occurred in the bunker before the explosives detonated. Further investigation determined that the failure of the bunker's heating system initiated a fire that caused the explosives in the bunker to detonate. Information gathered during the investigation revealed that an improperly installed damper partially obstructed the flow of heated air to the bunker. The air heating system was not equipped with over-temperature protection, allowing extremely hot air to enter through ceiling vents near the combustible plywood ceiling of the bunker. The only temperature sensor for the bunker heating system was a thermostat mounted approximately four feet from the floor. A 1/4-scale model of the explosives bunker was constructed to evaluate the thermal air stratification process that occurs inside the bunker when the airflow is deficient. Test results indicated that during low flow rate conditions, hot air entering the bunker stratifies, allowing the wood ceiling to be heated sufficiently to ignite before the thermostat would turn off the heater. The reinforced concrete construction of the incident and adjacent bunkers prevented a chain reaction of explosions in adjacent bunkers.

This incident highlights the fact that careful design and installation is necessary for even relatively standard systems such as HVAC units when handling reactive materials. In this presentation the importance of proper safety interlocks in heating system controls will be discussed.

1. Background

1.1 Incident Description

A commercial explosive was being stored in several adjacent explosives bunkers shown in Figure 1. The bunkers were constructed of reinforced concrete and buried underground. On the day of the incident, witnesses reported seeing smoke exiting exhaust vents on the front of one bunker. The area was evacuated, and eventually the explosives in one bunker detonated, destroying the bunker as shown in Figure 2.

Figure 1: Pre-incident photograph of explosives bunker. Bunker on far right of photograph involved in incident.

No personnel were injured in the incident. Although other bunkers received significant structural damage, the facility design prevented sympathetic detonation of explosives in neighboring bunkers.

Figure 2: Aerial photograph of explosives bunkers after incident. Crater visible in former location of explosive bunker. Neighboring bunkers to the left sustained some damage, but explosives did not detonate.

1.2 Pre-explosion Contents

The configuration of materials present in the bunker was recreated based upon witness statements and inspection of neighboring bunkers and components and is shown in Figure 3 and Figure 4. The bunker interior was 34 ft long, 20 ft wide, and 10 ft, 9 in high (10.4 m x 6.1 m x 3.3 m). Explosive was stored in both fiberboard drums on pallets and in larger aluminum bins. A heating system provided conditioning air through two diffusers present on the centerline of the ceiling of the bunker, with one near the front door of the bunker and one near the rear of the bunker. Exhaust vents were located on either side of the front door of the bunker just above floor level.

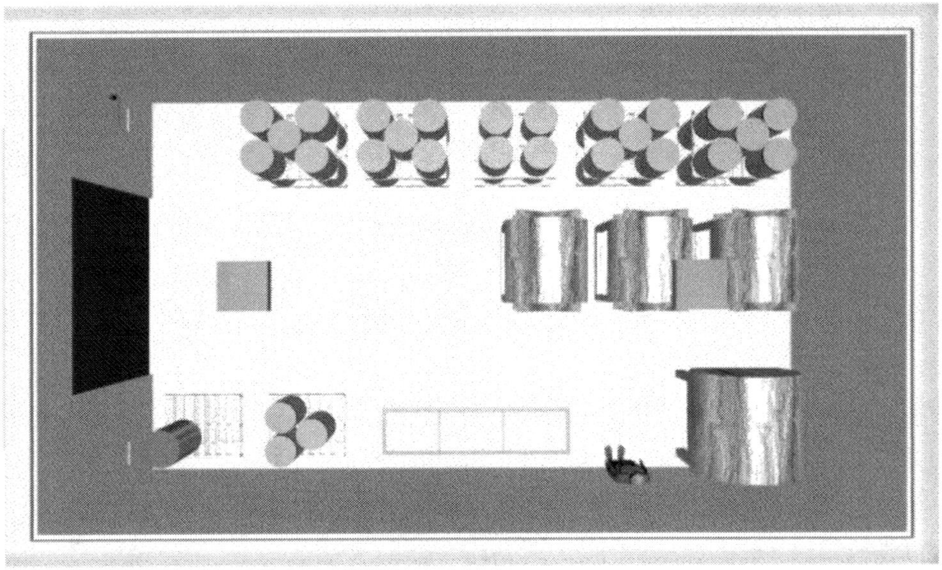

Figure 3: Plan view of bunker interior. Explosives were stored in both drums on pallets and in larger bins shown in the bottom right corner of photograph. Two diffusers located along the centerline of the bunker provided heated conditioning air. Exhaust vents and a door were present on the front of the bunker, shown in the left side of the figure.

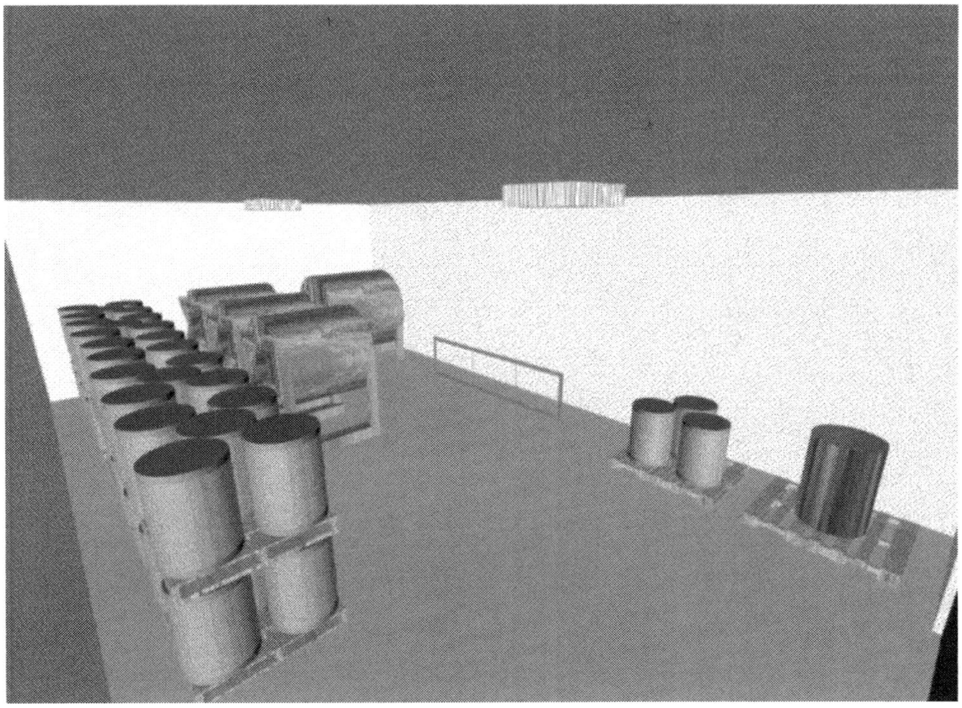

Figure 4: View looking from front corner towards back corner of explosives bunker. Two diffusers are visible on the bunker ceiling.

1.3 Initial Investigation

During the initial investigation, many potential causes of the incident were hypothesized. One hypothesis was that the explosive had become contaminated during processing, leading to self-heating and detonation of the explosive. A review of other materials that could have potentially contacted the explosive during handling revealed no incompatible materials. Additionally, witness statements and fire damage to the explosives bunker were consistent with a sustained fire in the bunker before detonation occurred. Witnesses had observed smoke coming from exhaust vents on the front of the bunker before the explosion. As shown in Figure 5, plywood that lined the interior of the concrete bunker walls was consumed by fire in some areas and deeply charred in other areas. The extremely short duration of the detonation could not have caused this fire damage to the bunker interior.

Figure 5: Photograph showing shattering of concrete bunker walls, and fire damage to plywood present on interior of bunker walls.

One potential cause of a fire in the bunker was a heating system that was used to maintain the bunker temperature at approximately 130°F. The remainder of this paper discusses analysis performed on the heating system to determine whether or not it was capable of causing a fire in the explosives bunker.

2. Bunker Heating System

2.1 Layout of Heating System

A schematic of the fresh air heating system for the bunker is shown in Figure 6. The heating system was located on the hillside above the explosives bunker. Fresh air was drawn through filters by a fan, then pushed through three electrical resistance heaters before entering the bunker through two diffuser boxes mounted on the ceiling of the bunker. The air exhausted the bunker through two exhaust ducts located on either side of the front door of the bunker, just above floor level. The system was designed to operate using a thermostat that would call for heat when the temperature dropped below 126°F at a temperature sensor located 4 feet above the ground on the front wall of the bunker. The system would cycle off when the temperature reached 140°F at that same location. Analysis performed after the incident indicated that the exhaust vents on the bunker were undersized and greatly restricted the flow through the heating system, decreasing the flow rate below the intended design value.

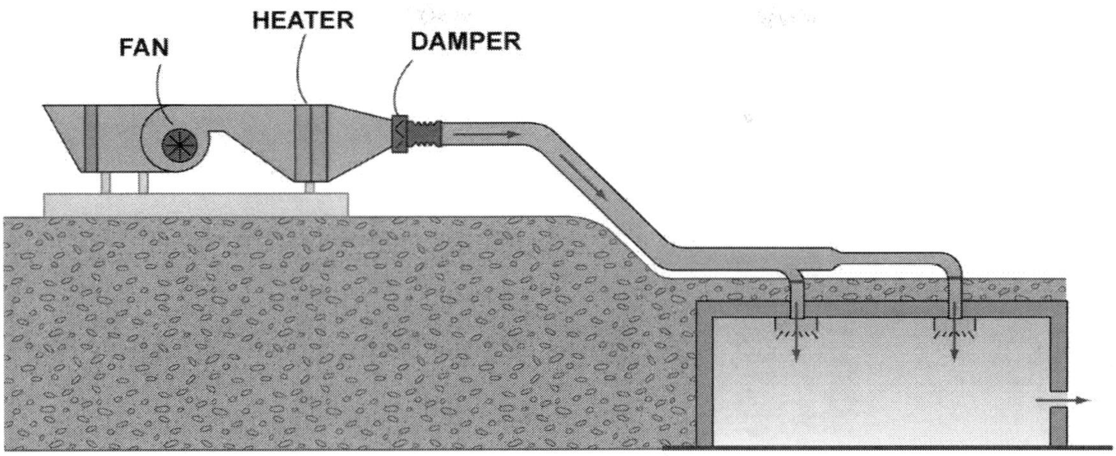

Figure 6: Schematic showing heating system for bunker. Heating system was located on hillside above explosives bunker and heated fresh outside air using electrical resistance heaters. This drawing is not to scale.

2.2 Modification of Heating System – Installation of Damper

During periods when the heating system was not operating, a chimney effect would occur. Driven by buoyancy, warm air would exit the bunker through the ceiling vents and continue out through the heating ducts, and cool air would enter the bunker through the exhaust vents near the floor level of the bunker. This would cool the bunker, causing much more frequent cycling of the heating system than intended.

To prevent this back flow of hot air through the heating system, the facility hired an outside contractor to install a damper downstream of the heating coils. The damper would open when the heating system was calling for heat and close when the system was satisfied, preventing the above chimney effect.

2.2.1 Observation of Damper and Actuator Post Incident

The motor actuator for the damper was found approximately 10% open after the incident. Although the damper was virtually destroyed by the explosion, pieces were recovered and reconstructed. It was discovered that the actuator had been misaligned during installation so that the damper linkage could become stuck in a mostly closed position.

2.3 Heating System as Cause of Fire

With the heating system damper stuck in a partially closed position, air flow rates through the heating system were well below design, causing abnormally hot inlet air temperatures into the bunker. Temperature as a function of heating system flow rate is shown in Figure 7, which demonstrates the theoretical high temperatures present at low flow rates. The installer of the heating system had failed to install a thermocouple downstream of the heating elements that could shutdown the system in the event of abnormal temperatures.

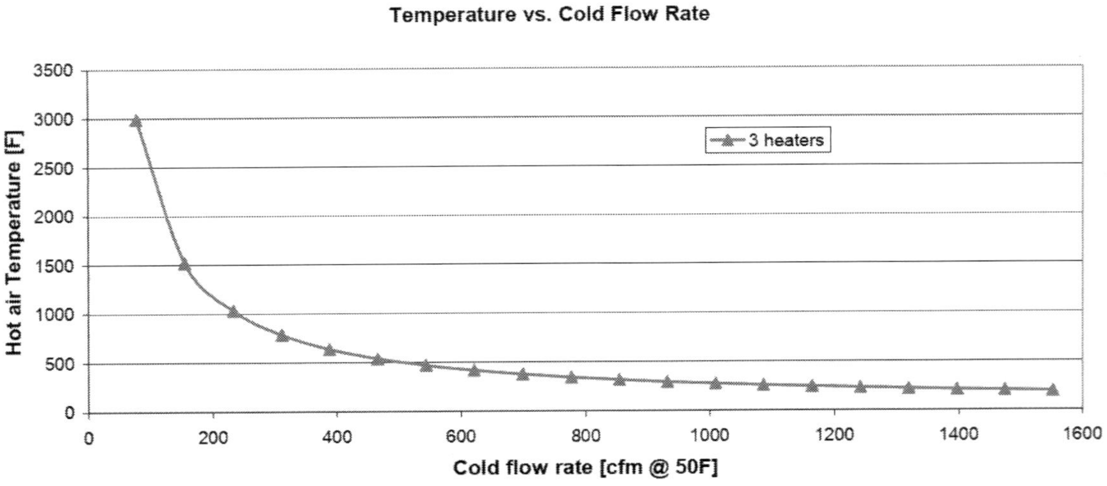

Figure 7: Temperature as a function of heating system flow rate.

If the air entering the bunker is sufficiently hot, it will tend to stably stratify in the bunker creating a layer of hot air against the ceiling over a layer of colder air below. The hot air could potentially heat the wood ceiling of the bunker above the ignition temperature of wood if the air were hot enough. The supply of hot air under this scenario would only go on for a finite duration. The thermostat in the bunker would shut down the heating system when the temperature at the thermostat on the front wall reached 140°F. A scale model of the bunker and heating system was created to determine the maximum temperatures that would occur near the ceiling of the bunker before the heating system thermostat would become satisfied and stop the call for heat. This testing is described in the next section.

3. Scale Model Testing

3.1 Dimensional Analysis

A ¼-scale model of the bunker was built and operated. This model was designed according to the scientifically accepted method of dimensional analysis and its application to model testing [1]. The governing physical equations of natural and forced convective flow and of heat transfer are the momentum equation and energy equation, respectively. Nondimensionalizing these governing equations yields three dimensionless parameters [2]:

$$\mathrm{Pr} = \frac{\nu}{\alpha}, \quad \mathrm{Fr} = \frac{\rho_{hot} u^2}{(\rho_{cold} - \rho_{hot})gL} = \frac{T_{cold} u^2}{(T_{hot} - T_{cold})gL}, \quad \mathrm{Re} = \frac{uL}{\nu}$$

The first parameter, Pr, is the Prandtl number and is the ratio of the momentum and thermal diffusivities of air, ν and α. The Prandtl number is a material property of air and varies by less than 1.5% over the temperature range 0°F to 1200°F. Thus, it is essentially identical between the model and full-scale bunkers at all temperatures as the model test was conducted with air.

The second parameter, Fr, is a modified Froude number and is the ratio of inertial forces of the hot air exiting the ceiling diffusers and gravitational buoyancy forces. This group depends on the air velocity, u, size of the bunker, L, gravitational acceleration, g, and hot and cold air densities, ρ_{hot} and ρ_{cold}, (or equivalently for an ideal gas at constant pressure, hot and cold air temperatures, T_{hot} and T_{cold}). This parameter captures the balance between the turbulent mixing created by the diffusers and the competing stabilizing effect associated with buoyancy. Since this experiment seeks to quantify the degree of stratification in the bunker, all variables in the model experiments were selected to match the modified Froude number of the model and the full-scale bunker.

The third parameter, Re, is the Reynolds number and is the ratio of inertia and viscous forces. This group depends on the air velocity of the air entering the bunker, u, the size of the bunker, L, and momentum diffusivity of air, ν. The Reynolds number of both the model and full-scale bunkers are large, indicating that inertia forces dominate viscous forces and the flow is turbulent.

In this type of scaling, one aims to match all dimensionless parameters, but that is not always possible. In this case, it was not possible to match the Reynolds while also matching the Froude number. The importance of the Reynolds number here is to ensure that flow regime out of the vents remains the same, namely turbulent. The geometry of fully turbulent jets exiting the diffusers into the bunker is independent of Reynolds number [3]. While the Reynolds number was smaller than in the actual application, it was maximized by operating the scale model at lower temperatures than the full scale model.

3.2.1 Scaling Test Variables to the ¼-Scale Model

The passage of time does not scale one-to-one between the model and the full-scale processes. The characteristic time to fill the bunker with hot air, τ, is proportional to the length of the bunker, L, divided by the average velocity of hot air flowing into the bunker, u:

$$\tau = L / u$$

The ratio of the fill time of the model, τ_m, to the fill time of the full-scale bunker, τ_{fs}, is given by:

$$\frac{\tau_m}{\tau_{fs}} = \frac{L_m u_{fs}}{L_{fs} u_m}$$

where the subscripts m and fs refer to the model and full-scale bunkers respectively.

When the modified Froude numbers are the same in both the model and full-scale bunkers, the ratio of the average hot velocities is:

$$\frac{u_{fs}}{u_m} = \sqrt{\frac{L_{fs}(T_{fs,hot} - T_{fs,cold})T_{m,cold}}{L_m(T_{m,hot} - T_{m,cold})T_{fs,cold}}}$$

Substituting the ratio of velocities into the expression for the ratios of times shows that, under conditions of equal modified Froude numbers, the ratio of model time to full-scale time is given by:

$$\frac{\tau_m}{\tau_{fs}} = \sqrt{\frac{L_m(T_{fs,hot} - T_{fs,cold})T_{m,cold}}{L_{fs}(T_{m,hot} - T_{m,cold})T_{fs,cold}}}$$

For the ¼-scale model used in experiments, this equation reduces to:

$$\frac{\tau_m}{\tau_{fs}} = \frac{1}{2}\sqrt{\frac{(T_{fs,hot} - T_{fs,cold})T_{m,cold}}{(T_{m,hot} - T_{m,cold})T_{fs,cold}}}$$

The experiments are run with a given initial temperature, which is also designated as the cold temperature, T_{cold}. As a result of the reduced flow, the inlet temperature into the duct reaches a steady hot value, T_{hot}. During the course of a model experiment, temperatures rise in the bunker with time. These temperatures were measured in the model experiments and were scaled to equivalent full-scale temperatures. To this end, a dimensionless temperature difference for both the model and full-scale bunkers is used:

$$\frac{T - T_{cold}}{T_{hot} - T_{cold}}$$

Equating the dimensionless temperature differences of the model and full-scale bunkers yields the expression for the equivalent full-scale temperature in terms of the temperature in the model bunker:

$$T_{fs} = T_{fs,cold} + \frac{T_m - T_{m,cold}}{T_{m,hot} - T_{m,cold}}\left(T_{fs,hot} - T_{fs,cold}\right)$$

3.2 *Testing*

3.2.1 Scale Model Construction

Photographs of the ¼ scale model are shown in Figure 8 to Figure 11. The interior walls of the model were constructed of non-combustible 1" thick marinate (calcium silicate). Scale models of explosive drums, pallets, and bins were also placed in the interior of the bunker as shown in Figure 9 and Figure 10. A variable speed blower and inline electric resistance heaters were used to control the flow rate and temperature of air into the bunker to match the thermal response of the full scale heating system.

Temperatures within the model bunker were measured by 40 Omega K-Type thermocouples that were recorded by computer through 2 Fluke Hydra data loggers at 8-second time intervals. The locations of the 40 thermocouples included:

- Six thermocouple trees in the bunker interior spanning from the floor to the ceiling.
- Thermocouples on the ceiling of the diffuser box and the metal duct entering the diffuser box.
- Thermocouples mounted on the ceiling
- Thermocouples mounted on the front wall including the location of the bunker thermostat and exhaust air.

Figure 8: Exterior View of ¼ Scale Model

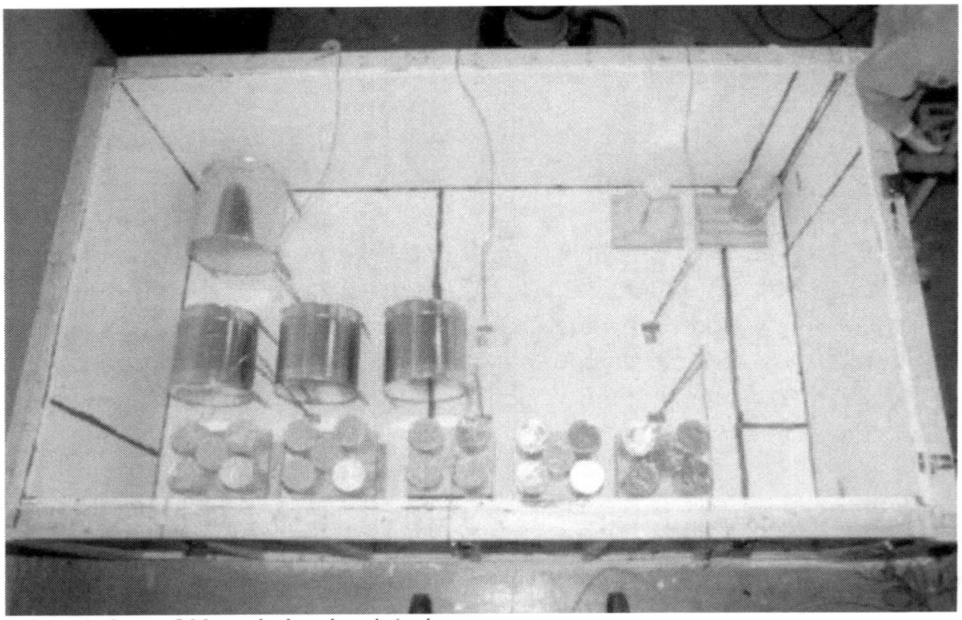

Figure 9: Overhead view of ¼ scale bunker interior.

Figure 10: Photograph of ¼ scale model interior looking towards front. Model ceiling is removed in photograph.

Figure 11: Photograph of ¼ scale model ceiling and hot air diffusers.

Test Results

Test results for two cases are shown below in Figure 12 to Figure 14. All results are shown transformed to the full-scale values. Data in Figure 12 corresponds to a condition where the air flow is approximately 20% of the nominal flow rate with the damper open, yielding an inlet temperature of 785°F. In this case the temperature at the location of the thermostat reaches a shutdown temperature of 140°F after approximately 1 hour. In the scale model the test was allowed to continue to run past this time. After 1 hour, the temperature of the exterior of the rear diffuser hot air duct, which was in contact with wood in the incident bunker was about 590°F. The ceiling temperature within the front diffuser box was about 525°F. Various sources provide a range of non-piloted or autoignition temperature of wood from about 375°F to 550°F [4,5], indicating that wood near the heating ducts in the bunker could ignite.

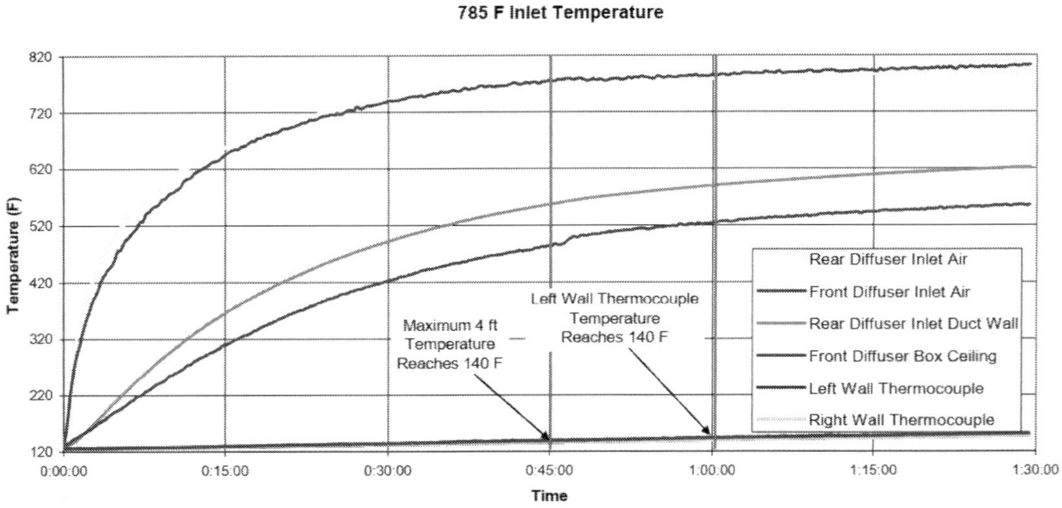

Figure 12: Transient temperatures with reduced airflow and 785°F inlet temperature.

Data in Figure 13 corresponds to an even lower airflow case with an inlet temperature of 1024°F. In this case the temperature at the location of the thermostat reaches a shutdown temperature of 140°F after 16 minutes. At 16 minutes the exterior temperature of the rear diffuser hot air duct was 685°F, and the temperature of the front diffuser box ceiling was 670°F. These temperatures are well above the non-piloted or autoignition temperature of wood.

Figure 14 shows the temperature profile from the floor to the ceiling in the center of the bunker for the 1024°F inlet temperature case. The temperature profiles show that stratification of hot air near the ceiling does occur. Notably the temperature of the ceiling is somewhat colder than the air several inches below it. This is due to heat loss through the ceiling and roof of the model. Analysis shows that the heat transfer loss through the ceiling of the scale model bunker is larger as a fraction of the total heating system power, than in the full-scale model.

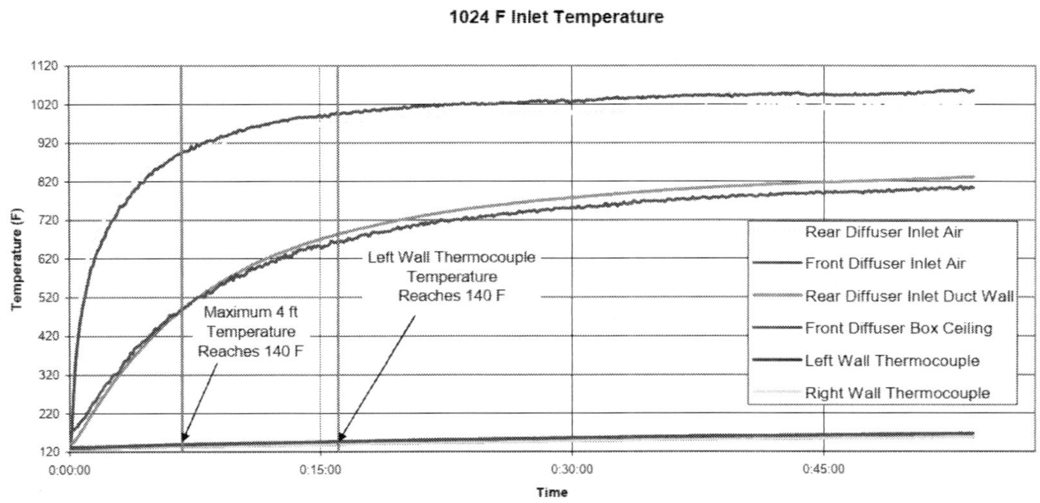

Figure 13: Transient temperatures with reduced airflow and 1024°F inlet temperature.

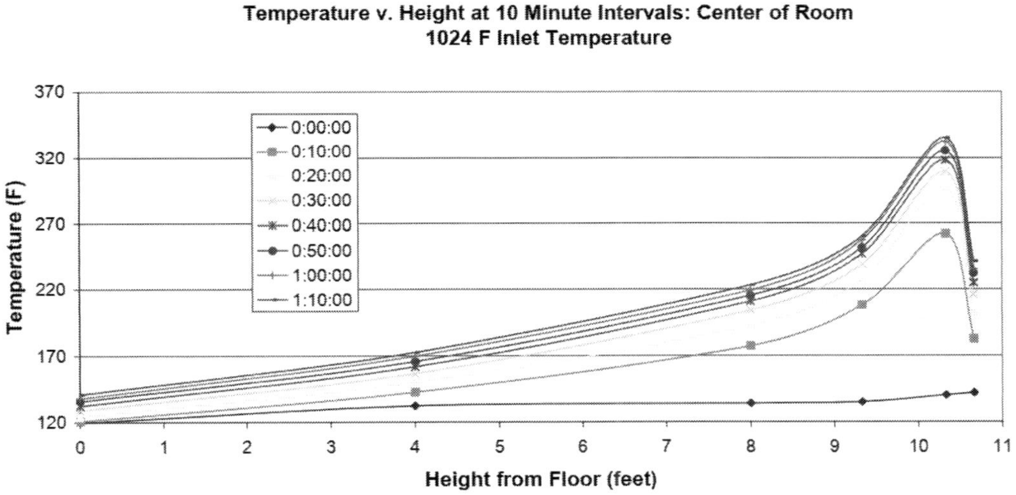

Figure 14: Temperature profile at center of room for 1024°F inlet temperature condition.

4. Lessons Learned

An explosive bunker was destroyed and neighboring bunkers were severely damaged due to a fire than occurred in the bunker. The fire was caused by a damper blocking airflow through a heating system, causing extremely high temperatures. Several specific lessons as well as more general lessons can be learned from this incident.

- Low flow through heating systems can cause excessively high temperatures.
- A single thermostat may have been sufficient to monitor the bunker temperature during normal operating conditions at high airflow rates and lower air temperatures. During upset conditions with low air flow rates and high air temperatures, however,

significant stratification occurred and a single temperature measured near the floor level of the bunker was not representative of the bunker temperature at higher elevations. Facilities should give careful consideration to the location and number of sensors for both normal and abnormal operating conditions.

- The reinforced concrete design and spacing of explosive bunkers was sufficient to prevent the sympathetic detonation of explosives in neighboring explosives bunkers.
- Blockage of airflow through a heating system caused a fire and ensuing detonation. When storing reactive chemicals or explosives, malfunction of even relatively standard equipment such as heating systems can cause significant incidents. Facilities should perform hazard analyses on these simple systems, much as they would for more complicated systems used in the production of reactive chemicals or explosives.
- A thermocouple that was supposed to cause shutdown of the heating system in the event of over temperatures was not properly connected. Facilities should test safety protection systems when equipment is originally installed, modified, and during the life of the equipment to ensure it is properly functioning.
- A damper was installed in the heating system to stop a chimney effect that drew hot air out of the bunker when the heating system was not functioning. This damper introduced a new failure mode into the system that was not identified. Facilities should consider performing hazard analyses whenever systems are modified.

5. References

[1] Avallone, E.A., Ed. and Baumeister II, T. Ed., Marks' Standard Handbook for Mechanical Engineers, Ninth Edition, McGraw-Hill Book Company, 1987.

[2] Incroprera, F.P., and DeWitt, D.P., Introduction to Heat Transfer, Second Edition, John Wiley & Sons, 1990.

[3] White, F.M., Viscous Fluid Flow, Second Edition, McGraw-Hill, Inc., 1991.

[4] O'Halloran, M.R., "Plywood In Hostile Environments, Physical Properties and Applications," APA, Report No. 132, December 1975.

[5] NFPA Fire Protection Handbook, Ninteenth Edition, NFPA, 2003, p. 8-36.

Expanding Known Process Safety and Risk Analysis Concepts to Manage Security Concerns

Mr. Donald W. Abrahamson—Celanese Chemical Corporation
Mr. Adrian L. Sepeda—A. L. Sepeda Consulting Inc.

A 10th PPSS Anniversary Paper

Abstract:

The generally accepted method for determining Risk by combining the probability that an incident will occur with the consequences of that occurrence can be upgraded and expanded to include security concerns with minor, yet significant, modifications in approach and avenues of analysis. For instance, when analyzing for process safety risk, the consequence and probability components of this equation are independent. But, when analyzing for security risks, they are interdependent. Another difference is in how Process Hazards Analysis (PHA) is used. While the basic PHA analytical components are essentially the same, the management approach (prevent and/or mitigate consequences) for an identified security hazard is different than the approach for managing a process hazard.

This paper identifies both the similarities and then the differences and needed upgrades in approach for analysis of security concerns versus process safety concerns. It focuses on the challenging aspects of security and helps the reader understand how to determine which upgrades are needed and how to populate the appropriate fields in those upgrades. It then frames the hazard assessment and risk management process with security as the prime focus. All this is done with familiar terms and well known, yet modified, hazard assessment and risk analysis procedures.

Introduction and Background

PHA Methodology and Process Overview

Process Hazards Analysis (PHA) techniques have been developed over time, proven in use and considered generally acceptable for analyzing the hazards associated with various types of processes and activities. Most of the more commonly used techniques are described in the CCPS publication "Guidelines for Hazard Evaluation Procedures, Second Edition." [1] Even though there are several acceptable PHA methodologies, they all contain the same basic components and end objectives—a drive to identify (and possibly quantify) hazards and potential paths toward management of those hazards.

Some PHA methodologies start the analysis with the resultant of failure. That is, they "look back" from the event you hope never happens and try to determine what sequence of events and actions led to that ultimate failure. Other PHA methodologies start at the opposite end of the sequence and list all events and actions that are anomalies from the original design intent. By following these to completion they might lead to the ultimate failure. Some PHA methodologies use a very structured approach to analyze for hazards. For instance, the Hazard and Operability (HAZOP) analysis approach asks specific questions built by combining "guidewords" such as high, low, etc. with specific operating parameters such as pressure, temperature, flow, etc. The resultant question, "What happens if the pressure is too low?" is than addressed by the PHA team to determine if it presents a hazard that should be addressed and to reach an understanding of the safety systems in place or needed. Other PHA methodologies are less structured in developing questions for the PHA team to consider. The "What-if" methodology, for instance, relies on creative thoughts by an appropriate PHA team to generate questions about a specific portion of the process. And, there are methodologies that graphically depict and guide the analysis process. Fault Tree, Event Tree, and Cause-Consequence are good examples of this type technique.

Regardless of where the analysis starts and how it progresses, the desired end result is an identification and definition of hazards and the structure of a plan for addressing those hazards. The purpose of that plan is to configure identified hazards into a format that can be prioritized and logically addressed. That plan often uses a risk consideration (ranking) approach to properly manage available resources for hazard/risk abatement. To convert a hazard identified in the PHA to a risk, two components must be developed (and quantified if reasonably possible)—"Consequence(s)" of failure and "Probability" that the identified hazard will result in that failure. When this is known, risks generated from different PHA findings or from within the same PHA can be compared and in some way ranked as to priority.

Based on the equation that Risk is related to a combination of Consequence and Probability, there are three approaches to risk management. One approach is to develop ways of reducing the Probability that the hazard will manifest itself as some type of failure. This approach is the Preventative approach in that its goal is to prevent (or at least make the likelihood so remote as to be tolerable) the failure scenario. Another approach is to develop ways of reducing the Consequence(s) of failure to tolerable levels. This is the Mitigative approach and attempts to reduce the consequences of failure. The third approach is a combination of both the Preventative and Mitigative approaches and is the more likely cost effective route. Regardless of the risk management approach and methodology, potential risk reduction efforts are generally called "safeguards" and are often compiled into "layers of protection." Sometimes they are analyzed for effectiveness singularly, but it is more appropriate to consider the entire suite of safeguards and layers of protection.

SVA Methodology and Process Overview

Security Vulnerability Analysis (SVA) is a process for identifying potential weaknesses in a facility that an adversary could exploit to cause damage and inflect fear in a community. An adversary is looking for targets with the greatest community impact (i.e. highest consequences) and the fewest countermeasures (i.e. safeguards) to circumvent. The SVA process is qualitative not quantitative, which challenges us to risk rank and compare security risks from facility to facility or, even more difficult, from company to company. To properly and efficiently utilize available resources, it is important to have a documented and well thought out process for measuring and comparing risks. A greater risk associated with a weakness can justify more resources to mitigate that weakness.

There are several types of SVAs. This writer utilized the method as described in the CCPS book "Guidelines for Analyzing and Managing the Security Vulnerabilities of Fixed Chemical Sites." [(2)] The SVA process is an effort to identify and then prevent a facility security weakness from being breached, which would allow an adversary to gain access to a chemical facility to cause an intentional release of a hazardous chemical. The SVA process is a systematic approach to identify security weaknesses. It is not the intention of this paper to go into the details of the SVA process. The CCPS book is an excellent example of how to perform a SVA. This paper suggests a method to compare SVA risks from facility to facility within a company so that the highest internal security risks can be recognized and appropriately prioritized.

The challenge with ranking of security risk is measuring and/or understanding the likelihood of the intentional action by an adversary. The likelihood or probability of an intentional action by an adversary is a function of the target attractiveness and the number of countermeasure to be circumvented. The risk of a successful attack by an adversary then becomes a function of the Target Attractiveness, Number and effectiveness of Countermeasures and the Consequence of the attack.

Similarities and Differences:

PHAs and SVAs start out with common goals—prevent a hazard from developing into an unwanted consequence. In fact, the initial analysis and the development of consequences are almost identical. The difference is in why failures resulting in (unwanted) consequences might occur and the relationship between consequence and probability.

In the conventional PHA analysis, failures are caused by unintentional situations or events. The resultant consequence and probability are independent—that is they have no common ground other than they are related to the same unintentional event. The consequence does not affect the probability that it will occur and the probability does not affect the consequence after it occurs. This allows three independent potential approaches to risk management—reduce the probability, reduce the consequence, or reduce both.

SVAs are different than conventional PHAs in that they address intentional situations or events! Because of the very nature of and reasons for SVAs, they do not have independence between probability and consequence. Because intention is the issue, probability and consequence are now interdependent, meaning that one is affected by and may be driven by the other. For instance, from a security standpoint, as the potential adverse consequences of an event increase, the attractiveness of the scenario to make that event occur may also increase. In other words, the likelihood that it will be the target of an attack may increase with the severity of the harm it could do.

The interdependence between consequence and probability is key to the development of security related risk management efforts and marks a difference in approach to addressing PHA findings and SVA findings. In the PHA world, risk related efforts usually result in adding safeguards and the development of layers of protection. For SVAs, risk related efforts usually result in the development of countermeasures to reduce the (apparent) attractiveness (including accessibility) of a target and/or ways to reduce the consequences of an intentional act all the while recognizing that the two are interrelated.

The Solution

Comparisons of PHA "Safeguards" to SVA "Countermeasures"

Over the years, PHA safeguards have developed in scope and sophistication. They have moved from single safeguards to layers of protection comprised of coordinated safeguards, often monitored and controlled by a computer system in the control room. Single safeguards still exist, a relief device for instance, but it is often monitored by some sensing device that sends a signal to the controlling computer system when it activates. We have also recognized that prevention is better than mitigation, but know that cost is a factor so a combination of both is often the most practical answer. The development of intertwined safeguards may be approaching the point of unmanageability unless interfaced with very sophisticated computer systems. Most safeguards have been in use for several years and are well understood by designers, maintenance personnel and operations. How they work and how they fail are common knowledge so there usually are no misunderstandings as to their goal, function or reliability.

On the other hand, SVA countermeasures are relatively new to the chemical industry. As such, the maturity and understanding level are not on par with PHA generated safeguards. Countermeasures are generally comprised of two distinct classes—1) those that come from or are closely related to PHA generated safeguards and, 2) those that are very specific to the prevention and/or mitigation of an intentional act.

Most PHA consequence related safeguards and layers of protection can be considered for use as a SVA countermeasure. PHA consequence safeguards function to mitigate the consequence(s) of an (intentional or unintentional) event. They work because they are independent of the initiating event. Therefore, the valid PHA consequence safeguards we have in place now are also valid SVA countermeasures.

However, the PHA generated preventative safeguards may not translate to SVA effective countermeasures. Certainly, a portion of them may be applicable, but those hazards posed because of the interdependence of consequence and probability (target attractiveness) as well as those hazards posed by an adversary may not be adequately addressed. These considerations form a new class of hazards specific to intentional (security related) acts.

The countermeasures for these intentional security related hazards are new to the chemical industry and have not reached the level of maturity and understanding that PHA safeguards enjoy. Most of these countermeasures fall in the prevention category in that they are in place to restrict access to or reduce the probability that an intentional action can cause unwanted consequences. They, in fact, reduce the attractiveness of the target because they make the target difficult to access. Because of "security" concerns, they may have been designed and installed by a contractor without input from local/plant employees. Explanations of how the newly installed countermeasures are to be used, their intended function, failure modes, repair procedures, etc. may not have been given to those responsible for the countermeasures daily use. These countermeasures may also be of a completely different style than familiar operational safeguards. They may be more in the vein of physical barriers than computer controlled systems and may actually cause inconvenience and confusion to facility employees unless their intended purpose is explained. They often are developed without the input of the site's direct workers, leaving them to feel not important enough to participate in the protection of their own jobs and livelihoods.

Time has shown industry how to develop independent safeguards for PHA discovered hazards. The problem with common-cause failure has been recognized and addressed with publications such as the CCPS book, Layer of Protection Analysis, Simplified Process Risk Assessment. [3] We are not there yet with SVA related countermeasures—party because of the interdependence between target attractiveness and consequence, and partly because of lack of experience.

Path Forward and Cautions

Identify critical scenario

All scenarios potentially leading to an unwanted event are not equal. The goal is to identify those that have the highest risk of progressing the event, in other words, the critical scenarios. To do that, first determine the consequences. Those unwanted events with the highest/most devastating consequences should be analyzed to determine if the existing safeguards are adequate from a security standpoint. If they are not, additional countermeasures should be considered. The first objective is to reduce the probability that this scenario will occur, typically by reducing the attractiveness and/or accessibility of the target. The existing safeguards should be analyzed to see if they can be enhanced to reduce consequences and thereby reduce the attractiveness of the target. The next step in reducing the attractiveness of the target is to restrict access and put in other measures (such as additional visual monitoring) to reduce the probability that someone can get close enough to cause the intentional event.

An Approach to Ranking Security Risks

The industry has ranked PHA defined risks for years. This risk ranking allows the development of action plans to address the highest process risks first. The ranking systems generally vary from company to company based upon a company's risk tolerance criteria. Risk ranking of security concerns also allows the prioritization of security risks. The CCPS book "Guidelines for Analyzing and Managing the Security Vulnerabilities of Fixed Chemical Sites"(2) provides a risk ranking example. This approach takes that example and attempts to more "quantitatively" integrate target attractiveness into the risk ranking process. This addition allows the likelihood to increase or decrease as the target attractiveness increases or decreases with a little less subjectivity.

Application of this approach requires information from a facility's SVA and it requires the development of risk ranking criteria. Necessary information from the SVA would include the Target Attractiveness (Attachment 5 – CCPS® Security Vulnerability Analysis Target Attractiveness Form) and Scenario Worksheet Form (Attachment 16 – CCPS® Security Vulnerability Analysis – Scenario Worksheet Form). Development of additional company specific risk ranking criteria is necessary and examples can be found in Tables 1, 2, 3 and 4, included below. These tables are for illustrative purposes only and would require further development based upon company specific criteria before being used at any specific company or facility.

Table 1 "Consequence Levels (1 - 5)" contains an example of a consequence table with consequence level ranking from 5 (highest severity) to 1 (lowest severity). There are several orders of magnitude difference between this consequence table and the typical consequence table used for risk ranking of PHA concerns. Most PHAs are trying to identify and mitigate hazards that could potentially impact few people. However, a malicious attack is intended to impact large numbers of people and the consequence table is scaled to reflect this intent.

In Table 2 "Example Likelihood Ranking Levels (1 - 5)" the likelihood estimation is impacted by the target attractiveness. Most PHAs would apply some type of layers of protection or fault tree analysis to determine the likelihood. Since a malicious event is intentional, it is difficult to apply the traditional layers of protection and there is little data available for the development of any type of fault tree. Additionally, the likelihood of the malicious event increases with the target attractiveness while the likelihood decreases with additional countermeasures. Target Attractiveness is split into five classifications from very high (51 to 60) to very low (15 to 23). Each of these five classifications has a likelihood ranking from 5 to 1 depending upon the number of existing countermeasures. Therefore, this table considers twenty-five different combinations of target attractiveness and countermeasure levels to generate a likelihood value from 5 to 1.

Table 3 "Example Risk Matrix" should look familiar to anyone who has done risk ranking of PHA concerns. It is a 5 X 5 matrix with severity on the "X" axis and Likelihood on the "Y" axis. This Matrix could be the same one that is used for PHAs.

Table 4 "Example Security Countermeasures List" is the last example. This table would be developed by a security expert and be provided to the SVA team. The SVA team would use this list to identify the existing countermeasures currently in place to mitigate a malicious attack. This list could also be used by the SVA team to identify recommendations to further reduce the likelihood of a malicious attack.

Consider as a hypothetical example, a facility that is manufacturing a toxic liquefied gas. The liquid gas is stored in a 200 ton pressure vessel until it is consumed on site or sent to a customer. The facility is in close proximity to a city with a population of 50,000. This would be a 4 consequence level according to Table 4 if the SVA team identified a credible malicious event.

The SVA team has raised the concern of an adversary gaining access to and breaching the storage tank resulting in total loss of containment. This liquefied gas has sufficient energy to evaporate 100 tons before auto-refrigeration slows the evaporation process. From Attachment 5 – CCPS® Security Vulnerability Analysis Target Attractiveness Form, the SVA Team has estimated a Target Attractiveness of 30, which is a low attractiveness according to Table 2. From Table 4 the SVA Team has determined there are 2 existing commeasures to prevent this malicious event. Based upon an attractiveness of 30 and 2 countermeasures, the likelihood for this scenario would be 5. The risk ranking value for a 4 Consequence and a 5 Likelihood is 2. The company has determined that a 2 security risk is not manageable. The SVA team has recommended the addition of 5 countermeasures to change the risk value to 7, which is consider a manageable level. Table 5 provides a summary of this example.

Table 1
Consequence Levels (1 - 5)

Severity Ranking	Example Consequence Level Description (This would be company specific)
5. (Highest)	**On Site or Offsite Impact:** ▪ **Toxic scenario impacting a population over 100,000** ▪ **Flammable/explosive scenario impacting a population over 10,000** **Environmental Impact:** ▪ **Extensive environmental impairment on and/or offsite (extensive clean-up).** **Financial Impacts:** ▪ **Extensive off-site property damage of more than $50 million** ▪ **Onsite property damage and/or loss of production revenue of more than $100 million**
4.	**On Site or Offsite Impact:** ▪ **Toxic scenario impacting a population of 10,000 to 100,000** ▪ **Flammable/explosive scenario impacting a population of 1,000 to 10,000** **Environmental Impact:** ▪ **Serious environmental impairment on and/or offsite—clean up lasting more than one year** **Financial Impacts:** ▪ **Onsite property damage or loss of production revenue of $25 million to $100 million** ▪ **Moderate offsite property damage of $5 million to $50 million**
3.	**On Site or Offsite Impact:** ▪ **Toxic scenario impacting a population of 1,000 to 10,000** ▪ **Flammable/explosive scenario impacting a population of 500 to 1,000** **Environmental Impact:** ▪ **Adverse environmental impairment on and/or offsite—clean up lasting less than one year** **Financial Impacts:** ▪ **Onsite property damage or loss of production revenue of $10 million to $25 million** ▪ **Offsite property damage of $1 million to $5 million**
2.	**On Site or Offsite Impact:** ▪ **Toxic scenario impacting a population of 100 to 1,000** ▪ **Flammable/explosive scenario impacting a population of 100 to 500** **Environmental Impact:** ▪ **Moderate environmental impact on and/or offsite—clean up lasting less than 30 days** **Financial Impacts:** ▪ **Damage or loss of production revenue of $1 million up to $10 million** ▪ **Offsite property damage of $25,000 to $1 million**
1. (Lowest)	**On Site or Offsite Impact:** ▪ **Toxic, Flammable or Explosive Scenario impacting a population of > 100** **Environmental Impact:** ▪ **Minor to no environmental impact on and/or offsite** **Financial Impacts:** ▪ **Onsite property damage or loss of production revenue of less than $1 Million** ▪ **Minor offsite property damage of less than $25,000**

Table 2
Example Likelihood Ranking Levels (1 - 5)

Target Attractiveness Rating (From CCPS SVA Attachment 5)	Available Countermeasures [1]	Likelihood Ranking
Very High 51 - 60	7 - 8	5
51 - 60	9 -10	4
51 - 60	12 - 13	3
51 - 60	14 - 15	2
51 - 60	16 - 17	1
High 42 - 50	5 - 6	5
42 - 50	7 - 8	4
42 - 50	9 -10	3
42 - 50	12 - 13	2
42 - 50	14 - 15	1
Medium 33 - 41	3 - 4	5
33 - 41	5 - 6	4
33 - 41	7 - 8	3
33 - 41	9 -10	2
33 - 41	12 - 13	1
Low 24 - 32	1 - 2	5
24 - 32	3 - 4	4
24 - 32	5 - 6	3
24 - 32	7 - 8	2
24 - 32	9 -10	1
Very Low 15 - 23	0	5
15 - 23	1 - 2	4
15 - 23	3 - 4	3
15 - 23	5 - 6	2
15 - 23	7 - 8	1

[1] This would be based upon a company specific list of countermeasures. It is understood that some countermeasures are more robust than others; subsequently it is important to specify the types of countermeasure that would be used as a credit towards reducing the likelihood. Examples of some types of countermeasures can be found in Table 4.

Table 3
Example Risk Matrix

		Highest	**Severity Ranking** (SVA Consequence Levels)			Lowest
		5	4	3	2	1
Highest	5	1	2	3	4	5
	4	2	4	6	7	8
Likelihood Ranking	3	3	6	7	8	9
	2	4	7	8	9	10
Lowest	1	5	8	9	10	10

1 is the highest risk to the facility/company and **10 is the lowest risk** to the facility/company. A company would need to determine an appropriate level of manageable risk.

Table 4
Example Security Countermeasures List

	Example Countermeasures in Alphabetical order [2]	Countermeasure reaction to adversary			
		Deter	Detect	Delay	Respond
1	Access control at risk areas			X	
2	Armed security / police	X			X
3	Barge/ship delivery inspections	X	X		
4	Barricades / bollards	X		X	
5	Chain link fence w/ barbed wire	X		X	
6	Chemical sensors	X	X		
7	Controlled access points	X	X		
8	Criminal background checks	X	X		
9	Employee security awareness training	X	X		
10	Fence-associated sensors		X		
11	Interior Motion sensors		X		
12	Lighting	X	X		
13	Local law enforcement				X
14	Package inspections	X	X		
15	Perimeter barriers	X		X	
16	Private Property / No Trespassing signs	X			
17	Portable barriers			X	
18	Proximity sensors		X		
19	Rail car inspections	X	X		
20	Retractable gates	X		X	
21	Screen visitors and deliveries	X	X		
22	Security master key system			X	
23	Security patrols	X	X		X
24	Setbacks / clear zones		X		
25	Special storage of sensitive / hazardous materials	X		X	
26	Structural barriers			X	
27	Subject to search signs	X			
28	Trenches / moats / bushes / dikes / knee walls	X		X	
29	Turnstiles			X	
30	Vehicle inspections	X	X		
31	Video surveillance	X	X		
32	Visible I.D. badges	X	X		
33	Visitor escorts	X	X		
34	Weapons policy	X			

[2] This is not intended to be a comprehensive list of potential security countermeasure. A company security expert should define this list

Table 5
SVA Example Summary

Scenario	Attractiveness Value	Consequence	Countermeasures	Likelihood	Risk Value
Without mitigation	30	4	2	5	2
With Mitigation	30	4	5 additional (7 total)	2	7

References

(1) *Guidelines for Hazard Evaluation Procedures, Second Edition with Worked Examples*, Center for Chemical Process Safety, AIChE, New York 1992

(2) *Guidelines for Analyzing and Managing the Security Vulnerabilities of Fixed Chemical Sites*, Center for Chemical Process Safety, AIChE, New York 2002

(3) *Layer of Protection Analysis—Simplified Process Risk Assessment*, Center for Chemical Process Safety, AIChE, New York 2001

A Series of Small Changes

John C. Wincek, CSP
Process Safety Manager
Croda, Inc.
8 Croda Way
Mill Hall, PA 17751
Phone: 570-748-7796
j.wincek@crodausa.com

Prepared for Presentation at
American Institute of Chemical Engineers
2008 Spring National Meeting
42nd Annual Loss Prevention Symposium
New Orleans, LA
April 6-10, 2008

UNPUBLISHED

42nd Annual Loss Prevention Symposium
AIChE 2008 Spring National Meeting
New Orleans, LA
April 6-10, 2008

A Series of Small Changes

By John C. Wincek, CSP
Process Safety Manager
Croda, Inc.
8 Croda Way
Mill Hall, PA 17751
j.wincek@crodausa.com

Abstract

A 6,000-gallon, stirred tank batch reactor was equipped with a one-inch, piston-type sample valve at the lower margin of the straight sidewall. This sample valve was originally installed with a two-inch long pipe nipple in the discharge port of the sample valve. Over a period of several years, recommendations from near miss reports and risk assessments, as well as operator concern with valve plugging resulted in a series of small changes to the sampling setup. Changes included lengthening the discharge pipe nipple, addition of a hand valve with position indication, and installation of heat tracing. These changes, while individually harmless, collectively created a hazard that resulted in the ignition of material as it was being sampled from the reactor.

A 6,000-gallon, stirred tank batch reactor was equipped with a one-inch, piston-type sample valve at the lower margin of the straight sidewall, just above the joint with the bottom dished head. When closed, the piston in the sample valve sits flush with the inside wall of the vessel. This type of valve is necessary to prevent plugging of the valve when high melt point products are produced in the vessel. As originally installed, the sample valve had a short pipe nipple threaded into the discharge port of the sample valve *[Figure 1]*.

A near miss report was submitted, stating that the sample valve discharging at approximately face level of the employee operating the valve created a hazard. Due to the high temperature products that were being sampled, serious burns could result if the product splashed onto the employees face. To minimize the hazard, a one-inch diameter, nine-inches long pipe nipple was threaded into the sample valve outlet, to extend the sample discharge point to the chest level of the operator obtaining the sample. Shortly after installation of the longer pipe nipple, operators discovered the nipple would frequently plug when sampling high melt-point products. When this occurred, the operator would unscrew the pipe nipple from the valve, and either rod it out or replace it with a new pipe.

Several months after this change, during a risk assessment of the vessel, it was recommended that all hand valves on the reactor be fitted with position indicators. These position indicators would be interlocked to hazardous materials charging valves on the reactor to prevent charging flammable and/or toxic materials to the vessel whenever a hand valve was inadvertently left open. For example, if the position indicator on the atmospheric vent valve indicated the valve was open, the automated charge valves for flammable and/or toxic materials would not open. This effort included installing a position indicator on the sample valve. However, because a position indicator was difficult to install on the piston-type sample valve, it was instead decided to install a quarter-turn ball valve on the end of the nine-inch sample valve extension pipe. This ball valve would later be equipped with a position indicator, such that even if the sample valve itself were left open, the operator could be assured that the ball valve was closed before flammable/toxic materials were permitted to the charged. This modification was reviewed and approved using the facility's Management of Change system.

Figure 1: Original Installation

Soon after installing the ball valve and position indicator, (as shown in Figure 2), the operators realized that they could no longer remove the extension nipple, as the position indicator on the ball valve would first have to be removed by an electrician. Episodes of sample line plugging became more frequent due to the addition of the ball valve on the end of the sample extension. Furthermore, plugged sample lines were more difficult to clear, because they could no longer be removed for cleaning. Operators resorted to applying steam, using a hand-held steam hose, to the sample assembly in order to melt the material plugging the line.

A department manager observed an employee unplugging a sample valve using a steam hose, and decided that steam tracing the valve and extension piping would be preferable. A work order was written, and ¼" steam trace tubing, supplied with 150 degree C steam, was installed on the sample valve and extension pipe. The installation of steam tracing was not reviewed using the Management of Change system.

Within a month of the steam tracing installation, while preparing to obtain a sample from the reactor, an operator opened the ball valve on the end of the sampling line, and witnessed a small, pressurized discharge of vapor. He then opened the sample valve to draw a sample into an 8-oz. glass sample jar. Two to three seconds after flow from the sample valve began, a six-inch diameter fireball erupted from the end of the sample extension pipe. The flash fire lasted less than one second, and diminished to a small (1" long) flame. The operator blew out the flame and closed the sample valve.

The ensuing investigation revealed that operators routinely charge a solution of 20 % sodium chlorite in water through the sample valve. 60 minutes before the fire, sodium chlorite had been charged through the sample valve. After the charge, both the sample valve and ball valve had been closed to allow for charging of a hazardous material. The investigation team theorized that residual sodium chlorite solution, trapped between the sample valve and ball valve, was heated by the 150-degree C steam tracing, boiling/evaporating off the water and leaving behind concentrated sodium chlorite. When the operator opened the ball valve, the observed discharge of vapor was the water vapor from the chlorite solution. When the sample valve was subsequently opened, the organic product contacted the concentrated chlorite and ignited, as foreseen by a warning on the sodium chlorite MSDS.

The conditions that permitted this incident to occur developed over a one to two-year period. A perfectly safe sampling arrangement was modified, in part with the "blessing" of an approved MOC, to make the reactor system safer. The MOC reviewers, however, did not realize the modification would increase the frequency of sample line plugging, and that installation of the position indicator would make clearing a plugged sample line more difficult.

Figure 2: Configuration at the time of the incident

An alert department manager noted that the second modification (created by installation of the ball valve and position indicator) created the hazard of melting out a plugged line with a live, open-ended steam hose. The facility had experienced a number of thermal burns, sensitizing the manager to potential burn hazards. The manager asked a subordinate to write a work order to install steam tracing on the sample valve. Neither the manager nor the subordinate thought the change was "worthy" of an MOC. The maintenance mechanics that installed the tracing had been trained to identify work orders that should have an MOC, but did not (perhaps justifiably?) flag this work order as requiring one. While it is easy to blame the lack of a MOC review as the causes of the incident, the four MOC approvers interviewed as part of the incident investigation indicated they did not believe they would have recognized the hazards of the steam tracing in combination with sodium chlorite charging.

The incident investigation team initially considered calling for a more robust review of MOCs. However, they also noted that the hazards of concentrated sodium chlorite were not well known in the facility, and in fact were not noted in the chemical hazards section of the facility's procedures. Further, an incident had occurred in the facility four years prior involving sodium chlorite residue igniting a wooden pallet. This incident, and the hazards of concentrated sodium chlorite, had not been communicated to the employees as the investigation recommended. The investigation team felt that had an MOC been written for this change, and had the hazards of sodium chlorite been well known in the facility, the potential for this incident to occur would have been recognized.

The final corrective actions from the investigation included:

Water-flushing lines and pumps used for sodium chlorite charging.

Water-flushing lines and pumps used for hydrogen peroxide charging (an oxidizer with similar to sodium chlorite).

Investigation of practices used for charging materials, including:

1. Should the procedures specify what specific method/equipment/nozzle to use for each charge?
2. Should the sample valve be used for charging?
3. Are chemical incompatibilities considered when charging multiple raw materials through the same nozzle/lines?
4. Is it acceptable to charge materials to the liquid phase of a product?

Conduct employee training on the hazards of sodium chlorite.

Conduct retraining on the types of small changes that require MOC review.

Combustible Dust Explosion Case History

J. Kelly Thomas, A.J. Pierorazio, D.C. Kirby & M.T. Edel
Baker Engineering and Risk Consultants, Inc.
3330 Oakwell Ct., Suite 100
San Antonio, TX 78218-3024
(210) 824-5960
kthomas@BakerRisk.com

Prepared for Presentation at
American Institute of Chemical Engineers
2008 Spring National Meeting
42nd Annual Loss Prevention Symposium
New Orleans, LA
April 7-9, 2008

UNPUBLISHED

Combustible Dust Explosion Case History

J. Kelly Thomas, A.J. Pierorazio, D.C. Kirby & M.T. Edel
Baker Engineering and Risk Consultants, Inc.
3330 Oakwell Ct., Suite 100
San Antonio, TX 78218-3024
(210) 824-5960
kthomas@BakerRisk.com

Abstract

This paper deals with a combustible dust explosion that occurred in 2003. The explosion and subsequent fires resulted in multiple fatalities, widespread injuries and damage to a processing area of the facility. The majority of the damage associated with the explosions was caused by a secondary dust explosion which resulted from the widespread accumulation of hazardous levels of combustible dusts throughout the processing area. The primary combustible dust involved in the explosion was phenolic resin, although similar damage would have been caused by any typical combustible dust allowed to accumulate in a similar manner. Although a number of potential initiating events have been postulated, the most likely initiating event was a gas explosion inside an oven associated with the manufacturing operation.

This event clearly demonstrates that allowing combustible dust to accumulate to hazardous levels within a building, room or similar enclosure creates the risk of a dust explosion. The accumulation of hazardous levels of combustible dust within an enclosure coupled with a competent initiating event can result in a dust explosion. The hazards associated with combustible dusts are widely discussed in the literature and addressed by national codes and standards (e.g., NPFA 654). Nevertheless, combustible dust explosions, such as the one described in this paper, continue to occur. It is therefore important to continue to disseminate information regarding combustible dust explosion hazards and case history information which illustrates that nature of these hazards.

42nd Annual Loss Prevention Symposium
New Orleans, LA
April 7-9, 2008

Session 7: Case Histories and lessons Learned

The Importance of Reliable Inertisation and Grounding of Ignition Sensitive Powder Handling Systems

Bill Roper
Ciba China Ltd.
18/F Xin An Building
99 Tian Zhou Road
Caohejing Hi –Tech Park
Shanghai 200233, P.R. China
0086 21 2403 2028
William_leonard.roper@cibasc.com

Prepared for Presentation at
American Institute of Chemical Engineers
2008 Spring National Meeting
42nd Annual Loss Prevention Symposium
New Orleans, LA
April 7-9, 2008

UNPUBLISHED

1. Introduction

This "learning from" will show how ignoring a small defect on a magnetic sieve together with unreliable inerting lead to the eventual increase of oxygen in a powder handling system to the point that an ignition occurred, probably through an electrostatic discharge, resulting in a fire..

The purpose of this paper is:

- To encourage management to establish a system of simple but regular controls of specific safety equipment, using their operation staff, in so doing reduce the probability of incidents.
  To this end a Safety Maintenance and Control systematic (SAMACO) will be presented.
- To bring to management attention the importance of an effective inertisation system which will provide the necessary level of safety for the people and equipment
- To show the consequences of not following the company safety standards recommending the use of a structured incident investigation technique which resulted in an incomplete analysis and a higher level of remaining risk. This highlights managements responsibility to ensure the safety standards, which are there as a provision and a protection for the people, are used

2. The Incident details

Incident Description

At approximately 23:30 on 17th September 2005 an alarm sounded in the milling room on the 4th floor of the powder handling plant C2 and when an operator investigated he found black and melted product residues scattered around various piece of equipment in the room. He smelt the odor of burning but saw no fire. He found a blackened hole in the material compensator above the mill from which smoke had emitted.

2.1 The powder handling equipment being used

The incident happened on the equipment shown on the following diagram.

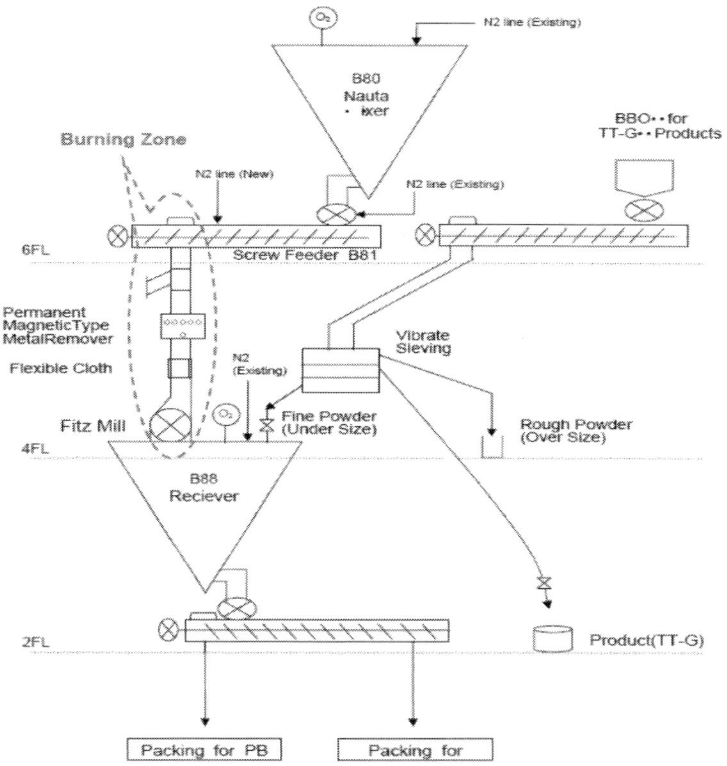

DIAGRAM 1

All the equipment is made of stainless steel apart from the flexible material compensator above the Fitz mill which is made of a nonconductive Poly Propylene

2.2 Description of the process

Product is transported using nitrogen gas as carrier into the Nauta mixer B80. Product from the Nauta mixer B80 is transported via gravity, rotary valve and screw feeder B81 through a permanent magnet metal remover to a Fitz mill. After milling the product passes into the receiver B88 and is ready for packaging.

At the time of the incident the equipment to the right (marked BBO for TTG products in the diagram 1 above) was not being used.

There is a gas balance line between B80 and B88 (not shown on diagram 1. The system is vented to the atmosphere via a dust filter mounted on B80 (not shown on diagram 1)

The atmosphere in this equipment is oxygen reduced to <6%v/v using nitrogen gas (introduced at the receivers B 80 and B88) and the oxygen concentration is monitored by on line analyzers whose sample ports are on receivers B 80 and B88. The oxygen analyzer detection principle is based upon "membrane barrier galvanic method".

2.3 Properties of the product

The product is an organic antioxidant additive used in the plastic industry. The following Fire and Explosion (FIREX) data for this product is known, see table 1.

Table 1

Characteristic	**Result**	**Comment**
Minimum Ignition Energy(MIE) (without induction)	<3mJ	Very low (electrostatically generated discharges could be an effective ignition source)
Powder particle size	<200 microns	Fine
Volume resistivity	>10^{14} ohm.m	Highly insulating
Solvent content	0.08 wt. %	≤0.5 wt. % considered solvent free
Melting point	120 Degrees C	
Burning Class @ 20 and 100 Degrees C	2	Not likely to sustain burning easily
Ignition temperature	410 Degrees C	

3. The incident investigation

Shortly after the incident a statement was made as to the probable causes of the incident and measures taken.
No root cause investigation technique was used an at that time the investigators recorded the following:

- Signs of a fire having taken place in the pipe line between the Fitz mill and the Nauta mixer B80 (see the dotted line on the Diagram 1). They took some photographs which show what they saw(see photo's)

Hole on the flexible cloth

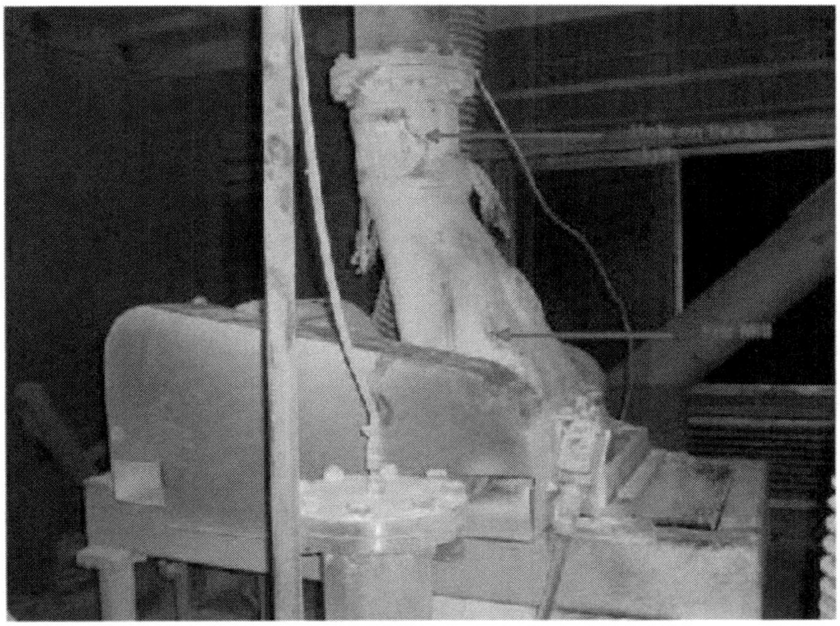

Black stuffs gushed from the hole on flexible cloth

Magnetic metal remover casing and black stuffs

- The maximum Oxygen concentration alarm for the measurement on line at receivers B 80 and B 88 which are set at 6% did not sound.
- The operator heard the Fitz mill alarm sound indicating a trip of the mill
- The Fitz mill had no visible scrape marks or colored spots on the metal indicating heating by friction.

Cutter for milling inside Fitz Mill
(We could not find any spark signs on this cutter)

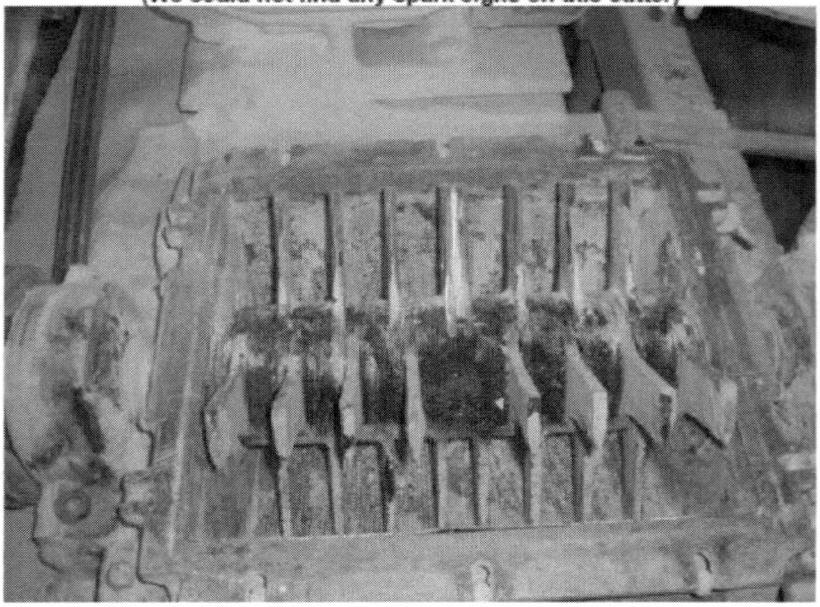

- The screw feeder B 81 did not have signs of friction visible

Black stuffs around screw feeder B81

- The operators had briefly opened the inspection cover on the screw feeder B 81 to check for powder flow
- The permanent magnet metal remover had 2 fasteners on the door of its casing and 1 of these was broken leaving a significant clearance between the door and the casing. This situation had been like this for some time

Black stuffs gushed from magnetic devices

- The resistance to ground for various pieces of equipment was checked and it was found that the magnet bars were insulated from the casing in which they were sitting. A measurement of 20 Mega ohms was recorded for the resistance of the magnet bars to ground.

3.2 The causes

The incident investigation isolated the following causes:

1. Air ingress into the system (primarily via the loose magnet door seal and secondly via the inspection cover opening to screw feeder B81) increased the oxygen concentration inside the magnet metal remover to a level >14%(1).
   - Management did not have a system in place to ensure changes(the loose magnet door seal) are corrected immediately

2. Electrostatic build up on a non grounded conductor (the magnet bars) due to the separation process of the powder falling between the screw feeder B81 and the Fitz mill.
   - Management did not have a system in place to ensure the integrity of the grounding of all conductive parts

3. Most probably a spark discharge between the insulated conductive magnet bars and their earthed casing ignited the ignitable dust cloud assumed to have a concentration >Lower Explosion Limit (LEL)

4. The Oxygen meters did not register the dangerously high oxygen concentration at the point where the ignition occurred (this cause was highlighted during the root cause analysis done ca. 2 years later in preparation for this paper)

3.3 The measures taken

1. The broken magnet door lid fastener was repaired to prevent air ingress. Inspection of the integrity of the magnet lid sealing system, using a Safety Maintenance and Control (SAMACO) program. If the inspection hatch is lifted on screw feeder B81 then nitrogen gas is purged in automatically at the rate of 100l/min to the screw feeder

2. Grounding of the bar magnet support rail through a grounding wire to earth.

3. Inspection of the integrity of the grounding regularly using a SAMACO program.
4. 2 additional Nitrogen purge points installed: 1 in the lid of the screw feeder B81 (see diagram 1) and the other in the down pipe leading to the magnet metal remover. Both points feed in 100 l/min Nitrogen gas when the Mill is running.

4. The Safety Maintenance and Control (SAMACO) systematic(2)

Equipment installed will only do its job if it works when required. One way to increase the probability of this happening is to continually control the conditions and the integrity of the used equipment and in doing so find any conditions that may cause failure before it actually happens.

Equipment failure prevention begins with the acknowledgment, by people, of the reality that equipment deteriorates over time and the conviction that their progressive action, from housekeeping to preventive maintenance, is the crucial factor for success in slowing down this inevitable wear process.

Safety Maintenance, Service and Control (SAMACO) is that part of preventive Maintenance which is especially concerned with simpler review, upkeep and inspection work not requiring special equipment and normally carried out by the production staff.

The SAMACO is part of preventive maintenance and is to be used for the early detection and elimination of failure, specifically for equipment installed to ensure safety.

4.1 Setting up a SAMACO Program

Implementing an effective SAMACO program should recognize the following 4 steps:

Step 1: Chose the crucial safety equipment you want to control and check

The first step is to set up a catalogue of all equipment and installations needing upkeep, inspection and function checks to ensure safety. Equipment identified from the risk analysis will be one of the inputs here. The compilation of such a catalogue requires a good and comprehensive knowledge of the plant with all its equipment, their function together and the consequences of malfunctions for safety.

In the catalogue for each piece of equipment the following is stated:

-The equipment unit number

And or

-The equipment description

Example of a page in the catalogue:

Equipment number	Name	control interval	Controller	The control instrucitions
R451 - X401	Inner temperature alarm	monthly	Operator	**> Set the temperature limit below the actual inner temperature**
R452 - X401				The visual temperature alarm blinks. Field 24 of the general alarm display lights up and horn sounds
R453 - X401				> Cancel the alarm at the vessel and at the lift
R545 - X401				The visual and audio alarms cease
	Agitator monitor alarm			**> Start the agitator**
R451 - X402			Operator	>Switch the agitator monitor on
R452 - X402				> Turn the agitator off
R453 - X402				The visual agitator alarm blinks. Field 24 of the general alarm display lights up and horn sounds
R454 - X402				> Turn the agitator monitor off
				The visual alarm at the vessel goes out
				> Cancel the alarm at the lift
				The general alarm display light goes out and the horn ceases

Step 2: Define what controls you want to do and how often

The second step involves choosing the exact control/inspection to be carried out on each piece of equipment and their frequency. It may help to ask the question, "how do we ensure that this piece of equipment will always be ready to work"? These controls and schedules will also be

prescribed from the various product risk analysis. Past experience about the equipment reliability, together with help from the equipment suppliers are resources that can be of value here.
In the catalogue for each piece of equipment the following is stated:

-The control/inspection instructions
-How often the control needs to be performed

Example of a control schedule

SAMACO programme		The Control Schedule																Month: Year:
		Week		1		Week		2		Week		3		Week		4		
Equipment number	The Control	Date	Sign	ok	not ok	Date	Sign	ok	not ok	Date	Sign	ok	not ok	Date	Sign	ok	not ok	Remarks
M451-X400	Potential equalisation check													25.1				
M452																		

Step 3: Choose who will do the controlling

The next step is to determine who is going to do the controlling, Controls and simple maintenance tasks can, with the respective training, be entrusted to the operating personnel of a plant. The plant personnel have an interest in the right functioning of the equipment under their control which make them the candidate of choice. Those jobs which cannot be done by plant personnel are delegated to the respective craftsman, some may need to be done by external firms.

In the catalogue for each piece of equipment the following is stated:

-The person/organization who carries out the required control or service

Step 4: Appoint a coordinator who will oversee the program

The next step is to determine who will administrate the control schedules. The program coordinator is responsible for the timely distribution of the control schedules and control instruction. He is also responsible for the filing of the completed control schedules.

4.2 Program effectiveness

Throughout the year additional equipment may be added to the Equipment Catalogue as new needs are identified and as new safety equipment is installed

It is recommended that at least1/year the plant management go over the past results with a view to determining how effective the program has been to reduce failure rates. Redefining inspection intervals, extending or shortening inspection instructions, and adding or subtracting equipment if necessary.

5. Ensuring an effective inertisation system

The following open question remains: Why did the Oxygen meters not warn of an increasing O_2 concentration in the system?

In reviewing the incident for this paper is became that this point had been neglected and the reasons could be a combination of the following:

- Pride- production and EHS thought they had it sewn up using quick and "dirty" methods!
- Production and EHS did not refer to the provision of the internal standard (Accident and Incident Reporting and Investigation)(3) requiring a structured investigation technique for such serious cases
- Misjudgment of the situation by the production and EHS staff

5.1 The hazard scenario:

In addressing the solution to the open point of an ineffective inertisation system the following hazard scenario can be stated:

The sealing of the equipment fails (an active failure)
AND
The Oxygen measurement system does not detect this change (passive failure) leading to the formation of an explosive air powder mixture in the system which is ignited causing an explosion injuring people.

<u>The RISK assessment</u>
Probability assumed to be Low
Severity assumed to be High

Possible reasons for the passive failure could include:

1. The oxygen measurement system is to slow to register changes in concentration
    a. The air sample point is in the wrong place eg. Next to the nitrogen inlet line
    b. The air sampling tube or its filter is blocked or partially blocked
    c. The air sampling tube from the sample point to the detector head is too long
    d. The oxygen measuring technique(electrolytic cell) is not optimal for the requirement
    e. The sample point geometry does not allow for a representative air sample to be taken
    f. Or combinations of the above

5.2. Verification of the above

At this point in time the above verification is under way but no results are known.

6. Summary

Simple controls and checks done regularly by the operation staff can greatly aid in ensuring the integrity of equipment as shown by the SAMACO example given

Ensuring compliance with company safety standards no matter how well they are written is a continual management responsibility which if neglected, as in this example, could lead to the same incident being repeated again

In this case management choose to accept the technical and organizational measures in place and take the responsibility for the residual risk until a reliable inertisation system can be found, implemented and verified.

7. References

[1] - Ciba Internal Guideline 16 "Safe Handling of Powders and Dusts", Edition 3 Revison 0 of 2003

[2] - Ciba Internal Guidance Note 11/6 "Safety, Maintenance, Service and Control", of October 1998

[3] - Ciba Internal Guideline 8," Accident and Incident Reporting and Investigation", of March 1998

Hydrochloric Tank Explosion

Chandrashekhar Chandwadkar,
National Petrochemical Company,
Nothern Sheikh Bahaei St.,
Mulla Sadra Ave.
Tehran, Iran.
c_chandwadkar@yahoo.com

Ghodaratollah Nasiri
National Petrochemical Company, Iran
nasir@nipc.net

Ali Seyfzadeh
National Petrochemical Company, Iran
hse@nipc.net

Prepared for Presentation at
American Institute of Chemical Engineers
2008 Spring National Meeting
42nd Annual Loss Prevention Symposium
New Orleans, LA
April 7-9, 2008

UNPUBLISHED

Hydrochloric Acid Tank Explosion

By
Chandrashekhar Chandwadkar,
National Petrochemical Company,
Nothern Sheikh Bahaei St.,
Mulla Sadra Ave.
Tehran, Iran.
c_chandwadkar@yahoo.com

Ghodaratollah Nasiri
National Petrochemical Company, Iran

Ali Seyfzadeh
National Petrochemical Company, Iran

ABSTRACT

Explosions in Hydrogen Chloride tanks are very rare. However, we have experienced one. The explosion was so powerful that the tank with a capacity 250 KL, 60% filled with 33% HCL, took off from the ground went up to 8-9 meters high in air and landed around 50 meters away. The contents of the tank sprayed all over the nearby area. Though the tank was surrounded by other plants and equipment, fortunately it landed in no-mans land! This averted loss of life and property that would otherwise might have resulted from a cascading effect.

1. Introduction

HCl tanks are generally considered to be safe from fire and explosion hazards; some times there are chances of contamination with flammables. Mainly this depends on the process being followed. Normally there is no problem if there is hydrogen and chlorine combustion and scrubbing with water (i.e. hydrochloric synthesis). However, when hydrochloric gas from the vinyl chloride monomer (VCM) process is used to produce HCl acid one has to be very careful to check the carry-over of flammables. The following case study describes an incident of an explosion in a HCl tank. In spite of nitrogen blanketing on the tank, some of the acts of commission and omission resulted in an explosion.

2. Narration of incident

It was a morning in late November at 6.40 am, near shift change time. Everyone was getting ready to handover the charge to the morning shift personnel and go home. The plant was running smoothly. One of the operators, who was changing his clothes, suddenly heard a loud bang. He just peeped through the window and to his surprise there was no HCl tank! After few moments again he heard loud thud, by this time he was out in open ground and saw a big metal structure in the near by area. It was the HCl tank that he was looking for! The explosion dislodged the HCl tank from its foundation. The tank, having 250 KL capacity, 60% filled with 33% HCl took off from ground went up around 8-9 meters high in air and landed around 50 meters away from its foundation, spraying contents all over near by area! Luckily it was shift change time and no one was outside in plant.

2.1 Photographs of the tank 603B after explosion & tank 603A after implosion:

Tank landed in no-man's-land

View from other side

Foundation of tank

Tank 603 A- Imploded

3. Brief history of earlier incidents:

There was history of minor incidents that should have provided adequate warning of the presence of flammables. In the year 1997 the transfer tank T-603 was emptied to perform welding. While performing welding on the tank a minor explosion was experienced. This was attributed to presence of hydrogen. After two years (i.e. in 1999) again in same tank T-603 another minor explosion was experienced, caused by hammering. This time the vent gas was analyzed and flammables such as acetylene, ethylene, etc were found. This led to the installation of nitrogen blanketing on all HCL tanks.

4. Routes of HCl Production:

4.1 Carbone Lorraine Process:

The SINTACLOR hydrochloric acid synthesis units continuously produce 37% concentrated aqueous solution of hydrochloric acid, from the combustion reaction of chlorine gas and hydrogen and subsequent absorption with de-mineralized water. , The hydrogen and chlorine combine in equal volumes according to an exothermal reaction to form gaseous hydrogen chloride. In practice 5% stoichiometric excess of hydrogen is employed to ensure the complete combustion of chlorine and avoid discharging chlorine waste into the atmosphere and to obtain pure acid.

The remainder of synthesis process involves:

Cooling of the gas formed.

Co-current absorption in the process water

Then cooling of the acid.

Like the combustion, these steps are highly exothermic, which implies the continuous use of cooling water to remove the heat generated by the system. Non absorbed gas is fed to Tail Tower (scrubber) and the weak acid formed is again fed to synthesis unit (Furnace + Absorber) to obtain pure acid.

4.2 HCl gas from VCM plant:

HCl gas produced with the pyrolysis of ethylene dichloride from the Vinyl Chloride plant is directly fed to chlorine line of hydrochloric synthesis unit as shown by the dotted line in fig. The synthesis unit is operated with the burner off and without supply of hydrogen.

4.3 Schematic for both the routes:

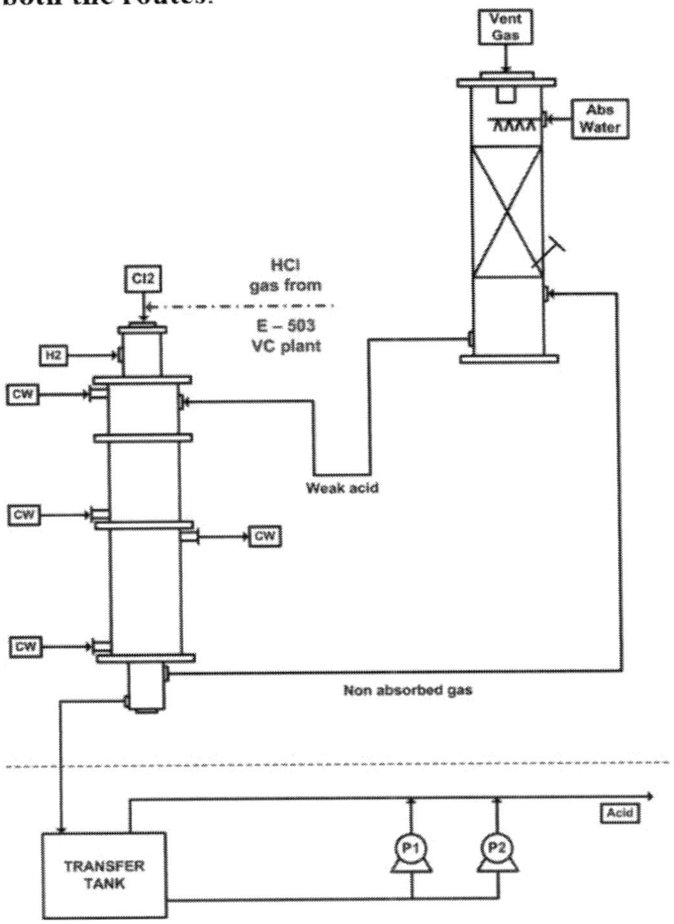

5. Observation & Analysis

- Acid production from SINTACLOR was in progress at the time of incident and just before the incident free chlorine was analyzed and found to be less than 10 ppm, so the possibility of a hydrogen and chlorine reaction is ruled out.
- Rubber lining of tank 605B was redone around 10 days back, so the possibility of hydrogen generation because of acid and tank body reaction is also ruled out.
- There was an interruption in production because of problem with the blower belt (25 Nov. 2006 at around 3 am). The plant was restarted at 2:30 pm on 25 Nov. 2006.
- HCl was being transferred to the tank at the time of the explosion and from the level chart it is clear that rate HCL removal for the five hours before the incident was high which can build static charge in rubber lined tank.
- The contents of the tank T-605 C was transferred in T-605 B. T-605 C contained HCl from the VCM plant gas. VCM plant was being run under capacity. C-502, the HCl Column in the VCM plant was also not functioning to its full capacity, as the plant was nearing annual shut down. This caused more flammable impurities in HCL gas in T-605 C.
- Black spots were observed on the inner surface of tank roof. Black spots imply that hydrocarbons were present.
- Earthing strips were found highly corroded & frailed. They may not be sufficient to take away the static charge produced in rubber lined tank.
- Earlier HCl tank vent gas analysis reports shows high traces of flammable in HCl gas from the VCM stream.
- Internal corrosion in nitrogen line discovered a few days before the implosion in tank T 605C partially/ completely blocked the nitrogen line.
- Common vent line of T-605 B & C was found open as Tank-C, damaged by implosion was being dismantled. This provided the ingress of air in tank, completing the fire triangle.

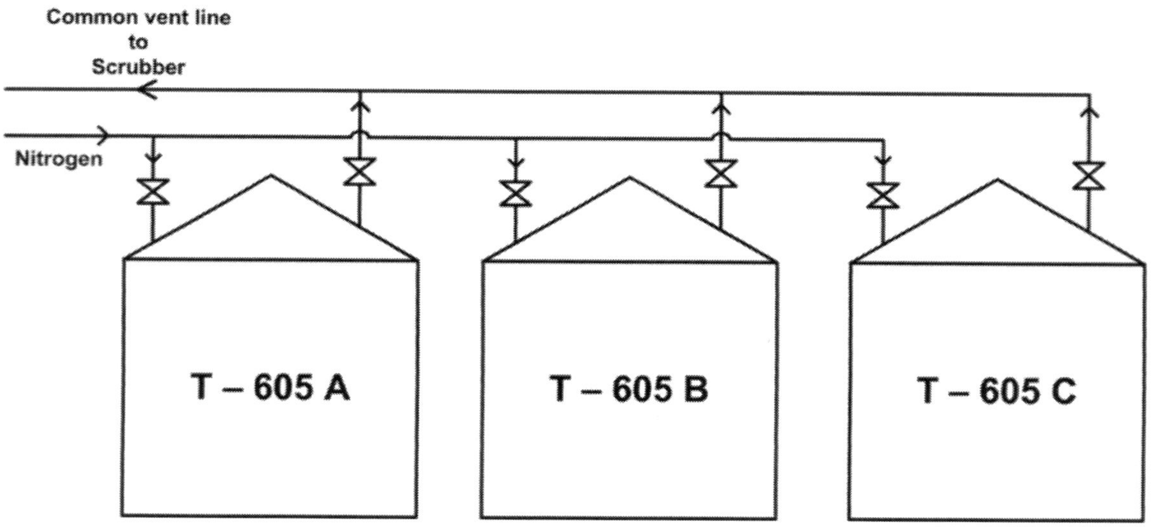

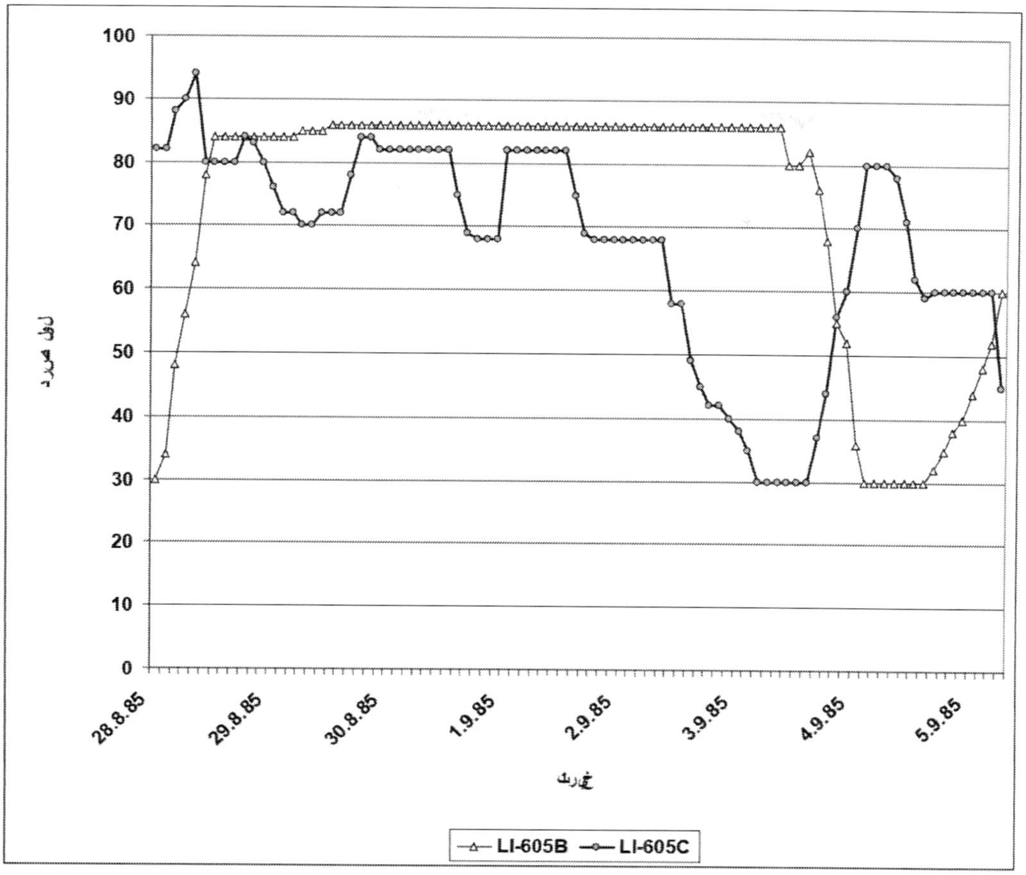

Tank Level Chart

6. Conclusion:

The carryover of flammable gases from the VCM plant HCl gas because of constrains in VCM plant, poor arrangements for static charge removal and the open vent line lead to the massive explosion that resulted in rocketing the 250 KL 60% full HCl tank around 9 meters high in air around 50 meters away from its foundation.

7. Recommendations:

- Limit of flow rate while drawing contents from tank.
- Install online sampling for flammables in VC plant HCl gas stream.
- Require regular sampling of tanks vent gas to find out concentration of flammables.
- Require nitrogen sampling for oxygen content in the tank head space.
- Provide efficient earthing for rubber lined tank.
- Change material of construction of nitrogen line 5 meters from the tanks of concern to FRP.
- While issuing work permit discuss and warn about the residual hazards, in this case the common vent line was opened and kept as it is.

8. References:

[1] Accident Investigation Report NPC.
[2] Documents from 'Carbone Lorraine' process.
[3] Process log sheets.

Synthron Runaway Reaction and Vapor Cloud Explosion

James Lay, PE
U.S. Chemical Safety and Hazard Investigation Board
2175 K Street NW, Suite 400
Washington, DC 20037
Jim.Lay@CSB.gov

Francisco Altamirano, CFEI
Johnnie Banks, CFEI
Jeffrey Wanko, PE, CSP
U.S. Chemical Safety and Hazard Investigation Board
2175 K Street NW, Suite 400
Washington, DC 20037

Prepared for Presentation at
American Institute of Chemical Engineers
2008 Spring National Meeting
42nd Annual Loss Prevention Symposium
New Orleans, LA
April 7-9, 2008

UNPUBLISHED

Synthron Runaway Reaction and Vapor Cloud Explosion

James Lay, PE
U.S. Chemical Safety and Hazard Investigation Board
2175 K Street NW, Suite 400
Washington, DC 20037
Jim.Lay@CSB.gov

Francisco Altamirano, CFEI
Johnnie Banks, CFEI
Jeffrey Wanko, PE, CSP
U.S. Chemical Safety and Hazard Investigation Board

Abstract

The CSB investigated a runaway chemical reaction and subsequent vapor cloud explosion and fire that killed one worker and injured 14 (two seriously) at the Synthron, LLC facility in Morganton, North Carolina, on January 31, 2006. The explosion destroyed the facility and damaged structures in the nearby community.

The accident occurred when plant managers increased the batch size for an acrylic additive by 12% and front-loaded all of the additional monomer into the initial batch-mode step in the production process. Subsequent laboratory testing showed that this process change more than doubled the heat release rate in the reactor. The increased thermal load exceeded the capacity of the reactor's condenser, causing a runaway reaction. Faulty bolting practices created a weak point in the pressure envelope, leading to a discharge of flammable solvent vapors into the manufacturing building. The resulting vapor cloud exploded.

The CSB identified numerous factors contributing to the incident, including an inexperienced staff, poor training, a lack of hazard awareness, poor oversight by the multinational parent company, and a lack of engineering safeguards and controls. Lessons learned are described.

1. Introduction

This paper describes a runaway chemical reaction and subsequent vapor cloud explosion and fires that killed one worker and injured 14 (two seriously). The explosion destroyed the facility and damaged structures in the nearby community. The incident occurred at the Synthron, LLC facility in Morganton, North Carolina, on January 31, 2006.

The CSB investigated the incident and published a Case Study in July, 2007 to emphasize the importance of implementing comprehensive safety management practices to control reactive hazards. The Synthron incident was also featured in the CSB's video, Reactive Hazards:

Dangers of Uncontrolled Chemical Reactions, released simultaneously with the Case Study. Both the Case Study and video are available for free download from the CSB website at http://www.csb.gov.

2. Incident Description

2.1 Summary

This incident occurred at Synthron, LLC's Morganton, North Carolina, facility. The company manufactured a variety of powder coatings and paint additives by the solution polymerization of acrylic monomers in a 1,500 gallon reactor.

The company had received an order for slightly more of an additive than the normal size recipe would produce. Plant managers scaled up the recipe to produce the required larger amount of polymer, and added all of the additional monomer needed into the initial charge to the reactor. This more than doubled the rate of energy release in the reactor, exceeding the cooling capacity of the reactor condenser and causing a runaway reaction.

The reactor pressure increased rapidly. Solvent vapors vented from the reactor's manway, forming a flammable cloud inside the building. The vapors found an ignition source, and the resulting explosion and fire killed one worker and injured 14. The blast destroyed the facility (Figure 1) and damaged off-site structures.

The U.S. Chemical Safety Board (CSB) found that the reactor lacked basic safeguards to prevent, detect, and mitigate runaway reactions, and that essential safety management practices were not in place.

Photo courtesy of Morganton Department of Public Safety

Figure 1: Synthron facility after the explosion

2.2 Description

The production department began preparing a 6,080 pound acrylic polymer batch the day before the incident, approximately 12 percent more material than would normally be made in a single batch. The plant superintendent determined the quantities of solvent, monomer, and initiator needed for the batch. The day shift operators then blended the solvents and used some of the blend to prepare the initiator solution. They added the balance to a 1,500 gallon reactor designated as vessel M1 (Figure 2). The second shift operators, in accordance with written instructions, added some of the monomer to the reactor and held back the remainder for use later in the reaction sequence. The day shift arrived on the morning of January 31 and added steam to the reactor jacket to heat the reactor to the temperature specified on the batch sheet, then shut off the steam.

The senior operator took the final step to start the reaction by pumping the initiator solution into the reactor. He then visually checked the flow of condensed solvent through the condenser sight glass to monitor the rate of reaction. While the reaction initially did not proceed as vigorously as he expected, the condensed solvent flow later increased and appeared to be in the normal range.

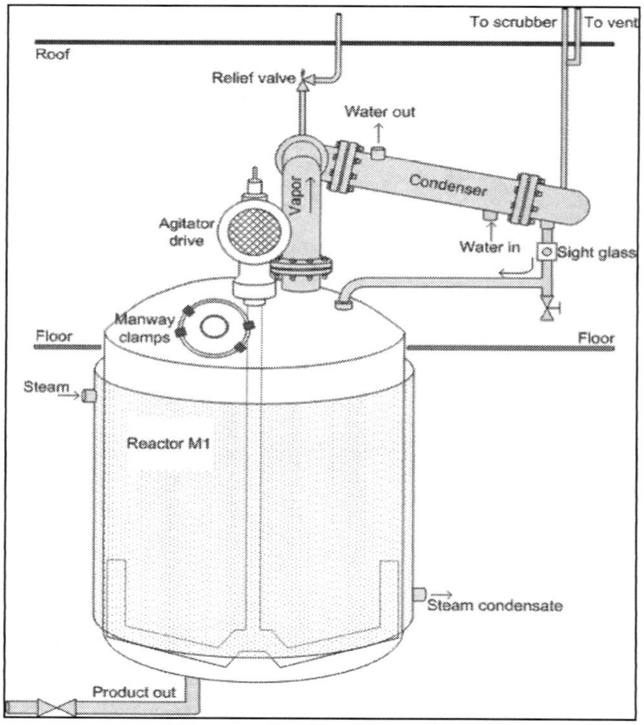

Figure 2: Reactor M1

Several minutes later, the senior operator heard a loud hissing and saw vapor venting from the reactor manway. The irritating vapor forced him out of the building. Three other employees were also forced from the building by the release. Joined by the plant superintendent and the plant manager, the employees gathered outside an upper level doorway (Figure 3). The senior operator re-entered the building wearing a respirator, and was able to start emergency cooling water flow to the reactor jacket. However, the building exploded less than 30 seconds after he exited.

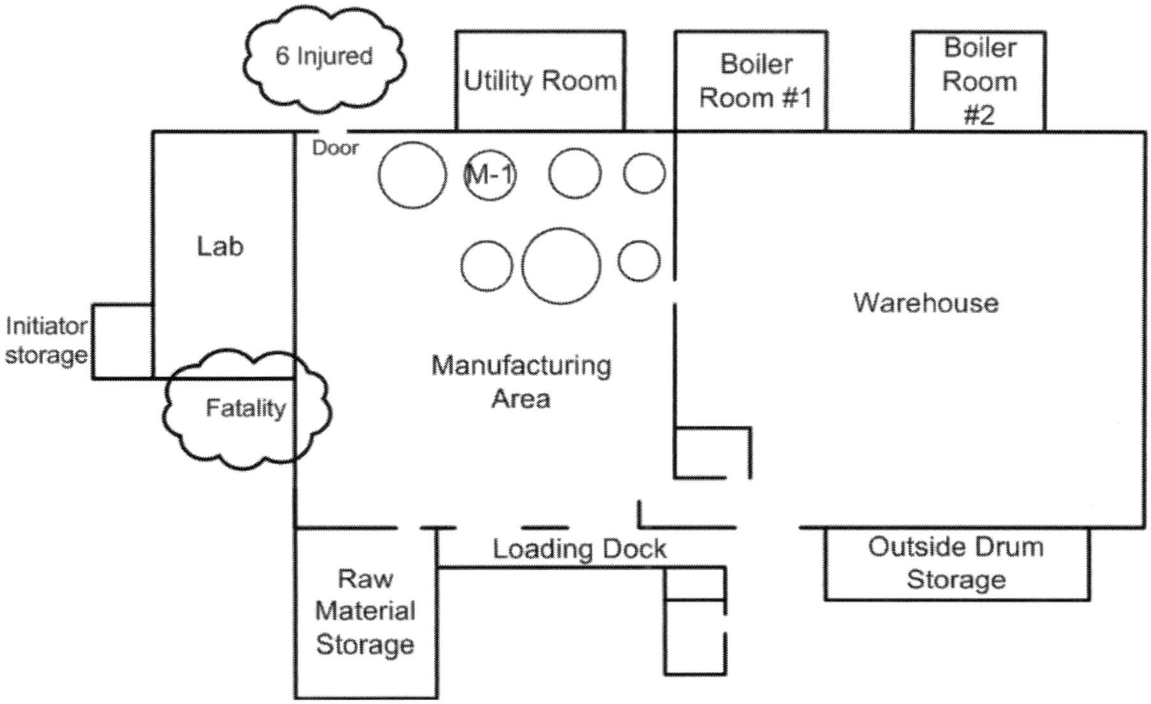

Figure 3: Synthron Facility Layout

The blast knocked down the personnel gathered outside the doorway. All were injured, and one required helicopter transport to hospital. Administrative personnel working in an onsite trailer also suffered minor injuries. The maintenance supervisor was near the lab on the lower level when the explosion occurred (Figure 3). He was severely burned over most of his body and was transported by helicopter to a regional burn center, where he died five days later.

The blast damaged structures in the nearby community. Two church buildings and a house were condemned, and glass was broken up to one-third of a mile from the site. Two citizens driving by the site were slightly injured.

The Environmental Protection Agency (EPA) took control of the site under the CERCLA (Superfund) regulation, remediated contamination from the incident, and eventually razed all of the structures on the property.

2.3 Emergency Response

The Morganton Department of Public Safety responded rapidly and called in mutual aid support from Burke County and nearby municipalities. Employees and Public Safety officers assisted injured employees. The fires following the explosion generated thick smoke, and local residents were asked to shelter-in-place for several hours. The fires were extinguished the next day.

3. COMPANY OPERATIONS

Protex International has owned a controlling interest in Synthron since 1972, and Protex' president is also president of Synthron. Protex, based in Paris, France, is a privately held company with over $100 million a year in sales, and operations in Europe, Asia, and North and South America. In addition to the Synthron facility, Protex operates chemical-related businesses in Massachusetts, New Jersey, and Florida.

Synthron's single U.S. facility was located in Morganton, North Carolina, approximately 70 miles northwest of Charlotte. The 17 Morganton employees included the company's vice president, who was responsible for day-to-day operations. Synthron had 2004 sales of approximately $3 million. Following the explosion, Synthron, LLC filed for bankruptcy under Chapter 7 of the bankruptcy laws.

Synthron produced acrylic polymers, primarily additives for the powder coating and paint industries. The polymers were produced by the free-radical polymerization of acrylic monomers in various flammable solvents. The product being manufactured on the day of the incident, Modarez MFP-BH, was a liquid acrylic polymer.

Polymerization was performed in a 1,500 gallon reactor, rated at 75 psig maximum operating pressure and designated as reactor "M1" (Figure 2). The reactor was located in a manufacturing area adjacent to the warehouse (Figure 3). In a typical reaction sequence, operators added a mixture of solvents and monomers to the reactor. They then injected steam into the reactor's jacket to heat the reaction mixture to a specified temperature, usually the expected mixture boiling point. The steam was turned off and initiator solution metered into the reactor to start polymerization.

The heat given off by the reaction boiled the solvent and monomer mixture, causing hot vapors to flow to an overhead water-cooled heat exchanger where they were totally condensed. Liquid solvent drained from the bottom of the condenser back into the reactor. The outlet of the condenser vented to the atmosphere through a small pipe, keeping the pressure in the reactor very near atmospheric under normal operation. In an emergency, water could be manually directed through the reactor's jacket to increase cooling.

A complete manufacturing cycle could include several reaction steps. The finished liquid polymer was stripped of solvent, cooled, and packaged into drums for shipment to customers.

4. Incident Analysis

The CSB investigators determined that the key factors leading to this incident included

- a lack of hazard recognition,
- poorly documented process safety information,
- ineffective control of product recipe changes,
- a lack of automatic safeguards to prevent or mitigate the effects of loss of control over the reaction,
- improper manway bolting practices,

- poor operator training,
- inadequate emergency plans drills, and
- inadequate corporate oversight of process safety.

4.1 Hazard Awareness

When performing reactive chemistry, companies should maintain a high degree of awareness of the hazards involved. Synthron combined monomers and reaction initiators in the presence of flammable solvent to produce polymer products, but failed to identify the hazards associated with this type of chemistry.

Texts such as Loss Prevention in the Process Industries (Mannan, 2005) and Bretherick's Chemical Reaction Hazards (Urban, 2000) describe appropriate means for characterizing industrial chemical reactions. These include determining the heat generation rate as a function of temperature, the available heat removal capacity, and the potential for excessive monomer or initiator accumulation in the reactor. Failure to control these critical characteristics can lead to severe upsets, including runaway reactions.

Companies can detect hazards by asking "What can go wrong?" to identify the potential consequences of inappropriate mixing, recipe changes, greater- and lower-than-normal heating, lack of mixing, etc. The answers can be used to develop improvements or changes that can prevent reactive upsets or mitigate their consequences. Synthron had not identified the hazards of its reactive chemical operations. No PHAs had been conducted to address "What could go wrong?" during reactor operations.

Furthermore, most of the management and operations personnel at Synthron had been on the job for less than a year, in some cases much less (Table 1), and lacked previous polymer manufacturing experience.1 In addition, Synthron's training program was informal and did not include reactive hazards training.

Table 1: Management and operations personnel tenure at Synthron

Employee	Time at Synthron	Polymer Experience?
Manager	9 months	No
Superintendent	8 months	No
Vice President	5 months	No
2nd Shift Op-1	3 months	No
2nd Shift Opt-2	3 months	No
Chemist	3 weeks	No

Personnel, including site managers, were thus poorly prepared to recognize potentially hazardous changes to product recipes, or to respond to an incipient runaway reaction. Additionally,

[1] Synthron's president had replaced the management team in an attempt to increase sales. A variety of causes had simultaneously contributed to high operator turnover. The senior operator had been brought back from retirement to provide a degree of continuity.

Synthron had no chemical or other engineers on staff, and none had been contracted to evaluate the hazards associated with reactive operations at the site.

4.2 Process Safety Information and Training

Synthron had minimal safety information on its polymerization process, even though this was the core of its manufacturing business. Synthron optimized product formulations to meet customer specifications. However, reaction characterization and calorimetry were not performed to establish process equipment performance requirements and operating limits for safe operations.

When scaling-up new products from the laboratory, the previous Synthron plant manager had typically estimated initial production batch sizes based on past experience. He then gradually increased batch quantities until the sight glass in the condensate return line showed that the condenser was close to flooding, or that another performance limit was being approached (e.g., the reactor pressure increased). Polymerization scale-up to their standard batch sizes had historically been done by simple trial-and-error. The condenser's cooling capacity was not documented, nor could the cooling load placed on the condenser be determined with the available instrumentation. As a result, information essential to the safe operation of the reactor was not available. Polymerization reactors can runaway with disastrous consequences if they are not carefully controlled.

Based on interviews the CSB conducted, Synthron employees and managers had little or no understanding of reactive hazards. They had not been trained on,[2] and did not understand, the margin of safety needed or available in their polymerization operations. Furthermore, they had little knowledge of the sensitivity of the reactor to changes in product recipes, batch sizes, or reaction conditions. Synthron's employees were thus unprepared to recognize and respond to the reactive hazards they faced the day of the incident.

4.3 Batch Recipe Changes

In planning the MFP-BH batch, Synthron managers made several changes that greatly increased the heat released by the reaction in reactor M1 and the potential for a runaway reaction. However, the changes were not effectively reviewed and the hazards went unrecognized. Synthron:

- increased the total amount of monomer to be charged to the reactor by 12 percent. The additional monomer, with a Normal Boiling Point Temperature (NBPT) of 147°C (297°F), was placed in the initial reactor charge.
- reduced the amount of aliphatic solvent, with an NBPT of 81°C (178°F), charged to the reactor by 12 percent.
- increased the amount of aromatic solvent, with an NBPT of 111°C (234°F), by 6 percent.

[2] Synthron's training program was informal, relying on unstructured on-the-job training. Testimony indicated that reactive hazards were not systematically addressed.

The customer had ordered 12 percent more MFP-BH than a standard batch would produce. To avoid the additional time and effort of running two half-size batches, the superintendent scaled up the recipe to produce the order in a single batch.

MFP-BH was produced in two stages. In the first, an initial charge of acrylic monomer and solvent was loaded into reactor M1, heated, and then reacted by adding an initiator over a brief period. In the second, the remaining monomer and initiator were co-fed into the reactor over an extended period. Unfortunately, the superintendent placed almost all of the additional monomer for the larger batch into the initial charge of chemicals loaded into the reactor.

According to the batch sheet, roughly equal amounts of aromatic and aliphatic solvent should have been added to the reactor. However, there was not enough of the lower boiling temperature aliphatic solvent available in storage. To compensate, the superintendent and manager decided to make up half the shortfall using the higher boiling aromatic solvent, and to run the batch with slightly less total solvent than specified in the recipe.

Together, these changes:

- increased the total amount of monomer in the reactor by 45 percent,
- increased the concentration of monomer by 27 percent, and
- increased the atmospheric boiling point temperature of the mixture by almost 5°C (9°F).[3]

Each of these changes would be expected to increase the rate of heat release in the reactor. When asked to review the changes in the solvent quantities, the plant chemist estimated that the boiling point of the solvent mixture would increase about 1°C (1.8° F). However, the chemist, manager, and superintendent did not recognize or address the potential impact of the increased monomer amount and concentration on the mixture boiling point, reaction rate, or total rate of heat release.

The combined effect of the changes was to increase the maximum heat output from the reaction to at least 2.3 times that of the standard recipe. Figure 4 illustrates the results of reaction calorimetry performed to simulate the early stages of the incident.

The lower curve shows the heat release rate (thermal power) versus time for the standard recipe, while the upper curve shows the heat release rate for the modified recipe.[4] The lower solid horizontal line is the estimated cooling capacity of the reactor condenser on the day of the incident. Once the heating curve exceeded the condenser cooling line, control of the reaction was lost, resulting in a runaway reaction.

[3] The CSB measured the boiling points of laboratory mixtures with the same compositions.

[4] Each curve was generated isothermally at the NBPT of the mixture, closely simulating the operation of the reactor up to the time when control was lost. Heat generation rates during the subsequent runway reaction were higher.

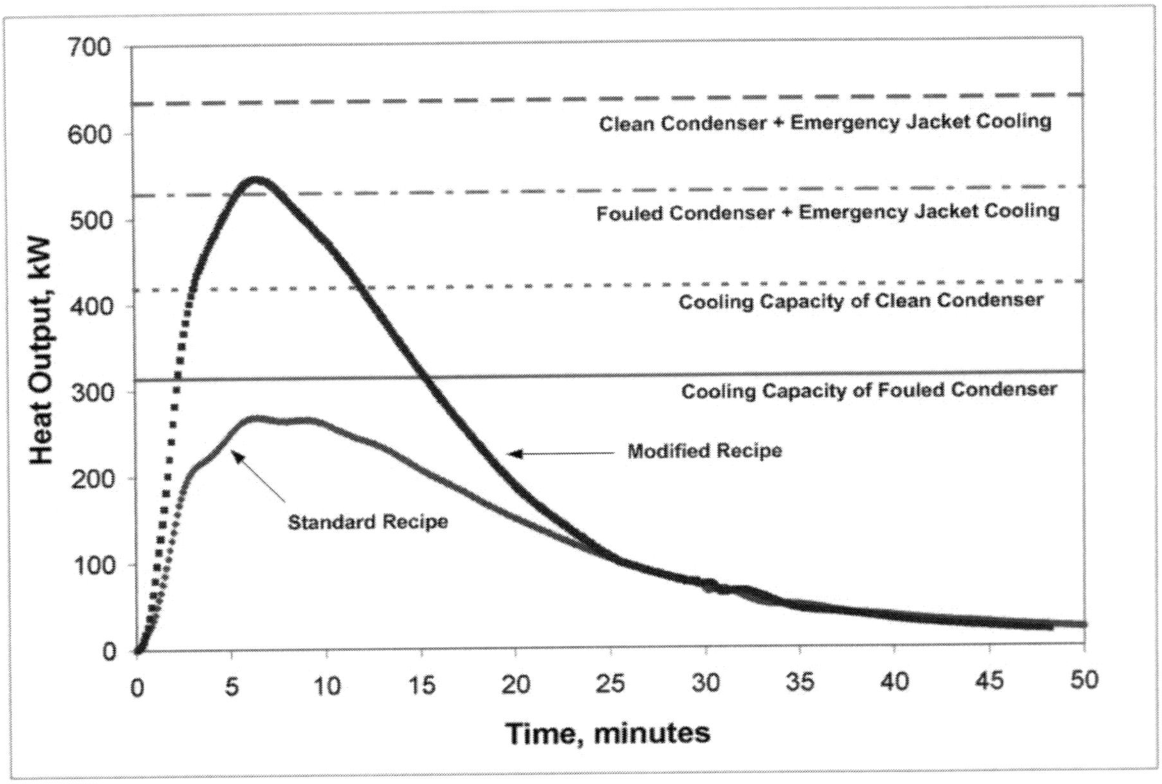

Figure 4: Reactor Energy Release and Cooling Capacity

4.4 Engineering Safeguards

Synthron relied primarily on a procedural safeguard to prevent loss of reactor control: the batch sheet, which was used as an operating procedure at the site. Procedures are essential for safety in chemical processing operation, but are the least reliable form of safeguard for preventing process incidents (CCPS, 2004).

Failures with potentially severe consequences, such as runaway reactions, should have multiple independent safeguards. Examples of safeguards that could have prevented or mitigated this incident, but were not installed at Synthron, include:

- high pressure alarms to notify operators of problems early in the incident when action to control the reaction might still be possible,
- automatic emergency cooling water flow to the reactor jacket,
- automatic shut-off of initiator feed,
- automatic or remotely operated injection of "short stop" solution to stop the polymerization reaction,[5] and
- automatic or remotely operated venting or dumping of the reactor to a safe location.

[5] Chemicals, such as phenothiazine, can be injected into a reactor to slow or stop free-radical polymerizations such as the reaction at Synthron.

Good practice is to review the adequacy of safeguards on chemical reactors using a structured method such as Layers of Protection Analysis (CCPS, 2001). Such reviews can help ensure that runaway reactions are prevented or are rapidly and reliably detected and controlled.

4.5 Manway Bolting Practice

Operators opened the reactor manway after every batch cycle to clean the reactor. Long-standing practice at the facility was to then close the manway and secure it using only four of the 18 clamps specified by the manufacturer (Figure 5).[6] The risk posed by this practice had not been recognized because the reactor normally ran at near-atmospheric pressure, for which four clamps were thought to be adequate.

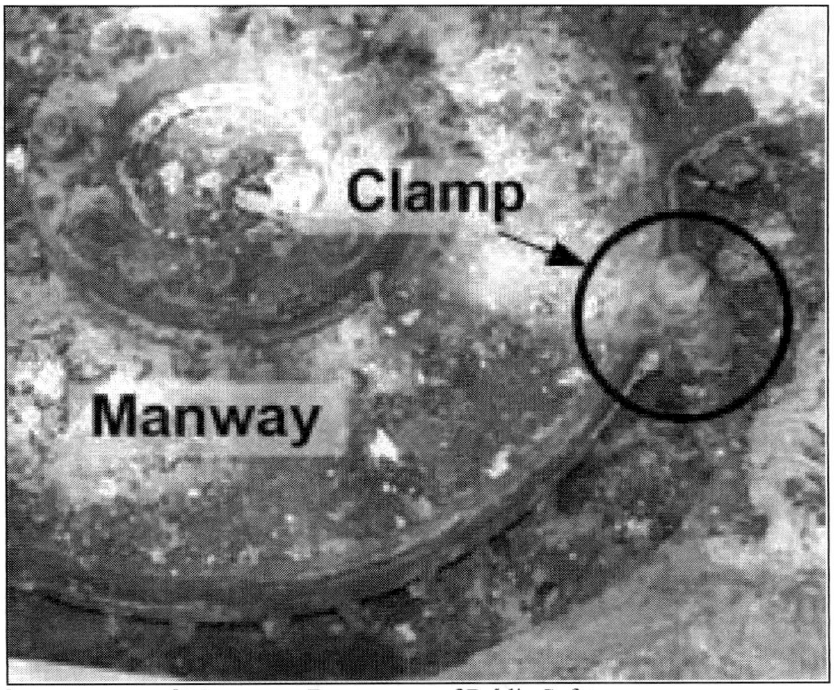

Photo courtesy of Morganton Department of Public Safety

Figure 5. One of four installed manway clamps

The CSB investigators calculated that, with only four clamps installed, the manway began to leak flammable solvent vapors when the reactor pressure reached approximately 23 psig, well below the reactor's maximum allowable working pressure (and likely relief valve set point) of 75 psig.[7] The vapor leak path is clearly visible in Figure 6 (arrow).

[6] The reactor manufacturer specified 18 clamps to maintain a tight seal at the reactor maximum working pressure of 75 psig. The clamps were not permanently attached to the manway, and no markings on the manway indicated the required number of clamps.

[7] The reactor M1 relief valve could not be recovered from the debris, and the relief valve records were destroyed in the explosion and subsequent fires. Other relief valves examined were set at the rated operating pressure of the vessels they protected.

Figure 6. Manway gasket and leak path (arrow)

4.6 Condenser Fouling

Synthron failed to establish procedures to maintain the performance of the M1 reactor condenser, an inclined shell-and-tube heat exchanger essential to safe operation of the reactor. In the condenser, solvent vapors flowed inside the tubes and were cooled and condensed by cooling water flowing through the shell (Figure 2). Inspection of the cooling water side of the condenser after the incident revealed that it was badly fouled, likely reducing condenser capacity at least 25 percent (Figure 7).

The condenser's design, with the tubes permanently bonded to plates at both ends of the condenser, made inspecting or cleaning the water side difficult.[8] Furthermore, the company had no program to systematically monitor and control water side fouling.[9] The CSB found no evidence that the water side of the condenser had ever been inspected or cleaned to remove the scale, rust, and sediment that had accumulated during 30 years of service. Synthron's employees lacked the expertise and experience to recognize the risk posed by water side fouling of the condenser.

[8] The condenser was a Tubular Exchanger Manufacturer's Association (TEMA) type BEM, fixed tubesheet exchanger with the process fluid on the tube side and cooling water on the shell side. The tubes could not be removed for inspection or cleaning. This design may require periodic chemical cleaning of the shell side to maintain good thermal performance.

[9] Chemical treatment of the cooling water is almost always required to prevent biological growth, fouling, and/or corrosion. The cooling water at Synthron was not treated.

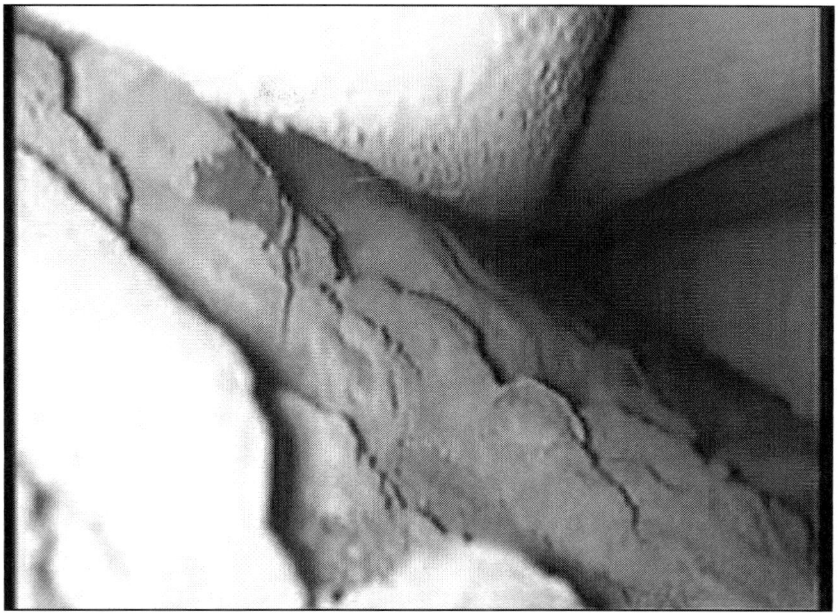

Figure 7: Deposits fouling the water side of the M1 reactor's condenser.

A clean condenser, combined with automatic emergency jacket cooling (Figure 4, upper dashed line), might have prevented the runaway reaction and subsequent explosion.

4.7 Emergency Evacuation Procedures and Training

Effective evacuation plans are important for minimizing injuries and fatalities in chemical emergencies. During this incident, none of the production employees evacuated to a safe location. At the time of the explosion, six employees, including the manager and superintendent, gathered outside a doorway on the upper level while a seventh was on the lower level by the lab. The employee on the lower level was killed, while all the employees outside the doorway were injured, two seriously.

Synthron was unprepared for an emergency; specifically:

- The facility Emergency Action Plan did not list events or describe situations that might necessitate a plant evacuation.
- Operating procedure did not specify employees' actions in the event of a chemical release or loss of reactor control.
- Employees were not trained on the Emergency Action Plan and evacuation drills were not conducted.
- The facility was not equipped with an emergency alarm system.

In this incident, there was sufficient time after the release began to evacuate employees to a safe location. Adequate emergency response planning by Synthron could have prevented the death and serious injuries caused by the explosion.

4.8 Corporate Safety Oversight and Support

Synthron was part of Protex International, a much larger organization. However, the parent company provided little safety oversight or support to Synthron.

While Synthron performed quality control testing and limited product development work in Morganton, the procedures, equipment, and trained personnel needed to characterize reactive hazards were not in place.

However, Protex had the ability to perform detailed reaction characterization work, including reaction calorimetry, at its European facilities. It is good practice (CCPS, 2003) to ensure that reactive safety programs at small facilities are adequately supported by technically qualified resources. Protex did not provide adequate reactive safety support to Synthron.

Furthermore, in the summer and fall of 2005, Synthron's president (also the president of Protex) hired a new vice president, plant manager, and plant superintendent for the Morganton site. While two of these key management employees had degrees in chemistry, they had no previous polymerization experience and were not trained on the parent company's process safety procedures or its testing capabilities. Having little reactive chemistry background or training, they did not recognize the reactive hazards at the Morganton site.

CCPS (1995) stresses that expertise in managing process safety must be ensured when making staffing changes. Synthron's president failed to ensure that the new team he installed at the Morganton site had the requisite training, knowledge, and experience to operate the facility safely.

Finally, Protex did not comprehensively audit or review Synthron's safety program. Such a review should have identified the absence of an effective reactive hazard control program.

5. Regulatory Analysis

Synthron had not implemented a Process Safety Management (PSM) system consistent with the requirements of the Occupational Safety and Health Administration's (OSHA) PSM regulation (29 CFR 1910.119). The site had last been inspected by North Carolina OSHA (NC-OSH) in 1996. Following the explosion, NC-OSH proposed numerous PSM citations, which Synthron contested.

The batch being processed at the time of the explosion, had it been completed, would have contained in excess of the OSHA PSM threshold of 10,000 pounds of flammable liquid. However, only 4,500 pounds of material were in the reactor at the time of the incident, well below the PSM threshold quantity. Nonetheless, a catastrophic incident occurred. Synthron's experience demonstrates that companies working with reactive chemical quantities less than the PSM flammables' threshold still need to consider and guard against potentially catastrophic accidents.

6. The CSB Reactives Study

In its comprehensive 2002 report, "Improving Reactive Hazard Management" (CSB, 2002), the CSB found that reactive incidents are a serious problem in the United States, and that both management system and regulatory improvements are needed to help facilities control reactive hazards. This report studied reactive incidents, causal factors, and preventive measures,[10] and outlined the screening, hazard identification, hazard review, operating procedures, and training needed to prevent reactive incidents.

The report documented 167 serious reactive incidents in the United States between January 1980 and June 2001 that resulted in 108 deaths, hundreds of injuries, and significant public impacts. Ongoing monitoring by the CSB indicates that reactive incidents, such as the Synthron explosion, continue to occur.

The CSB report also found that 70 percent of reactive incidents occurred in the chemical manufacturing industry, with 35 percent due to runaway reactions, such as that which occurred at Synthron.[11] While 42 percent of reactive incidents resulted in fires and explosions, another 37 percent caused toxic emissions. Many reactive incidents occurred at small manufacturing sites such as Synthron.

More than 50 percent of the 167 incidents documented in the CSB report involved chemicals not covered by existing OSHA PSM (29 CFR 1910.119) or EPA Risk Management Program (40 CFR Part 68) regulations. The CSB recommended better coverage of reactive hazards by these regulations. EPA now requires reporting of reactive chemical incidents under RMP reporting rules. OSHA has taken steps to increase industry awareness of reactive hazards, but has not fully implemented the CSB recommendations.

The CSB report identified many valuable sources of good practices for managing reactive hazards. Since the report's publication, additional guidance has become available. The Center for Chemical Process Safety (CCPS) published *Essential Practices for Managing Chemical Reactivity Hazards* (CCPS, 2003), which can be downloaded at no cost from http://info.knovel.com/ccps/ (January 2007). This book lays out the basic steps manufacturers should take to protect against reactive hazards.

Other valuable sources of guidance are available. The EPA's Chemical Emergency Preparedness and Prevention web page http://yosemite.epa.gov/oswer/ceppoweb.nsf/content/ap-book.htm (May 2007) contains useful links. The United Kingdom's Health and Safety Executive also offers an informative web page, http://www.hse.gov.uk/pubns/indg254.htm (May 2007) and the comprehensive booklet, "Designing and Operating Safe Chemical Reaction Processes" (HSE Books, 2000).

---

[10] Available for download at http://www.csb.gov.

[11] Of reactive incidents, 25 percent originate in reactors, with the rest occurring in a wide range of equipment.

7. Conclusion

This incident provides important lessons for manufacturers with operations involving reactive chemistry.

7.1 Identify and Control Reactive Hazards

Manufacturers should take a comprehensive approach, and:

- identify and characterize reactive hazards;
- systematically evaluate what can go wrong, including mis-charging of reagents, loss of cooling, instrument malfunction, and other credible failure scenarios; and
- implement, document, and maintain adequate safeguards against the identified failure scenarios. Multiple, independent safeguards may be needed to reliably ensure the safety of the reactive process.

7.2 Control Change

Chemical manufacturers and others with reactive chemistry operations should control changes to batch recipes, including key operating conditions, such as:

- the quantities, proportions, and sequencing of reactor feeds,
- reaction temperature,
- conditions that could cause initiator or monomer accumulation, and
- conditions that could affect the deactivation of monomer inhibitors or stabilizers.

7.3 Maintain Process Equipment Capability

Manufacturers with reactive chemistry operations should:

- document the performance requirements and capabilities of process equipment, such as the reactor condenser at Synthron; and
- periodically inspect and service process equipment, including the water side of heat exchangers, to maintain appropriate safety margins.

7.4 Train Personnel on Hazards and Procedures

Manufacturers should ensure that worker training includes:

- the nature of the reactive hazards, including process safety margins; and
- operating procedures, including appropriate cautions and warnings, the consequences of deviations, recognition of deviations and abnormal operations, and the appropriate responses to control or mitigate their effects.

7.5 Prepare for Emergencies

Manufacturers should:

- implement an effective emergency plan,
- train employees on the plan,

- install an evacuation alarm system that is audible and/or visible throughout the facility,
- conduct regular exercises to help ensure rapid evacuation to a safe location in an emergency, and
- coordinate their emergency planning with offsite response organizations and the Local Emergency Planning Committee (LEPC).

The CSB made a recommendation to Protex International to establish a program to ensure that reactive hazards at Protex' U.S. facilities are managed in accordance with good industry practices.

8. Disclaimer

This paper has not been independently approved by the Board and is published for general informational purposes only. Any material in the paper that did not originate in a Board-approved report is solely the responsibility of the authors and does not represent an official finding, conclusion, or position of the Board.

9. REFERENCES

[1] Health and Safety Executive (HSE) [Online], *Chemical Reaction Hazards and the Risk of Thermal Runaway*, www.hse.gov.uk/pubns/indg254.htm, London: HSE, (accessed 2007)

[2] HSE, Designing and Operating Safe Chemical Reaction Processes, Rugby, UK: HSE Books, 2000

[3] Center for Chemical Process Safety (CCPS), Guidelines for Hazard Evaluation Procedures - With Worked Examples, 2nd ed, New York: American Institute of Chemical Engineers (AIChE), 1992

[4] CCPS, Plant Guidelines for Technical Management of Chemical Process Safety (Revised Edition), New York: AIChE, 1995

[5] CCPS, Layer of Protection Analysis - Simplified Process Risk Assessment, New York: AIChE, 2001

[6] CCPS, Essential Practices for Managing Chemical Reactivity Hazards, New York: AIChE, 2003

[7] CCPS, Inherently Safer Chemical Processes, A Life Cycle Approach, New York: AIChE, 2004

[8] Mannan, S., Lees' Loss Prevention in the Process Industries, 3rd ed., Burlington, MA: Elsevier Butterworth-Heinemann, 2005

[9] Urban, P. G., Ed., Bretherick's Handbook of Reactive Chemical Hazards, 6th ed., St. Louis, MO: Elsevier Science and Technology Books, 2000

[10] The United States Chemical Safety and Hazard Investigation Board (CSB), Improving Reactive Hazard Management, Washington, DC: CSB, 2002

Statistical Analysis of Pressure Relief Valve Proof Test Data: Findings and Implications

Dr. Julia V. Bukowski
Villanova University
Villanova, PA, USA

Dr. William M. Goble
exida
Sellersville, PA, USA

A 10th PPSS Anniversary Paper

ABSTRACT

This paper reports on our statistical analysis of pressure relief valve (PRV) proof test data for the failure mode, fail-to-open, i.e., the PRV remains closed when actual pressure reaches or exceeds 150% of set pressure. Three data sets, from Fortune 500 operating companies, which met the intent of the quality assurance of proof test data as documented by the Center for Chemical Process Safety Process Equipment Reliability Database (CCPS PERD) initiative, were analyzed. Although the original intent of our analysis focused solely on estimation of the failure rate during the useful life of the equipment, it became apparent that the probability of failure upon initial installation or reinstallation after proof test, and the need to address what constituted end of useful life were very significant. This paper provides three important findings that are summarized as follows.

1) The statistical analysis of each data set predicted a 1% – 1.6% probability of initial failure (PIF) where initial failure is understood to be at the time of initial installation or reinstallation after a proof test. This implies that most of the failures found during the useful life via proof test are pre-existing failures from the time of installation or reinstallation rather than failures that occurred randomly after installation or reinstallation of the PRV.

2) Our calculations, based on the three independent data sets, led to consistent estimates of PRV useful-life failure rates between 10^{-8} and 10^{-7} failures/hour. Additionally, we compared our estimates from data analysis to the prediction of useful-life failure rate for a particular PRV model using the Failure Modes Effects and Diagnostics Analysis (FMEDA) method. The prediction was consistent with the data estimates.

3) The data further indicated that the low useful-life failure rate was not supported beyond a 4 to 5 year proof test interval as the threshold of wear out seemed to be approached.

The importance of these finding cannot be overestimated. When taking credit for a PRV in a risk assessment and in calculating the probability of failure on demand (PFD), both the initial probability of failing to open, as well as the probability of PRV failure due to the useful-life failure rate, must be taken into account. (The paper discusses how to do this.) Even then, the results are only defensible when it can be demonstrated that the proof test occurs before wear-out (i.e. end of useful life) begins.

1. INTRODUCTION

Pressure relief valves (PRV) are safety devices used extensively in the chemical process industry to reduce the risk of accidents caused by overpressure events. In order to quantify a PRV's contribution to risk reduction, it is necessary to know the device's failure modes and failure rates. This paper focuses on the PRV failure mode "fail-to-open," also known as "stuck-shut," and analyzes data to determine, for this failure mode, the PRV failure rate, useful in-service life before maintenance is required, and PIF.

The "stuck-shut" failure mode occurs when the PRV remains closed at pressures greater than or equal to 150% of set pressure. This failure mode is especially important because the PRV is normally closed. Therefore, the "stuck-shut" failure mode cannot be detected while the PRV is installed within the process being protected. Identifying this failure mode requires a proof test. In a proof test, the PRV is removed from the process and is pressurized to determine the pressure at which the PRV opens. This type of testing is costly and, while it identifies failed PRV, it does not provide an indication of *when* the failure occurred prior to the proof test – information normally needed to estimate failure rate. The nature of the proof test data requires a special form of statistical analysis known as quantal response analysis which is described in this paper.

The failure rate, $\lambda(t)$, of a PRV changes over time. Normally, risk assessment is interested in the useful-life failure rate which is approximately constant and represented by λ. In order to estimate λ from proof test data, it is first necessary to ascertain what time interval encompasses useful life as opposed to longer time intervals when wear out is first evident and failure rates begin to rise. Thus, the first analysis described in the paper is how we obtained a useful life of 4 – 5 years. Once the useful-life interval is identified, the analysis of proof test data from this time interval is used to estimate λ. Both the full data analysis and the useful-life data analysis give estimates for the PIF. Finally, we address how PIF is incorporated into the calculation of PFD and PFDavg. We close with a discussion of the results and suggestions for further work.

2. NOTATION

FMEDA	failure modes, effects, and diagnostics analysis
PFD	probability of failure on demand
PFDavg	probability of failure on demand averaged over $[0, T_P]$
PIF	probability of initial failure; 1-R(0)
PRV	pressure relief valve(s)
q_i	percentage of proof tests resulting in failure in the ith interval
R(0)	initial reliability; 1 - PIF; may be less than 1

R(t)	reliability function for $t > 0$
T_i	equivalent failure time associated with q_i
T_P	length of a particular PRV time in-service until proof test
λ	useful-life failure rate, a constant
$\lambda(t)$	failure rate as a function of time; may be a constant

3. QUANTAL RESPONSE ANALYSIS

The theory and practice of quantal response analysis as applied to proof test data are well described in [1] and [2]. Here we provide sufficient background to support understanding of the analysis in the remainder of the paper.

In quantal response analysis, the in-service operating times between installation and first proof test, or between re-installation and subsequent proof tests, for each PRV are arranged in ascending order without regard to the outcome of the test. The ascending operating times are then grouped into *m* non-overlapping time intervals such that each interval includes a small number of proof tests that resulted in failure. This grouping process is not arbitrary and requires some understanding of the assumptions and mathematical details of quantal response analysis. For the ith interval, $i = 1, \ldots m$, two quantities are calculated: q_i, the fraction of failed proof tests for the interval, and T_i, calculated as either the average operating time for all PRV in the interval or the average operating time for just the failed PRV in the interval. There is rarely a significant difference due to different methods of calculating T_i and, in general, either method is acceptable.

Once the q_i and T_i calculations are complete, a plot is made of the points $-\ln(1-q_i)$ vs T_i and two different curves are fit to these points. First, a power curve of the form $At^n + B$ is fit to all of the quantal response points generated by a given data set. The curve fit is valid if $A > 0$ and $B \geq 0$. Normally, a valid curve will rise approximately linearly at first but then will reach a time when the ascent is more rapid. This time region of rapid ascent indicates the end of the useful life and the need for proof testing and maintenance. Furthermore, the value of B from a valid curve fit is, for $B < 0.05$, an estimate of the PIF. Secondly, once the useful-life region is identified, the data points for just the useful life are fitted with a straight line of the form $mt + b$. The curve fit is valid if $m > 0$ and $b \geq 0$. The slope of the straight line, m, is an estimate for the constant useful-life failure rate, λ, given in failures/year. This can be converted to the more familiar failures/hour by dividing by 8760 hours/year. If $b < 0.05$, b is another estimate for the PIF. The values of B obtained for a power fit and b obtained for a straight line fit should be similar for a given data set.

4. DATA SETS ANALYZED

Three independent data sets, obtained from two Fortune 500 operating companies, were analyzed by three independent analysis groups. Two of the analysis groups were from within the operating companies. The third analysis was independent of either operating company. The analyses produced the information contained in the Appendix

Tables A1-A3. Data Set I consisted of 3403 proof tests performed on 1949 individual PRV resulting in 48 "fail-to-open" test results. Data Set II consisted of 2578 proof tests which included 57 failures. Data Set III consisted of 3282 proof tests which included 24 failures. Data Set III is unique in that the 3282 proof tests include 2377 proof tests that were performed prior to initial installation and these tests include 10 initial failures.

5. RESULTS OF DATA ANALYSIS

Determining the Useful Life Interval

Fitting power curves of the form $At^n + B$ to the data in Tables A1-A3 results in the plots in Figure 1. The annotations in each graph indicate the estimated values for A, n, and B with their standard deviations (for Data Sets I and II) along with the mean squared and root mean squared errors.

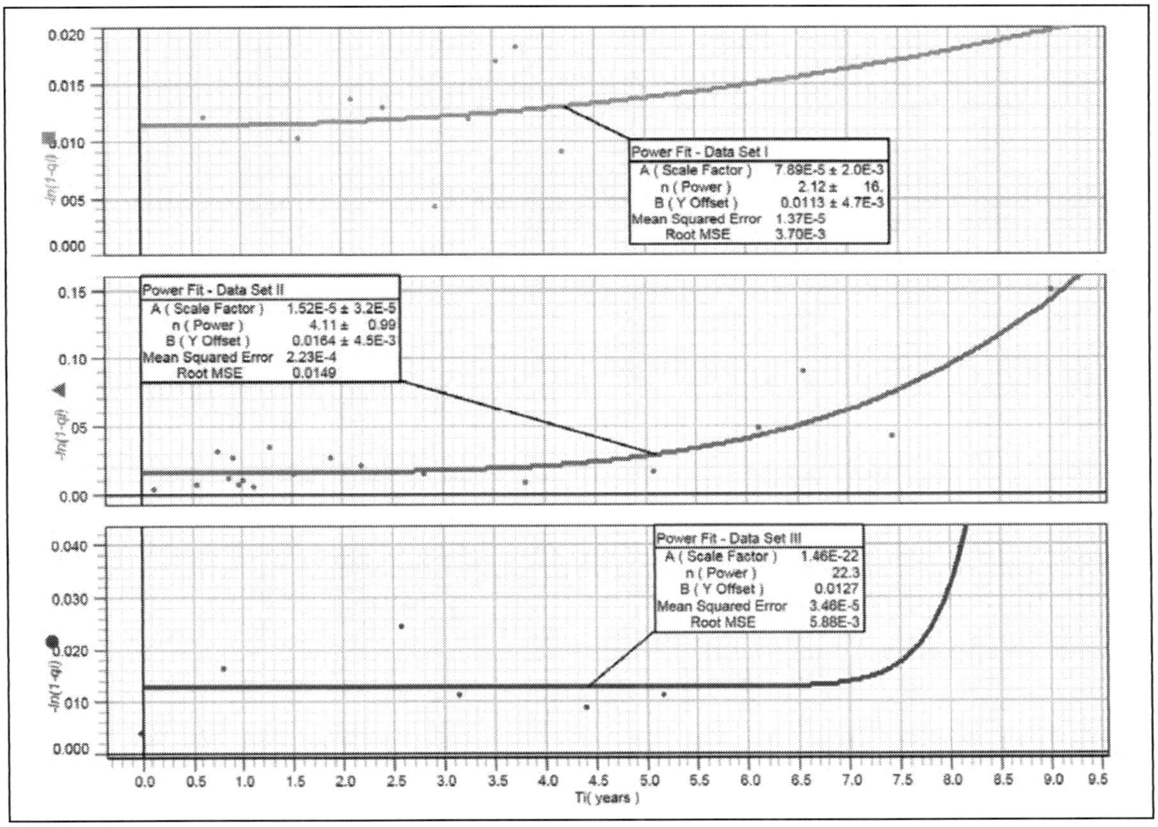

Figure 1. Plots of -ln(1-q_i) vs T_i along with power curve fits for the three data sets.

First note, that in each case, A and B are positive, indicating valid curve fits. For Data Set I there is no significant increase in rate of ascent within the region for which we have data. This indicates that wear-out has not yet become evident at the time of the last data point at approximately 4.2 years. For Data Set II, a change in rate of ascent occurs between 4 and 5 years indicating the first evidence of wear-out. In Data Set III, although there is no significant change in rate of ascent until just after 7 years, it is also true that there are no data points between 5.2 years and 8.1 years to shape the curve. Absent this information, we must conservatively conclude that there is no evidence of wear-out prior to 5.2 years but beyond that, there is insufficient information to support a useful-life interval longer than about 5.2 years. Based on analysis of these three data sets, we conclude that PRV useful life extends to about 4 - 5 years at which time proof testing and maintenance are required to extend the PRV useful-life. If proof testing and maintenance are not performed by this time, we must consider the PRV to be beyond its useful life for purposes of reliability and risk analysis.

Determining λ, the Useful-Life Constant Failure Rate

Using linear regression, a straight line of the form $mt + b$ is fit to the useful-life data for each data set, resulting in the plots shown in Figure 2. The annotations in each graph indicate the estimated values for m (in failures/year) and b with their standard deviations along with the correlation coefficient, r, and the mean squared and root mean squared errors.

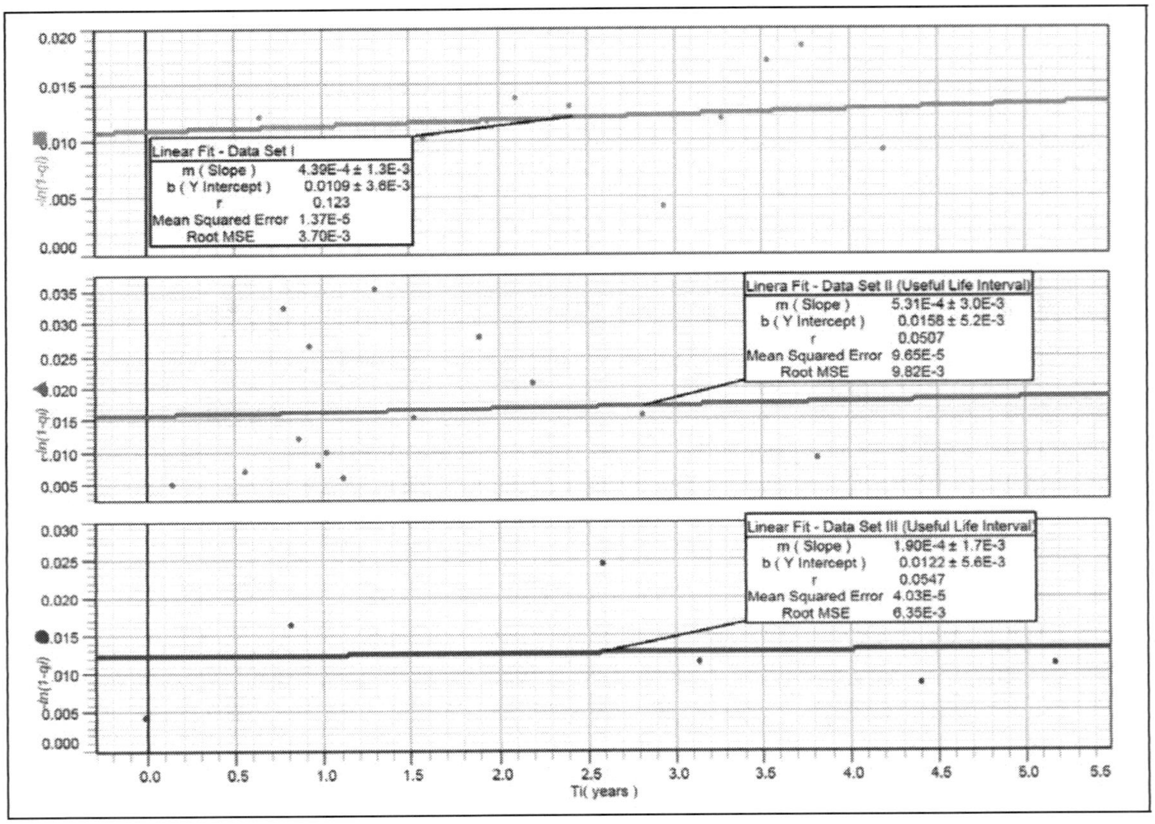

Figure 2. Plots of -ln(1-q_i) vs T_i along with linear fits for the three data sets.

The values for m are all of the same order of magnitude and, when divided by 8760 hour/year, correspond to failure rates between 10^{-8} and 10^{-7} failures/hour as summarized in Table 1 below. It is noted that the r values are small and this may lead some readers to question the appropriateness of a linear fit to this data. However, as is explained in detail in [1] along with corroborating simulation evidence, the small values for r result from the fact that the PIF is substantial and initial failures, in fact, account for most of the failures discovered in proof test. To further support the appropriateness of the linear regressions and the information derived from them, we performed a FMEDA analysis [3] on a particular PRV model. The FMEDA predicts a failure rate based on a database of failure rates [4] for each part used to construct the PVR. The database is designed to generate conservative estimates of PRV failure rate. For the specific PRV model analyzed, the FMEDA predicted a failure rate of $8.4*10^{-8}$ failures/hour [1], consistent with the estimates for failure rates obtained from our analyses based on data aggregated over many PRV manufacturers and models.

Summary of Data Analysis Results

Table 1. Summary of Results along with General Conclusions Drawn

Data Set	Estimated Useful-Life Interval	Estimated PIF from *B*	*b*	Estimated λ failures/hour
I	4.2 years	1.13%	1.09%	$5.0*10^{-8}$
II	4-5 years	1.64%	1.58%	$6.1*10^{-8}$
III	5.2 years	1.27%	1.22%	$2.2*10^{-8}$
General Conclusions	4 - 5 years	1 - 1.6%		10^{-8} - 10^{-7}

Note that for each data set the PIF estimated as B from the power fit and as b from the linear fit is about the same.

6. INCORPORATING PIF INTO PFD

PFD is a time varying quantity that measures the probability that the PRV is in a "stuck-shut" state when an over-pressure event occurs, i.e., the probability that the PRV is failed when a demand occurs. By definition,

$$PFD(t) = 1 - R(0)R(t). \quad (1)$$

If the PRV is known to be in working order when the process is started, $R(0) = 1$. However, the data clearly show that $R(0) < 1$. To incorporate this into the calculation of PFD, we write

$$PFD(t) = 1 - (1 - PIF)\, R(t) = 1 - (1 - PIF)\, e^{-\lambda t}. \quad (2)$$

For λ as large as 10^{-7} and t as large as 5 years (43,800 hours), λt is less than 0.005. Hence, $e^{-\lambda t}$ is well approximated by $(1 - \lambda t)$, allowing us to approximate (2) as

$$PFD(t) \approx 1 - (1 - PIF)\,(1 - \lambda t). \quad (3)$$

Usually, we calculate not PDF(t) but rather PFDavg defined as

$$PFDavg = (1/T_P) \int_{T_P}^{0} PFD(t)\,dt \qquad (4)$$

where T_P is the in-service time of the PRV from last installation/reinstallation until the current proof test. Using (3) to approximate PFD(t) in (4) and integrating results in

$$PDFavg \approx PIF + (1 - PIF)^* \lambda T_P/2. \qquad (5)$$

In (5) we see that the longer we wait before proof test, the greater the value of PFDavg. To determine the extent to which PIF influences PFDavg, consider the ratio PIF/PFDavg. This ratio is plotted in Figure 3 for several values of PIF with $\lambda = 10^{-7}$ over a range of T_P from 0 to 5 years. As would be expected, the PDFavg is initially due solely to the PIF and thus the ratio is 1. As time progresses without a proof test, random failures during the useful life of the PRV contribute to PDFavg and the ratio becomes smaller. However, note that for a PIF as little as 0.5% the ratio dips just under 0.7 after 5 years. A PIF of only 0.5% accounts almost 70% of the total PDFavg at 5 years. Clearly the PIF plays a strong role in reducing the safety that could otherwise be contributed by the PRV.

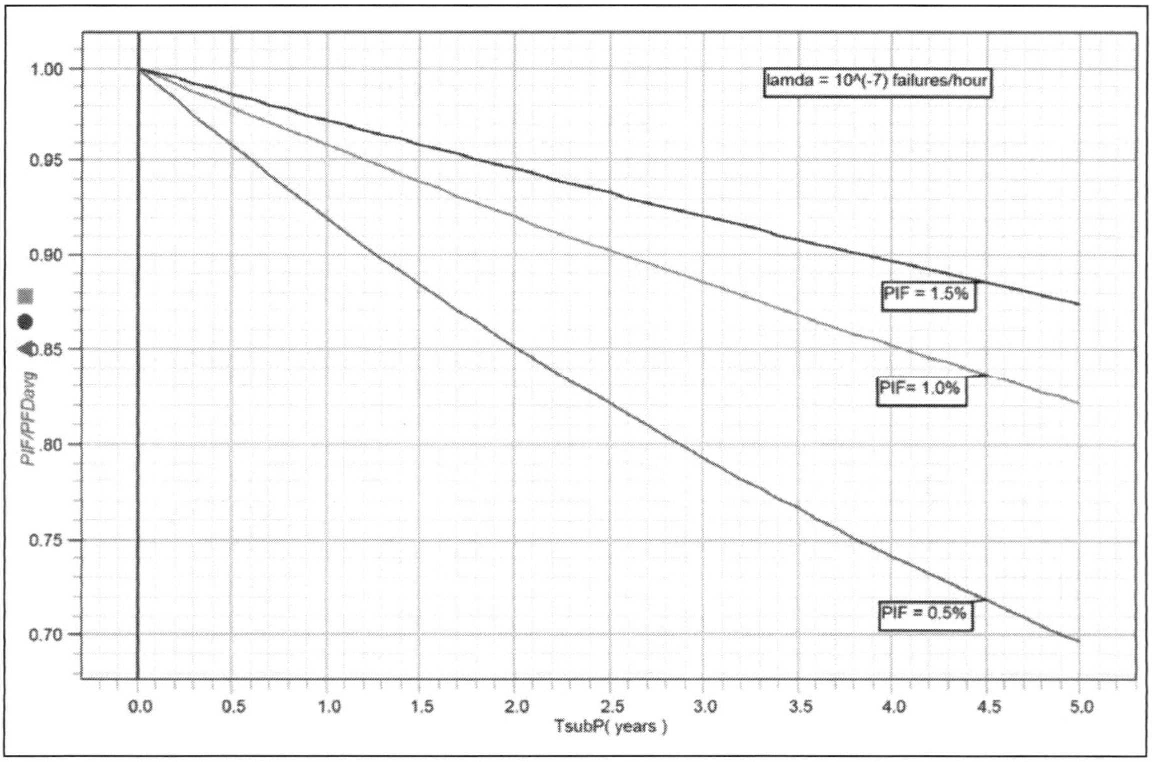

Figure 3. Plot of PIF/PFDavg for various values of PIF

7. DISCUSSION AND SUGGESTIONS FOR FURTHER WORK

The main findings and implications of this research are appropriately captured in the ABSTRACT section at the top of this paper and are not repeated here. This section discusses a few additional points and suggests areas for further work.

FMEDA analysis is an appropriate tool for estimating useful-life failure rates for a given PRV model. It does not require extensive data collection and is the only viable method for predicting failure rates of new designs either not yet built or without sufficient numbers and in-service times to allow for data analysis. However, FMEDA analysis cannot predict PIF, an important parameter to be considered in risk assessment. Thus, this research clearly shows the need for adequate and appropriate data collection and analysis.

Given the dominant effect of PIF on the process safety provided by the PRV, it is crucial to identify and address the underlying causes of initial failures. If PIF can be significantly reduced, process safety can be significantly improved while reducing the frequency of (but not eliminating the need for) proof testing. Of course, to establish that PIF is, in fact, being reduced will require adequate and appropriate data collection and analysis.

8. REFERENCES

1. Bukowski, J. V. (2007), "Results of Statistical Analysis of Pressure Relief Valve Proof Test Data Designed to Validate a Mechanical Parts Failure Database," Technical Report, September, exida, Sellersville, PA.

2. Sheesley, J. H., Thomas, H. W. and Valenzuela, C. A. (1995), "Quantal Response Analysis of Relief Valve Test Data," *ASQC 49th Annual Quality Congress Proceedings*, pp. 741-748.

3. Goble, W. M. and Brombacher, A. C. (1999),"Using a Failure Modes, Effects and Diagnostic Analysis (FMEDA) to Measure Diagnostic Coverage in Programmable Elecronic Systems," *Reliability Engineering and System Safety*, Vol. 66, No. 2, November.

4. *Electrical & Mechanical Component Reliability Handbook*, exida, Sellersville, PA, 2006.

5. Gross, R., Harris, S. (2008), "Analysis of Safety Relief Valve Proof Test Data to Optimize Lifecycle Maintenance Costs", (Draft H), *Proceedings of the 2008 Annual Reliability and Maintainability Symposium*, Los Vegas, NV, January.

9. ACKNOWLEDGEMENTS

The authors wish to acknowledge the contributions and helpful suggestions of Harold Thomas of Air Products & Chemicals, Inc.

10. APPENDIX

The appendix contains the data tables generated by quantal response analysis of Data Sets I, II and III, which were reported in [1], [2], and [5] respectively. Note that in [5], the original data set included an interval with no failures. This is not permitted in quantal response analysis so the interval containing no failures was combined with the next interval to produce the data in Table A3 at $T_i = 0.82$ years.

Table A1. Quantal Response Data for Data Set I

T_i (years)	$-\ln(1-q_i)$
0.65	0.0121
1.59	0.0102
1.96	0.0117
2.11	0.0136
2.42	0.0130
2.96	0.0042
3.28	0.0118
3.55	0.0169
3.74	0.0182
4.21	0.0090

Table A2. Quantal Response Data for Data Set II

T_i (years)	$-\ln(1-q_i)$
0.16	0.0052
0.57	0.0072
0.79	0.0322
0.88	0.0120
0.93	0.0267
0.98	0.0080
1.03	0.0100
1.13	0.0061
1.31	0.0351
1.53	0.0155
1.91	0.0277
2.21	0.0207
2.83	0.0158
3.83	0.0090
5.10	0.0166

6.15	0.0483
6.58	0.0902
7.46	0.0415
9.04	0.1495

Table A3. Quantal Response Data for Data Set III

T_i (years)	$-\ln(1-q_i)$
0	0.0042
0.82	0.0163
2.60	0.0244
3.15	0.0113
4.41	0.0087
5.18	0.0112
8.12	0.0392

2007 NPRA National Safety Conference
May 2-3, 2007
Woodlands, Texas

PASSIVE DEVICE TECHNOLOGY
THE TREND AWAY FROM RELIANCE UPON VALVING SYSTEMS
TO ADDRESS LINE FAILURE

Andrew Abrams, Thomas Steinbach, Joseph Abrams
Smart Hose Technologies, 2536 South 59th Street, P.O. Box 116828, Philadelphia, Pennsylvania, 19142. 215-730-9000

Abstract

Since the landmark incident in Sanford, North Carolina to the more recent tragedy in Riverview, Michigan, reliance upon excess flow valve technology to stop the potentially catastrophic consequences of hose and line failure has been eroding. Through legislation requiring "passive device" technology to formal warnings against this reliance issued by all of the major regulatory agencies in the United States to special permits that provide regulatory relief for users of passive device technologies, a growing trend has commenced towards eliminating this over-reliance upon excess flow valves in the event of transfer line failure. The basis of our paper is to identify this trend, explain its genesis and provide an understanding of the benefits to safety, cost and regulatory compliance through the growing trend and regulatorily encouraged use of passive devices for line and hose failures.

Introduction

For more than a decade, safety concerns have been proliferating throughout industry – none more so than the liquefied petroleum gas (LPG) industry. Incidents have spawned greater concerns over "weak links" in the safety chain of LPG transfer systems. Incidents have caused countless millions of dollars and claimed many lives. On September 8, 1996, in Sanford, North Carolina, during delivery of propane to a bulk storage facility by an MC 331 bulk transport, more than 35,000 gallons of propane were released. The discharge hose separated from its hose coupling at the delivery end of the hose. Most of the transport's 9,800 gallons of propane and more than 30,000 gallons from the storage tanks were released. If this quantity of released propane had ignited, local authorities estimated that about 125 emergency response personnel could have been seriously injured or killed.

On June 3, 1997, in Caro, Michigan, while unloading propane into a storage tank at an industrial facility, the transport delivery hose ruptured. The ensuing fire and a series of explosions seriously burned the driver, destroyed four vehicles and extensively damaged the facility. Initial estimates of property damage were at least $2.0 million.

The history of major accidents in the transportation of anhydrous ammonia is similar to that involving the transportation of liquefied petroleum gases. Pulmonary injuries are more significant with ammonia while fire damage is more significant with liquefied petroleum gases. An example of a major accident involving the release of ammonia is an incident that occurred, in Houston, Texas. The driver of an MC 331 transport lost control while negotiating an interstate exit ramp. The cargo tank motor vehicle released its entire cargo of 7500 gallons of anhydrous ammonia. The driver was killed in the crash. Five other persons were killed and 78 others were hospitalized, all due to inhalation of ammonia. Another 100 persons were treated for less severe injuries. Favorable wind conditions prevented the vapor cloud from reaching a nearby elementary school.

These incidents as well as many more in other petrochemical industries all pointed to a fundamental flaw in the fluid transfer system - over reliance on the excess flow valve. This recognition spawned a landmark Rulemaking issued by the United States DOT-RSPA in which they declared formally that excess flow valves are not reliable to address hose failures and new technology must be developed and identified to provide a greater degree of safety and reliability. Thus the term, **passive technology** was defined as shutdown equipment that has the capability to shut off the flow of product without human intervention within 20 seconds of an unintentional release caused by complete separation of a delivery hose. This definition and requirement thereby changed the safety landscape. This rulemaking marked the first, of several important, steps that have since been taken to require alternative means to address the consequences of transfer line failure by relying upon systems other than excess flow valves.

Excess Flow Valve Technology and The related problems of Over Reliance

An excess flow valve is, typically, a metallic device inserted into the interior piping of a tank car, just below the valve(s) used to load and unload the car. In the event that the valves are sheared off in a railroad accident, there will be a sudden rush of product out the opening thus created. With nothing to impede the flow of fluid product, the excess flow valve will move toward the opening and seat, thus sealing off the opening. In response to concerns that the then-current regulatory provision for excess flow valves might be ambiguous, in 1985 the DOT published a notice of proposed rulemaking to amend the tank car specifications by adopting what is now the contemporary standard. Proponents of the clarification stated that tank-mounted excess flow valves are not intended to substitute for adequate excess flow equipment in plant loading systems. ``The only use of such valves is for protection against loss of lading due to shearing of external closure during transit." \1\

The hazardous materials regulations (HMR) are quite clear that excess flow valves are limited in purpose and scope: An excess flow valve as referred to in this specification, is a device which closes automatically against the outward flow of the contents of the tank in case the external closure valve is broken off or removed during transit \2\

Excess flow valves, by their nature, must encounter a high-volume, surging flow of product to be activated. If that were not the case, they might function in unintended situations, such as when a tank car is being unloaded with the aid of a strong pump. As designed, essentially any apparatus attached to the outside of the external closure valve will create sufficient internal friction (whether hose or pipe) that the flow of product will not be sufficient to activate the excess flow valve. Excess flow valves, by both design and regulation, are intended to function only when the external closure valve is sheared, broken off, or otherwise removed during transit. These devices may also function as a back-up flow control device during tank car loading or unloading activities.

Despite these clear and obvious functional limitations, in the absence of newer passive device technology, industry has placed great reliance on these types of valves to address a scenario that they are not designed for – line/hose failure. (See NTSB Report in which, inter alia, six of nine companies surveyed rely on tank car excess flow valves as a method of stopping or limiting a leak in the transfer equipment.).

In the face of this limitation and over reliance, what would the regulatory bodies do when the next major incident occurs?

Rulemaking HM-225 and 49 CFR 173.315

New regulations **HM-225A** in effect since July 1, 1999, imposed new requirements on operators of cargo tanks used to transport liquefied compressed gases and new procedures for unloading liquefied compressed gases from cargo tanks. The new regulations seek to prevent unintentional releases during unloading of liquefied compressed gases, assure prompt identification of unintentional releases, and reduce the consequences of unintentional releases. Specific provisions include: New inspection, maintenance, and testing requirements for cargo tank discharge systems. Revised unloading requirements for liquefied compressed gases, including revised attendance requirements. New requirements for emergency discharge control equipment, such as passive systems that will shut down unloading without human intervention and remote control devices that enable an attendant to stop the unloading process at a distance from the vehicle.

The new requirements for emergency discharge control equipment on cargo tanks used to transport liquefied compressed gases are keyed to the degree of risk associated with the transportation of specific liquefied compressed gases. The regulation **49 CFR 173.315** specifies two types of emergency discharge control equipment - **Passive Shut-**

down Equipment. Passive shutdown equipment must shut off the flow of product without human intervention within 20 seconds of an unintentional release caused by complete separation of a delivery hose.

This landmark legislation marked the first time that any regulatory body not only identified the risks associated with excess flow valve reliance but further regulated against their use in lieu of other technology. Since this regulation went into effect several systems have been developed and no recorded incidents have occurred.

These same standards were adopted in Canada. **Safety standard CSA B620** has been revised to introduce off-truck remote or **passive** emergency discharge control systems for highway tanks or portable tanks during the unloading of liquefied compressed gases. This applies to highway tanks and portable tanks of the type TC 331 and TC 51, respectively that are used to transport for example, liquefied petroleum gases (LPG) or anhydrous ammonia, the most common of these gases transported. This revision applies equally to compressed gases with a poison inhalation, refrigerant gases and other compressed gases that pose similar risks. It does not apply to Class 2.2 without a subsidiary risk, for example, gases like argon, helium, etc.

The standard calls for new highway tanks or portable tanks to be equipped with these emergency discharge control systems by January 1, 2003. The standard also includes a five-year retrofit period for the installation of these off-truck remote or **passive** systems on existing tanks consistent with the five-year pressure re-inspection date. This retrofit period will begin January 1, 2005, and will end December 31, 2010. The emergency shutdown systems are also subject to test procedures. Appendix D introduces acceptable test procedures to test the off-truck emergency shutdown systems.

The remote capability introduced in CSA B620 enables operators attending the unloading operation to shut off the flow of product when away from the motor vehicle during delivery. These remote systems are associated with TC 331 tanks that are in metered delivery service, more commonly referred to as Bobtail units. The **passive** emergency discharge control equipment (sometimes referred to as a **Smart hose**) is designed to shut down unloading operations without human intervention. These **passive** systems are found on larger trailer type units.

Post-Rulemaking
Atofina Accident and The NTSB's Conclusions

About 3:45 a.m. on July 14, 2001, at the **ATOFINA** plant in Riverview, Michigan, a pipe attached to a fitting on the unloading line of a railroad tank car fractured and separated, causing the release of methyl mercaptan, a poisonous and flammable gas. About 4:09 a.m., shortly after the Riverview Fire Department chief arrived on scene, the methyl mercaptan ignited, engulfing the tank car in flames and sending a fireball about 200 feet into the air. Fire damage to cargo transfer hoses on an adjacent tank car resulted in the release of chlorine, a poisonous gas that is also an oxidizer. The fire was extinguished about 9:30 a.m. Three plant employees were killed in the accident. There were several other injuries; most of the injured were treated for respiratory symptoms and released. About 2,000 residents were evacuated from their homes for about 10 hours. Two tank cars, railroad track, and plant equipment (including hoses and fittings) were damaged in the fire. /3/

The Safety Board determined that the probable cause of the accident was a fractured cargo transfer pipe that resulted from (1) the failure of **ATOFINA** to adequately inspect and maintain its cargo transfer equipment, and (2) inadequate Federal oversight of unloading operations involving hazardous materials. Contributing to the accident was **ATOFINA**'s reliance on a tank car excess flow valve to close in the event of a leak from cargo transfer equipment and the company's failure to require appropriate safety equipment for employees involved in tank car loading and unloading operations.

While **ATOFINA** could not provide data or records to confirm whether or when the transfer pipe that failed in this accident had last been inspected under the mechanical integrity program, the erosion-corrosion that was found within the failed pipe indicated that the program had clearly not been effective. The Safety Board concluded that **ATOFINA**'s failure to implement effective procedures for inspection and maintenance of its unloading pipes and fittings allowed the transfer pipe in this accident to gradually deteriorate and ultimately fail.

Both OSHA and the Environmental Protection Agency (EPA) required ATOFINA to develop and document safety plans for the Riverview facility that included safeguards intended to reduce the risk and consequences of catastrophic releases of hazardous materials. ATOFINA's risk management plan (mandated by the EPA) and process hazard analysis (mandated by OSHA) included an accident scenario that involved the failure of a flexible hose on the unloading apparatus for a methyl mercaptan tank car - a scenario similar to this accident. Under both plans, ATOFINA indicated that the release of methyl mercaptan would be stopped by the automatic closure of the tank car's excess flow valve. Further, **ATOFINA**'s risk management plans explicitly noted that excess flow valves on the tank car would activate in the event of a pipeline or unloading hose rupture. **However, when the transfer pipe/hose failed on July 14, 2001, the excess flow valve on the tank car did not close and stop the release of the methyl mercaptan.**

Calculations made by Safety Board engineers and parties to the investigation indicated that the flow rate of methyl mercaptan through the broken transfer piping was insufficient to cause the excess flow valve to close. Excess flow valves are designed to close and stop the release of product from the tank car in the event a tank car valve or fitting is broken or sheared off during transit. Attaching cargo transfer apparatus to a tank car can change product release rates and flow rate characteristics and can prevent the excess flow valve from closing in the event of an emergency. As noted by the Chlorine

Institute in its *Chlorine Manual Pamphlet 57 and* by the Safety Board in its investigation of a July 30, 1983, accident at the Formosa Plastics plant in Baton Rouge, Louisiana, tank car excess flow valves are not designed to act as an emergency shutoff device during cargo transfer. /4/

To determine whether reliance upon tank car excess flow valves as safety mechanisms during transfer operations is restricted to **ATOFINA** or is a broader problem, Safety Board investigators interviewed a sampling of domestic chemical companies. Interviews with personnel responsible for company safety plans revealed that six of nine companies surveyed rely on tank car excess flow valves as a method of stopping or limiting a leak in the transfer equipment. Only one company reported having remotely operated shutoff valves on the unloading piping just outside the tank car dome. (The other two companies did not respond to the Safety Board's inquiry.) Although the Safety Board's sampling was limited, the results suggest that the **inappropriate use of tank car excess flow valves may be a widespread practice in the chemical industry.**

As a result of its investigation of the Riverview accident, the National Transportation Safety Board therefore made the following safety recommendation to the Occupational Safety and Health Administration: Assist the U.S. Department of Transportation in developing safety requirements that apply to the loading and unloading of railroad tank cars, highway cargo tanks, and other bulk containers that address personal protection requirements, emergency shutdown measures (passive device technology), and the inspection and maintenance of cargo transfer equipment.

Festus, Missouri and the Continued Consequences of Over Reliance on Excess Flow Valves

A chlorine repacking plant receives 4 railcars contained nearly 250,000 lbs of chlorine to be repacked into cylinders and other distribution. After 59 days, a hose wrongfully identified as a chlorine hose catastrophically failed spilling 48,000 pounds of Chlorine over 3 ½ hours. The root cause in addition to the hose failure is that backup systems (ESD'S) as well as Excess Flow Valves (EFV's) failed to work. /5/

DPC Chemical, Festus, Missouri August 2003

Below is the section of the hose that failed after three workers on break heard a loud "pop", which clearly was this hose failing. The hose was made of stainless steel not suitable for Chlorine uses, although it was identified as hastelloy by the manufacturer, a material that is suitable for Chlorine usage. There is no visual difference between stainless steel and hastelloy, therefore the chlorine repackagers employees relied upon the manufacturers representation that they were indeed shipped the correct hose. Moreover, the facility relied upon the four valves (ESDs and Excess Flow Valves) in case of this precise type of error.

Ruptured chlorine transfer hose with its Teflon inner layer exposed;
when the hose burst it ejected pieces of Teflon as well as other components of the hose.

The illustration below (Figure 1) shows the positions of all valves and hoses as it relates to this incident. Determining the actual positioning of the five ESD valves at tank car station #3 led the United States Chemical Safety Board's investigators to conclude that the chlorine flowed through the partially opened ESD valve #3, to the attached intact liquid CTH line, into the plant side piping, and back through the ruptured hose (see arrows denoting product path). Closing tank car valve A did not stop the release because valve C continued to supply chlorine through the intact liquid CTH line. Even if these events replayed themselves the product flowing through the building because of the failure of the ESD's would have been stopped known passive device technology on the inboard side . The pressure of the product itself would have kept the plunger/valve sealed and limited loss of product to what was in the hose, likely below the RQ (reportable quantity).

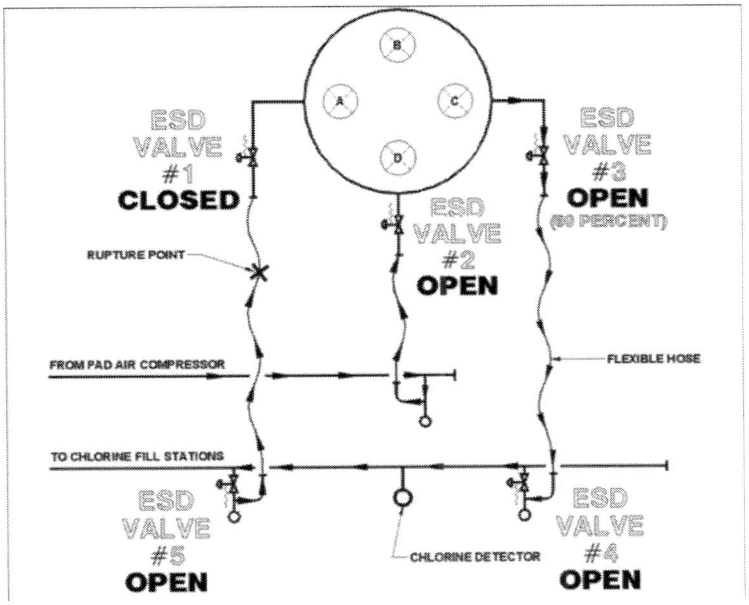

Figure 1

CSB investigators concluded that it is all but impossible to determine all possible routes of escape for hazardous materials or the infinite numbers of LPHC (Low Probability/High Consequence) scenarios. It is however abundantly clear that the hose is the weakest link in the chain and that devices, training and other precautions have proven to be inadequate in combating hose failures. In most reports of hazardous material release incidents we submit that one constant seems to be that employees do not follow regulations, rules and precautions that management, organizations and agencies have mandated and reasonably assume are being followed. Moreover, the valving technology that relies upon excess flow principles has been vastly over relied upon to address these incidents and governmental regulatory bodies have taken notice through he issuance of warnings and landmark Special permits encouraging the use of the passive device system mandated in 49 CFR 173.315.

Regulators Begin their Proliferation of Passive Device Technology Benefits

A Special Permit (previously referred to as an Exemption to Regulation) is temporary regulatory relief from one or more regulations given to a person or class of persons subject to the regulations, or who intend to engage in an activity that would make them subject to the regulations. (b) A Special Permit provides the person or class of persons with relief from the regulations for up to two years, and may be renewed. (c) Special Permits may only be granted from one or more of the requirements contained in the certain sections of the various regulations. To apply for such a Permit, the applying party must identify how they would ensure that they could achieve a level a safety that is **equivalent to, or greater than, the level of safety that would be obtained by complying with the regulation.** Thus the issuance of such a permit indicates that the regulatory body considers the procedure or technology to be of a greater safety value than the regulations promulgate. This is precisely the conclusion that the United States Federal Railroad Administration reached when they issued the landmark Special Permit 12325-SP. (See attached FRA Analysis)

Exemption/Special Permit 12325-N provides that the use of a Smart-Hose Technologies hose assembly will exempt the user from 49 CFR 173-177-Railcar Unloading Attendance Requirements. This regulation normally provides that rail tank car unloading operations must be performed by a reliable person who has been properly instructed in unloading hazardous materials and attended by the unloader until unloading is completed. Section 177.834 requires that a cargo tank motor vehicle be attended by a qualified person (usually the driver) at all times while it is involved in transfer operations to ensure the safe loading or unloading of hazardous materials being loaded or unloaded. The purpose of the attendance requirement in section 174.67 is to ensure the safe loading or unloading of hazardous materials and that, in the event of an emergency; such processes can be quickly halted. The Permit provides that the attendance requirements in the above regulations are relaxed if the rail car operator is using a Smart-Hose hose assembly.

The loading and unloading of tank cars requires the driver to stand and monitor the discharge of gas and fluids from the vehicle. The DOT recognized that this monitoring does not always occur particularly in the cold wintertime when the drivers may retreat from the cold to their vehicles. Air Liquide, a current user of the Smart-Hose passive device technology effectively petitioned the DOT to create a Special Permit 13484 to provide for relaxed attendance requirements for MC 330, MC 331 and 338 cargo tankers for division 2.2 gases.

Warnings are Issued

Nearly two years after the National Transportation Board made its recommendations; all major regulatory bodies in the United States issued warnings. In April of 2004, the United States Environmental Protection Agency issued the following warning: "Notify all facilities that are required to submit risk management plans to the

Environmental Protection Agency that tank car excess flow valves cannot be relied upon to stop leaks that occur during tank car loading and unloading operations and that those companies that have included reliance on such valves in their risk management plans should instead identify and implement other measures that will stop the uncontrolled release of product in the event of a transfer line[hose] failure during tank car loading or unloading."

In, September 2003, the **United States Department of Transportation – Federal Railroad Administration** issued the following warning: FRA is issuing Safety Advisory 2003-02 advising all persons involved in loading and unloading products from railroad tank cars that they cannot rely on internal excess flow valves to stop the flow of product except under the limited conditions for which these valves were designed and installed.

As early as 1980 **OSHA had issued warnings** indicating that in Anhydrous Ammonia service, it is emphasized that the system is not adequately protected without devices designed specifically to protect the system in case of rupture of the applicator tank liquid withdrawal hose. It is strongly stressed that excess-flow valves or other protective devices installed in bulk storage tank feed hose outlet lines are not designed to protect the system in case of an applicator tank withdrawal hose rupture. These devices are required to protect the lines that they serve and are designed specifically for this purpose only. For example, a manufacturer of anhydrous ammonia equipment advised that an excess flow valve installed in a bulk storage tank hose outlet (designed to accommodate a one-inch fill hose that is 15 feet in length) will be spring-loaded to trip should the hose be severed at any point along the 15 foot length of the hose; however, it is doubtful that adequate protection would be provided beyond the 15 foot point. Thus in the case of the applicator tank equipped with the combination valve, the hose outlet excess flow valve provides protection of the system from the bulk tank to the applicator fill/withdrawal valve with doubtful effectiveness in protecting the system beyond this point. It should be noted that the applicator tank involved in the accident was provided with a 3/4 inch liquid withdrawal hose. A reduction in the diameter of the hose line on the applicator tank further reduces the possibility of the bulk tank hose outlet excess flow valve tripping in case of severance of the applicator tank withdrawal hose -- especially in view of the fact that excess flow valves function on the basis of gallons per minute (GPM) flow rates and pressure differentials, i.e., a sudden drop of pressure in the line.

The preceding discussion has been limited to liquid lines and the filling of an applicator tank from a bulk storage tank. However, the same principles apply when filling the applicator tank from a nurse tank. In addition, it is equally important that protective devices be installed in the vapor lines to protect the system in case of vapor line rupture. It should also be noted that the bulk tank that was used to fill the applicator tank involved in the accident was not provided with an excess flow valve on the one-inch hose outlet on the bulk tank.

HM-223 - the Jurisdiction and Perspective Changes

Under the newly enacted Rulemaking (98-4952), otherwise known as HM-223, RSPA clarified the applicability of the Hazardous Materials Regulations (HMR) to specific functions and activities, including hazardous materials loading and unloading operations and storage of hazardous materials during transportation. They also listed in the HMR pre-transportation functions to which the HMR apply. They stated that pre-transportation functions are functions performed to prepare hazardous materials for transportation in commerce by persons who offer a hazardous material for transportation or cause a hazardous material to be transported. They determined that transportation functions are functions performed as part of the actual movement of hazardous materials in commerce, including loading, unloading, and storage of hazardous materials that is incidental to that movement. For purposes of applicability of the HMR, they concluded that ``transportation in commerce" begins when a carrier takes possession of a hazardous material and continues until the carrier delivers the package containing the hazardous material to its destination as indicated on shipping papers or other shipping documentation. Thus, in promulgating this rule, they effectively transferred the regulatory oversight of loading and unloading railcars to the Occupational Safety & Health Administration ("OSHA") and, to a lesser extent, the Environmental Protection Agency.

In light of the National Transportation Board's recommendations to OSHA to include these warnings in their assumption of jurisdiction and related regulator interpretations, will OSHA utilize this opportunity to further the trend away from excess flow valve reliance towards passive device technology? It would appear at first review, the answer is "Yes". In a recent submission seeking a Letter of Interpretation, the following questions were asked and verbally answered in the affirmative with formal comment expected to be issued with 30 days of this submission:

1. Does OSHA have regulations that specify tank car unloading similar to 49 CFR 174.67 (i) (j) and (k) which provide: (i) Tank cars may not be allowed to stand with unloading connections attached after unloading is completed. Throughout the entire period of unloading, and while car is connected to unloading device, the car must be attended by the unloader.(j) If necessary to discontinue unloading a tank car for any reason, all unloading connections must be disconnected. All valves must first be tightly closed, and the closures of all other openings securely applied. (k) As soon as a tank car is completely unloaded, all valves must be made tight by the use of a bar, wrench or other suitable tool, the unloading connections must be removed and all other closures made tight.

2. In the absence of these regulations, will OSHA's General Duty Clause (Section 5a1) which requires each employer to "furnish to each of his employees employment and a place of employment which are free from recognized hazards that are causing or are likely to cause death or serious physical harm" be the defining standard for these procedures?

3. Will the use of Smart-Hose assemblies, *(a passive device, as defined under 49 CFR 173.315, and proven hose technology designed to counteract the hazardous effect of hose rupture or failure during fluid or gaseous transfer operations through its unique, patented and patent pending design that eliminates the potential for disaster through the use of an internal cable connected to specially designed, normally unseated valve "wedges or plungers" located on each end of the cable. In the event of hose separation, stretching to the point of an unsafe condition or coupling-to-hose separation, the valve "wedges or plungers" are released and instantly seat stopping the flow in both directions.)*, in conjunction with the railcar loading and unloading requirements of 49 CFR 174 and the specific requirements of Special Permit 12325 (see below excerpts and complete copy is attached hereto and made a part hereof) satisfy the General Duty provisions of 5a1 thereby making the user compliant with OSHA's railcar loading and unloading regulations and related interpretations?

The direction that has been recommended by the National Transportation Safety Board and promulgated through warnings by the Environmental Protection Agency, the Department of Transportation- Federal Railroad Administration and the US-Department of Labor's Occupational Safety and Health Administration is away from over reliance on excess flow valves towards the regulatorily established Passive Device technology.

Conclusion

Given the numerous, detailed and accurate warnings from manufacturers, advocacy groups, government agencies and from accidents themselves, it is not reasonable for anyone be they industry or agency to be surprised when the next hose bursts and an EFV fails to stop the catastrophic consequences. These warnings are advice to beware to professionals charged with the safety of facilities, communities and individuals and are intended to be actionable.

The "passive device" addresses directly the deficiencies of the excess flow device. These deficiencies are not an indictment of the EFV device, the device behaves exactly as designed when used and installed properly. Rather, where the fault lies is in the properties and attributes that industry affixes to these devices to do a job for which they were not designed. The EFV is designed to trigger at a preset flow rate exceeding normal operation and contemplating the "shearing off" of valves or other catastrophic hardware event. In the event of a transfer line failure, in truth the EFV does not fail it simply works as designed, as the flow through a transfer line is likely within the parameters of normal operation and a rate at which the EFV set to accept.

The "passive device" is designed to shut off the flow of product in both directions in the transfer hose in the event of a failure between the couplings. This device provides safety at flow levels that are common in instances that sometimes are not sufficient to activate the EFV. While the EFV triggers at only high rates a "passive device" will trigger in incidents not only at high and low rates but even when a system is on standby (Festus) making the device not only complimentary but redundant to the EFV. When the next transfer lines fails and the call goes to the responsible party rather than wondering if they have dodged a bullet with an EFV the first question will be, "What did the hose do?"

A proven solution is at hand and industry and government are moving inexorably towards adaptation of passive device technology. The aforementioned warnings are not intended to provide cover to those that ignore them, as they are now so widely known. The fact is that given this knowledge and "best practice" the consequences to those who do take notice and action are likely severe.

Through the use of Passive Device Technology, sanctioned and in some cases mandated, the LPG industry can provide a level of safety for our operators, a level of compliance increasingly mandated and further confidence to our customers that safety does count and we embrace it.

Acknowledgment

We want to thank the various nongovernmental organizations that have supported this effort. We appreciate the contributions by the National Transportation Safety Board, the United States Chemical Safety Board, The United States Department of Transportation, The United States Environmental Protection Agency and The United States Department of Labor – OSHA. We further appreciate the World LP Gas Association for engaging us to research and develop a comprehensive paper on the impact of excess flow valve technology on the LP gas industry.

References

\1\ Docket HM-166W, NPRM at 53 FR at 36418, September 19, 1988; Final Rule adopting the amendment as proposed, 54 FR 38790, September 20, 1989.

\2\ 49 CFR 179.100-13(d).

National Transportation Safety Board, *Hazardous Materials Release From Railroad Tank Car With Subsequent Fire at Riverside, Michigan, July 14, 2001*. Hazardous Materials Accident Report NTSB/HZM-02/01 (Washington, D.C.: NTSB, 2002).

\3\ National Transportation Safety Board, *Safety Recommendation,* Norman Minetta, July 16, 2002.

\4\ The Chlorine Institute, Pamphlet 57, Emergency Shut-Off Systems for Bulk transfer of Chlorine, Edition 3, October 1997 section 3.1.

\5\ The Chemical Safety Board, Investigation Report Chlorine Release, DPC Enterprises, May 2003, incident Festus, Missouri August 14, 2002

2 National Transportation Safety Board, *Vinyl Chloride Monomer Release From a Railroad Tank Car and Fire, Formosa Plastics Corporation Plant, Baton Rouge, Louisiana, July 30, 1983*, Hazardous Material Accident Report NTSB/HZM-85/08 (Washington, D.C.: NTSB, 1985).

Although excess flow valves are routinely used as safety mechanisms in the piping systems of fixed facilities, those excess flow valves are designed and constructed for specific piping systems and the properties of the material flowing through the pipe.

Correspondence dated July 16, 2002, from Marion C. Blakely, Chairman, NTSB, to Allan Rutter, Administrator, FRA, summarizing the Board's investigation, including a public hearing, of this incident.

RISK BASED METHOD TO ESTABLISH INSPECTION INTERVALS FOR PRESSURE RELIEF DEVICES

Philip A. Henry, P.E.
The Equity Engineering Group, Inc.
Shaker Heights, OH

Valerie L. Magyari
The Equity Engineering Group, Inc.
Shaker Heights, OH

ABSTRACT
Properly functioning pressure relief devices (PRDs) are essential to protect plant personnel and equipment, as unexpected overpressure events can cause equipment damage and loss of containment, and result in costly plant shutdowns. PRDs are considered to be the ultimate layer of protection of pressurized systems.

The inspection and testing of relief devices is needed to ensure PRD functionality on demand. But when and how often should PRDs be inspected? Many plants apply inspection intervals to PRDs in general agreement with API 510 and NBIC, without considering historical performance or the potential hazards associated with the relief device application.

Previously, the primary focus of Risk-Based Inspection (RBI) programs has been to create inspection plans that reduce risk and optimize total maintenance and inspection expenses for pressurized static equipment. This approach has been expanded to include a quantitative methodology for PRDs. The PRD RBI technology provides a system approach (includes the damage state of the protected equipment) and evaluates the probability (reliability) and the consequence (criticality) of pressure relief valve failure and sets the appropriate inspection frequency based on risk. Results of recent studies have demonstrated the benefits of the new approach. Compared to a typical API510 approach, the risk based plan can significantly reduce the overall risk in the plant while at the same time a significant reduction in inspection costs can be realized.

A case study of implementation in a refinery process unit illustrates the methodology has been successfully used to establish risk-based inspection intervals for PRDs.

Chemical Safety Database - Application for Flammability Limits of Biogases

Maria Molnarne and Volkmar Schröder

Federal Institute for Materials Research and Testing (BAM)
12200 Berlin, Germany
E-mail: maria.molnarne@bam.de

ABSTRACT

The chemical safety database CHEMSAFE® contains safety characteristics for more than 3000 flammable substances and approximately 1000 mixtures of gases, liquids and dusts. The data is evaluated by experts of two Federal Institutes in Germany, BAM and PTB. The main advantage of CHEMSAFE® is that the data based on own tests of these institutes. The temperature and pressure influence on safety properties are also taken into account. Flammability limits are measured up to 400 °C and 10 MPa. In this paper an application of the database is shown for calculation of flammability limits of biogases. Biogas consists mainly of methane and carbon dioxide, with low levels of hydrogen sulfide and other gases. For the explosion protection in biogas plants, it is necessary to know the flammability limits of the process gases in the air. The composition of biogases from agricultural plants is varying very strongly, so that for each mixture composition the flammability limits should be determined experimentally. This is associated usually with considerable time and effort. A method is presented here which shows that the flammability limits of a biogas with known composition can be assessed with the help of two explosion diagrams. In addition to the methane and carbon dioxide content of the biogas a portion of water vapor is included in the calculation method and the influence of oxygen content in the system is considered.

1. INTRODUCTION

In the recent years biogas has become an important source of renewable energy. During the last three years some serious accidents, accompanied with injuries and fatalities, have happened in biogas production plants in Germany. These agricultural plants must also be controlled to avoid pollution and the risk of explosion. According to the European Directive 1999/92/EC (ATEX 137) the operator of biogas plant in Europe must meet explosion protection measures, such as the prevention of explosive mixtures (primary explosion protection) and ignition sources (secondary explosion protection). For this purpose an explosion protection document must be created. Such a document is based on the safety characteristics of hazardous substances. For biogas plants this safety characteristic is determined by the flammability limits of biogas in air. The biogas consists of mainly methane and carbon dioxide with small percentages of nitrogen, oxygen, hydrogen sulfide and other trace gases.

2. AGRICULTURAL BIOGAS PLANTS

A simple plant, as it can be found in a large number of farms, is shown in Figure 1. The biogas is present in the following operation units: in the digester, the connected gas storage tank and the combined heat and power unit (CHP) with possibly connected emergency flare. For larger plants, which have the permission to feed the gas into the natural gas network, there is a reprocessing plant with gas scrubbing, etc., mostly equipped with a downstream desulfurization unit. Especially in these operating units - with the appropriate vessels, valves, pipes and compressors - a failure may occur that results in the formation of an explosive gas-air mixture.

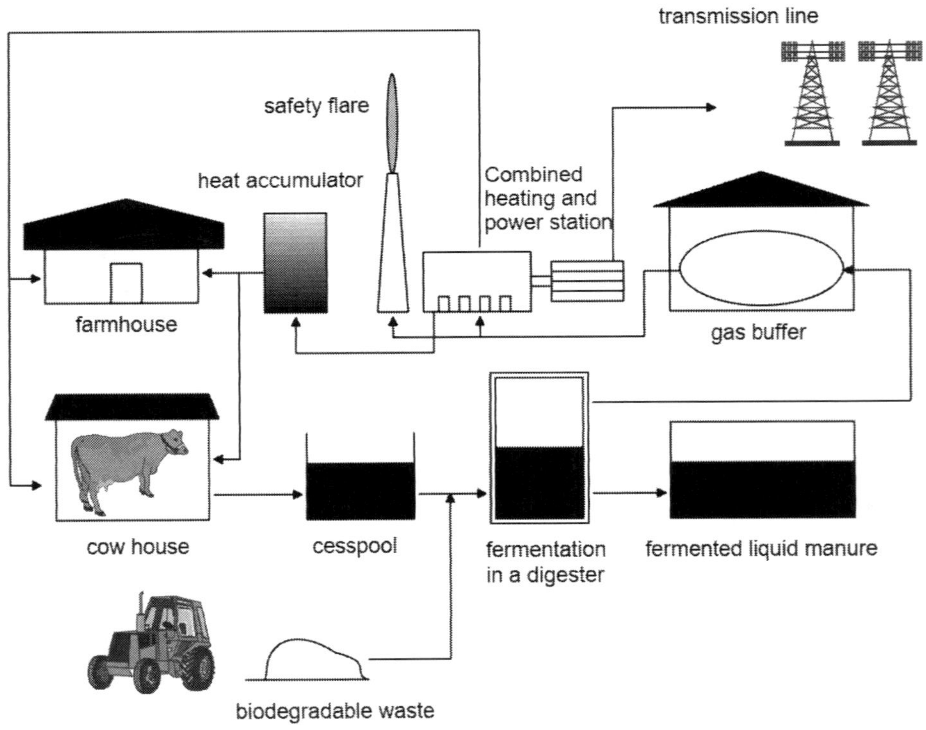

Fig. 1: A simple biogas plant in a farm.

Biogas is a naturally produced gas which is available from decomposition of organic materials in an anaerobic digestion process. The composition varies strongly depending on the introduced organic materials. The raw biogas usually consists of the following components:

- methane (45 mol% to 75 mol%),
- carbon dioxide (25 mol% to 55 mol%),
- water vapor (0 mol% to 12 mol%)
- nitrogen (0 mol% to 5 mol%)
- oxygen (0 mol% to 2 mol%)
- hydrogen sulfide (0.05 mol% to 0.5 mol%), and traces of ammonia, hydrogen and higher hydrocarbons.

The composition is given in mol% which is the same as volume fraction in percent at atmospheric conditions.

In addition, the composition of the process gas depends on the place, the part of the plant, where the gas is located. The biogas in the digester is saturated with water vapor at temperatures of up to 50 °C. On the other hand, the water vapor portion in the gas is significantly lower at ambient temperature. If the gas is fed into the natural gas network, it must be desulfurized, and the proportion of methane has to be increased. For the desulfurization usually some air is added.

The upper (UFL, fuel-rich) and lower flammability limit (LFL, fuel-lean) are referred to the mole fractions (concentration) of the biogas in air at which flame propagation after spark ignition no longer occurs, the last non-ignition point. The range between the lower and upper flammability limits is called as explosion range. In European standards and in this paper the flammability range is called explosion range to emphasize the explosive behavior of the mixtures and set it apart from the flammability of pure combustible substances. The area restricted by the lower and upper flammability limit curves in a fuel-air-inert gas system is called the explosion region. Mixtures with a composition in the explosion region are explosive and dangerous. The flammability limits of pure substances, such as methane, can be taken from data books and safety databases. For a mixture like "biogas" this is not readily available, so that in practice often the flammability limits of methane (4.4 mol% to 17.0 mol%) [1] are applied. At the lower flammability limit this is allowed, because the mixtures are in the "safe" area. At the upper flammability limit, such as by air inlet into biogas, this can lead to dangerous conditions.

It is better, therefore, if the composition is known, to estimate the flammability limits of biogas with the help of so-called explosion diagrams. The explosion ranges can be represented in different types of flammability diagrams. Most frequently diagrams are used with triangular or Cartesian coordinates. Both types of representation allow several possibilities for defining the axis. Unfortunately in the safety literature no uniform definitions can be found. For the present work an axis assignment has been selected, as it is used in the Federal Institute for Materials Research and Testing (BAM). A collection of such flammability diagrams of fuel-inert gas-air mixtures, as measured according to the German standard DIN 51649, is in [2].

3. EXPERIMENTAL DETERMINATION OF THE EXPLOSION REGIONS

The flammability limits, in case of the diagrams used in this paper, have been determined at atmospheric pressure with an apparatus in accordance with DIN 51649 [3]. Since 2004 this standard has been replaced by the DIN EN 1839 [4] tube method. Both standards recommend almost identical test apparatus. The new European standard describes in the tube method a little larger inner diameter for the ignition vessel (80 mm instead of 60 mm). The larger diameter does not affect the results for the mixtures of methane in question here. It was demonstrated by comparing measurements in [5].

In Figure 2 an apparatus can be seen for determination of the flammability limits according to DIN EN 1839 (DIN 51649), as it is used in the BAM. The gas mixture is metered by thermal mass flow controllers at different nominal volume flow rates. For the corresponding gas the mass flow controller must be calibrated by a precision gas meter.

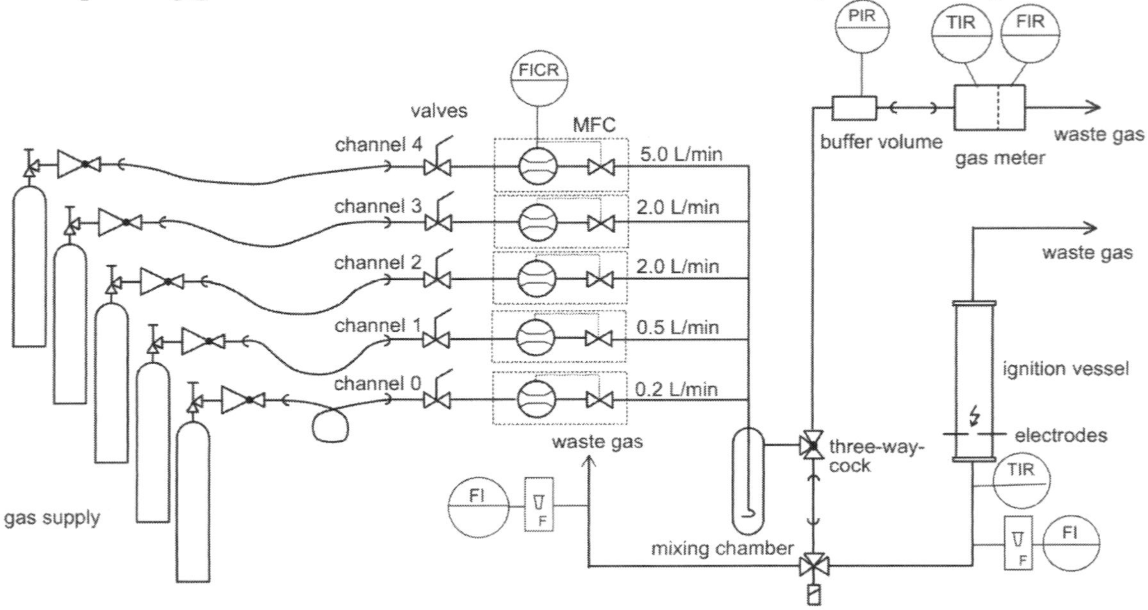

Fig. 2: BAM apparatus according to DIN 51649 and DIN EN 1839 „Tube method", respectively.

Thereafter the gas is piped through a mixing vessel into the ignition vessel. After purging the ignition vessel the gas is ignited by using high voltage sparks and observed visually whether a flame is detaching from the electrodes and spreading out through the ignition vessel in test mixture. The ignition vessel consists of a vertical glass tube with a length of 300 mm and an inner diameter of 60 mm (DIN) and 80 mm (DIN EN), respectively. The apparatus is tested for functionality by a reference gas facing a test series. With the help of gas chromatography, the composition of the mixtures is randomly tested. The explosion regions of gas mixtures have been determined by varying the percentages of flammable gas, inert gas and air, respectively. The increment in the determination of the flammability limits was 0.2 mol% methane, and for determination of limiting oxygen concentration a step size of 0.5 mol% was used.

4. THE EXPLOSION REGION OF METHANE-CARBON DIOXIDE-AIR MIXTURES

The explosion regions of ternary mixtures can be represented most clearly in the so-called triangular diagrams. Here each side corresponds to a binary mixture and the three corners correspond to the respective pure substances. The composition of a mixture can be read directly from the diagram. The explosion diagram of methane-carbon dioxide-air in Figure 3 is an example of this type of diagrams.

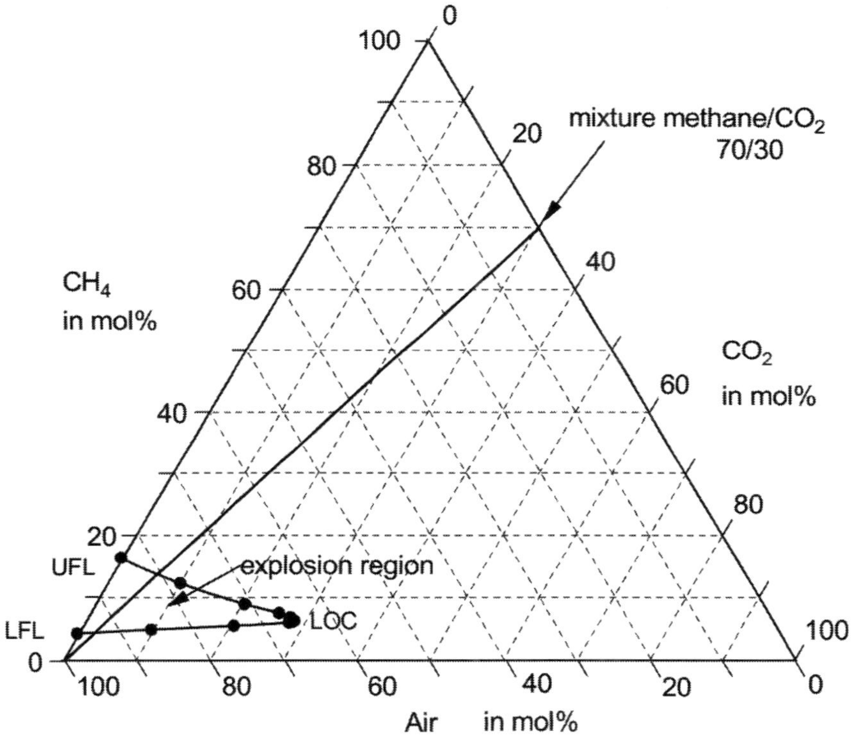

Fig. 3: Explosion region of methane-carbon dioxide-air mixtures [2], measured according to DIN 51649 at 20 °C and atmospheric pressure. The points represent the measured values.

The lower (LFL) and upper flammability limits (UFL) of methane in air are read off on the left side of the diagram. The values are LFL = 4.3 mol% and UFL = 16.3 mol% in this test. The air fraction at the apex of the explosion region is used to determine the limiting oxygen concentration (LOC). The air fraction is found with 65 mol%. This represents an oxygen concentration of 13.6 mol%.

Some basic statements can already be made by using these key data. In case of leakages and the accidental release of biogas in the surrounding atmosphere an explosive mixture can only be formed, if the lower flammability limit of methane is exceeded. Furthermore, it is only possible to obtain an explosive mixture in case of ingress of air into a biogas vessel, if the LOC is exceeded. The LOC of 13.6 mol% refers to a dry biogas, in which the water vapor portion is negligible. For the digesters, in which the biogas has a major portion of water vapor at temperatures up to 50 °C, an LOC of 12.4 mol% can be conservatively used. This value is determined from a methane-water vapor-air explosion diagram.

The straight line shown in Figure 3 corresponds to a biogas with a methane content of 70 mol% and a carbon dioxide content of 30 mol%. By constant addition of air to this mixture the composition of the mixture moves from the right side along the straight line by intersecting the explosion region to the lower left corner which corresponds to pure air. The flammability limits of the biogas (70/30) can be obtained also from the intersection points of the line with the explosion region.

5. CALCULATION OF THE FLAMMABILITY LIMITS OF METHANE-CARBON DIOXIDE-AIR MIXTURES

The lower percentages of other fuel gases which are present in biogas can be added to the methane portion for the calculation procedure. The flammability limits can be determined from the explosion region in Figure 3 for dry biogas with negligible portions of water vapor and oxygen. For better representation of the intersection points with the explosion region a diagram with Cartesian coordinates is often applied. This representation has the advantage that the axes are scalable, and the explosion region can be displayed slightly in optimum size. Moreover, most graphic and spreadsheet programs do not deliver triangle coordinates by default. A disadvantage of this Cartesian type diagrams is, however, that the representation of the axis for the third component is missing. Therefore the portion of air in a mixture can be gained by subtracting the portions of methane and carbon dioxide from 100 mol%. In Figure 4 two biogas compositions with methane-carbon dioxide (70/30) mol% / mol% and (50/50) mol% / mol% are exemplary represented. The flammability limits of biogases are calculated with the help of this diagram by using the mole fractions of methane and carbon dioxide according to equation (1) and (2).

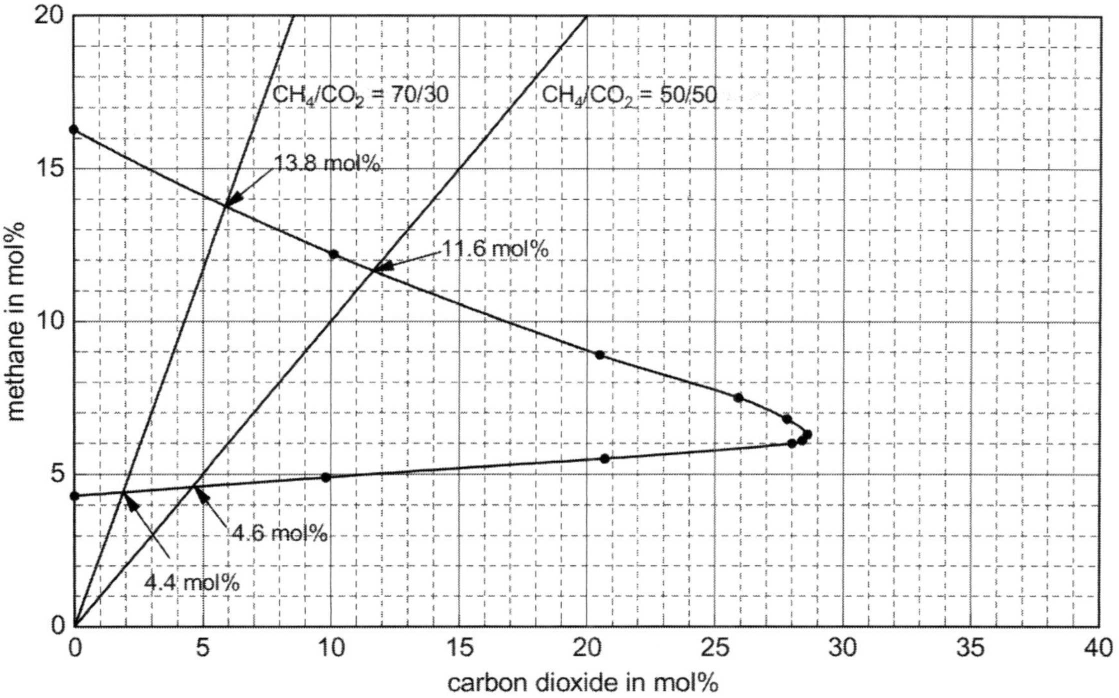

Fig. 4: Explosion region from Figure 3 in Cartesian coordinates.

$$LFL_{Biogas} = \left(1 + \frac{x_{CO2}}{x_{CH4}}\right) \cdot LFL_{CH4,CO2} \tag{1}$$

$$UFL_{Biogas} = \left(1 + \frac{x_{CO2}}{x_{CH4}}\right) \cdot UFL_{CH4,CO2} \tag{2}$$

x_{CO2}, x_{CH4} fractions of the biogas components
$LFL_{CH4,\ CO2}$, $UFL_{CH4,\ CO2}$ methane fraction at the intersection point with the limit curves

These equations resulted in the following lower flammability limits

$LFL_{Biogas\ 70/30}$ = (1 + 30/70) 4.4 mol% = 6.3 mol%
$LFL_{Biogas\ 50/50}$ = (1 + 50/50) 4.6 mol% = 9.2 mol%

and the upper flammability limits for the sample mixtures

$UFL_{Biogas\ 70/30}$ = (1 + 30/70) 13.8 mol% = 19.7 mol%
$UFL_{Biogas\ 50/50}$ = (1 + 50/50) 11.6 mol% = 23.2 mol%

If these calculations are carried out for multiple configurations, a diagram can be constructed as shown in Figure 5. From this diagram the flammability limits of the biogas immediately can be read, if the methane content is known.

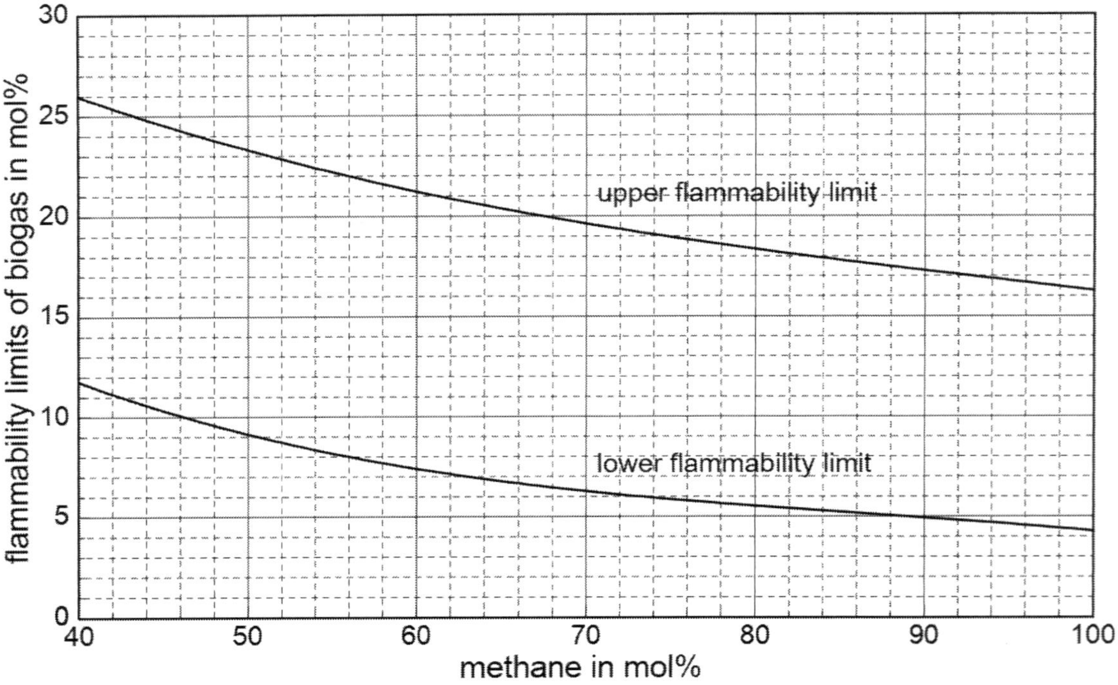

Fig. 5: Flammability limits of dry biogas at known methane content.

6. THE SIGNIFICANCE OF THE FLAMMABILITY LIMITS FOR MONITORING EQUIPMENTS

The first case to be considered is an accidental leakage and release of biogas into the surroundings. The key safety characteristic for this case is, in order to avoid of an explosive biogas-air mixture, the lower flammability limit. It indicates the percentage of biogas above which a biogas-air mixture becomes explosive. For this purpose monitoring equipments (gas warning devices) will be mounted on the units of biogas plants which detect the portion of methane entering into the atmosphere. Here it is important to mention that gas warning devices used in the explosion protection should be tested for functionality by an authorized testing laboratory. The accuracy of the measurement function must be ensured under the operating conditions (temperature, humidity, cross sensitivities).

From Figure 4 and from the calculation examples it is shown that, although the LFL of biogas with different CO_2 portions is quite different, the proportion of methane in the LFL remains nearly constant. For a biogas containing 30 mol% CO_2 the proportion of methane in the LFL is 4.4 mol% and for a 50 mol% biogas it increases only to 4.6 mol% methane (see Figure 4). It also shows that it is acceptable to use the LFL of pure methane as a basis for alarms in gas warning devices. A calculation of the LFL of biogas is therefore in this case not necessary.

The situation is quite different in the upper flammability limit. This case can occur by ingress of air into a part of biogas plants or even during the proper addition of air, such as by desulfurization. Because of an oxygen deficiency the air concentration has the priority in the consideration. It is useful here to apply oxygen sensors for monitoring. The allowable air concentration is calculated from the percentage of methane and carbon dioxide by subtraction from 100 mol%. In our example the biogas with 30 mol% carbon dioxide has an UFL of 19.7 mol% according to equation (2) and with 50 mol% CO_2 an UFL of 23.2 mol%. This corresponds to an air portion of 80.3 mol% and 76.8 mol%, respectively. The permissible oxygen concentration is 16.8 mol% or16.1 mol% calculated from the air concentrations. Using the upper flammability limit of pure methane in air is the oxygen concentration of 17.5 mol%. Using this oxygen percentage the resulting biogas –air mixture can become explosive.

7. CONSIDERATION OF WATER VAPOR IN THE DIGESTER

Before drying and compressing of biogases water vapor is present in different units of the biogas plant in phase equilibrium. The portion of the water vapor in the total gas mixture can be obtained from the temperature dependence of the vapor pressure. According to Table 1 there are significant water vapor contents only at higher temperatures, for example in the digester.

Tab. 1: Vapor pressure of water [6]

Temperature in °C	Water vapor content in mol% at 101.3 kPa
20	2.3
25	3.2
30	4.1
35	5.5
40	7.3
45	9.5
50	12.2

If air enters the digester, the flammability of these mixtures can be conservatively estimated from an explosion diagram of methane-water vapor-air (see Figure 6). This is acceptable, because the water vapor has slightly lower molar heat capacity than carbon dioxide. It does not lead to any explosion hazards, if the carbon dioxide proportion like water vapor portion is considered.

The following example illustrates this approach. The conditions in the reactor after an air access are: T = 50 °C, p = 101.3 kPa, the composition of dry biogas: 70 mol% methane and 30 mol% carbon dioxide. The detected oxygen percentage is 12 mol%.

At 50 °C is a water vapor content of 12.2 mol% according to Table 1.

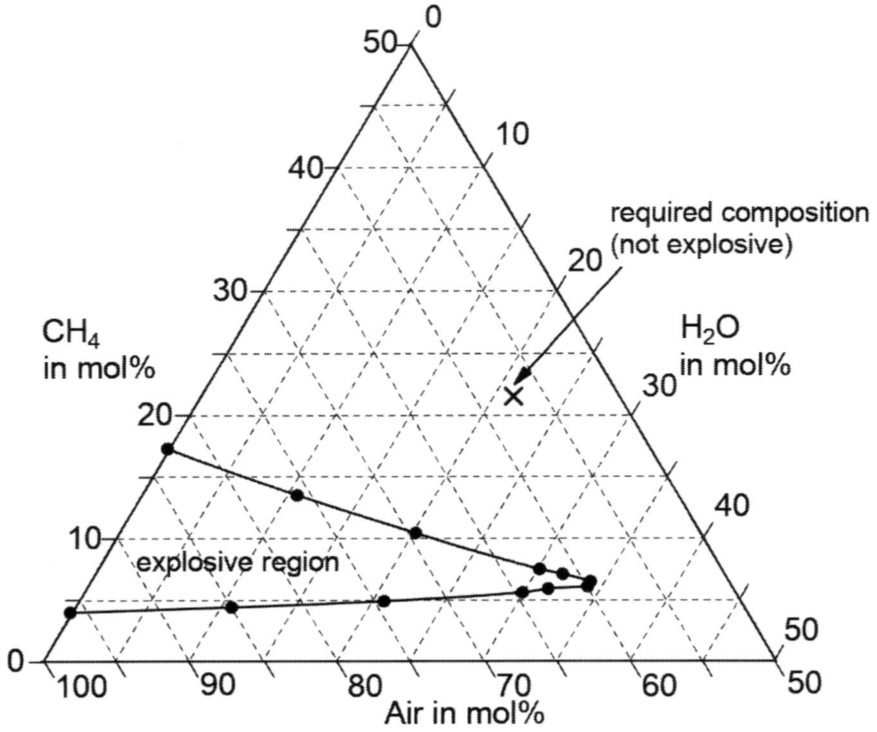

Fig. 6: Explosion region of methane-water vapor-air mixtures [2], measured according to DIN 51649 at 100 °C and atmospheric pressure.

First, the percentage of air is calculated by equation (3) from the measured oxygen content.

$$x_{Air} = x_{O2}/0.21 = 0.571 \triangleq 57.1 \text{ mol\%} \quad (3)$$

The next step is the calculation of the sum of the inert gases (CO_2 + H_2O) by equation (4).

$$x_{inert} = (1\text{-}x_{Air}\text{-}x_{H2O})*0.3 + x_{H2O} = 0.214 \triangleq 21.4 \text{ mol\%} \quad (4)$$

By using these data, the composition of the mixture can be placed into the explosion diagram in Figure 6 and it can be seen that this point is outside of the explosion region. That means that the mixture in our example - in the digester - is not explosive.

8. CONCLUSION

For the indication of the flammability limits of biogas in air it is important to distinguish whether these refer only to the methane content, or to the whole biogas. Both can be found in the safety literature and confusion may lead to dangerous situations.

It is sufficiently safe, to use the lower flammability limit of methane at atmospheric conditions as a warning alarm in biogas plants.

If air is entering into parts of biogas plants, such as by shut downs and start-ups, repairs, etc., the upper flammability limit of biogas should be used to estimate the risks of explosion. The use of the upper flammability limit of pure methane as a reference point for monitoring equipment is not allowed here.

In these cases with the proposed calculation method or from the diagram in Figure 5 of the remaining allowable air content will be determined. Gas warning devices which are applied for measuring the oxygen content can then be set up to these values. For gas alarms the LOC value (see Figure 3) can also be used here, but it overestimates the actual risk.

The influence of temperature on the flammability limits is low. Up to the temperatures of 50 °C, e. g. in the digester or by sunlight, it can be neglected. The influence of higher pressures, such as by compression of biogas in CHP unit may occur, however, must be observed and separately considered. For this purpose explosion diagrams for higher initial pressures are needed.

Avoiding ignition sources as the sole explosion protection measure is usually not enough. Experience has shown that in case of complex systems, usually it is not possible all possible ignition sources to eliminate.

9. REFERENCES

1. DECHEMA, BAM und PTB, *CHEMSAFE® - Database for evaluated Safety Characteristics*, Update 2007, STN International, Karlsruhe (http://www.dechema.de/en/chemsafe.html)
2. Molnarne, M., Schendler, Th. and Schröder, V., *Sicherheitstechnische Kenngrößen Band 2: Explosionsbereiche von Gasgemischen,* Bremerhaven, Wirtschaftsverlag NW - Verlag für neue Wissenschaft GmbH, 2003.
3. DIN 51649: *Bestimmung der Explosionsgrenzen von Gasen und Gasgemischen in Luft,* Berlin, Beuth Verlag, 1986.
4. DIN EN 1839: *Bestimmung der Explosionsgrenzen von Gasen und Dämpfen,* Berlin, Beuth Verlag, 2004.
5. Schröder, V. and Daubitz, R., *Evaluation of Standard Test Methods for the Determination of Explosion Limits of Gases and Vapors*, Proceedings of 11th Int. Symp. of Loss Prevention, Praha, Czech Rep., 2004.
6. Poling, B. E., Prausnitz, J. M. and O'Connel, J. P., *The Properties of Gases and Liquids*, Fifth Edition, New York, McGraw-Hill, 2001.

Development of a Methodology for Toxic and Flammable Gases Sensors Positioning in Oil Platforms

Giordhanne Bruno Carpaneda Gimenes

Mechanical Engineer, CHEMTECH - A Siemens Company
Rio de Janeiro, RJ, Brazil
giordhanne.gimenes@chemtech.com.br

Aristeu da Silveira Neto, PhD.

Federal University of Uberlândia -FEMEC/LTCM
Uberlândia, MG, Brazil
aristeus@mecanica.ufu.br

Solidônio Rodrigues Carvalho, D.Sc.

Federal University of Uberlândia -FEMEC/LTCM
Uberlândia, MG, Brazil
srcarvalho@mecanica.ufu.br

Valter César de Souza, D.Sc.

Consultant, CHEMTECH - A Siemens Company
Rio de Janeiro, RJ, Brazil
valter.souza@chemtech.com.br

Patrícia Carneiro dos Santos

Process Engineer, CHEMTECH - A Siemens Company
Rio de Janeiro, RJ, Brazil
patricia.santos@chemtech.com.br

ABSTRACT

Computational fluid dynamics (CFD) and mechanical systems optimization concepts are currently quite important tools for engineering problems analysis. A typical application of such tools is gas dispersion studies, a theme of great importance for the oil and gas industry. An important safety item for oil and gas production installations (especially oil rigs) is the use of flammable and/or toxic gas sensors to detect gas leaks capable of starting fires, explosions and causing toxic exposure. The

Importance of these sensor rules may be evaluated if one considers all potential costs of compensations, environmental impacts, life loss and profit loss.

This work presents the development of a computational tool that supplies optimum position and number for gas detectors in an oil platform. Associating CFD simulation results with coherent mathematical treatment, this methodology presents improvements in relation to other studies of gas dispersion and sensor positioning. This mathematical treatment supported by this computational tool is responsible for all calculation methods and for the display of graphical results of the locations to place detectors both for toxic (hydrogen sulfide, for instance) and flammable (like methane) gases.

Gas detection projects are usually based on standards and technical notes that do not supply all necessary information for proper sensor positioning. In the same way, although qualitative methodologies are used intensively and are quite accepted, they are not as accurate because, in most cases, the final decision is based on personal analysis and previous experience. Although experience will always remain an important asset in this kind of service, development of such a computational tool as the one described in this paper will reduce project schedules, besides providing more accurate and less intuitive results.

1. INTRODUTION

One of safety items for oil installations is the use of concentration sensors for flammable and/or toxic gases due to high risk proportionated by leaks (fires, explosions, intoxications and others). Consequences of an accident will result in considerable setbacks including compensation, environmental impact and profit stagnation. Furthermore they damage operator's image before investors and society. Pioneering companies in safety studies developed methods for the determination of the positioning of these sensors; however such methodologies are not published because of their strategic value.

Nowadays, gases detection project are usually based on norms and technical notes that don't supply all the necessary information for the correct positioning of the sensors. In the same way, qualitative methodologies are used intensely, however, in spite of being accept; they are not so good because, most of the time, final decision is based on the personal analysis and the engineer's experience.

Due to the mentioned reasons, a new methodology of easy application and in agreement with the norms and specifications in vigor is being developed by CHEMTECH Services of Engineering and Software Ltda. in partnership with the Laboratory of Heat and Mass Transfer and Fluid Dynamics (LTCM) (College of Mechanical Engineering (FEMEC) of the Federal University of Uberlândia (UFU)).

This methodology is based on the use of data resultant from gas dispersion simulations through CFD – Computational Fluid Dynamics. These data are treated with statistical methods aiming to identify the necessary places for the installation of gas detectors. This treatment is executed with the support of a computational tool that is responsible for the process calculation and supplies graphical results of most indicated areas for the installation of detectors.

2. METHODOLOGY

The methodology in development can be used both to toxic gas (sulphidric gas, for instance) and to flammable gas. In the present article, however, it will just be broached the detection of flammable gas, through the example of methane (CH_4). This gas is predominant in the natural gas composition (Table 1) and so it deserves special attention in the studies of gases dispersion and allocation of detectors in oil facilities.

ORIGIN	% VOLUME						Density	Superior Calorific Power
Country / Field	Methane CH_4	Propane C_3H_8	Ethane C_2H_6	C_4 and Higher	CO_2	N_2		(MJ / Nm³)
USA/Panh.	81,8	5,6	3,4	2,2	0,1	6,9	-	42,7
USA/Ashlaw	75,0	24,0	-	-	-	1,0	-	46,7
Canada	88,5	4,3	1,8	1,8	0,6	2,6	-	43,4
Russia	97,8	0,5	0,2	0,1	0,1	1,3	-	39,6
Australia	76,0	4,0	1,0	1,0	16,0	2,0	-	35,0
France	69,2	3,3	1,0	1,1	9,6	0,6	-	36,8
Germany	74,0	0,6	-	-	17,8	7,5	-	29,9
Netherlands	81,2	2,9	0,4	0,2	0,9	14,4	0,640	31,4
Persia	66,0	14,0	10,5	7,0	1,5	1,0	0,870	52,3
North Sea	94,7	3,0	0,5	0,4	0,1	1,3	0,590	38,6
Algeria	76,0	8,0	3,3	4,4	1,9	6,4	-	46,2
Venezuela	78,1	9,9	5,5	4,9	0,4	1,2	0,702	47,7
Argentina	95,0	4,0	-	-	-	1,0	0,578	40,7
Bolivia	90,8	6,1	1,2	0,0	0,5	1,5	0,607	38,8
Chile	90,0	6,6	2,1	0,8	-	-	0,640	45,2
BRAZIL								
Rio de Janeiro	89,44	6,7	2,26	0,46	0,34	0,8	0,623	40,22
Bahia	88,56	9,17	0,42	-	0,65	1,2	0,615	39,25
Alagoas	76,9	10,1	5,8	1,67	1,15	2,02	-	47,7
Rio Grande do Norte	83,48	11	0,41	-	1,95	3,16	0,644	38,54
Espírito Santo	84,8	8,9	3,0	0,9	0,3	1,58	0,664	45,4
Ceará	76,05	8,0	7,0	4,3	1,08	1,53	-	52,4

Table 1: Composition of the natural gas.

Source: www.gasnet.com.br

Methane is an odorless and colorless gas. It has low solubility in water and when dispersed into the air (in certain ratios) it becomes a mixture with high explosive power. Some important physical-chemistry properties for this gas are in the Table 2.

<table>
<tr><th colspan="3">METHANE</th></tr>
<tr><td>NOMENCLATURE</td><td>MOLECULAR FORMULA</td><td>MOLECULAR WEIGHT</td></tr>
<tr><td>Methane, natural gas, gas of the swamps</td><td>CH_4</td><td>16,04</td></tr>
<tr><th colspan="3">PHYSICAL-CHEMISTRIES PROPERTIES</th></tr>
<tr><td>Physical state and risk classification: flammable, simple suffocating gas</td><td colspan="2">Color and scent: Colorless of soft and sweetened scent.</td></tr>
<tr><td>Solubility: 2,86x10^{-3}g/L @ 20ºC</td><td colspan="2">Reactivity: Forms explosive mixies with air and oxidizers agents. Incompatible with halogenous and interhalogenous.</td></tr>
<tr><th colspan="3">METHANE
FLAMMABILITY</th></tr>
<tr><td>Lower Flammable Limit: 5 % v/v</td><td colspan="2">Extinguishers agents: CO_2, dry chemical pownder, water vapor for small focuses.</td></tr>
<tr><td>Upper Flammable Limit: 15 % v/v</td><td colspan="2">Incompatible extinguishing agents:
Separately used water can be inefficient.
Not to extinguish the flames before eliminating the emptying source, therefore the new ignition can be explosive.</td></tr>
<tr><td colspan="3">Toxic products of combustion: He is not toxic, but its partial combustion can generate toxic gases (CO).</td></tr>
<tr><th colspan="3">OBSERVATIONS</th></tr>
<tr><td colspan="3">It can be flamed by heat, flashes or open flames
Cause asphyxia for oxygen reduction. It can provoke giddiness and asphyxia without acknowledgment.
In the liquid phase, it can generate burning in the skin and eyes for freezing.</td></tr>
</table>

Table 2: Some properties of methane.

Source: Crowl et al (2002), Cote (2003), Eltschlager et al (2001) e Perry (1997).

According to technical specifications usually employed in the oil and gas industry, alarms in the central control room should go off whenever levels from 20% to 60% of Lower Flammable Limit (LFL) for combustible gases are detected, where LFL is the minimum concentration of gas (% in volume) dispersed in the air that, in contact with an ignition source, is capable to cause combustion of the mixture. In a similar form, the Upper Flammable Limit (UFL) is the maximum gas concentration (% in volume) dispersed in the air that, in contact with an ignition source is capable to cause combustion of the mixture (Figure 1).

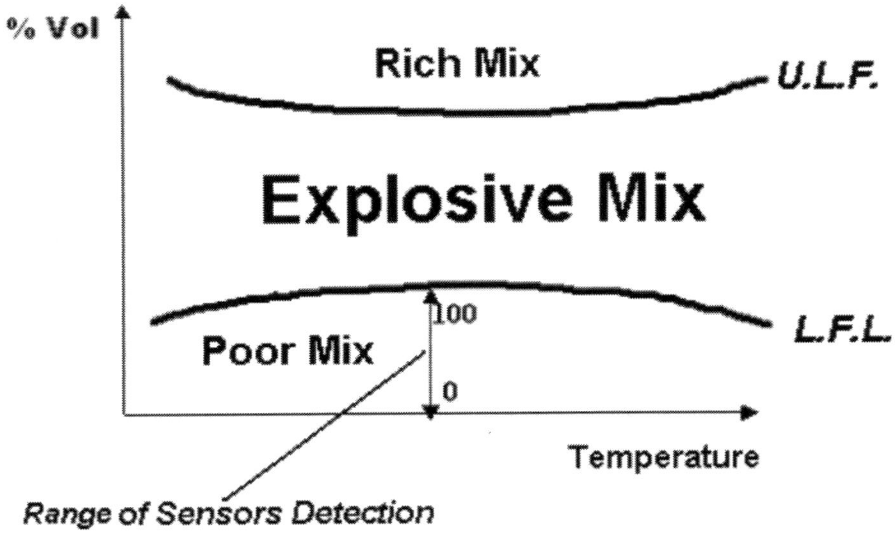

Figure 1: Strips of inflammability of a mixture gas-fuel

Source: Own elaboration

Lower Flammable Limit for methane is 5% in volume in air (Table 2), based on standard conditions of temperature and pressure (20°C and 1 atm). Therefore, the range of detection for this gas will be between 1% in volume (corresponding to 20% of the L.F.L) and 3% in volume (corresponding to 60% of the L.F.L).

In light of this information, three-dimensional gas dispersion simulations are carried out using tools of Computational Fluid Dynamics (CFD) considering geometric aspects, meteorological conditions and gas properties at leak conditions. Scenarios of leaks are simulated with several directions and speeds of wind, including wind calms. Meteorological conditions are important variables in models of gas dispersion (Crown et al, 2002), because gases are dispersed by wind. Atmospheric instability favors dispersion of the gaseous pollutant in the atmosphere. Particularly, in this study developed by the partnership CHEMTECH-UFU, the CFD tool used is the software PHOENICS (Parabolic Hyperbolic Or Elliptic Numerical Integration Code Series) - version 2006, developed by the British company CHAM (Concentration, Heat and Momentum Limited). CHEMTECH is the PHOENIC's licensor in Brazil. More details can be found in the website http://www.cham.co.uk.

In the computational tool in development by CHEMTECH-UFU the user should enter the data referring to leak conditions (probability of leak occurrence in a certain equipment or component) and meteorological aspects (probability of occurrence of each wind direction considered in the simulations). The reference data giving the probabilities of leaks can be extracted from historical reports supplied by the proper operator or from reports published by specialized institutions in health, safety and environment, such as the Health & Safety Executive (HSE), the agency responsible for health and safety regulation in Great Britain. The data for the probabilities of occurrence of winds in the considered directions can be obtained from existing meteorological stations in the area where the plant will be installed. With this information, the probabilities of occurrence of the flow configurations will be calculated. This probability is general and it takes into consideration both the leak occurrence as the wind occurrence and the analyzed direction, synthesizing in a unique value these two variables.

The probabilities calculation, is based on Bayes's Theorem. According to the magazine Ciência Hoje (2006) Thomas Bayes's reasoning is nowadays considered a new form to view the world, as the basis of a true revolution in different fields of the knowledge, from genetics to theology.

Bayes's Theorem is based on one of the basic beginnings of the probability: the probability that an event A occurs simultaneously with an event B, is given by the Equation (1):

$$P(A \cap B) = P(A/B).P(B) \tag{1}$$

Where P(A) and P(B) they are the occurrence probabilities of the events A and B, respectively, and $P(A/B)$ is the probability of occurrence of A considering that B has happened. The development of the Bayes's Theorem is available in specialized literatures in Statistics. The final form of this theorem is given by the Equation (2):

$$P(A/B) = \frac{P(B/A).P(A)}{\left[P(B/A).P(A)\right] + \left[P(B/A^c).P(A^c)\right]} \tag{2}$$

Where A^c is the complementary event of A, also called of $not\,A$. The bayesian reasoning is not intuitive, being conceptually a 'opinion modifier', in other words, the probabilities are adjusted with the knowledge of new evidences.

After the calculation of the probabilities, the next step is to load the output from CFD simulations (in the case of PHOENICS, extension .phi), separating the values of the variable of interest, in this case, concentration (variable C_1). Based on loaded files, reminding that the number of loaded files is the same to the number of simulations, a new file will be generated (C_2). In this file they will just be printed the computational volumes (Figure 2) that contain concentrations among $20\% \le C_1 \le 60\%$ of L.F.L. (for the methane, $1\% \le C_1 \le 3\%$ in volume). In the computational volumes out of this range, the value "zero" it will be attributed. At the end of this stage a file will be generated one C_3 file for each file .phi loaded.

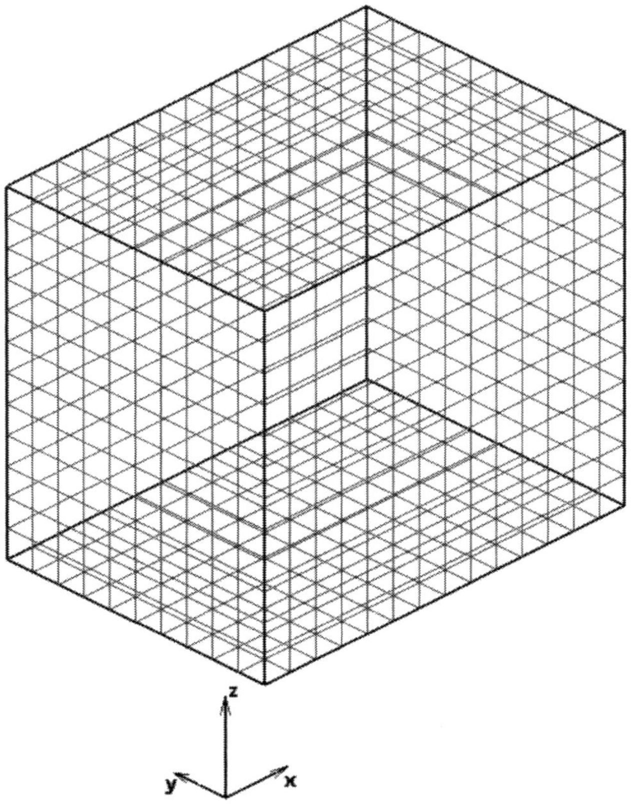

Figure 2: Three-dimensional domain divided in small computational volumes.

Source: Own elaboration.

Soon after, multiply each C_2 file for the respective occurrence probability (resultant of the calculation of the Bayes's Theorem), generating C_3 files. Then, all the files are compared, selecting the highest value for each volume, saving them in an unique file (C_4).

Finally, these values are normalized so that their range varies between 0 (zero) and 1 (one). This final file supplied by the computational tool in development by the partnership CHEMTECH-UFU, may be opened in a graphic software or in the own post-processing module of the CFD tool. This way, it is possible to visualize the field that shows importance index by colors the of the possible areas for allocation of sensors, among 0 (without need) and 1(priority). The Figure 3 summarizes these stages through a schematic flowchart of sequence of files created by this computational tool in development.

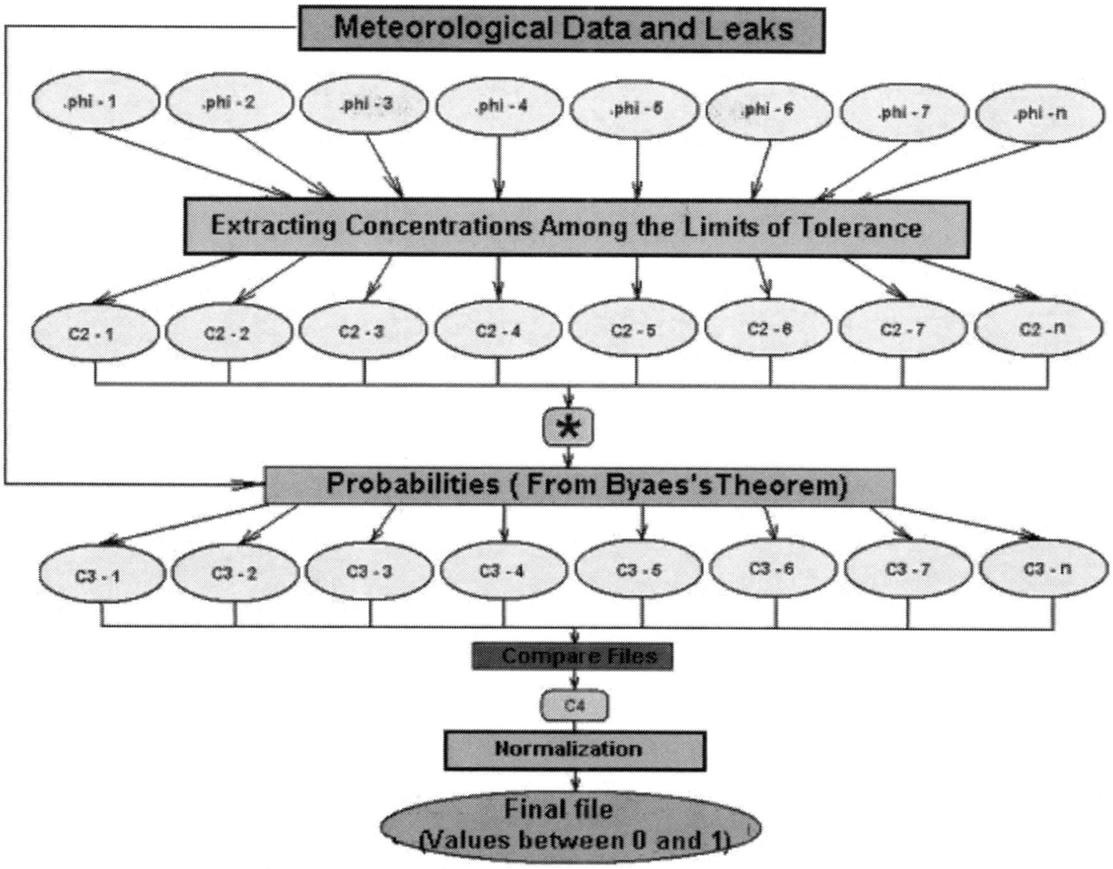

Figure 3: Flowchart of the computational tool.

Source: Own elaboration.

Therefore, at the end of the process, an index will be generated between 0 and 1 indicating probabilities of having gas with concentrations between $20\% \leq C \leq 60\%$ of L.F.L. in the computational volume, considering the probabilities of occurrences of leaks and wind for all the configurations (leak and wind).

It is important to note that is not necessary to know the exact gas percentage in each computational volume. The final file created only considers the existence of gas among $20\% \leq C \leq 60\%$ of L.F.L. this way, to know if the percentage is 21% or 30% adds a little in relation to the needs of the detection system, because any percentage between 20% and 60% of L.F.L. will be detected by the sensors, activating the alarms.

For each computational volume, a number of 0 (zero) to 1 (one) is attributed, that represents in probabilistic terms the classification of importance for each volume in the space potentially receiving a sensor. The methodology offers the advantage of being independent of opinion due to the mathematical treatment explained in this section.

3. RESULTS

Thin computational tool in development was applied to a previous sensor location project executed by CHEMTECH in 2003, in order to validate the proposed methodology. The results obtained in 2003 (through the qualitative methodologies) were compared with the results provided by the new methodology. Figure 4 shows the sensors positioned by CHEMTECH in the earlier study.

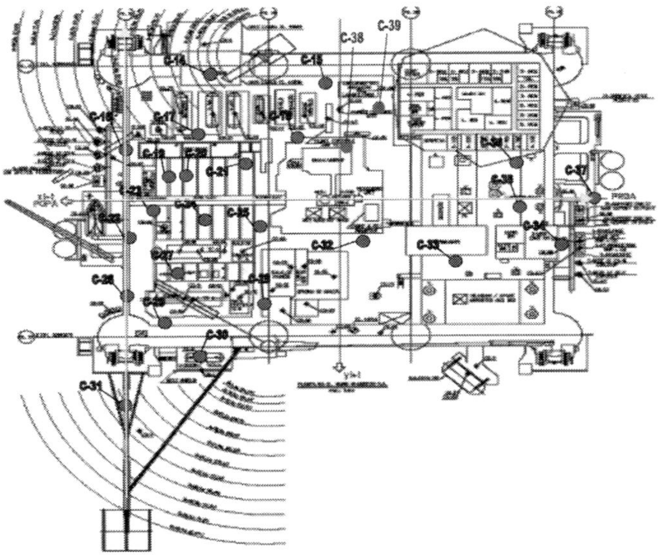

Figure 4: Sensors positioned by CHEMTECH in 2003

Source: Own elaboration

Figure 5 shows the comparison between the original location and the results obtained by the new methodology.

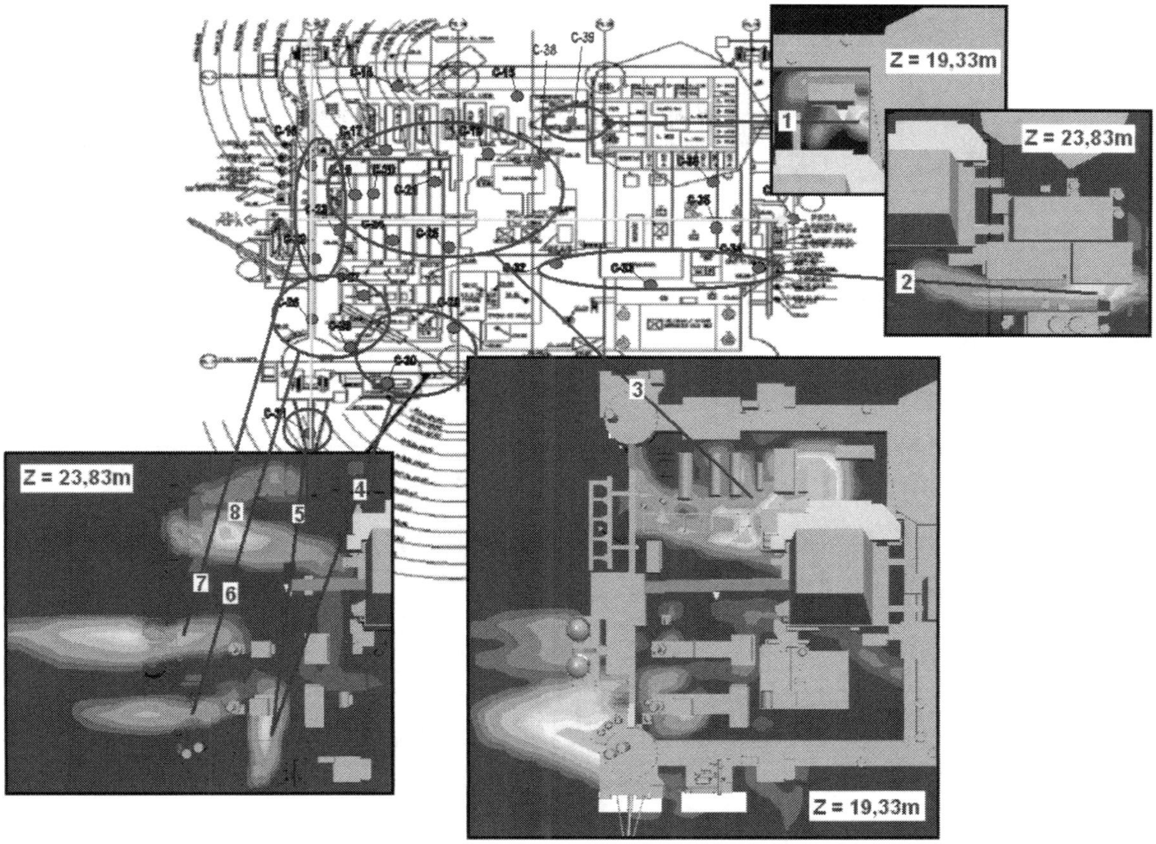

Figure 5: CHEMTECH 2003 versus the new methodology

Source: Own elaboration

The last figure shows that the results provided by the new methodology are similar to the results obtained in 2003. The scale of colors indicate the best regions for the installation of detectors, where "z" is the elevation in meters. With the support of this new tool, it was possible to take into consideration the analysis of many plumes with gas concentration between 20% and 60% of L.F.L. much more quickly. Qualitative methodologies usually analyses only plumes in the threshold gas concentrations for positioning sensors (20% and 60% of L.F.L.) due to many scenarios to be studied.

4. CONCLUSION

This methodology presents improvements in relation to others studies of gas dispersion and positioning of sensors, forming an alliance between CFD simulation results and coherent mathematical treatment. Development of the computational tool described in this document will reduce the time for conception of the projects, besides supplying more necessary and less intuitive results. It is important to remember that, in spite of the great progress that this methodology provides, researches will continue, in order to advance in aspects related to an optimum number of sensors.

5. REFERENCES

1. Ciência Hoje, *Revista de Divulgação Científica da SBPC*, Vol. 38, Julho 2006.

2. Cote, A. E. *Fire Protection Handbook* – Volume II. 19ª Edição. Editora Editora National Fire Protection Association. Estados Unidos, 2003.

3. Crowl, D. A.; Louvar, J. F. *Chemical Process Safety – Fundamentals with Applications.* 2ª Edição. Editora Prentice Hall. Estados Unidos, 2002.

4. Eltschlager, K. K.; Hawkins, J. W.; Ehler, W. C.; Baldassare, F. *Technical Measures for the Investigation and Mitigation of Fugitive Methane Hazards in Areas of Coal Mining.* US Department of Interior - Office of Surface Mining Reclamation and Enforcement. Pittsburgh, PA. Setembro, 2001.

5. HSE, Hid Statistics Report, *Offshore Hydrocarbon Releases Statistics and Analysis, 2002.*

6. Kelsey, A., Hemingway, M.A., Walsh, P.T. and Connolly, S., *Evaluation Of Flammable Gas Detector Networks Based On Experimental Simulations Of Offshore, High Pressure Gas Releases*, Institution of Chemical Engineers, Vol. 80, Part B, March 2002.

7. Perry, R. H. *Perry's Chemical Engineers' Handbook.* 7ª Edição. Editora McGraw-Hill. Estados Unidos, 1997.

8. Pupe, Camila Gonçalves; Soares da Silva, Ályson – *O Uso de CFD na Previsão da Dispersão de Gases em Plataformas Ofsshore*, 2006.

9. Yudkowsky, *An Intuitive Explanation of Bayesian Reasoning, 2003.* http//yudkowsky.net/bayes/bayes.html

A graphical approach for representing hazard scenarios

Rafael Batres

Toyohashi University of Technology
E-mail: rbp@pse.tut.ac.jp

Yukiyasu Shimada

Chemical Safety Research Group, National Institute of Occupational Safety and Health
E-mail: shimada@s.jniosh.go.jp

Tetsuo Fuchino
Tokyo Institute of Technology
E-mail: fuchino@chemeng.titech.ac.jp

ABSTRACT

This paper describes "hazard scenario graph" or HSG which is a visual representation technique for performing hazard identification. HSGs explicitly represent all the intermediate and final consequences in the scenario. HSG is based on ontologies and knowledge representation techniques with which knowledge extraction is possible. The method explicitly represents causality, substances, equipment, physical and chemical transformations, plant operations and personnel, facilitating the integration with other engineering tools such as CAD systems.

1. INTRODUCTION

Several methods for hazard identification have been developed, including checklists, What If Analysis, FMEA, and HAZOP. Among these, HAZOP is probably the most common method for hazard identification of a detail design. Broadly speaking, once a potential hazard scenario is generated, engineers capture the resulting information into two parts: initial causes and consequences. Causes of a scenario normally refer to a single initial cause. However, consequences can be recorded in multiple ways. Sometimes, as a single final consequence, such as an explosions or damage to facilities, but also as several consequences (either final or intermediate) grouped together in a single paragraph. As a result, intermediate parts of the scenario remain in the minds of the team or are hidden in the analysis report (typically in a tabular form). Furthermore, information is often captured in text format which limits the reuse of information offered by CAD and other plant information systems.

This paper describes "hazard scenario graph" or HSG which is a visual representation technique as an aid to hazard identification. HSGs explicitly represent all initial causes, intermediate and final consequences in the scenario. HSG is based on ontologies and knowledge representation techniques with which knowledge extraction is possible. The method explicitly represents causality, substances, equipment, physical and chemical transformations, plant operations and personnel, facilitating the integration with other engineering tools such as CAD systems.

This paper is structured as follows. The HSG notation is presented in section 2. Next, the method for its construction is presented in section 3. The proposed method is illustrated with the hazard analysis of a DAP (Diammonium Phosphate) production process in section 4. Finally, section 5 presents the conclusions and discussion.

2. HAZARD SCENARIO GRAPHS

A Hazard scenario graph (HSGs) is a visual representation technique based on concepts defined in the ISO 15926 standard that specifies definitions for the representation of knowledge about process plants [1].

A hazard scenario graph is composed of the following elements from the ISO 15926 standard:

1. Activities. An activity is a possible individual that brings about change by causing an event. Classes of activities include physicochemical processes, plant operations, and abnormal situations. Examples of activity are "Level increasing" and "Closing feed valve". Activities are shown as square boxes.

2. Events. An event is a possible individual that has zero extent in time, which means that it occurs at an instant in time. Events are shown by ovals.

3. Physical objects. A physical object is a distribution of matter, energy, or both. Examples of physical object are a tank, pump, a stream of material, a person. Participating entities are enclosed by hexagons. For example, tank T-01 is a participating entity in the "level increase in T-01" activity..

4. Causal relations are represented as solid arrows identified by the phrase *cause of event*

5. Temporal relations (beginning and ending) are represented as solid line with a filled circle identified by either the words *beginning* or *ending*

6. The participation relation is represented as a solid line with an empty circle identified by the word *participation*. A participation relation indicates that physical objects are resources, instruments, or performers of an activity

2.1 Causality and time-related information

Causality refers the links between the initial cause and the final consequences of the hazard scenarios passing through a series of intermediate consequences. Causality in HSGs is captured by means of the *cause of event* relation. A cause of event indicates that an event (the caused) is caused by an activity (the causer).

Another important element of hazard scenarios is the precedence-in-time relationships, represented by beginning and ending relations. A beginning relation marks the temporal start of a possible individual (an activity or a physical object). An ending relation marks the end of a possible individual.

Causality in HSG follows the properties described by Shoham [2] some of which are listed here:

1. A cannot cause B if B is the cause of A.
2. A cannot cause itself.
3. Causes cannot succeed their effects in time. If A causes B, A precedes B in time.
4. Activities are temporally bounded. For example, explosions, runaway reactions, mixing operations, all have a beginning and an ending. Similarly, a reactor during a runaway reaction has also a beginning and an end.
5. The property of transitivity [3]. If A causes B and B is the cause of C, then A is also the cause of C.

3. CONSTRUCTING HAZARD SCENARIO GRAPHS

Here we describe the construction of a hazard scenario graph when used together with a HAZOP study of a continuous process. We assume that the HAZOP team has already identified the system boundary, decomposed the system into subsystems or nodes.

Step 1. Select the subystem to be analyzed

Step 2. Select a deviation from the design intent of the subsystem. If the resulting deviation is temporally bounded, the deviation is represented as an *abnormal* activity. In this case, the deviation is identified by a verb or verb phrase. Examples include temperature increase, pressure decrease. If the resulting deviation is not temporally bounded, the deviation is represented as a *normal* activity ending in an event. In this case, the activity is a normal behavior (the intended behavior). For example, a "no flow" deviation is represented as an activity for the specific flow in the intention and event for the cessation of flow linked by the *ending* relation.

Step 3. Identify all possible causes of the deviation. Create a HSG for each cause and represent the cause as an activity. Classify the activity as equipment malfunctions, undesired physicochemical phenomena, abnormal operations, or external events (impacts, earthquakes).

Step 4. For each cause (in a single HSG) identify a possible propagation scenario based on the topology of the plant. An activity causes an event that is the beginning of another activity.

During the construction of the scenario part-whole relations can be used to indicate that activities or events occur during the time boundaries of another activity. For example, an "abnormal" no-flow situation in a pipe can only occur when material is flowing in that pipe.

4. CASE STUDY: HSG OF A DAP PROCESS

This case study is based on the DAP reaction system found in [4]. This is a process in which ammonia and phosphoric acid react to form diammonium phosphate (DAP), a non-hazardous product. The DAP flows from the reactor to an open-top storage tank. Low ratios of phosphoric acid/ammonia result in an off-specification product, but the reaction is safe. If both, ammonia and phosphoric acid increase, the rate of energy release may accelerate, resulting in increase in temperature and pressure and reactor failure. If too much ammonia is fed to the reactor, unreacted ammonia may flow into the DAP storage tank, causing ammonia release and consequently personnel exposure. The schematic is shown in Figure 1.

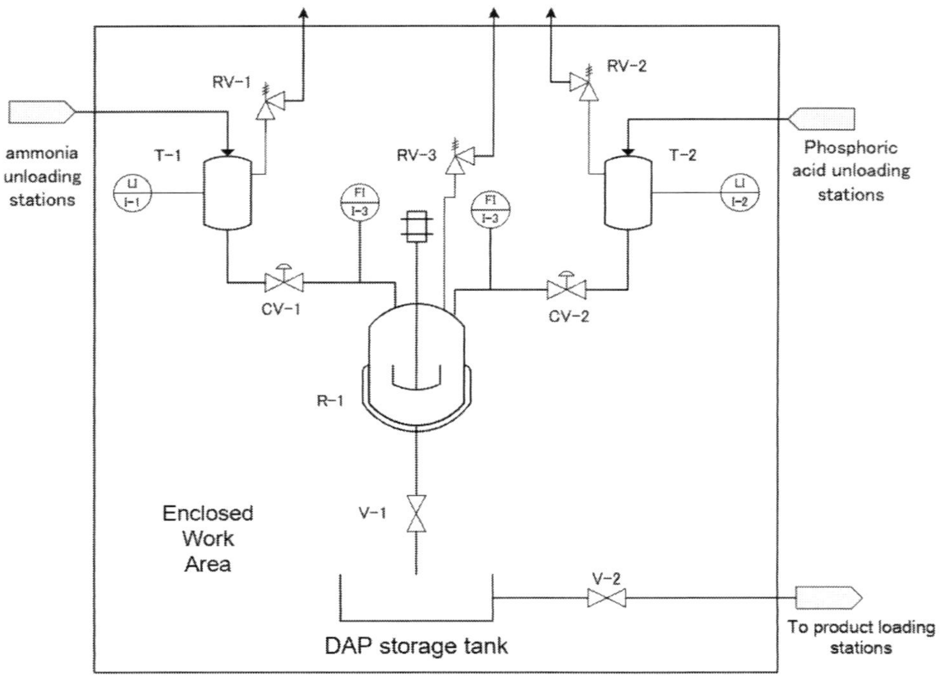

Figure 1. Schematic of the DAP reaction system

Step 1. Subsystem to be analyzed: Phosphoric Acid Feed Line to the DAP Reactor.

Step 2. A deviation is proposed by combining the guideword *no* with the parameter *flow*. The resulting deviation (no flow in Phosphoric Acid Feed Line) is not temporally bounded for which it is represented as an event (that ends the activity of normal flow).

Step 3. Identify all possible causes of the no flow deviation of Step 2. One of such causes is "line leaks". This cause is temporally bounded so it is written as a verb phrase and represented as an activity.

Step 4. Identify possible propagation scenarios for "line leaks". Two of the possible scenarios are shown in Figure 2, leading to "personnel exposure" and "loss of DAP production".

The resulting HSG is shown in Figure 2. Note that the deviation selected in Step 2 is shown explicitly in the diagram, facilitating the process of deducing that one of the consequences is the "loss of DAP production".

The same diagram could be obtained using an alternative approach. The initial cause is obtained by finding equipment malfunction, undesired physicochemical phenomena, and external events. For the *Phosphoric Acid Feed Line*, equipment malfunctions include no flow of ammonia in and out the tank; undesired physicochemical phenomena include corrosion, erosion; abnormal operations include inadequate maintenance. External events include external impacts. Figure 2 shows how the diagram that results from corrosion in the line (an undesired physicochemical phenomenon).

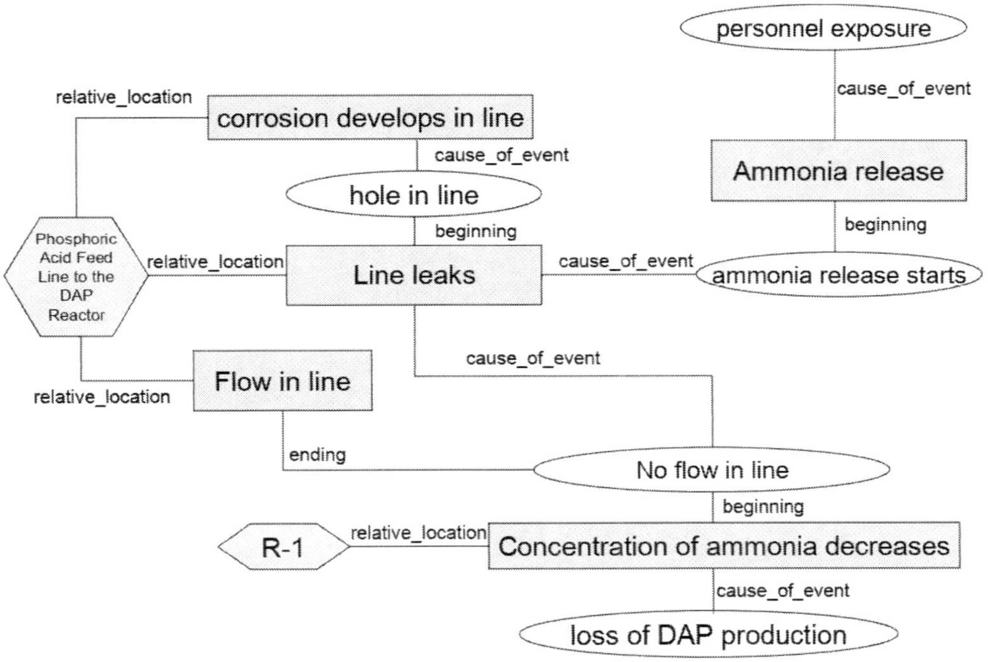

Figure 2. A Hazard Scenario Graph of the DAP reaction system

5. CONCLUSIONS

This paper introduced a graphic approach for representing hazard scenarios, which can be used as together with HAZOP or other hazard identification methods.

One advantage of the HSG representation is that with the use of adequate software tools engineers can search and combine scenarios in an automatic way. For example, if we were interested in finding the causes of ammonia release, the computer environment would combine several scenarios including those that apparently do not lead to ammonia release but lead to corrosion in the line.

6. REFERENCES

1. Batres, R., West, M., Leal, D., Price, D., Katsube, M., Shimada, Y., Fuchino, T. An upper ontology based on ISO 15926. *Comp. & Chem. Eng.*, 31, 5-6, 2007, pp. 519-534
2. Shoham, Y. *Reasoning about Change*, Cambridge, Massachusetts, MIT Press, 1988
3. Findler, N. V., Bickmore, T., *Applied Artificial Intelligence* 10 (1996) 455-487
4. CCPS, *Guidelines for Hazard Evaluation Procedures, with Worked Examples*, AIChE 1992

Effects of Polyethylene Size on Ignition of Electrostatics Discharge

LIU Yi, ZHAO Dong-feng, LU Shuai, ZHANG Lin, YIN Fabo

(College of Chemistry and Chemical Engineering, China University of Petroleum, Dongying 257061, China)

ABSTRACT

Minimum ignition energy of coarse polyethylene granule and screened polyethylene dust is tested with a standard dust explosion testing equipment of Hartmann tube. By comparing the minimum ignition energy and different types of electrostatic discharge energy, the effects of polyethylene size on ignition of electrostatics discharge has been investigated. The results show that, the minimum ignition energy of polyethylene has an exponential relationship with maximum particle size while the particle size is less than 2 mm. No types of electrostatics discharge can ignite the coarse polyethylene granule with average diameter above 2 mm. For the polyethylene dust with diameter below 0.5 mm, electrostatics discharge of corona discharge, brush discharge and cone discharge are excluded from the potential ignition source. Sparks and propagation brush discharge are the possible ignition source. Preventing the occurrence of sparks and propagation brush discharge is an effective measure to defend against the dust explosion of polyethylene from an industrial view.

1. INTRODUCTION

With the high-speed development of petrochemical industry in China, both numbers of polyethylene (PE) production unit and its throughput are all increased. Now the maximum throughput of high-density PE for a single production unit is 200,000 tons per year in China. However, the explosion accident potential has also increased along the process of PE manufacture. Based on our investigation, sixty-eight PE explosion accidents have happened in China during the last several years. The accidents caused a great loss of life and damage to property, which was also a big disturbance to the manufacture and management for the enterprise. The accident investigations showed that most accidents occurred at the silo and attributed to electrostatics discharge.

To assess the hazard of PE electrostatics discharge, the following steps should be included. First, is it possible to form combustible mixture? Second, is it possible for the electrostatics charge to accumulate? Third, what types of electrostatic discharge will occur? Finally, is the energy of electrostatic discharge big enough to ignite the combustible mixtures? During the process of PE manufacture, PE is conveyed with air, and the concentration of PE is above its upper explosive limit, so it is easier for the mixture of PE and air to form combustible mixture. In addition, the volume resistivity of PE is larger than 10^{14} Ω • m, and the electrostatic charge because of friction, collision and segregation could not be eliminated by grounding. Electrostatic would accumulate with a high speed. When the field strength is larger than the breakdown field strength of air, different types of electrostatic discharge will occur. As a result, the key step to evaluate the hazard of electrostatic discharge during the process of PE manufacture is to assess the ignition ability of electrostatic discharge.

Most of previous studies focus on the types of electrostatic discharge, energy of electrostatic discharge and method of evaluation [1-6]. However, more detailed information about the effects of PE size on ignition of electrostatics discharge, which is important to assess the hazard of electrostatic discharge of PE, has not been sufficiently explored.

In this study, the effect of PE size on ignition of electrostatics discharge has been conducted experimentally.

2. EXPERIMENTAL

2.1. EXPERIMENTAL APPARATUS

The experimental system is schematically shown in Fig. 1, which is composed of a Hartmann tube, dust dispersion part, ignition part and control part. To observe the ignition of PE granule conveniently, the upper part of Hartmann tube is made of Plexiglas, which volume is 1.2 L. The dust dispersion part is composed of a gas-supplying unit, a solenoid valve, a gas nozzle and a sample dish. The ignitions part consists of a pair of copper wire electrodes with 0.4 mm diameter. The electrodes were positioned in the middle of the tube, which locates 7.6 cm above the bottom of the tube. The distance between the electrodes tips is approximately 5 mm. The startup time of the ignition time and on-off time of the solenoid valve were set and controlled by the programmable logic controller.

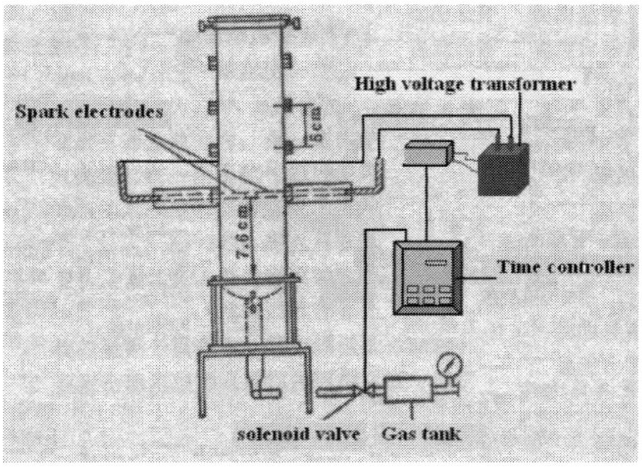

Fig. 1 Experimental system

2.2. EXPERIMENTAL CONDITIONS AND PROCEDURE

The experiments were carried out at atmospheric pressure with the upper end open and the lower end equipped with a dust dispersion part. A specified mass of PE granule weighed by a balance was put on the sample dish evenly. The dust was dispersed into the tube with pressurized air. The PE granule used in the test was made of PetroChina Company Limited, which is a product of high-density PE. The size distribution of the original product is shown in Table 1. Before each experiment, the sample was stored in a drier of 50℃ for 24 h to eliminate the moisture.

Table 1 PE particle size distribution

Particle average diameter/ μm	<2000 μm/ %	<1000 μm/ %	<125 μm/ %	<80 μm/ %
1352	100	28.6	0.5	0.2

The detailed experimental conditions are listed as follows:

Injection time: 100 ms, injection pressure: 0.5 MPa, interval between injection end and ignition: 120 ms.

Experimental procedure is listed as follows:

(1) Inject air into the gas tank until the pressure in the gas tank is 0.5 MPa.

(2) Connect the experiment system and all the other devices as shown in Fig. 1. Open the solenoid valve connected to the gas tank.

(3) Put 1 g PE granule on the sample dish evenly.

(4) Charge the high voltage transformer to a certain value and initiate the control unit to trigger the solenoid valve and ignition part.

(5) Judge whether the PE-air mixture is ignited or not in a range of spark energy. If the PE-air mixture is ignited, and then adjust the voltage and capacitance to lower the spark energy and repeat the procedure from step 1 until the mixture cannot be ignited by 20 times.

2.3. EXPERIMENTAL METHOD

At first, the PE granule was ignited with high voltage circuit discharge, where the voltage and capacitance are 9000 V and 0.1 μF respectively. If the PE granule could not be ignited, the method of the high voltage breakdown and low voltage arc circuit continued to be used. The principle of the high voltage breakdown and low voltage arc circuit is introduced as follows. Break down the spark gap at 10 kV voltage first, where the spark energy is less than 10 mJ and could be neglected, and then the low voltage capacitance discharge in the electric arc between the electrode gaps. The ignition energy was changed by adjusting the voltage and capacitance. Due to the low voltage, the energy consumed on the resistance is small, so the discharge energy can be calculated simply by the difference before and after the electric discharge based on the following formula.

$$E=0.5C\times(U_1^2-U_2^2) \quad (1)$$

Where E is the minimum ignition energy (MIE) of the sample particle, C is the capacitance. U_1 and U_2 are the voltage before and after the electric discharge respectively.

3. RESULTS AND DISSCUSSIONS

3.1. EXPERIMENTAL RESULTS

Based on on-site measurement, most of suspended PE granule diameter is less than 0.5 mm. The MIE of the original PE granule and the particles with diameter below 0.5 mm were tested. The results are shown in Table 2. To compare with results of standard PE dust, the MIE of 75 μm PE dust is also listed in the table [7].

Table 2 MIE of the tested PE particles

Sample	MIE (mJ)
Original PE granules	$>10^7$
PE granule with diameter less than 0.5 mm	656
PE dust with diameter less than 75 μm	10

3.2. DISCUSSIONS AND ANALYSIS

3.2.1 RELATIONSHIP BETWEEN MIE AND PARTICLE SIZE

Base on the experimental results, if the MIE of original PE granule is assumed to 10^7 mJ, the relationship between the MIE and maximum particle size is plotted as shown in Fig. 2. The curve shows that there is a linear relationship approximately between the normal logarithms of PE's MIE and its maximum particle size, where the correlation coefficient is 0.996. As a result, when the particle size of PE is less than 2.0 mm, the MIE has exponential relationship with the maximum particle size.

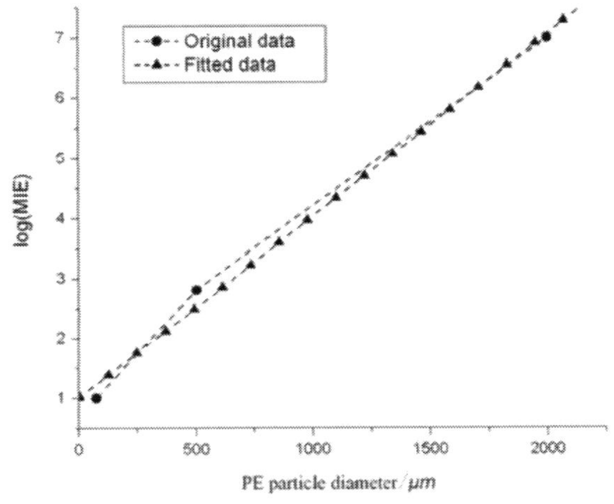

Fig.2 Relationship between MIE and maximum particle diameter of PE

3.2.2 COMPARISON BETWEEN MIE OF PE AND MAXIMUM DISCHARGE ENERGY OF DIFFERENT ELECTROSTATICS DISCHARGE TYPE

During the process of production, transportation and storage, different types of electrostatics discharge will occur while the PE particle collides with pipeline or other devices. Electrostatics discharge type includes corona discharge, brush discharge, cone discharge, spark discharge and propagation brush discharge. Different electrostatics discharge has different maximum discharge energy [8-9]. Corona discharge can just ignite some flammable gases, which has a lower hazard for PE granule. The maximum energy of brush discharge is smaller than 3.6 mJ, which could ignite most flammable gas and some of dust with lower MIE. Cone discharge is also called Maurer discharge, whose maximum discharge energy is 10 mJ, which could ignite most gas and fine dust. Spark discharge generally occurred between two conductors, whose maximum discharge energy is up to 1 J. Propagation brush discharge has maximum discharge energy of 10 J. Almost all the dangerous flammable gas, flammable liquid vapors and the mixture of flammable dust can be ignited by propagation brush discharge.

Comparing the maximum energy of different electrostatics discharge with the MIE results under different particles size, we can conclude that the original PE granule with MIE larger than 10 kJ cannot be ignited by any type of electrostatics discharge. When the maximum particle diameter of PE is less than 0.5 mm, its MIE is larger than the maximum discharge energy of corona discharge, brush discharge and cone discharge, which is smaller than the maximum energy of spark discharge and propagation brush discharge. Under this conditions, much attention has to be paid on spark discharge and propagation brush discharge, and some necessary measures to prevent the two electrostatics discharge has to be taken. When the particles diameter of PE is smaller than 75 μm, its MIE is larger than the energy of corona discharge and close to the maximum discharge energy of brush discharge and cone discharge. In this situation, we should prevent all the electrostatics discharge except corona discharge.

The above analysis confirms that the PE particle size has much effect on the ignition probability. The smaller the diameter of PE is, the more hazards the PE particles are. With the increase of PE particle diameter, the MIE of PE will increase approximately exponentially. Moreover, the hazard of PE dust explosion mainly comes from the suspended dust, so the key index to assess the hazard of on-site electrostatics discharge is the MIE of particles with diameter less than 0.5 mm.

4. CONCLUSIONS

For better understanding the effects of PE size on ignition of electrostatics discharge, the MIE of different diameter PE granules was measured and compared with the maximum electrostatics discharge energy. The conclusions obtained are as follows.

(1) For the PE granule used in the study, the MIE has exponential relationship approximately with the maximum particles diameter of PE when the PE particle diameter is less than 2 mm.

(2) The electrostatics hazard during PE manufacture can be assessed by comparing the MIE of suspended PE particles with the maximum discharge energy of different electrostatics. Some effective measures to prevent PE dust explosion can be taken based on the comparison results.

(3) The smaller diameter has more hazards of PE dust explosion. No kinds of electrostatics can ignite the original PE particles with MIE of approximately 10^7 mJ. When the PE particle diameter is less than 0.5 mm, corona discharge, brush discharge and cone discharge are impossible to ignite the mixture of PE particles and air. Spark discharge and propagation brush discharge should be the possible ignition source. Much attention has to be paid to prevent spark discharge and propagation brush discharge.

5. REFERENCES

1. Dahn, James C., Electrostatic hazard evaluation methods of dust/hybrid mixtures-current & future, *Proceedings of the 4th International symposium on hazards, prevention, and mitigation of industrial explosions*, France: Bourges, 2002.
2. Glor, Martin, Hazards due to electrostatic charging of powders, *Journal of Electrostatics*, 1985, 16(2-3): 175-191.
3. Glor, Martin, Luttgens, G., Maurer, B., and Post, L, Discharges from bulked polymeric granules during the filling of silos-characterization by measurements and influencing factors, *Journal of Electrostatics*, 1989, 23(1): 35-43.
4. Glor, Martin, Ignition hazard due to static electricity in particulate processes, *Powder Technology*, 2003, 135-136(1):223-233.
5. Glor, Martin, Schwenzfeuer, K., Direct ignition tests with brush discharges, *Journal of Electrostatics,* 2005, 63(6-10):463-468.
6. Hoppe, Tom, Jaeger, Norbert, and Terry, John, Safe handling of combustible powers during transportation, charging, discharging and storage, *Journal of Loss Prevention in the Process Industries*, 2000, 13(3-5):253-263.
7. Eckhoff, Rolf R., *Dust explosions in the process industries*, London: Gulf Professional Publishing, 2003.
8. Pratt, Thomas H., *Electrostatic Ignitions of Fires and Explosions*. New York: Center for Chemical Process Safety/AIChE, 2000.
9. Wilson, Norman, Luttgens, G., *Electrostatic Hazards*, Oxford: Buttenuorth-Heinemann, 1997.

Chemically Unstable Gases – Flammability of Ethylene Oxide Mixtures in Sterilization Processes

Volkmar Schroeder and Enis Askar

Federal Institute for Materials Research and Testing (BAM)
Berlin, Germany
E-mail: volkmar.schroeder@bam.de

Aydan Acikalin and Joerg Steinbach

Technical University Berlin
Berlin, Germany

ABSTRACT

For the assessment of explosion hazards by industrial sterilization processes with ethylene oxide (EO), the flammability regions of 3-component systems EO/nitrogen/air, EO/carbon dioxide/air and EO/water vapor/air were determined. The tests were performed at temperatures of 20 °C and 100 °C and at pressures of 40 kPa and 100 kPa in accordance with the standard test method EN 1839-B.

The observed flammability regions are similar in shape and typical for mixtures with ethylene oxide. According to the molecular heat capacities the regions get larger with nitrogen and smaller with carbon dioxide. They become larger with increasing pressure and increasing temperature.

Using experimental data a semi-empirical model was created that allows the calculation of flammability limits of process gases in sterilization processes. Such process gases can consist of EO, nitrogen, carbon dioxide, water vapor and air.

The model is based on the assumption that the adiabatic flame temperatures along the boundary curves of a flammability region have a certain temperature profile that is nearly independent of the type of the inert gas. The adiabatic flame temperatures were calculated by using the "Gaseq" Code.

Using a temperature profile calculated from only one experimental system EO/inert gas/air it is possible to predict the flammability limits of systems with other inert gases or of process gases containing several inert gases.

1. INTRODUCTION

Ethylene oxide (EO) is among the top 3 % of high volume chemicals. EO is both, a flammable and a chemically unstable gas that can be explosive without any amount of oxygen or air. During sterilization procedures the used EO can easily form explosive mixtures. In the past, EO already caused a number of explosions in sterilization industries. One example is the explosion accident happened in Ontario, California, 2004 [1]. Subsequently, producers and users of sterilization plants in Germany required exact knowledge about the flammability limits and explosiveness of process gases created by sterilization processes.

During such sterilization cycles gas mixtures are being created that consist of EO, nitrogen (N_2), carbon dioxide (CO_2), water vapor (H_2O) and air. For safety reasons it is recommended to run the process in such a way that the mixtures do not become explosive. Therefore certain limiting concentrations of EO and air must not be exceeded. The accurate knowledge of the flammability limits under process conditions is a prerequisite for those protection measures.

Both, the upper and lower flammability limit (UFL and LFL, respectively) are understood as the mole fractions (concentration) of EO in the gas mixture at which flame propagation does not occur. They are the last non-ignition points. This European definition is somewhat different to the one in the United States where according to ASTM E 681 the mean value between ignition and non-ignition point is taken. Nevertheless, because of the small step sizes the difference is not very important in practice. The range between the LFL and UFL is the so-called flammability range. If inert gases are present in a 3-component system the term flammability region or explosion region is being used. Inside this region the process gas is explosive and can be ignited by ignition sources.

Flammability regions of 3-component systems are given in [2, 3]. However, most of the studies were performed at atmospheric conditions. For higher temperatures and non atmospheric pressures only a few data is available.

The flammability region of EO mixtures is strongly influenced by the type of the inert gas, the temperature and the pressure. In sterilization processes mixtures with more than one inert gas can be created at different temperatures and pressures. Hence, the flammability limits had to be measured for all these conditions. As the matter of fact, this is not feasible due to the high experimental effort.

Therefore, the goal of this study was the systematic experimental determination of basic data on sterilization processes and the development of a reliable calculation method by means of this data.

2. EXPERIMENTAL

The flammability limits were investigated systematically for the systems EO/N_2/air, EO/H_2O/air and EO/CO_2/air at temperatures of 20 °C and 100 °C and pressures of 40 kPa and 100 kPa. The experimental set-up is schematically shown in Fig. 1.

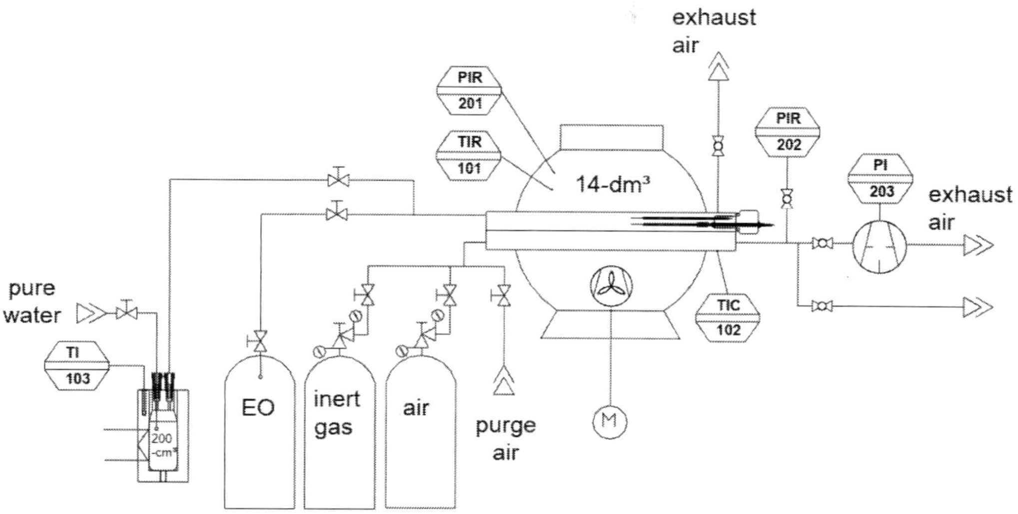

Fig. 1: Experimental set-up according to the standard DIN EN 1839, bomb method.

The ignition tests have been performed following the bomb method of DIN EN 1839 [4]. The ignition vessel used was a heatable spherical steel autoclave with an inner volume of about 14 dm^3. The ignition source was located in the center of the autoclave. It was a 0.12 mm fusing NiCr-wire (exploding wire) soldered across two electrodes at a distance of 5 mm. The ignition energy was controlled by an electronic switching unit that allows the adjustment of the ignition energy between 1 J and 50 J. According to DIN EN 1839 the energy was set between 10 J and 20 J for all tests.

Filling and explosion pressures were measured by piezoresistive pressure gauges (from Keller, type PAA-10, sensitivity 10^{-4} of full scale, precision 0.5%). They were calibrated prior to the experiments. After ignition the pressure-time courses were recorded and stored in a computer.

The temperatures of the autoclave and the gas phase were measured using 0.5 mm NiCr-Ni thermocouples.

The gas mixtures were prepared according to a manometric method on the basis of the partial pressures. Subsequently, the test mixtures were ignited in the ignition vessel. The overpressure caused by the reaction was measured in order to characterize the explosibility of the test mixture. As criterion for a successful ignition served the pressure rising factor F (ratio of explosion pressure p_{ex} and initial pressure p_0) with $F = p_{ex}/p_0 > 1.05$. In case of the lower initial pressure of 40 kPa the criterion was increased to $F > 1.10$ because of a possible influence of the ignition source on the explosion pressure.

The flammability limits were determined by variation of the EO concentration in the mixture. The step size was 0.2 mol% near the LFL and 0.5 mol% near the UFL.

3. TEST RESULTS AND DISCUSSION

The experimentally determined flammability regions (explosion regions) for the investigated systems EO/N_2/air, EO/H_2O/air and EO/CO_2/air at different pressures and temperatures are shown in the following Fig. 2 – 4.

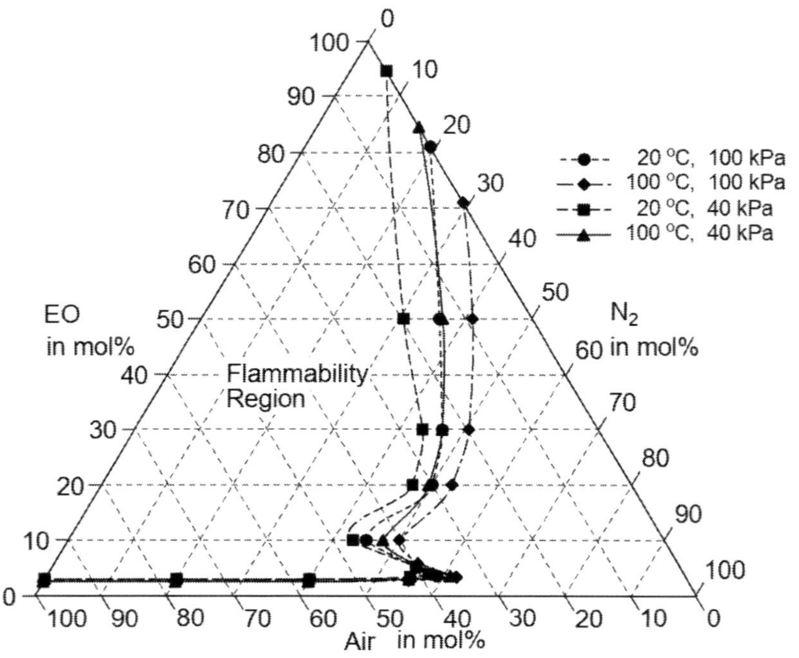

Fig. 2: Experimentally determined flammability regions for the system EO/N_2/air at different temperatures and pressures.

At lower EO fractions, i.e. in the fuel poor range, the run of the boundary lines that enclose the flammability regions is similar to those of other common flammable gases. Here, the oxidation reaction of EO with air dominates the decomposition of EO. At about 10 mol% EO a reversal point can be observed. Approaching the fuel rich range, the decomposition reaction of EO becomes more influential and causes the typical expansion of the flammability region.

With increasing pressures and temperatures a further expansion of the explosion region is observed. The influence is still rather low in the range of LFL. However, in the range of UFL the influence of initial pressure and temperature on the decomposition is particularly strong.

Also different inert gases have an effect on the size of the flammability region. The region decreases with an increasing molar heat capacity of the inert gases in the following order: N_2, H_2O, CO_2. This corresponds to the theory of Burgess and Wheeler [5]. They assumed that a minimum flame temperature that is largely independent of the type of inert gas shall be reached. In case of inert gases with higher molar heat capacities more energy is needed to heat up the system. Therefore more amounts of the flammable gas have to be burnt resulting in a smaller flammability region.

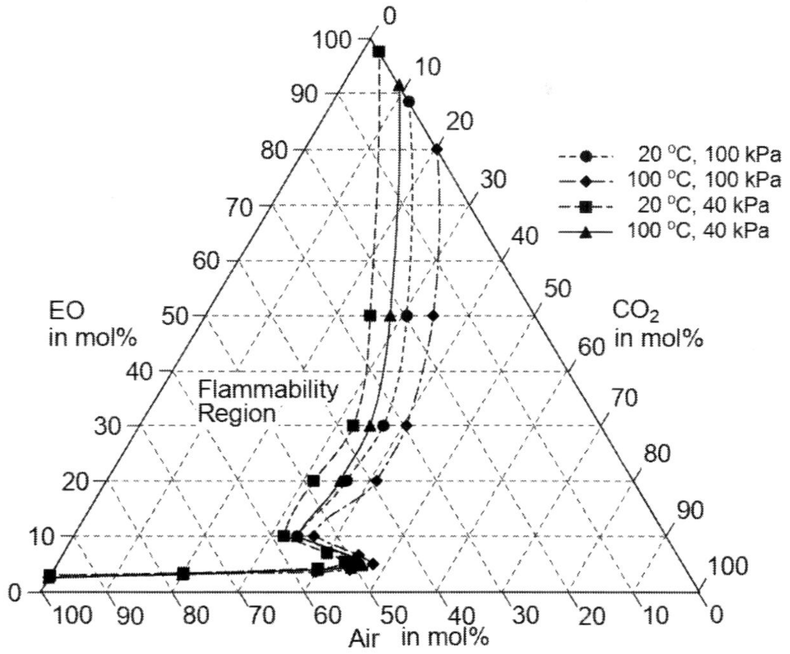

Fig. 3: Experimentally determined flammability regions for the system EO/CO_2/air at different temperatures and pressures.

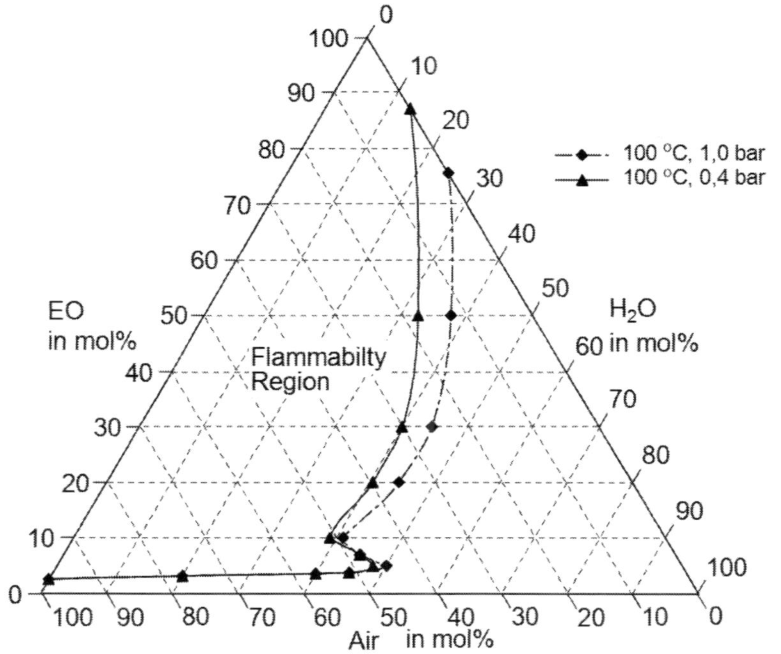

Fig. 4: Experimentally determined flammability regions for the system EO/H_2O/air at different pressures.

4. CALCULATION OF FLAMMABILITY LIMITS

The complete numerical computation of flammability limits is a complex and ambitious task. For many substances not feasible as the complete kinetics of the combustion process and all transport and radiation phenomena had to be taken into account.

Practically, it is much easier to apply semi-empirical methods. Such a model was developed in the 1990s at the University Paderborn, Germany, for the calculation of flammability regions of flammable gas/inert gas/air mixtures [6]. The applied model was based on the assumption of Burgess and Wheeler of "constant minimum flame temperatures". At first the adiabatic equilibrium flame temperatures are calculated for the experimentally determined LFL and UFL. At this stage no inert gases are considered. In the following steps different amounts of inert gases are included and the flammability regions are computed. For this purpose the flame temperature is regarded to be constant.

Using the code "Gaseq" [7], the intention of the present study was to prove that the assumption of Burgess and Wheeler can be used to predict correct flammability regions for EO mixtures, too.

"Gaseq" has a Windows graphical user interface and is easy to run. It allows the computation of equilibrium concentrations and adiabatic temperatures for gas reactions at constant volume or pressure. All calculations were performed for EO and its reaction products by means of NASA thermodynamic data sets [8]. The reaction kinetics were not considered. The solution of the system of equations is based on the minimization of Gibb's energy G according to equation (1):

$$\frac{G}{RT} = \sum_{i=1}^{nSp} \left(\frac{y_i G_i}{RT} + y_i \ln \frac{y_i}{\sum y_i} + y_i \ln p \right) = Min \qquad (1)$$

On the product side the following species were considered: C, CH_4, CO, CO_2, C_2H_4, C_2H_6, C_3H_8, C_2H_4O, CH_3, CH_2O, H, OH, H_2O, O, O_2, CH_3CHO, C_2H_2, C_3H_6, C_4H_8, C_4H_{8CIS}, C_4H_{8TRANS}, N_2, HCN, CN, NH_3, NO, HCO, NH_2, NH, N, CH, CH_2. The formation of carbon black was not taken into account because only small amounts of carbon black were observed during the experiments.

In addition it was necessary to modify the calculation procedure. Especially in those cases where the "inert gases" water and carbon dioxide take part at the equilibrium reactions extremely high amounts of H_2 and CO were calculated. This led to calculated equilibrium temperatures that were much lower compared to the case with nitrogen. Furthermore gas analysis resulted in much lower fractions of H_2 and CO. Therefore the program was modified in such a way that the inert gas fractions were kept constant on the product side. Only the combustion products of EO took part in the equilibrium reactions. Based on this assumption the calculated temperatures at the flammability limits were similar for all types of inert gases following the theory.

Furthermore it was observed that the calculated temperatures along the boundary line of a flammability region are not constant for the whole curve but varied between 1300 K and 1600 K (see Tab. 1). However, the general shape of the temperature profiles found was similar for all three types of inert gases.

Tab. 1: Calculated adiabatic flame temperatures (equilibrium temperatures) for EO/N_2/air mixtures at 20 °C and 100 kPa (see also Fig. 2)

Flammability limits EO in mol%	N_2 in mol%	Flame temperatures in K
3.0	0.0	1369
3.0	20.0	1371
3.0	40.0	1376
3.2	55.0	1437
3.6	59.0	1389
5.2	55.0	1355
10.0	45.0	1485
20.0	50.0	1381
30.0	46.5	1409
50.0	36.0	1478
81.0	19.0	1539

For a better modeling the mean temperature profiles of the three investigated inert gases were taken in order to calculate the flammability regions. Now, by providing arbitrary but specific flame temperatures and using the "Gaseq" code, it was possible to calculate the corresponding EO concentrations (flammability limits) for different amounts of inert gases. It shall be emphasized that this can be done for a gas mixture containing a single inert gas but also for process gases containing more than one inert gas and for different initial temperatures and pressures.

In Fig.5, an experimental explosion region is compared with a calculated one. The differences are very small. A first application of the new calculation method was the verification of the explosiveness of a specific process gas consisting of EO, N_2, CO_2, H_2O and air in sterilization chambers at 100 kPa and 60 °C. Our client also required to know the maximum allowable EO concentration in such a gas mixture. The ignition tests showed that the process gas was not explosive. The flammability limit was found to be 43 mol% EO. Applying the new calculation method the flammability limit was determined to be 42.5 mol%. The calculation method worked very well.

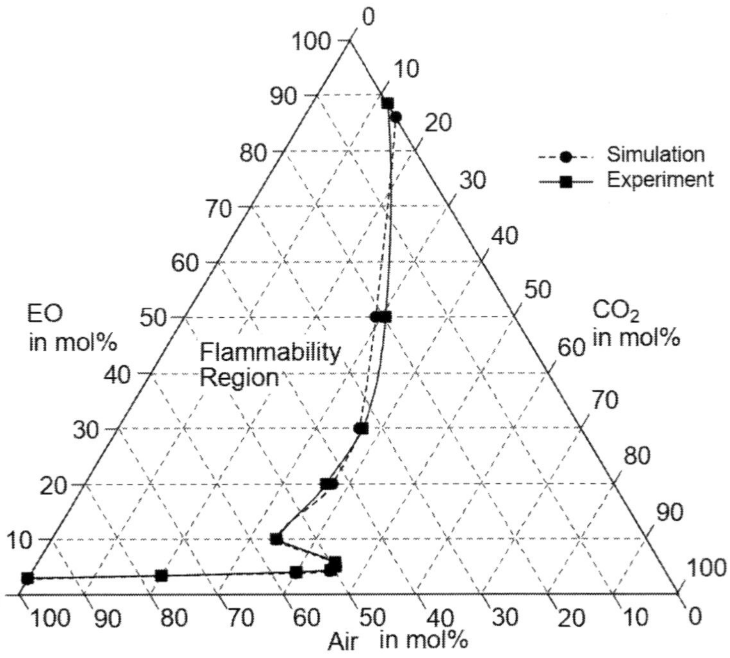

Fig. 5: Comparison between simulation and experiment for the system EO/CO_2/air at 20 °C and 100 kPa

5. REFERENCES

1. U.S. Chemical Safety and Hazard Investigation Board, *Investigation Report No. 2004-11-I-CA, Sterigenics,* 2006
2. Molnarne, M., Schendler, T., Schröder, V., *Sicherheitstechnische Kenngrößen Band 2: Explosionsbereiche von Gasgemischen,* Verlag für neue Wissenschaft GmbH, Bremerhaven 2003
3. DECHEMA, BAM und PTB, *CHEMSAFE® - Database for evaluated Safety Characteristics*, Update 2007, STN Karlsruhe
4. DIN EN 1839: *Bestimmung der Explosionsgrenzen von Gasen und Dämpfen,* Beuth-Verlag, Berlin 2004
5. Burgess, M. J., Wheeler, R. V., *J. Chem. Soc.*, 1911, 99, S. 2013-2030
6. Thiel-Böhm, A.., *Explosionsgrenzen methanhaltiger Brenngasgemische - experimentelle Bestimmung und Simulation; Dissertation*, Universität-GH Paderborn, VDI Verlag, Fortschrittberichte, VDI Reihe 3, Nr. 258, Düsseldorf 1991.
7. Morley, C. "GasEq – A Chemical Equilibrium Program for Windows", online auf http://www.gaseq.co.uk/, Stand: 2007
8. Gordon, S., McBride, B. J., *NASA Reference Publication 1311,* 1994

How to Apply Process Safety Management to Capital Projects

Hunter Shang

3M China Lt., Co.
Shanghai, P.R. China
hshang@mmm.com

ABSTRACT

As a project progresses from beginning to end, how is good Environmental, Health and Safety (EHS) control maintained? When are EHS gate reviews needed during the progress of a capital project? How does one verify and confirm all related requirements have been met before starting the next phase of the project? As a practice example, the "7 Phases and 6 Gates" system integrates project requirements, local laws / governmental requirements, and the 14 elements of the OSHA PSM standard into the overall progress measurement of the capital project, while at the same time, uses an electronic-database to carry out all approval flows, action item tracking and houses all documentation.

1. Background introduction

A capital project normally involves multiple functions within an organization and numerous internal and external requirements. It is difficult to ensure the project progresses smoothly without any EHS and/or operations issues. In the following example, a plant is setting up a new coating production line. The following issues will be considered:

i) Process scope and manufacturing requirements;
ii) Compliance to local laws and regulations;
iii) Hazard and risk evaluation;
iv) Operability;
v) Cost or spending;
vi) Supplier verification and certification;
vii) Contractor management …

What needs to be done, and when? At the same time, how does it meet related requirements or receive confirmation by each function member? So, a clear flow and a multi-function communication platform is necessary. In this case, based on OSHA PSM standard and local law requirements, we separate the overall project process into seven continuous phases, then, apply thread-up safety tools of MOC, PHA, PSSR, etc. to control project progress with function team gate review. Please see the F-1 chart: *Overall Flowchart of the "7 Phases and 6 Gates" System.*

F-1: Overall Flowchart of the "7 Phases and 6 Gates" System

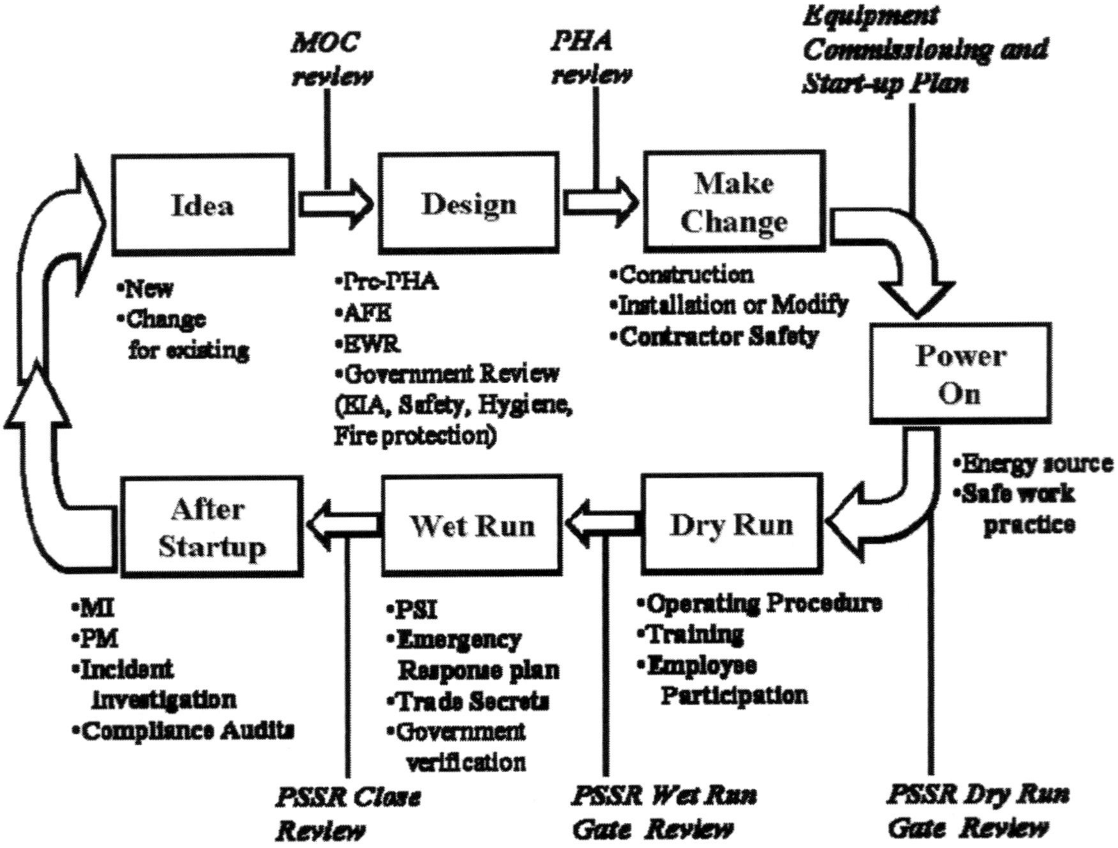

2. "7 Phases and 6 Gates" system Introduction

2.1 Project Phases – "7 Phases"

A capital project usually goes through the following seven phases from first to last. We call it "7 Phases". Please see the F-2 chart: *Project Phases*.

F-2: Project Phases

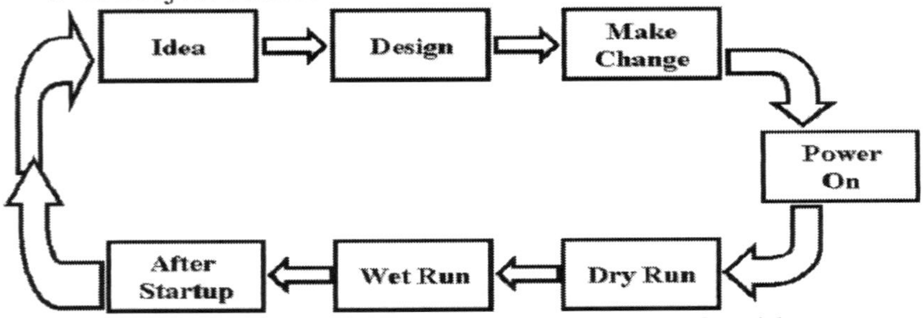

First, any one project will originate because of an idea or a request. It may be a new requisition, or a change for an existing situation. We call the first phase the "Idea" phase.

Then, it needs to be designed based on process requirements or the owner desirability in the light of related regulations. The result is commonly a drawing, a specification or a model, etc. We call the second phase the "Design" phase.

After finishing the formal design, we may need to construct related equipment or facilities by suppliers, and install or modify it on site. We call the third phase the "Make Change" phase.

After basic installation or modification is finished, we attach all kinds of energies; electric, compressed air, water, etc. We call the forth phase the "Power On" phase.

Next, we need try to run the equipment to check if the related functions are right. We call the fifth phase the "Dry Run" phase.

After basic functional testing, we'll try to run the whole equipment or process with materials or loads to check if all overall functions of the process meet the original design requirements. We call the sixth phase the "Wet Run" phase.

Finally, we enter into the normal production or using phase after we verify all functions and related requirements. We call the seventh phase the "After Startup" phase.

Of course, during the "After Startup" phase (normal using phase), we may repeat the entire cycle from the first phase again if we have any more new ideas.

2.2 EHS Gate Review – "6 Gates"

As a project progresses from beginning to end, how is good EHS control maintained? It's a really tough topic! As a core value, EHS should be considered throughout whole project process. In this example, we set up six EHS gate reviews for the whole project process. We call it "6 Gates" review. Please refer to F-1 chart.

The first gate review is the "MOC review" between the "idea" phase and the "design" phase of the project. When we have a new idea, we usually need go through the Management of Change (MOC) system before the formal project design. Generally the goal is a feasibility study of the project. Except business factors, it will focus on big hazards and risks identification of EHS aspects based on related laws / regulations and corporate requirements. So, this is a vital step for the whole project, it is also a basis of other successful gates reviews and EHS systems applications, for example, PHA, PSSR, contractor management, etc.

MOC is one of the 14 elements of the OSHA PSM standard. It should be organized by a multi-functional team, and all results of MOC should be approved by the team and related authorized people before the change implementation. As an actual project communication platform, normally we also discuss and define the future safety control plan and flow of the project in the MOC review meeting.

The second gate review is the "PHA review" before the "make change" phase. The goal of Process Hazard Analysis (PHA) is to complete comprehensive and detailed study to identify potential process hazards/risks and define safe guards and protection action plans based on the actual process design information. PHA is also one of the 14 elements of the OSHA PSM standard, and usually it uses some combination of tools such as: HAZOP, what-if, checklist, FMEA and LOPA etc. Of course, except the formal PHA review, project team will have lots of related work in the "design" phase of the project. For example: Authority for Expenditure (AFE) application; Government review and pre-evaluation for Environmental, Health, and Safety (EHS); Pre-PHA review, etc.

The third gate review is the "Equipment commissioning and start-up plan" review before the "Power On" phase. The goal is to make sure the team has a clear and orderly trial run schedule and no EHS issues happen during testing. During machine/project commissioning normal equipment operations do not exist and a number of potential hazards may be present, and each individual takes responsibility to be informed of such hazards. So, this step will define and inform all related EHS issues of the project trial run phase, meanwhile, energy source control and safe work practice (hot work permit, confine space entering permit, etc.) management will be critical.

At the same time, Pre-Startup Safety Review (PSSR) will be also initiated in this phase. A PSSR should be required before startup of new or modified equipment or facilities. Normally, a PSSR will include an organized checklist of key EHS and equipment requirements. Here we divide the whole PSSR process into three parts ("dry run" phase, "wet run" phase and the "after startup" phase) It is mainly used to determine:

- Whether new and modified facilities have been constructed in accordance with engineering design specifications;
- Safety and training requirements have been met;
- Checks that operating, maintenance, and emergency procedures are adequate and documented.
- Whether all action items (from MOC, PHA, etc.) have been completed

The fourth gate review is the "PSSR Dry Run Gate Review" before the "Dry Run" phase of the project. The goal is to make sure all related EHS issues of the "dry run" phase have been addressed before the project equipment has been started. By getting the approval of the team and related authorized people, the project team can start to dry run the equipment. From the beginning of the "dry run" phase, "Employee Participation" practice will be increased, involving lots of production people and engineers. The "Operating Procedure" will be formed, and the "Training" also need be done.

The fifth gate review is "PSSR Wet Run Gate Review" before the "Wet Run" phase of the project. The goal is to make sure all related EHS issues of the "wet run" phase have been addressed before the project equipment has been started to wet run. At this time, new chemicals or materials will be involved to the production line, increasing the likelihood of potential risks of fire, explosion and toxic hazards. So, this is a vital step of whole PSSR process. In this phase, local government verification (such as: fire protection system certification, explosion-proof equipment certification, etc.) need be confirmed; Process Safety Information (PSI) and Emergency Response Plan (ERP) should be also drafted and employees trained on their use.

The Sixth gate review is "PSSR Close Gate Review" after the "Wet Run" phase of the project. The goal is to make sure all related action items of whole project have been completed. The step is usually executed as a deliverable between the project team and the manufacturing owner. Of course, it also needs the approval of the team and related authorized people. After the equipment startup, the manufacturing team will finally define the procedures and plans of "Prevention Maintenance" (PM), "Mechanical Integrity" (MI), and the "Compliance Audits". "Incident Investigation" also needs to be implemented during the normal production phase.

3. E-Database and Documentation Management

From above "7 Phases and 6 gates" system, new problems will undoubtedly appear. How do we conveniently go through all gate approval flows of all function teams? How can we easily track the status of so many action items? How do we document and keep all related files / drawings for the life of the equipment / process? An Electronic-Database is a good solution which may help resolve those issues.

The following is a main menu of a plant PSM database for your reference:

Main Menu of PSM Database
1. Procedure sub-database
2. MOC sub-database
3. PHA sub-database
4. PSSR sub-database

5. PSI sub-database
6. MI sub-database
7. Incident Investigation sub-database
8. Training manual
9. Contractor safety sub-database
10. Safety Permits sub-database
11. EHS communication conforming sub-database
12. Action items sub-database

Each of the 12 sub-databases listed above are linked to each other in the total PSM database using the overall flow of "7 Phases and 6 gates" system. Each one has its own functions: Creating and editing the request forms, store function, email and approval function, editing action items function, etc. The PSI sub-database may store all projects and product information (including equipment, utilities, processes, chemicals, etc.). The action items sub-database may directly send separate action item to related responsible people, and it will clearly state the following points in a separate email: What's the action? Who is the responsible person? When is the target date? Who is the final approver?

These functions working together help us to integrate all requirements, flows, actions and documents into the same database, improving project implementation efficiency and operability.

4. Acronyms

AFE – Authority for Expenditure
CFR - Code of Federal Regulations
EWR – Engineering Work Request
EIA – Environmental Investigation agency
MI - Mechanical Integrity
MOC - Management of Change
NFPA - National Fire Protection Association
OSHA - Occupational Safety and Health Administration
PHA - Process Hazardous Analysis
PSI – Process Safety Information
P&ID - Piping and Instrument Drawing
PM - Preventive Maintenance
PSM - Process Safety Management
PSSR - Pre-Startup Safety Review

5. REFERENCES

1. Regulations (Standards -29 CFR), *Process Safety Management of highly hazardous chemical. – 1910.119,* Occupational Safety & Health Administration, U.S. Department of Labor, http://www.osha.gov

Specialized Working Procedure

I Made Sukrajaya
BP West Java
Jakarta, Indonesia
sukrajim@bp.com

Yenny Bambang
BP West Java
Jakarta, Indonesia
yenny.bambang@bp.com

ABSTRACT

Everyone knows that standard operating procedures (SOPs) are used for providing clear instructions for safely conduction "critical" or "routine" tasks. But do we forget that non-routine tasks should also have written procedures, called specialized working procedures (SWPs), which should provide clear instructions for safe operations and maintenance?

On closer look, SOP and SWP must be treated the same in terms of purposes, performance requirements and documentation. We should not forget that a majority of risk is due to human factors during non-routine activities and that the task risk assessment or job safety analysis may be insufficient to address all hazard and risk reductions. Risk assessment is used to define the hazard, risk and mitigation plan, but need to be converted to a written procedure to ensure the operator doing the task has clear instructions. By preparing an SWP, it is intended to reduce the frequency of incidents due to human error. An example of SWP is tie-in procedure for plant modification, turn-around operation procedure, non-routine maintenance, etc.

Issues to be addressed when developing an SWP, similar to an SOP; should at minimum include:

- *Overview of work description*
- *Expert personnel that know well the work activity*
- *Responsible person(s)*
- *Tools used for doing all required tasks*
- *Potential Hazard (before and when activities are executed)*
- *Input on Experience Based Hazards from the field*
- *Action steps / accountabilities for key points which must be address if hazards and risks are assessed outside of tolerable limits*
- *Knowledge Test (optional)*

This paper will describe in more detail requirements of SWPs, how to develop, what the key points are which must be addressed, what procedure steps must be considered, the MOC which must be followed-up on and how the SWP should be documented.

1. Introduction

Since 1980s, a number of major industrial accidents have spurred more and more industrial personnel to be aware of the importance of process safety. A major accident is commonly modeled as the iceberg: only a small portion of the iceberg is visible above the waterline, but the tip is supported by much larger and invisible part underneath. Similarly, a major accident is represented by the tip of the iceberg and is caused by many minor and seemingly trivial incidents.

Most of those incidents are usually seen as "easy" tasks; either because they are regular or routine activities, or just "simple" activities such as tightening or loosenening stud nuts. Additionally, some workforce personnel become over-confident with their "used-to-do" activities, such as cable pulling, scaffolding erection, lifting, etc, which they consider these activities as a habit and begin to lose caution during performing the job.

Several practices, such as risk assessment, job safety analysis, work permit and operating procedures are implemented to reduce minor incidents. Although these methods do not overcome the problem completely, they do, with an appropriate approach, reduce the occurrence of minor incidents. Nevertheless, a significant number of cases involve non-routine activities. Hence, a good safety program must not overlook the process of addressing non-routine task.

This paper focuses on discussing the Specialized Working Procedure (SWP) because minor incidents typically occur during performing non-routine activities. It is necessary because of the following considerations:

- there is often no standard procedure for non-routine activities
- there are usually very few person that really understand about the particular non-routine activities
- there is assumption that it is not important to develop standard procedure for non-routine activities, either because the activity is considered trivial or the occurrence is infrequent

Non-routine task can be defined as tasks that are performed apart from the usual work routine. It also refers to an activity which Operating Procedures do not exist or when the Operating Procedure is out-dated or inadequate.

The intentions of the SWP are basically the same as for the Standard Operating Procedure:

- comply with HSE requirements
- reduce process hazards and risk that may occur during procedure preparation until execution of the works. Human error is one primary aspect that needs to be considered.
- establish consistency in developing working procedure
- improving training program to Operator & person that will do the work

2. Hazard Identification and Hazard Analysis

Prior to starting any jobs, routine or non-routine, identifying hazard around the work place is a must. By doing so, the operators are informed of the potential hazard, are encouraged to think several options to perform the work safely, and are aware of the risk behind each options, thus allow them to select the best method to perform the work safely. Hazard identification can be done by walk-around inspection or observation program. A lesson learned from previous similar activity is also a source for identifying the hazard of a task.

After the hazards have been identified, they need to be analyzed. Hazard analysis is a task of analyzing the safety of a performing a job in advance, to ensure that the hazards are fully understood and cautiously tackled and mitigated prior to commencing the job. Commonly, this activity is called Job Safety Analysis (JSA).

Every non-routine task is subject to a thorough job safety analysis. During its preparation, each team member should look carefully at the hazard in the area and determine the best way to eliminate or reduce the risk. Job Safety Analysis should also be reviewed frequently to ensure that it has covered all scope of work and is capable of being followed. Any changes in process, scope of work, technology requires the Job Safety Analysis to be reviewed and changed accordingly.

Although Job Safety Analysis is a good safety program for ensuring safe working environment, it may be insufficient to address all hazard and risk reductions. A majority of workers who did jobs unsafely did so because they believed that it was the safest to do the job. Likewise, some of them did not know how to perform the work, either because the job has never been performed before or because the workers lacked proper training. Another obstacle to safely performing a non-routine job is that frequently there is no experienced person who fully understands and is capable of performing the job safely.

Considering the above explanations, a written procedure for non routine task is required to be developed to ensure adequate information is communicated to personnel involved in performing the work. The written procedure is also intended to provide clear instructions for the operator doing the task. The written procedure is known as Specialized Working Procedure (SWP).

2.1. Human Error Consideration

SWP is intended to provide guidance for consistency hence reducing human error during performing a job. Thus several human error factors should be considered during SWP development and also when the SWP is used as guidance for performing the work on site. The list of human aspect include :

- Human Error for understanding the facility
- Human Error for understanding the hazard and the risk that may occur.
- Human Error due to lack of concern
- Frequency of Error
- Basic approaching to review human error at facility
- Understanding Lesson Learnt from previous experience
- Basic Process of Human Error at Company Based Experience
- Lack of basic knowledge on Risk Based Process Safety

3. What is a Procedure?

A Procedure is a set of steps to be performed to obtain a specified outcome. Generally, a procedure is comprised of action steps, decision steps and repetition steps. Action steps consist of what should be done to perform the job. Usually it is written in imperative sentences and starts with an action verb. The subject of an action steps is always implied of "you". Decision steps are points which a person has to make a choice based on the information available at that time, usually written in yes-no question in an if-then format. Repetition step is a special command or imperative sentence that tells the user to repeat a set of actions and/or decisions until some desired condition is achieved. The repetitive steps must specify exactly which action steps or decision steps to repeat and define the condition that will end the repetition. A procedure also has a starting point defining initial conditions or inputs and a goal-defining output. In summary, a procedure can be described as three elements: starting point, components and goal.

An example of a procedure that most people are familiar with is Standard Operating Procedure. An SOP can be defined as a set of written instructions or steps to allow someone performing a task in a standardized manner, which is the preferred and safest method. SOP is usually intended for a specific task, usually a routine task, and for providing assistance to the operators for maintaining the quality control and ensuring compliance to regulation. An SOP is developed to minimize variance and ensure consistence implementation even if there are temporary or permanent personnel changes.

Similar to SOP, a Specialized Working Procedure is a written instruction for non-routine tasks that allow an operator to perform the task in a standardized manner by the preferred and safest method. Taking the importance of providing clear instruction and ensuring good communication, especially for non-routine work, into consideration, SWP should be treated the same as SOP in terms of purpose, performance, requirement and documentation.

4. How to Develop Specialized Working Procedure

There are a lot of procedures, either Standard Operating Procedure or Specialized Working Procedure, available throughout industry, but few people understand how to develop a procedure, what should be included in the procedures and what a procedure should be look like. This section presents elements which should be included in a procedure. Application in different industries and types of work may necessitate appropriate modification.

A procedure is not just comprised of a set of instructions. Several points are essential in a Specialized Working Procedure document so that the procedure can accomplish its purpose of communicating hazard of non-routine work, and outlining the preferred and safest method, as well as ensuring consistency of performing the non-routine work.

Issues to be addressed when developing an SWP is similar to an SOP, which consist of the following:

- Overview the work description
- Expert personnel that know well the work activity

- Responsible person
- Tolls used for doing all required tasks
- Potential Hazard (before and when activities are executed)
- Input or Experience Based Hazards from the field
- Action steps / accountabilities for key points which must be addressed if hazards and risks are assessed outside of tolerable limits
- Knowledge Test

4.1. Overview of work description

The overview of work description provides an explanation of what objective is to be achieved. It consists of the purpose of developing the SWP, the particular intended user and process and/or regulation standard. Clear boundary definition of work to be covered by the SWP must also be explained in order to provide guidance for the writer to focus and providing clarification for the reader and the user.

Definition of any specialized or unusual terms is also necessary to be described in separate section. The reasons are to familiarize the reader with those terms and to make it easy-to-find when the definitions are needed. Sometimes it can be in the form of a list of glossary provided in the beginning of the document. Which format is chosen depends on its convenience and the writer's preference.

4.2. Expert personnel that know well the work activity

Who should develop the SWP needs to be carefully considered. Usually a SWP is developed by a person who has good knowledge of the task, preferably one who has performed the particular task. Consultation with additional experts is required under some circumstances, for example if there is no experienced person in the organization of this particular task. The name of the preparer of SWP should be clearly stated.

Regardless level of competency, it is encouraged that personnel involved in performing the job also be involved in preparing the SWP. This will help engaging the operators and ensuring that the procedure is to be properly followed.

4.3. Responsible Person

One of the most significant issues to be considered is the person responsible for the performance a particular task. This section should provide complete list of the roles and responsibilities of all personnel for undertaking each step in the SWP. The SWP should specify an appropriate level of responsibility for each of the working team members. Minimum required competencies for performing specific task are encouraged to be defined. By providing clear roles and responsibility of everyone involved, team members are more aware of their task and the limits of their own competence. Additionally, this is also ensures that in the absence of a team member, others should not have increased responsibility or do the work whose competence and knowledge they are unsure of.

It is also important to define the accountability of each task to ensure that each personnel are professionally accountable and carefully consider all decisions they are making.

4.4. Tools used for doing all required task

Any equipments or tools required for performing the entire required task specified in the SWP should be listed. To simplify the list, it is preferable using a general terms for common tools and equipment. This section also includes required certification and inspection to be carried out to all tools and equipments prior to commencing the work. Required personal protective equipment for performing the task is also necessary to be stated.

4.5. Potential Hazards

Prior to writing action steps of the procedure, it is critical to understand about the potential hazard and associated risks of the task. This information can be simply obtained by conducting hazard identification and hazard analysis. Through job safety analysis, preventive action to eliminate or reduce the risk can be identified.

However, not only the hazard before starting the work that needs to be considered. But any potential hazard that may arise during working period should also be assessed and properly mitigated. For example, during hot work activity, there is possibility of potential hydrocarbon release within working location.

For certain activities, such as hot work, the environment where the task is to be performed should be clearly defined, such as required good ventilation, special working location, etc. Each permitted or prohibited area should be specified. Additionally, environmental conditions should also be clearly defined; for example is the work is to be carried out during daylight only or during sunny conditions.

4.6. Input or Experience based Hazard from the field

Risk assessment alone is not considered adequate for identifying all hazards for performing specified tasks. Lesson learned inputs from similar activities are one of the best options for identifying the hazard. Other source is from experience based hazard identification from the field. This can be carried out by having walk-through inspection accompanied by experience person or by simulating the implementation of the procedure in the actual working condition. Through walk-through inspection, the personnel with basic understanding can ask question to enhance its knowledge and confidence for doing the work. By field testing, each action steps can be challenged to be the preferred and safest method.

4.7. Action Steps

Action Steps are clear and simple step by step instructions described in sequential order for performing the particular task in a preferred and safest manner. The instructions should be written as clear as possible to avoid ambiguity and misunderstanding. In case of any two steps to be done at the same time, the sentence should be to differentiate the action to the other.

A SWP should also provide clear guidance for exceptional circumstances if hazards and risks are assessed outside the tolerable limits.

4.8. Reference

A good procedure needs a complete reference section. This section consists of complete list of standard and regulations used for developing the SWP. It also provides source for the reader and user if they need to seek additional information or if they are required to clarify an issue.

4.9. Knowledge Test

A Knowledge Test is an optional feature of a SWP. This test is intended to ensure basic understanding of the involved personnel to grasp and comprehend the potential hazard of the work, the risk associated in the work and key point of the action steps. This test may also be a useful measure of successful hazard communication to all involved personnel.

Besides the above issues, one needs to remember that Specialized Working Procedure is aimed for a specific job and intended to provide guidance for specific users. Hence, it should provide sufficient detail to ensure that the operators with limited experience or knowledge of the procedure can successfully perform the work as described in the procedure without being supervised. To avoid misunderstanding, the procedure should be written in a concise, logical, step-by-step and easy-to-read format.

SWP should be developed based on detail study such as review of process information, lesson learned from previous activity, and observation on how the operator tendency to perform the work.

5. Training

A written procedure, besides ensuring quality and consistency of performing a task, is also intended to reduce the frequency of incidents due to human error. SWP provides a simple set of instructions to perform the work in the safest method and provides role clarification. Having that assistance, confusion over who does what and the actions to do for completing the job can be avoided.

To achieve those objectives, SWP should be prepared as concise and as simple as possible so that a person with basic understanding and limited experience can understand and follow the SWP accurately. Nevertheless, not all operators have adequate basic understanding about a particular task. Thus, training on the task and a review of the procedure is necessary to be carried out prior to commencing the work.

The objective of the training is not simply to familiarize the operator with the procedure or to ensure that they can follow the steps in the procedure, but emphasize understanding of the reason for developing the procedure and to assess where the procedure may not be sufficient. Simulation is one useful concept for an effective training for non-routine jobs.

6. Documentation

SWP should be properly documented and numbered accordingly. Date of preparation and/or date of review should also be precisely stated. SWPs must be readily available at all times. It does not mean that they are physically available, but more that they are up-to-date at all times to ensure it relevancy. So it is necessary to outline updating process for a certain period.

It is also critical to record any deviation from the recommended procedures and what has been done to tackle those deviations. Any changes of standards, regulations and technologies should be followed by reviewed the SWP and updated the SWP accordingly.

Any changes or revision should also be properly communicated to relevant users. For ensuring that every revision is recognized, a section in the procedure should be dedicated for recording any revision: the initiator, date of the revision occurred, the approval authority and details of changes.

7. Study Case

This section describes Hot Work Procedure at Offshore Facilities of BP West Java as one example of Specialized Working Procedure. The purpose of this procedure is to describe the required safeguard when performing hot work, such as welding, cutting, and grinding.

Since 2001, BP West Java limits Hot Work Activities for Offshore facilities. By having issuing this policy, Hot Work Activities should be done at an onshore workshop. Except for special circumstances, for example during total shut down of Turn Around Program, Hot Work Activities are permitted to be performed offshore but with thoroughly consideration and –high-level management approval. As Hot Work Activities considered as risky task and it is infrequently performed, a Specialized Working Procedure is required for performing this work.

Hot Work Procedure of BP West Java comprised of the following:

1. Purpose / Scope
   This section describes the purpose of developing this procedures and the scope of activities covered by this procedure.
2. Definitions
   Several specific terms are explained in this section, as well as providing scope boundary of those terms. For examples, Hot Work is defined as welding, burning, cutting, grinding, hot tapping or work with open flame.
3. General Requirements
   The General Requirements sections illustrate Work Description and Work Location.
4. Key Responsibilities, Inspection and Certifications
   This section describes each the roles and responsibilities of each personnel, for example the Person-in-charge, Issuing Authority, Fire Watch and Personnel Doing the Work. It also contains the required certification, such as Welder Certification and Equipment Certification. Required inspection on welding machine and associated equipment is also necessary to be defined.

5. Procedures
The procedures are required for the Pre-Plan, Site Preparation and Inspection, including Energy Isolation. This section also contains Permitting Process and required Training.
6. Documents, Tools and References
Revision of the procedure should always be documented properly, complete with the person initiating revision, date, authority approval and detail revision.

8. Conclusion

Specialized Work Procedures are powerful tools for seizing control of non-routine activities. They define the subtle details that maintain consistent implementation, ensure compliance to regulation, and complete the activity safely. In addition, well-written SWPs act as effective communication tools that contribute to worker understanding and safely job completion.

It is important to remember that SWP should have clear and specific title, reference source, revision dates, approval and detail, intended user, purpose, list of tools and parts, definitions of terms and concepts, safety warnings and cautions, detail and simple step by step instructions for performing the job.

SWP development process also provides significant performance improvements. Development process of a SWP, when properly and fully carried out, allows worker, engineers and managers working together for ensuring safe working condition. As the result, everyone focuses their abilities on doing the best job possible as well as improving working condition.

However, the essential purpose of SWP should be enhanced with assessment of hazard and risk and strong behavior-based safety program, as well as with continuous improvement through observation and training.

9. REFERENCES

1. Stewart, Maurice I, *Risk-Based Operating Procedures and Safe Work Practices*, Indonesia, LDI Training, 1997.
2. Center for Chemical Process Safety (CCPS), Guidelines for Risk Based Process Safety, 2007
3. *BP Indonesia HSE Handbook 6th edition,* Indonesia, 2005

Case Study of a Hydrogen Explosion in an Electrical Panel

Timothy J. Myers
Exponent, Inc.
Natick, MA
tmyers@exponent.com

Alfonso F. Ibarreta
Exponent, Inc.
Natick, MA
aibarreta@exponent.com

Prepared for Presentation at
American Institute of Chemical Engineers
2008 Spring National Meeting
42nd Annual Loss Prevention Symposium
New Orleans, LA
April 7-9, 2008

UNPUBLISHED

Case Study of a Hydrogen Explosion in an Electrical Panel

Timothy Myers, Ph.D.
Exponent, Inc.
Natick, MA
tmyers@exponent.com

Alfonso Ibarreta, Ph.D.
Exponent, Inc.
Natick, MA
aibarreta@exponent.com

Abstract

A chemical facility generated nitrogen for use in its processes by removing oxygen in air using pressure swing adsorption (PSA). In order to remove residual oxygen in nitrogen from the PSA process, hydrogen was added to the nitrogen and the mixture was then passed through a catalyst bed, converting the remaining oxygen to water. The water was then removed in a drying or dehumidification process yielding high purity nitrogen. During what had become a routine process upset at the facility, a worker opened an electrical panel to disconnect a pneumatic line, which in turn caused a solenoid valve to open and reset the system. As the system reset, an explosion occurred in a neighboring electrical panel and the panel door struck the worker in the head, causing serious injuries. A subsequent investigation revealed that the explosion was caused by the ignition of hydrogen that had migrated into the panel.

As is often the case, multiple factors contributed to the cause of the incident, any one of which could have prevented the incident. In this paper we discuss the root causes and contributing factors of the incident and lessons learned that can prevent similar events from occurring in the future. Issues discussed include: identification of scenarios during process hazard analyses, initial electrical classification of the area, design and maintenance of electrical conduit systems, maintenance of safety shut-off valves, management of change, procedures for working on electrical panels, and response to process upsets.

1. Background

An accidental release of hydrogen caused a small explosion in an electrical panel at a chemical facility, causing the electrical panel door to fly outward and strike a worker in the head. The incident occurred during a valve switching operation at a nitrogen-generating and purification processing facility, injuring one man. Investigation of the incident indentified several factors that contributed to the incident.

2. Process

The process equipment involved in the incident is shown in Figure 1 and consisted of three sections: (1) a pressure swing adsorption unit that separated nitrogen from air, creating crude nitrogen, (2) a catalytic oxidation unit which reacted residual oxygen in the nitrogen with hydrogen, and (3) a drying section that removed water from the nitrogen. A continuous oxygen analyzer monitored the final product quality.

2.1 Pressure Swing Adsorption (PSA) Generator

In the PSA, filtered air is passed through a Carbon Molecular Sieve (CMS) material and oxygen in the air is adsorbed. The CMS is then regenerated in a desorption stage by venting to atmospheric pressure, releasing adsorbed oxygen. Two adsorption vessels are used to alternate between the adsorption and desorption cycles while providing a continuous stream of nitrogen-rich exhaust. The crude nitrogen stream exiting the PSA process typically consists of about 0.5% oxygen content.

2.2 Catalyst Oxidation Unit

After the PSA stage, hydrogen is added to the nitrogen stream and the mixture is then circulated through a catalyst bed, in order to convert the remaining oxygen to water. The purification is a continuous process, with hydrogen being injected at the proper flowrate to maintain a slightly higher than stoichiometric (2 to 1) ratio of hydrogen to oxygen. The ratio is maintained with a ratio controller that adjusts the hydrogen addition rate based upon measurements of oxygen in the raw nitrogen stream and hydrogen in the mixed stream. The hydrogen from a pressurized source passes through a pressure regulator, a solenoid valve, the control valve and then a manual shut-off valve (as shown in Figure 1) before being introduced into the main nitrogen stream. The stream then passes through a mixer and a catalyst bed, where the oxygen and hydrogen are converted to water. This process produces a final nitrogen stream with 5 ppm maximum oxygen content.

2.3 Dryer

Once the excess oxygen is consumed in the catalytic reactor, the generated water is removed first cooling the stream and removing condensed water, and then by passing the stream through a desiccant dryer. Two dryers are present in parallel and flow alternates between the two dryers while one dryer is regenerated by heating. The final product stream is dry nitrogen gas.

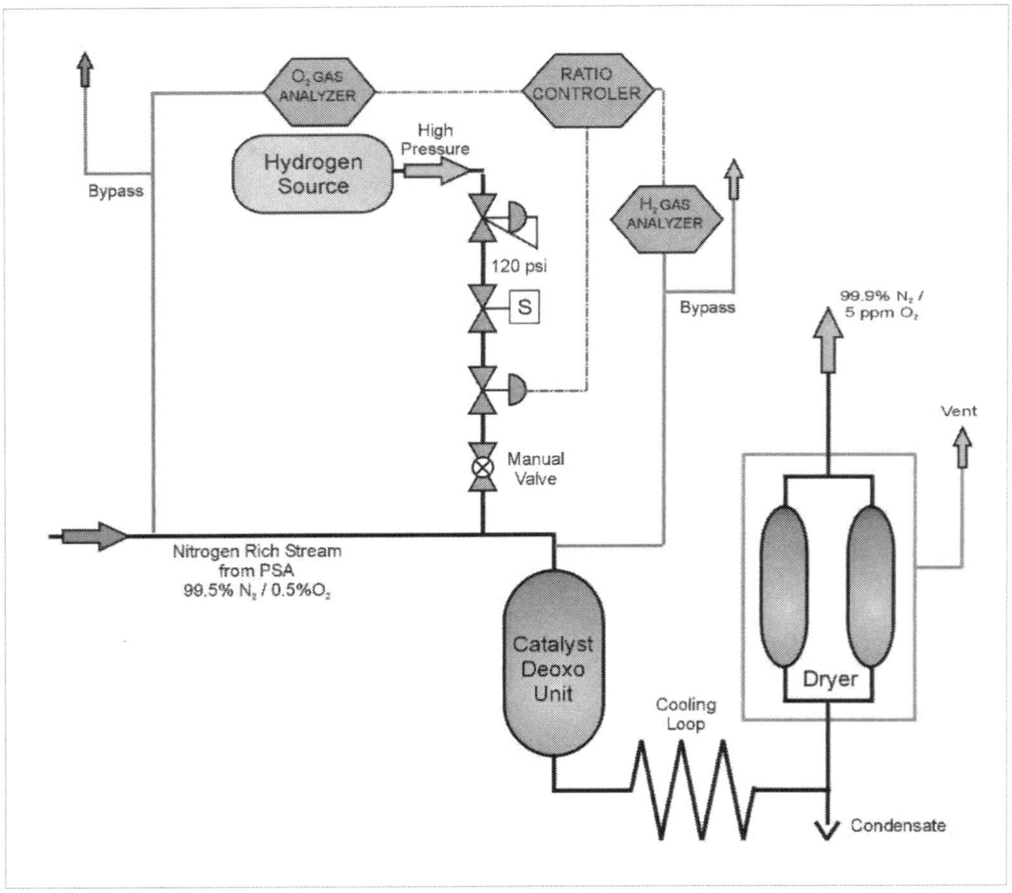

Figure 1: Flow diagram of selected components of nitrogen generation unit

3. Events Preceding the Incident

3.1 Initial Process Hazard Analysis

Prior to the installation of the nitrogen generation process, the facility owner had performed a limited hazard analysis to determine the electrical classification of the area where the nitrogen generation system would be installed. The system was going to be installed in an area with a roof, one wall, and three open sides. Because of the relatively open environment, and the low density of hydrogen, the facility deemed that in the event of a release hydrogen would accumulated near the top portion of the equipment. The facility therefore decided to require gravity ventilation at the high point of the roof, and only classify regions one foot below the roof and higher as Class I, Group B, Division II [1]. All other regions were classified as general purpose, and did not require special electrical equipment. The facility also specified that the hydrogen pipe header should be welded construction and any flanged connections should be at least 8 feet above floor level.

In addition to the facilities requirements, the manufacturer of the hydrogen system used electrical equipment suitable for a Class I environment [1] on the hydrogen line, and placed seals in the electrical conduit between this equipment and other electrical equipment, consistent with practices for a Class I area. Other electrical equipment in the unit, however, was not designed for a Class I environment because of the electrical classification determined by the facility.

3.2 Changes to System Prior to Incident

Prior to the incident, a solenoid shut-off valve on the hydrogen feed line (shown in Figure 1 with an 'S') was replaced. Because it was part of the hydrogen feed flow system, the original solenoid valve was explosion proof and rated for a Class I, Division 1 area. The facility replaced the valve with a general purpose valve from the same manufacturer. During the investigation it was determined that the replacement valve had been installed backwards. Like many solenoid valves, when installed backwards the valve would not properly seal in the closed position, but would leak when the pressure drop across the valve was sufficient to counter the solenoid spring force – approximately 35-40 psi. The hydrogen pressure upstream of the valve was regulated to 120 psi; providing enough pressure to open the valve and allow hydrogen to leak through the system. A hydrogen flow control valve was also present in the system, but as is common with many flow control valves, it was not designed to be leak tight or to serve as a shut-off valve.

3.3 Initial Failure and Operator Response

On the morning of the incident there was a process shut down for four to five hours. The shut down was caused by a failure of a valve to open and release the gas from one of the desiccant drying vessels. This, apparently, was a frequent event at this facility, and operators had devised a process for resetting the system. The system was reset by disconnecting a pneumatic line that provided nitrogen as a pilot to a valve on the dryer vent line. The operators would disconnect the pneumatic line inside a low-voltage control cabinet which would cause the valve to open. After a few minutes of delay, a relay in an adjacent high-voltage cabinet activates a switch and resets the process.

3.4 Final Catastrophic Failure

Immediately after the reset procedure on the day of the incident, a pocket of leaked hydrogen gas ignited in the high voltage electrical cabinet, blowing the cabinet door away from the box. After the explosion, an injured worker was found with severe head injuries due to impact from the door of the high voltage electric cabinet. A photograph of the cabinet, with bulged walls an damaged hinges, is shown on the left in Figure 2. The door to the cabinet, shown to the right in Figure 2, was found approximately 20 feet away where it had impacted a tank. The worker's hard hat and safety glasses had been damaged and knocked off his head. The low voltage electrical cabinet was found open, with the pilot line disconnected, indicating that the worker was in the process of resetting the system using the method commonly used at the facility. Junction box gaskets on electrical conduit near the high voltage electrical box were also bulged outward indicative of internal pressurization.

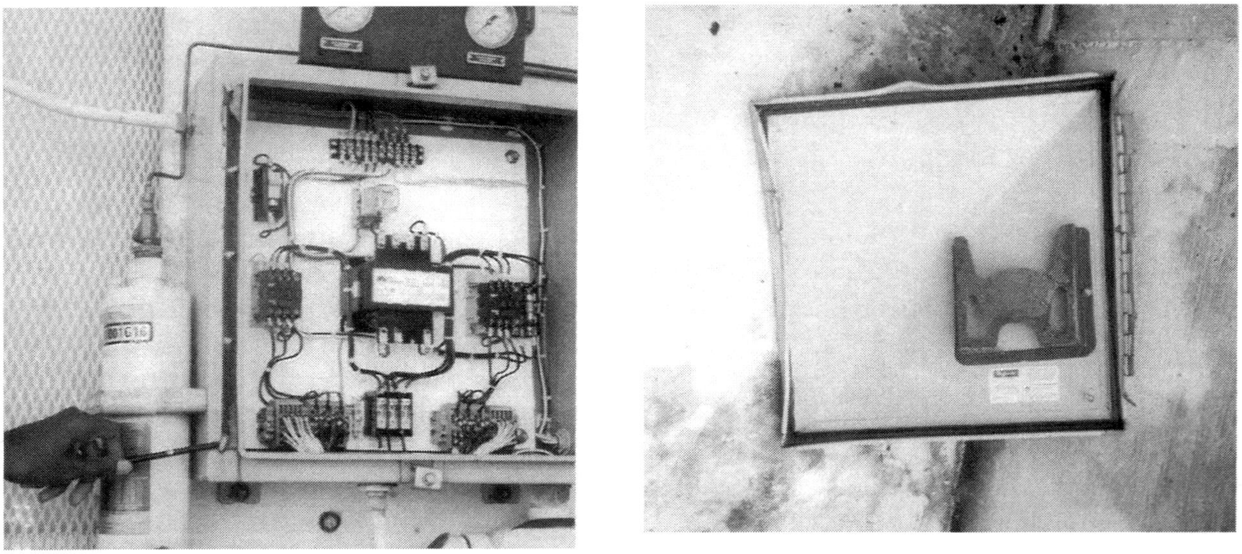

Figure 2: Photographs showing damage to electrical box and door.

4. Incident Investigation

4.1 Source of Energy to Pressurize Cabinet

Investigators originally identified three possible scenarios that could have led to the explosion of the electrical cabinet.

1. Nitrogen Pressurization - Pressurization by process nitrogen by a leak through conduit between the desiccant driers and the high voltage electrical cabinet.
2. High Voltage Electric Arc – A high voltage electric arc in the high voltage cabinet could have led to pressurization of the cabinet.
3. Hydrogen Combustion – Hydrogen migrating into the high voltage electrical cabinet, mixing with oxygen and igniting by a spark from a switch in the electrical cabinet.

The first two scenarios were ruled out. Pressure testing indicated that there was no leak path between the desiccant heater and the electrical panel, and inspection of the electrical cabinet showed no evidence of electric arcing. The third hypothesis could not be ruled out and further testing and analysis were performed to determine whether hydrogen could have migrated into the electrical cabinet

4.2 Source of Hydrogen Leak into Cabinet

Pressure testing was used in an effort to identify leaks through the electrical conduit from the electrically actuated hydrogen shut-off valve and a pressure switch to the high voltage electrical cabinet. The valve and pressure switch were both designed in a manner that did not allow gas to migrate from the hydrogen process side to the electrical conduit. Pressure testing verified that no leak path existed between the hydrogen feed line and electrical conduit. Thus, no direct leakage path from the hydrogen line to the high voltage cabinet was identified.

Other possible hydrogen leak paths existed because the hydrogen shut-off valve was installed backwards and would allow hydrogen to continue to flow through the system even during a shutdown. The system design also enabled an unforeseen leak path of hydrogen through the processed gas exhaust. During a shutdown there was no nitrogen flow through the system. Thus, during the shutdown, portions of the system downstream of the hydrogen addition point that normally handled low concentrations of hydrogen in nitrogen would fill with pure hydrogen. One investigator hypothesized that hydrogen could have been leaking inside the electrical cabinet of a hydrogen analyzer and then migrated through electrical conduit to the high voltage cabinet. This was not tested experimentally, but appears unlikely because gas entered the cabinet at a very low flow rate and pressure, and the cabinet was not air tight so gas leaking in the cabinet would have preferentially leaked out of the cabinet rather than through the tortuous path of the electrical conduit to the high voltage panel.

Another possible leak path into the high voltage electrical cabinet was identified. The drying system was a pre-assembled system purchased from a third-party. Since the dryer was intended for nitrogen service, its manufacturer configured it to use the dried process stream as the pilot gas for solenoid valves associated with the dryer. During the shutdown, eventually the desiccant drier would have filled with hydrogen. Thus, when the pilot line was disconnected to reset the system, it would have released hydrogen and not nitrogen.

Because, this connection was frequently disconnected and reconnected, it may have been leaking before the worker disconnected it as well. This gas could then leak into the neighboring high voltage electrical panel through electrical conduit where it was ignited. The system had been reconfigured before this potential leak path could be tested by experiments.

Although the exact source of hydrogen entering the high voltage electrical cabinet could not be identified, two possible sources were ultimately identified: a leak through electrical conduit from the hydrogen analyzer, or a leak from the pilot line from the drier.

5. Lessons Learned from Incident

Even though the exact leak source of hydrogen into the high voltage cabinet was not identified, multiple factors were identified that contributed to the cause of this incident and could have caused other incidents. The mitigation of any of these factors could have prevented the final incident. This particular case therefore offers multiple lessons to be learned.

5.1 Failure to Identify the Hazard

The original hazard analysis focused solely on external hydrogen leaks from the hydrogen header that could potentially enter electrical equipment. It did not consider possible leak paths that would allow hydrogen to migrate from process piping or vessels to electrical cabinets.

Additionally, the hazard analysis failed to identify upset conditions where high concentrations of hydrogen could enter the downstream portions of the process and potentially migrate into electrical equipment. During normal operation combustion of hydrogen once mixed with nitrogen was not possible because hydrogen was being added to a nitrogen stream at concentrations of approximately 1%, well below the 4% LFL of hydrogen, and the oxygen concentration was approximately 0.5% well below the 5% minimum oxygen concentration of hydrogen. However, as this incident demonstrated, failure of the hydrogen shut-off valve would allow much higher concentrations of hydrogen to exist during a shutdown. When mixed with air, the hydrogen would be ignitable. Had this been identified, the facility may have taken additional measures to seal or purge electrical conduit and cabinets.

When performing hazard analyses or determining electrical classification of areas, facilities should consider potential upset conditions as well as normal operation and determine whether extra protection is needed to address these upset conditions.

5.2 Incorrect Installation and Maintenance of Equipment

Automatic shutoff of hydrogen to the system relied on proper functioning of a single solenoid valve. During routine maintenance, however, the valve was replaced and installed backwards. Like many solenoid valves, the valve was designed to be installed in one direction so that a higher pressure upstream than downstream would act to close the valve when it was not powered. When installed backwards a higher pressure upstream than downstream acts to open the valve. In this case a pressure drop of 35 to 40 psi was sufficient to overcome the force of the spring in the valve and cause the valve to open.

When replacing or performing maintenance on safety related valves such as a shutoff valve, facilities should consider having a second person verify that the valve is properly installed through either visual inspection, or where possible, through testing.

5.3 Redundancy of Shutoff Valves

This incident also demonstrates that if the hydrogen shutdoff valve failed open due to other reasons, hydrogen would continue to flow into the system during shutdowns. Facilities should consider redundant shutoff valves or double block and bleed combinations for shutoff valves with safety implications.

5.4 Use of Process Stream from Desiccant Dryer for Instrument Pressure

Because the desiccant dryer was designed to treat nitrogen, the dryer had been designed to use the process stream as a pressure source for pilots on solenoid valves. The process stream was then routed through the electrical cabinets. Using an independent source of instrument air, rather than the process stream would have prevented hydrogen from flowing through gas lines within electrical cabinets and being released when the pilot line was disconnected. When using process streams in place of instrument air, facilities should consider the possible composition of the process stream during upset conditions as well as the intended composition. Kletz describes a somewhat different incident where solvent vapors in a drier process stream entered electrical equipment causing an explosion [2].

5.5 Improper Handling of Process Upset

The facility had experienced this same process upset, the failure of a drier exhaust valve to open, on previous occasions. Rather than addressing the failure that was causing the upset, employees had developed an ad hoc procedure for resetting the system. No management of change procedures or hazard analysis were followed to develop a safe procedure for addressing the upset. The manual for the desiccant drier included specific instructions for troubleshooting this upset that did not require disconnecting a line in an electrical box. The manual recommended using a "push-to-test button" on the solenoid valve, and if that did not open the valve, manually disconnecting the pilot line at the valve. Both of these steps would have been performed remote from the electrical panel. The troubleshooting procedures provided by the manufacturer would have also identified the cause of the reoccurring process upset allowing the problem to be resolved and preventing future upsets.

When possible, facilities should consider implementing management of change procedures or performing process hazard analyses before addressing process upset conditions. These hazard analyses should result in specific procedures to use to address routine process upsets. The procedures should also not counter the instructions or recommendations set forth by equipment manufacturers. Additionally, where possible the causes of routine upsets should be addressed to stop or minimize future upset conditions.

5.6 Improper Alarm Management

During this event, several alarms would have been activated. It is known that the reset being performed by workers was a result of a system shutdown alarm. At the time of the incident, the high hydrogen alarm would have been also have been present, indicating that a significant process upset had occurred. The implications of these alarms should have been considered before addressing the upset. Workers should have considered closing manual hydrogen shutoff valves and de-energizing equipment before performing work.

6. Conclusions

As often is the case, this incident required several simultaneous failures of equipment and improper actions by workers. The original process design failed to identify hazards during process upset conditions, a solenoid valve had to be installed backwards, operators had to miss the improper installation, and hydrogen alarms had to have been ignored by operators. Several lessons were learned through the investigation of this case that can prevent future incidents from occurring.

7. References

[1] NFPA 70, National Electrical Code, National Fire Protection Association.

[2] T. Kletz, Learning from Accidents, Third Edition, (2001) pp. 22-31.

Robust and reliable modelling for a distillation column

Yahya CHETOUANI

Université de Rouen, Département Génie Chimique, Rue Lavoisier, 76821 Mont Saint Aignan Cedex, France. Tel: (0033)235528466, Fax: (0033)235147191, Yahya.Chetouani@univ-rouen.fr

ABSTRACT

The main aim of this paper is to establish a reliable model of a process behavior both for the steady-state and unsteady-state regimes. The use of this accurate model allows distinguishing a normal mode from an abnormal one. Therefore the neural black-box identification by means of a NARX (Nonlinear Auto-Regressive with eXogenous) model has been chosen. It shows the choice and the performance of the neural network. The model is implemented by training a Multi-Layer Perceptron Artificial Neural Network (MLP-ANN) with input-output experimental data. After describing the system architecture, a realistic and complex application as a distillation column is presented in order to illustrate the neural model reliability.

1. INTRODUCTION

In the last few years ever-growing interest has been shown in production quality standards and pollution phenomena in industrial environments. However process development and continuous request for productivity led to an increasing complexity of industrial units. The dynamic nature and the nonlinear behavior of such units pose challenging control system design when products of constant purity are to be recovered.

In chemical process control, the processes typically exhibit nonlinear behavior. The intrinsic highly nonlinear behavior in the industrial process, especially when a chemical reaction is used, poses a major problem for the formulation of good predictions and the design of reliable control systems [1, 2]. Similar problems arise also from the uncertainty for the parameters of the process, such as the reaction rate, activation energy, reaction enthalpy, heat transfer coefficient, and their unpredictable variations.

Most large-scale process models derived from first principles are represented by nonlinear differential–algebraic equation (DAE) systems. Since such models are often computationally too expensive for real-time control. Even if global nonlinear models are available, they may not be appropriate for control and monitoring or fault detection (FD) purpose [3].

On the other hand, when fundamental process knowledge is unavailable or incomplete, or when the resulting model may not be particularly suitable for FD applications, the input/output models identified from plant data may be more useful. Nonlinear autoregressive models have been used [6], as well as models that use a set of polynomial basis functions (e.g. Volterra functions). Also ANNs provide an excellent mathematical tool for dealing with severe nonlinear problems. Another attractive property is self-learning ability. As a result, nonlinear systems can be modeled with a great flexibility. These features allow one to employ artificial neural networks to model complex, unknown and nonlinear dynamic processes.

Distillation is one of the most important separation processes used in many chemical industries. It has extensively been studied [4, 5]. There are essentially two approaches by which nonlinear models can be developed for a distillation column; from first principles by using the process knowledge or empirically from input/output data. The advantages and disadvantages of each approach are well known. In industrial practice, it is not always possible in general to obtain accurate first principles models for high-purity distillation columns.

Most industrial columns are used to separate multi-component mixtures whose constituent elements are often not known completely; the fundamental thermodynamics of multi-component vapor-liquid equilibrium, the physical property data, and other essential constitutive relations required for the successful development of a first principles model are not always available. And even when such knowledge is available, the resulting models usually occur in the form of a very large system of coupled nonlinear ordinary differential equations, and may therefore not always be the most convenient for controller or fault detection (FD) design.

The purpose of this study is to obtain a reduced and reliable model which allows reproducing the dynamics of a nonlinear process as a distillation column. This reliable model enables to reproduce the process dynamics under all operating conditions i.e. steady-state or unsteady state. The present study focuses on the development, and implementation of a NARX neural model for the forecasting of the distillation column dynamics. Experiments were performed in a distillation column and experimental data were used both to define and to validate the model. The performance of this neural model was then evaluated using the performance criteria. Results show that the NARX neural model is representative for the dynamic behavior of this nonlinear process. The experimental set-up, modelling procedure, and prediction results are described in the following sections.

2. EXPERIMENTAL SET-UP

The feed tank (fig. 1) contains a mixture to be separated (Toluene-Methylcyclohexane) with a mass composition at 23 % methylcyclohexane. The operation in continuous mode involves charging the still with the mixture to be separated, bringing the column to equilibrium under total reflux. The product is introduced through the optimal feed tray so that the light components are volatilized, while the heavy part goes down again with the reflux in the column reboiler. The quality of the collected top product of the column depends on the reflux flow rate. The reflux ratio is varied through the magnetic valve by changing the relative quantities of material returning to the column and flowing to product storage. Feed preheating system is constituted by three elements of 250 W each one. In addition it has a low liquid level switch in order to avoid the running if the level is excessively low. The reciprocating feed pump is constituted by a membrane allowing firstly the suction of the mixture and the discharge towards the tank with a flow capacity $F = 4.32$ L.h^{-1}. The column has also a reboiler of 2 liters hold-up capacity, an immersion heater of a power $Q_b = 3.3$ kW and of a level liquid switch sensor which allows the automatic stop of heating if the level is insufficient. The stirring of the mixture in the reboiler is ensured by the boiling mixture. The internal packing is made of Multiknit stainless 316L which enhances the mass transfer between the vapor and liquid phases. In order to approach the adiabatic conditions, heat-insulation made of glass wool is laid around the column. A condenser is placed at the column overhead in order to condense the entire vapor coming out from the column. The cooling medium used in exchangers is water. The heat-transfer area of the total overhead condenser is 0.08 m^2. Moreover the reflux timer (R_t) allows to control the reflux ratio (R_r). It is monitored by the overhead product temperature (T_d). When the required distillate temperature (T_d) is attained, the reflux timer opens. In the opposite case, it remains closed. Distillation supervision control system allows to modify the parameters and to follow their evolution such as the pressure drop (ΔP), the flow or the temperatures at different points of the distillation column. This control system, therefore, must hold product compositions as near the set points as possible. The thermocouples are coupled to a calibrated amplification circuit (4-20 mA, 0–150°C) whose signals are inputted to the computer on-line, which permits the bottom and top temperatures to be obtained. The unit has twelve sensors which measure continuously the temperature throughout the column.

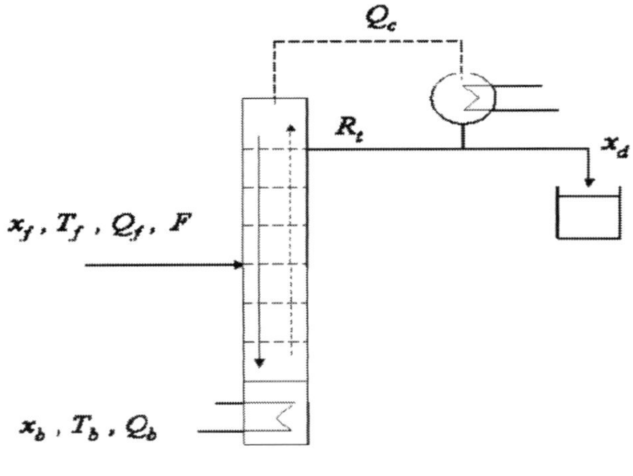

Figure 1. Experimental device: Distillation column

3. IDENTIFICATION OF NARX MODELS

The NARX neural model used to describe accurately the process behavior is the classical MLP-ANNs [7, 8] with one layer of hidden neurons. The NARX model of a finite dimensional system [9] with order (n_y, n_u) and scalar variables y and u is defined by:

$$y(t) = \phi(y(t-1),..., y(t-n_y), \quad u_i(t-n_k),...,u_i(t-n_k-n_u)) \qquad 1 \le i \le m \qquad (1)$$

where $y(k)$ is the Auto-Regressive (AR) variable or system output; $u(k)$ is the eXogenous (X) variable or system input. n_y and n_u are the AR and X orders, respectively. m is the number of the used inputs. n_k is the time delay between u and y. ϕ is a nonlinear function.

The feed-forward MLP-ANN is used in this study and consists in a large number of highly connected nonlinear simple neurons. The log-sigmoid transfer function is used on the hidden layer and a linear transfer function on the output layer. A back-propagation training function for feed-forward networks using momentum and adaptive learning rate techniques is used.

3.1. Calculation of the NN output

Though the applicability of neural networks to solve several nonlinear complex problems has been amply demonstrated, the time taken to train neural networks off-line can be quite excessive [10, 11]. The following steps explain the calculation of the ANNs output based on the input vector.

1. Assign $\hat{w}^T(k)$ to the input vector $x^T(k)$ and apply it to the input units where $\hat{w}^T(k)$ is the regression vector given by the following equation:

$$\hat{w}^T(t) = [y(t-1),\dots,y(t-n_y),\;\; u(t-1-n_k),\dots,u(t-n_u-n_k)] \qquad (2)$$

2. Calculate the input to the hidden layer units:

$$net_j^h(k) = \sum_{i=1}^{p} W_{ji}^h(k)x_i(k) + b_j^h \qquad (3)$$

where p is the number of input nodes of the network, i.e. $p = n_y + n_u + n_b$; j is the jth hidden unit; W_{ji}^h is the connection weight between ith input unit and jth hidden unit; b_j^h is the bias term of the jth hidden unit.

3. Calculate the output from a node in the hidden layer:

$$z_j = f_j^h(net_j^h(k)) \qquad (4)$$

where f_j^h is the log-sigmoid transfer function.

4. Calculate the input to the output nodes:

$$net_l^q(k) = \sum_{j=1}^{h} W_{lj}^q(k)z_j(k) \qquad (5)$$

where l is the lth output unit; $W_{lj}^q(k)$ is the connection weight between jth hidden unit and lth output unit.

5. Calculate the outputs from the output nodes:

$$\hat{v}_l(k) = f_l^q(net_l^q(k)) \qquad (6)$$

where f_l^q is the linear activation function defined by:

$$f_l^q(net_l^q(k)) = net_l^q(k) \qquad (7)$$

3.2. Back-propagation training algorithm

The error function E is defined as:

$$E = \frac{1}{2}\sum_{l=1}^{q}(v_l(k) - \hat{v}_l(k))^2 \tag{8}$$

where q is the number of output units and $v_l(k)$ is the lth element of the output vector of the network. Within each time interval from k to k+1, the back-propagation (BP) algorithm tries to minimize the error for the output value as defined by E by adjusting the weights of the network connections, i.e. W_{ji}^{h} and W_{lj}^{q}. The BP algorithm uses the following procedure (Eqs.9, 10, 11, 12):

$$W_{ji}^{h}(k+1) = W_{ji}^{h}(k) + \alpha\Delta W_{ji}^{h}(k) - \eta\frac{\partial E}{\partial W_{ji}^{h}(k)} \tag{9}$$

$$W_{lj}^{q}(k+1) = W_{lj}^{q}(k) + \alpha\Delta W_{lj}^{q}(k) - \eta\frac{\partial E}{\partial W_{lj}^{q}(k)} \tag{10}$$

where η and α are the learning rate and the momentum factor, respectively; ΔW_{ji}^{h} and ΔW_{lj}^{q} are the amounts of the previous weight changes; $\partial E/\partial W_{ji}^{h}(k)$ and $\partial E/\partial W_{lj}^{q}(k)$ are given by:

$$\frac{\partial E}{\partial W_{ji}^{h}(k)} = -\left[z_j(k)(1-z_j(k))x_i(k)\right]\sum_{l=1}^{q}\left[(v_l(k)-\hat{v}_l(k))\hat{v}_l(k)W_{lj}^{h}(k)\right] \tag{11}$$

$$\frac{\partial E}{\partial W_{lj}^{q}(k)} = -(v_l(k) - \hat{v}_l(k))z_j(k) \tag{12}$$

The implementation of the ANN for forecasting is as follows:

1. Initialize the weights using small random values and set the learning rate and momentum factor for the ANN.
2. Apply the input vector given by Eq. 2 to the input units.
3. Calculate the forecast value of the error using the data available at (k−1)th sample (Eqs.2, 3 4, 5, 6, 7).
4. Calculate the error between the forecast value and the measured value.
5. Propagate the error backwards to update the weights (Eqs. 9, 10, 11, 12).
6. Go back to step 2.

For weights initialization, the Nguyen-Widrow initialization method [12] is best suited for the use with the sigmoid/linear network which is often used for function approximation.

4. RESULTS AND DISCUSSION

In order to select the optimal number of hidden neurons, tests were performed by varying the number of neurons between 5 and 25 for the temperature (T_d) modelling. The minimal number of inputs is avoided to ensure the model flexibility. Also, the maximum number of inputs is excluded to avoid the over-fitting.

$$T_d(t) = \phi(T_d(t-1), R_t(t-2), \Delta P(t-3), Q_f(t-4), T_f(t-6), F(t-8), T_d(t-1) - \hat{T}_d(t-1))$$

The evolution of the loss function is given in fig. 2 in the training and the validation database. This figure shows that the neurons number in the hidden layer in training can be 7, 10 or 21. However this number can slightly change and becomes 8, 10 or 14 in the test.

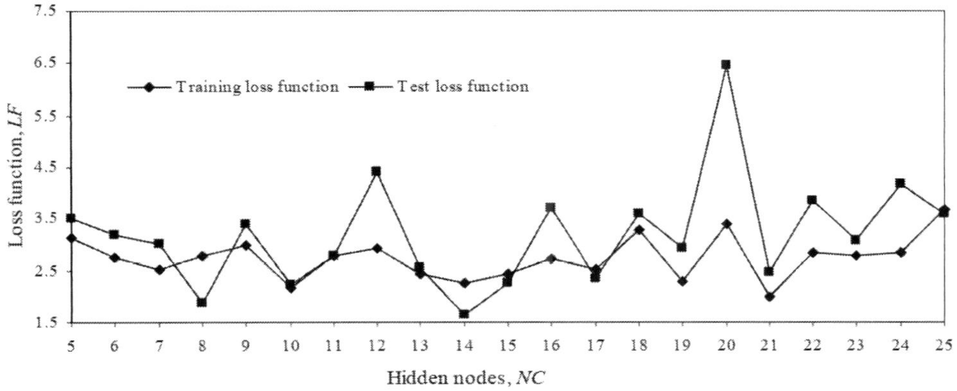

Figure 2. Loss functions

The models which have a structure (7-8-1 and 7-10-1) exhibit the acceptable *LF*. Based on engineering judgment, the model 7-8-1 would be preferred without significant loss of accuracy. Fig. 3 shows the difference between the experimental overhead product temperature and those simulated by the neural model.

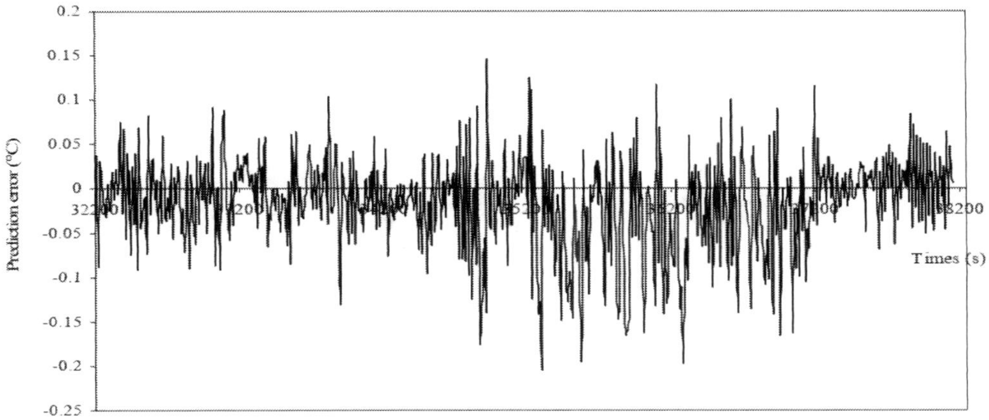

Figure 3. Prediction error of the temperature (T_d)

After analyzing this figure, it emerges that the NARX model 7-8-1 for modelling the temperatures (T_d) ensures satisfactory performances as it is indeed able to correctly identify the dynamics of the distillation column. Therefore, the reduced neural model is considered as reliable one for describing the dynamic behavior of this column. There is a good agreement between the learned neural model and the experiment in the validation phase.

In conclusion the main advantage of the neural approach consists in the natural ability of neural networks in modelling nonlinear dynamics in a fast and simple way and in the possibility to address the process to be modeled as an input-output black-box, with little or no mathematical information on the system.

5. CONCLUSION

This work aims to identify process dynamics by means of a NARX model. The identification of the system dynamics by means of input-output experimental measurements provides a useful solution for the formulation of a reliable model. In this case, the results showed that the model is able to give satisfactory descriptions of the experimental data.

The developed neural models are used in a recursive scheme in order to test their ability to perform the behavior prediction. Finally, the identified neural models will be useful for the detection and the isolation (FDI) of faults which can occur through the process dynamics.

6. REFERENCES

1. Chetouani, Y., *Use of Cumulative Sum (CUSUM) test for detecting abrupt changes in the proccess dynamics*, International Journal of Reliability, Quality and Safety Engineering, 14, 65-80, 2007.
2. Cammarata, L., Fichera, A., Pagano, A., *Neural prediction of combustion instability*, Applied Energy, 72, 513-528, 2002.
3. Huang, Y., Reklaitis, G.V., Venkatasubramanian, V., *Dynamic optimization based fault accommodation*, Computers and Chemical Engineering, 24, 439-444, 2000.
4. Eden, M.R., Koggersbøl, A., Hallager, L., Jørgensen, S.B., *Dynamics and control during start-up of energy integrated distillation column*, Computer Chemical Engineering, 24, 1091–1097, 2000.
5. McAvoy, T.J., *Modeling chemical process systems via neural computation*, IEEE Control Systems Magazine, 3, 24-29, 1990.
6. James, S., Legge, R., Budmann, H., *Comparative study of black-box and hybrid estimation methods in fed-batch fermentation*, Journal of Process Control, 1, 113–121, 2002.
7. Chen, S., Billings, S.A., *Representation of nonlinear systems-The NARMAX model*, International Journal of Control, 49, 1013–1032, 1989.
8. Narendra, K.S., Parthasarathy, K., *Identification and control of dynamical systems using neural networks*, IEEE Trans. on Neural Networks, 1, 4–21, 1990.
9. Ljung, L., *System Identification, Theory for the User*, Prentice Hall, Englewood Cliffs, New Jersey, 1999.
10. Chetouani, Y., *Modelling and prediction of the dynamic behavior in a reactor-exchanger using NARMAX neural structure*, Chemical Engineering Communications Journal, 5, 691-705, 2007.
11. Fung, E.H.K., Wong, Y.K., Ho, H.F., Mignolet, M.P., *Modelling and prediction of machining errors using ARMAX and NARMAX structures*, Applied Mathematical Modelling, 27, 611-627, 2003.
12. Nguyen, D., Widrow, B., *Improving the learning speed of 2-layer neural networks by choosing initial values of the adaptive weights*, International Joint Conference of neural networks, 3, 21-26, 1990.

Process Safety Incident

Hydrogen Compressor Panel Explosion

Date & Time : 24.01.2007 at 2:30 AM

Presentor :
Shantanu Date
Reliance Ind. Ltd , India

Description:

While stopping H2 compressor from the panel small explosion took place.

Both doors of the control panel were thrown out, which struck the operator resulting in minor injury to his head

Hydrogen Compressor Panel

Hydrogen Compressor Panel cable termination Area

Hydrogen Compressor Panel cable termination Area

Possible causes of Hydrogen ingress to the Panel

- Contamination of purge air going to panel by hydrogen .

- Ingress of Hydrogen to panel from atmosphere in case of purge system failure.

- Entry of hydrogen through cables entering the panel.

Probable cause # 1

Possible cause include ingress of hydrogen to local instrument air header through damage to actuator of integral actuator type recycle/decompression valves.

LEL checked at Instrument Air header with hydrogen pressure applied to these valves. No leak observed.

Probable Cause # 2

Probable causes of ingress to panel include

- Ingress from atmosphere in case of Purge failure.
  - Sealing of panel checked and found to be ok.
  - Panel purge system was also found to be working.
  - No ‘ Loss of Pressurisation ’ alarm

Probable Cause # 3

Probable causes of ingress to panel include

- Ingress though cable entry.

Panel LEL was checked by lining up Hydrogen to compressor.

- Observed 40% LEL in panel.
- Observed 100% LEL in Junction box on skid
- Individual inputs were checked and cause was narrowed down to Suction pressure switch diaphragm.

Hydrogen Compressor System

Area specification : Zone 2 Gr II C

- Panel – E Exdep (iia)
- P - Purged with Inst air as media
- E - Enhanced safety through purge interlock which isolates the total power to panel incase of purge failure detection
- d - Purge controller Power supplies protected with Ex-Proof JBs
- (iia) - All field instruments through barriers.

- E/I signals at voltage level of 24VDC not through barriers.

H2 Compressor

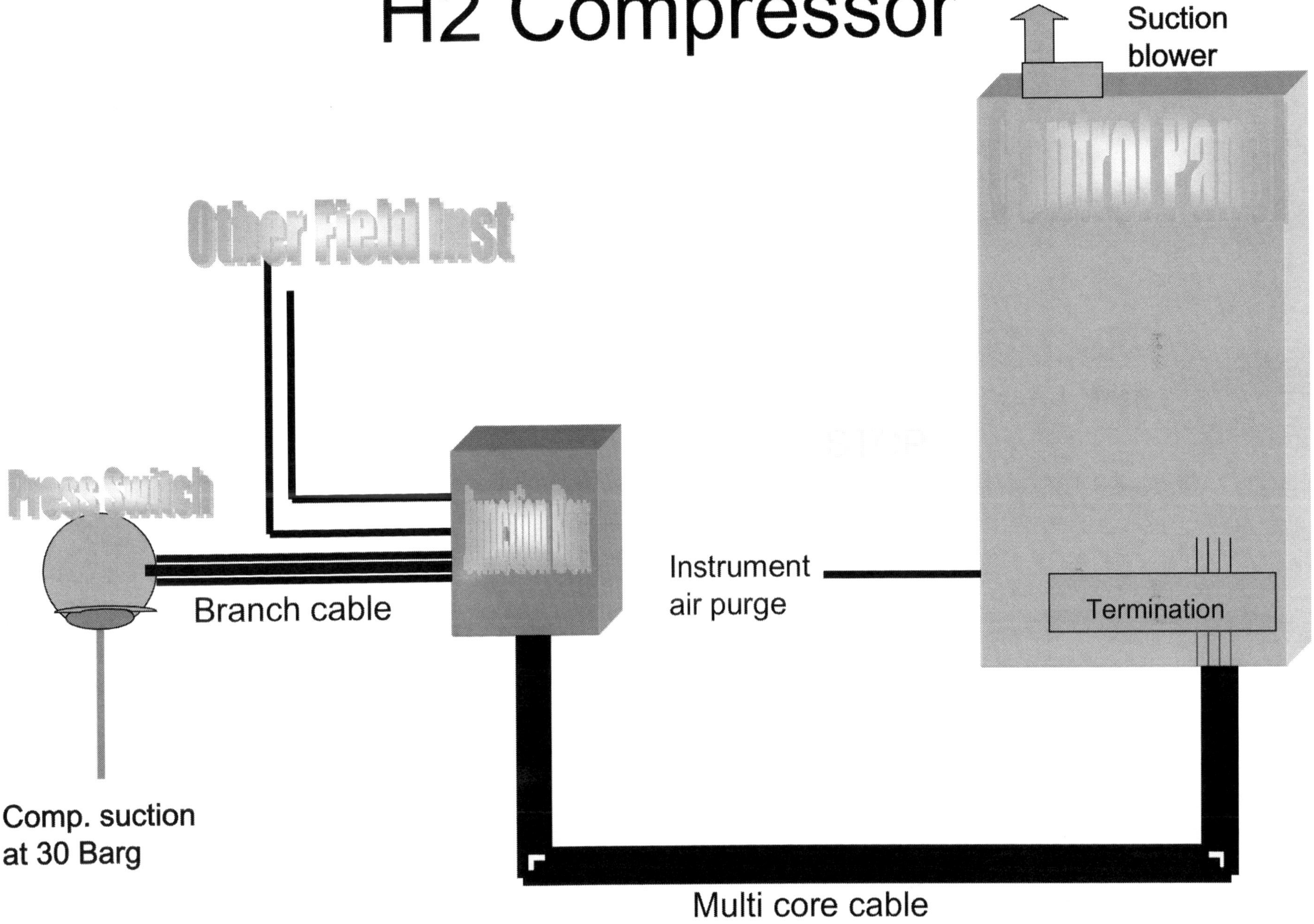

PTA2 H2 Panel Explosion Area

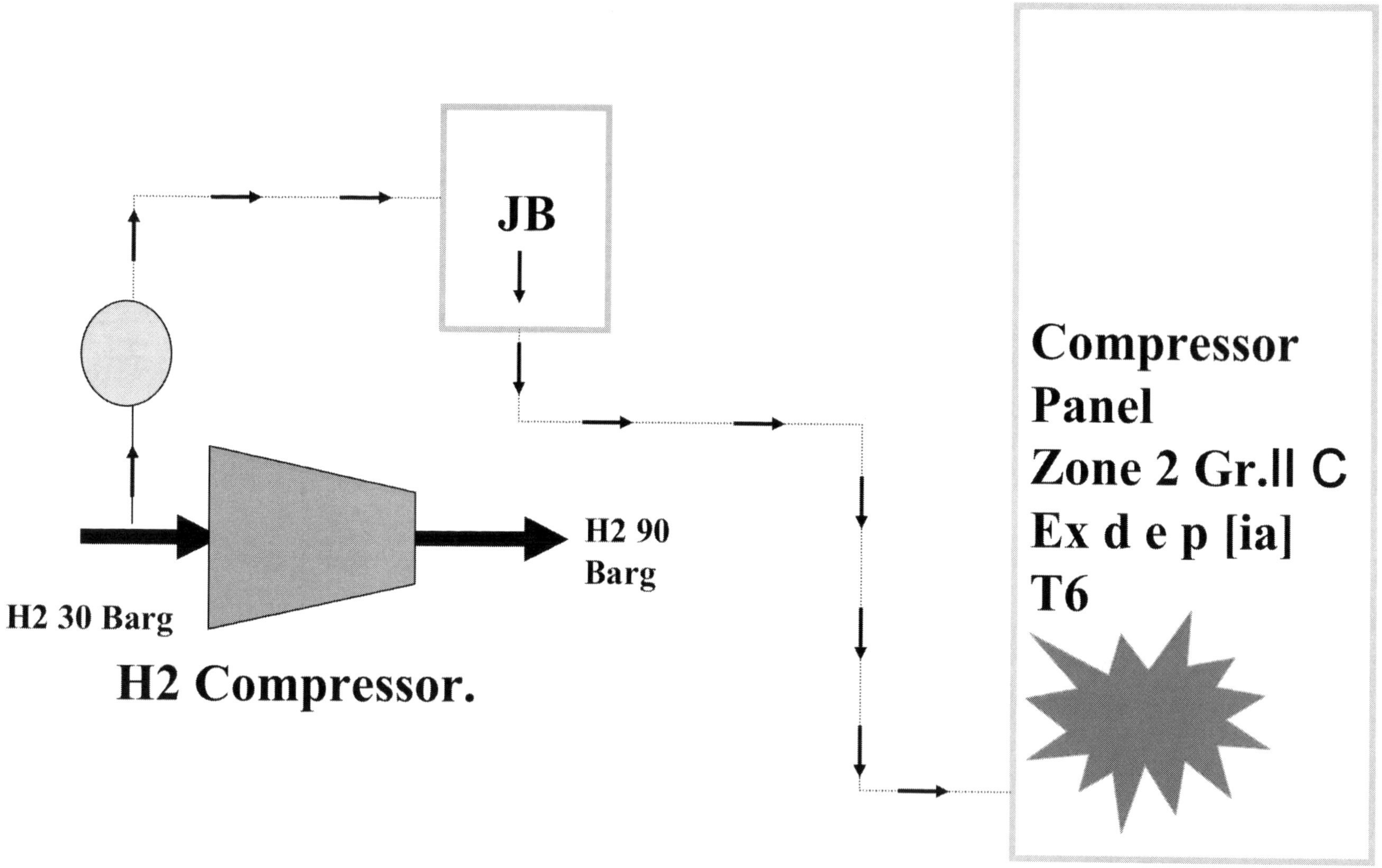

Panel Inside Area classified as safe if pressurized.

Observations :

1. Pressure switch impulse tube contained water
2. Multicore instrument cable from junction box to panel had no seal.

Root Causes :

1. Pressure switch diaphragm got dislocated due to pressurization by water , backed up due to passing NRVs during plant shutdown when compressor was depressurised.
2. Allowed hydrogen ingress into pressure switch Absence of seal in inst. cable allowed hydrogen entry to panel via cable .

Applicable Standards

- ISA 12.04.01 – 2003 – Electrical apparatus for explosive gas atmosphere – Pressurized Enclosure

- ISA 12.01.01 – 1999 – Definition & Information pertaining to Electrical instruments in Hazardous Locations.

ISA12.04.01-2003

- **Type px pressurizing**

  pressurization that reduces the classification within the pressurized enclosure from zone 1 to unclassified non-hazardous

- **Type py pressurizing**

  pressurization that reduces the classification within the pressurized enclosure from zone 1 to zone 2.

- **Type pz pressurizing**

  pressurization that reduces the classification within the pressurized enclosure from zone 2 to unclassified non-hazardous.

Standard for Pressurization:ISA 12.04.01
H2 compressor Panel –Type px Pressurization

Table 1 — Determination of protection type

Flammable substance in the containment system	External Zone classification	Enclosure contains ignition-capable apparatus	Enclosure does not contain ignition-capable apparatus
No containment system	1	Type px [a]	Type py
No containment system	2	Type pz	No pressurization required
Gas/vapor	1	Type px [a]	Type py
Gas/vapor	2	Type px (and ignition-capable apparatus is not located in the dilution area)	Type py [b]
Liquid	1	Type px [a] (inert) [c]	Type py
Liquid	2	Type pz (inert) [c]	No pressurization required [d]

Standards For Pressurization.ISA12.04.01
H2 Compressor Panel protection – IP65

Table 2 — Design criteria based upon protection type

Design criteria	Type px	Type py	Type pz with indicator	Type pz with alarm
Degree of enclosure protection according to IEC 60529 or IEC 60034-5	IP4X minimum	IP4X minimum	IP4X minimum	IP3X minimum
Verifying purge period	Requires a timing device and monitoring of pressure and flow	Time and flow marked	Time and flow marked	Time and flow marked
Door/cover requiring a tool to remove	Warning, see 5.3 and 6.2 b) ii)	Warning, see 5.3 (Note 1)	Warning, see 5.3 and 6.2 b) ii)	Warning, see 5.3 and 6.2 b) ii)

Standard for Pressurisation ISA 12.04.01
Safety Provision & Safety devices

Table 3 — Safety devices based upon protection type

Design criteria	Type px	Type py	Type pz
Safety device to detect loss of minimum overpressure	Pressure sensor, see 7.9	Pressure sensor, see 7.9	Indicator or pressure sensor, see 7.9 d)
Safety device(s) to verify purge period	Timing device, pressure sensor, and flow sensor at outlet, see 7.6	Time and flow marked, see 7.7 c)	Time and flow marked, see 7.7 c)
Safety device for a door or cover that requires a tool to open	Warning, see 6.2 b)	Warning, see 6.2 b)	No requirement
Safety device for a door or cover that does not require a tool to open	Interlock, see 7.12 (No internal hot parts)	Warning, see 6.2 b) (No internal hot parts)	Warning, see 6.2 b)
Safety device for hot internal parts when there is a containment system (see Clause 15)	Alarm and stop flow of flammable substance	Not applicable for protection type since internal hot parts not permitted	Alarm (normal release not permitted)

ISA 12.01.01 –1999 Wiring Methods

Seals are installed in conduit and cable systems to

- confine an explosion occurring in an enclosure or a conduit system to only that enclosure or that portion of the conduit system.
- minimize the passage of gases, vapors, or liquids and prevent the passage of flames through the conduit or cable system from a classified to an unclassified area, or from one enclosure to another;
- prevent *pressure piling* the buildup of pressure inside conduit systems (ahead of an explosion's flame front) caused by precompression as the explosion travels through the system. Exploding precompressed gases may reach pressures that would exceed the design pressure of the enclosures.

ISA 12.01.01- 1999 Wiring Methods

- Seals and drains shall be installed in each conduit or cable connection to devices installed on process lines that contain flammable fluids and depend upon a single compression seal, diaphragm, or tube (such as a Bourdon tube) as a barrier between the process fluid and the conduit or cable.

This is to prevent flammable process fluids from entering conduit or cable systems and being transmitted to unclassified areas.

- Typical examples of such devices are solenoid valves and pressure, temperature, and flow switches or transmitters. This requirement applies even in unclassified areas.

ISA 12.01.01 –1999 Wiring Methods

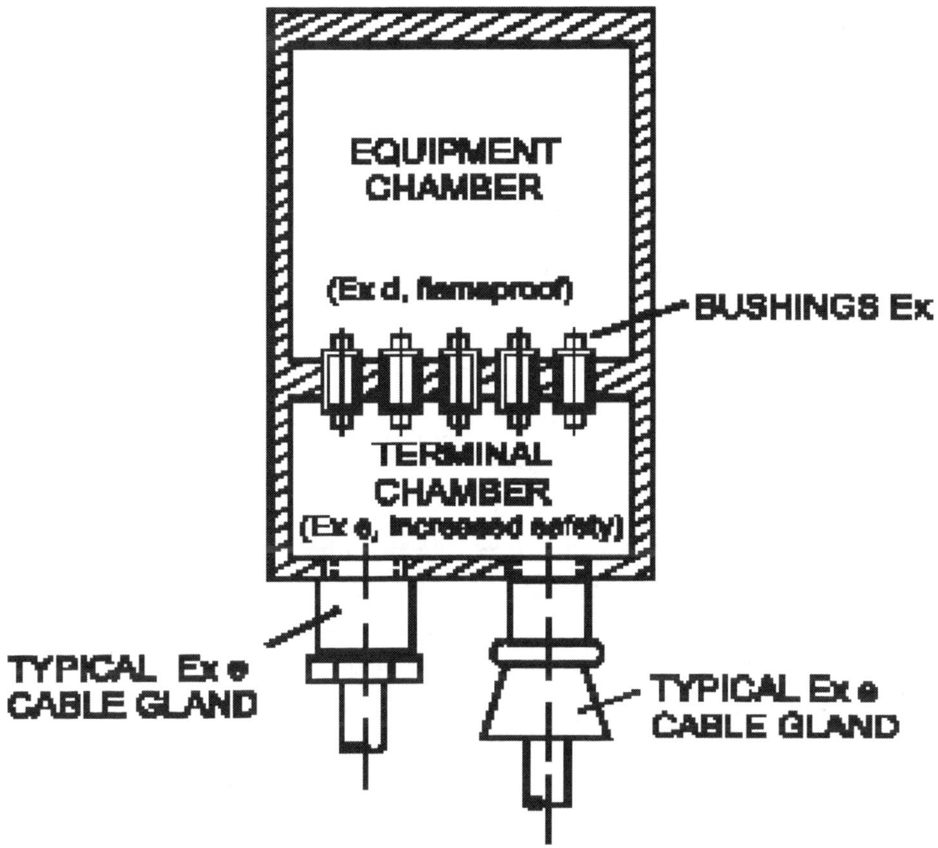

Figure 5 — Cable system (indirect entry)

ISA 12.01.01-1999 Wiring methods

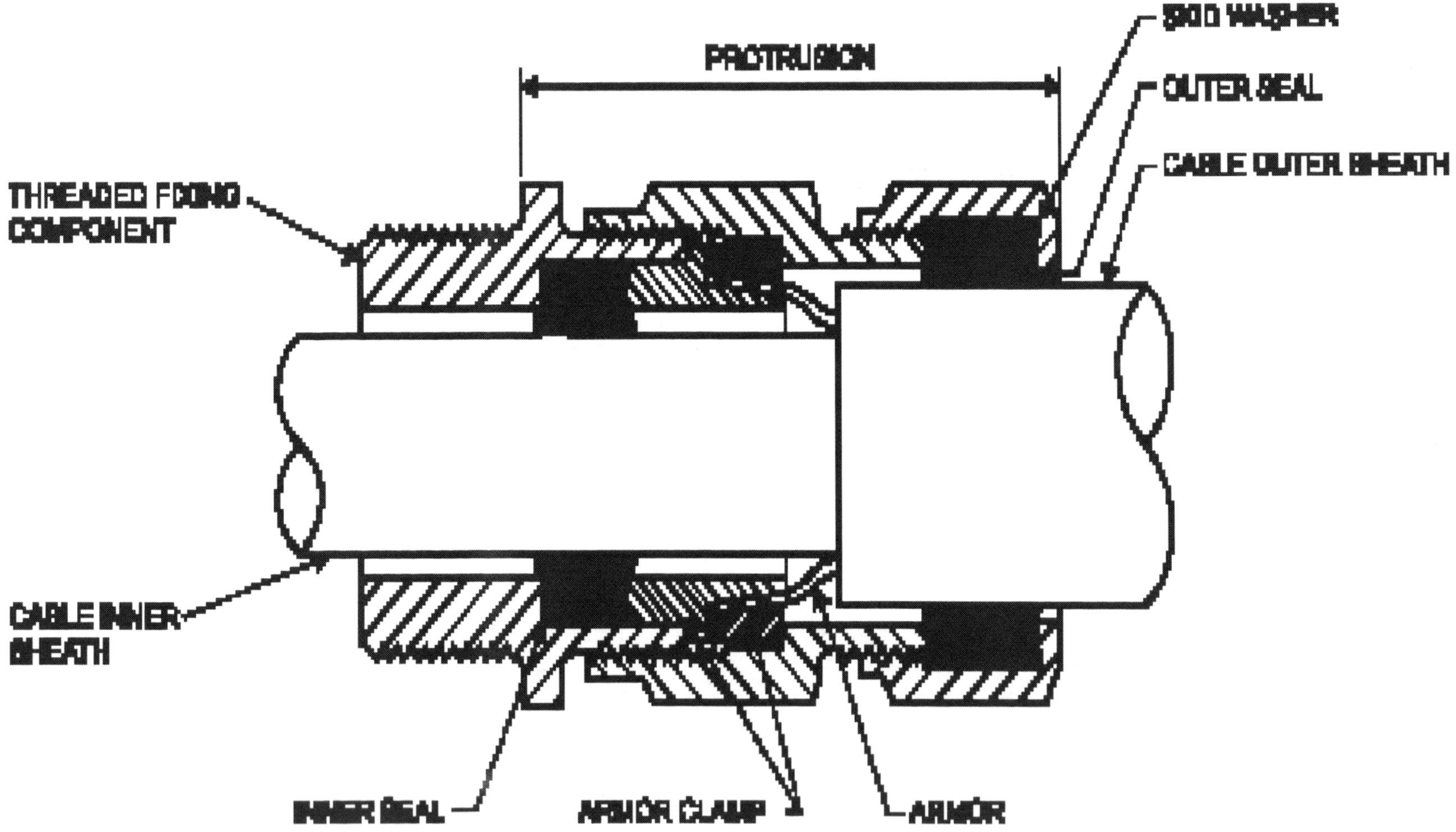

Figure 6 — Cable gland (indirect entry)

ISA 12.01.01 –1999 Wiring Methods

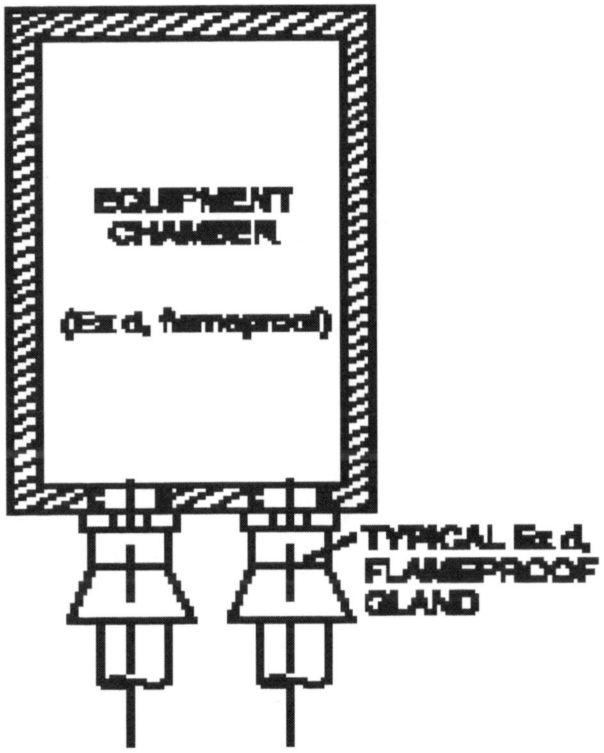

Figure 7 — Cable system (direct entry)

ISA 12.01.01-1999 Wiring Methods
Type of seal to be used at Instrument End

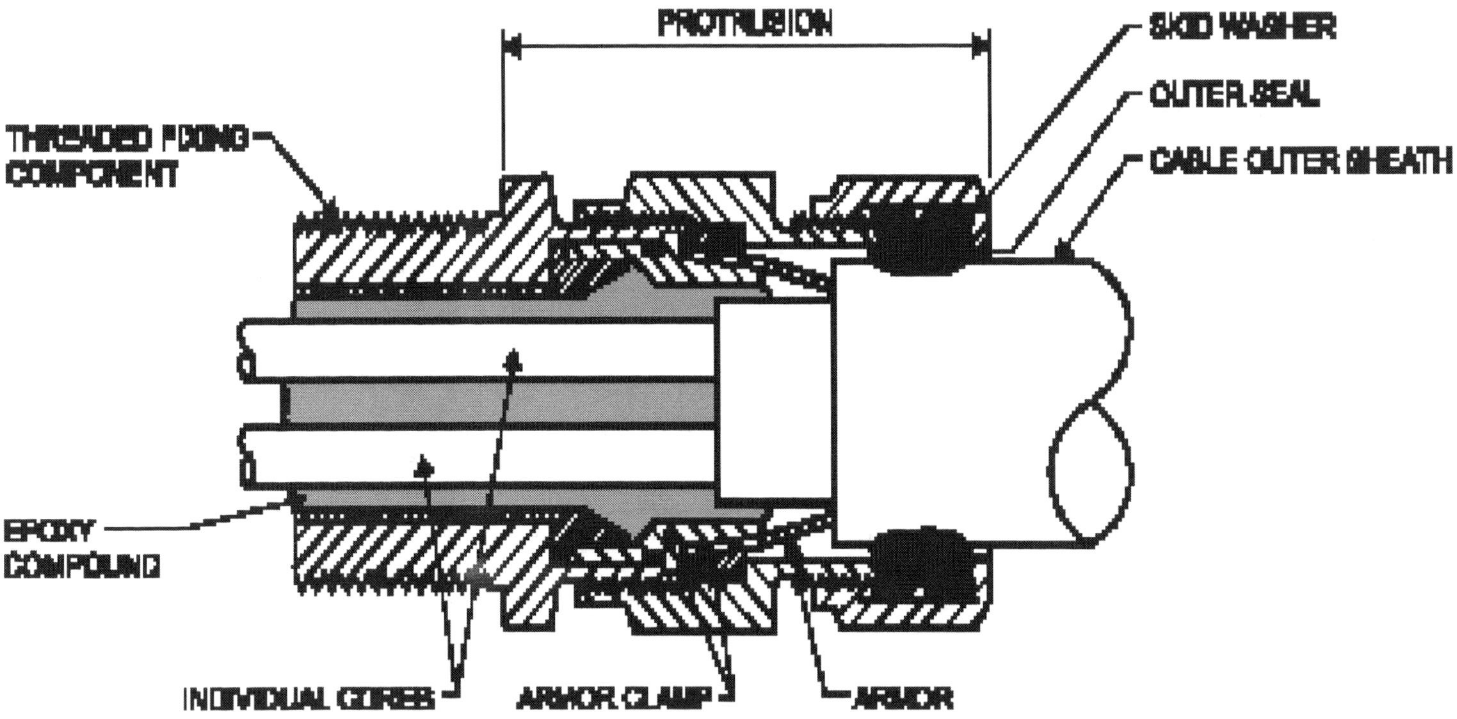

Figure 8 — Cable gland (direct entry)

Systemic Causes Responsible

- There is a practice in the organisation to order cables in bulk. Procurement is done as per applicable codes & standards.
- In this case it was established that though compressor skid was supplied by renowned vendor , supply scope excluded cables as per company's approach.
- Correct cables were procured as per vendor specification & applicable codes

BUT -- - -

During project execution , correct cable was not used. It went undetected as there was no leak in Pressure switch that time.

AFTER 10 YEARS OF OPERATION - -

- Led to unsafe incident when another cause (back up of process water due to passing of Non Return Valves) created leak in pressure switch & combined with unsafe condition already present due to systemic cause .

Domino Effects in the Safety Assessment of Petrochemical Enterprises

Zhao Dongfeng Zhang Bo Wang Wendong

(Center of Environment and Safety Technology, China University of Petroleum (East China), Dongying , China)

Abstract: Petrochemical units are normally evaluated independently for the purpose of optimization, but they could easily be involved in chain accidents which we define as domino effects. Domino effects can cause serious consequences in petrochemical units. However, the existing safety assessment methods analyze equipment separately and ignore the influences of the neighbor units. It is of great urgency and importance that safety assessments take the domino effects into consideration. Based on the domino effect analysis model presented by V.Cozzani, we conceived the framework of domino safety assessment method. Domino safety assessment is a safety assessment method in which domino effects should be analyzed and evaluated. It evaluates the interaction between the surrounding units qualitatively and quantitatively to determine the possibility of happening of a domino effect, so that the results of the safety assessment would be more objective and more close to the facts. Three kinds of factors which could cause domino effects presented by V.Cozzani including thermal radiation, overpressure and explosion missiles have been included in the framework of domino safety assessment method. An industrial example has been studied and it explains the accidents phenomenon well. Domino safety assessment method is a useful supplement to the existing safety assessment method and it has great value for research.

Key words: domino effect domino safety assessment method petrochemical enterprises

1. Introduction

Heinrich describes the occurrence of accidents as a fall of domino chain which contains four elements: P—human error; H—unsafe actions of men or unsafe conditions of machines; D—accidents happen; A—injuries. Once a domino falls, a chain reaction will occur; the dominos behind it will fall one by one. The domino theory intuitively shows the causality of the occurrence of accidents, points out that the accident analysis should be carried out from the phenomena of accidents and should be prevented from the very beginning. However, the disadvantage of the classic domino theory is also obvious. It sticks to the domino chain too absolutely and describes the causes and effects of accidents too roughly. In order to evaluate the falling possibility of the domino chain reaction more accurately, we need to consider the probability of the secondary accident, and then establish an evaluation method which can evaluate the interactional effects of the domino chains.

Petrochemical installations are normally evaluated independently for the purpose of optimization and it is good for the process design. However, due to domino effects, installations would be affected by each other in chain accidents, neighboring equipments or units would be involved. Domino effects can be found everywhere, especially when it happens, the domino chains will fall one by one, and then cause more serious consequences. Nowadays, many countries have introduced the assessment of domino effects to safety assessment, but in China, the safety assessment methods are still focusing on the accidents effects simulation for independent equipments or units and have not considered the effects from the condition of equipments or units which are closely surrounding the assessment units. It means that we have not thought much of the analysis and evaluation of domino effects. So, it's of great importance to bring forward and study domino safety assessment method.

2 .Domino Safety Assessment

Domino safety assessment is a safety assessment method in which domino effects should be analyzed and evaluated. It evaluates the interaction between the surrounding units qualitatively and quantitatively to determine the possibility of occurrence of a domino effect [(1)]. So that the results of the safety assessment would be more objective and more close to the facts. The difference between domino safety assessment and existing safety assessment methods is that domino safety assessment focuses on the influence among all neighboring assessment units and emphasizes on the analysis of the probability of the secondary accidents. Domino safety assessment needs to firstly determine the source of significant hazards which cause accidents which including the potential hazards of installations or equipments. Secondly, determine the assessment range and chose a domino effect assessment method according to the hazards types. Thirdly, carry out an assessment for each evaluation unit and at last, point out the prevention and contingency measures to minimize the influence of domino effects. There are many factors which could result in domino effects including thermal radiation, overpressure, missiles of explosion, toxic substances leakage, and a combination of several or all of these factors could lead to second accidents. In these factors, toxic substances leakage mainly relates to the impact of human beings and environment, and researches have been developed before. Our study mainly focuses on the three factors which could primarily cause domino effects: thermal radiation, overpressure and explosion missiles. [2,3]

2.1. Domino Effects Caused by Thermal Radiation

Thermal radiation produced by fire is a major factor leading to domino effect. Flammable gas or liquid leakage comes out from installations could catch fire or explode when provided with certain ignition energy. The targets of thermal radiation could be people, equipments, facilities, buildings and so on. The detailed assessment can carry out according to the following steps: choose a thermal radiation model, calculate the thermal flux a target received, and then use the probability model to assess the damage. The death probability of human beings and ignition probability of wood due to thermal radiation has already been developed by the existing safety assessment method. So our study focuses primarily on the probability model of different types of equipments damage caused by thermal radiation. Valerio Cozzani developed a probability mode for domino effect caused by thermal radiation. It can assess the equipments damage probability according to the amount of thermal radiation. The models are as follows:

$Y = 12.54 - 1.84\ln(ttf)$, where Y is the damage probit.

For the atmospheric and vertical cylindrical containers,

$$\ln(ttf) = -1.128\ln(I) - 2.667\times10^{-5}v + 9.887;$$

For high pressure containers,

$$\ln(ttf) = -0.947\ln(I) + 8.835v^{0.032},$$

where

ttf is "the time to fail" (s),
I is the thermal flux of the equipment targets (Kw/m^2),
V is the equipment capacity (m^3). [4]

2.2 .Domino Effects Caused by Overpressure

The overpressure theory points out that once the overpressure caused by a blast wave reaches a certain value, it will cause a certain degree of damage to the targets. It is another main factor which could cause domino effects. When explosion occurred in the petrochemical installations, the overpressure can reach to several atmospheric pressures. According to experiments, when two same kinds of explosives with different amount explode, if the ratio of distance from explosion source equals to the cube root of the ratio of the explosive amount, they would produce the same overpressure. According to this model and checking from the blast wave overpressure table of the 1000Kg TNT explosion in the air, we can calculate the overpressure received by the target equipments from the explosion source.

Valerio Cozzani developed the following model for the quantitative assessment of the equipment damage caused by blast wave:

$$Y=K_1+K_2Ln(\Delta P^0).$$

Where:

Y—equipment failure probit,

ΔP_0 —over pressure (Pa);

K_1, K_2 assumes different values according to four different types of containers,:

for atmospheric pressure vessel, K_1=–18.96, K_2=2.44;

high Pressure Vessel, K_1=–42.44, K_2=4.38;

long vertical equipment, K_1=–28.07, K_2=3.26;

small equipment, K_1=–17.79, K_2=2.18.

The probit analysis is a wide-used method which could assess the human body reflection to toxic substances and radiations, and it also can be used to assess the damage probability of equipment caused by the thermal radiation, overpressure damage and other factors. The relation of the probability F and the probit Y agrees with the normal distribution: $F=\frac{1}{\sqrt{2\pi}}\int_{-\infty}^{Y-5}e^{-u^2/2}du$. Through the relationship, we can make the conversion between probit and the probability, and make the quantitative assessment. [5,6]

2.3 .Domino Effects Caused by Explosion Missiles

As pressure vessels or storage tanks explode, they release enormous energy. The vessels or tanks could be torn into several pieces or debris and be ejected at high speed. These missiles with high speed or in great quantity have enormous energy when hitting the equipments or people around, could result in serious accidents, which is another major factor that could cause domino effects. Currently the domino effects triggered by the explosion missiles could be taken by the initial quantitative evaluation, which is specifically divided into the following steps: determine the probability distribution of the direction of the fragments' initial projectile and the velocity, the debris path, the probability of collision between fragments and target equipments, determine the probability of damage to equipments. The key point is to establish the initial parameters, which can impact fragments flying projectile. Great study must be undertaken in this field. [7]

3 .Case Study Example

We take a petrochemical plant as an example, the occurrence of a series explosion in a petrochemical plant, which has the thermal radiation, overpressure, explosion missiles can lead to domino effects. This study will focus on the domino effect caused by the overpressure as an example to illustrate the domino effect assessment method. We take a tank explosion which contains 25 tons of nitrobenzene as an example. According to the worst accident evaluation guidelines recommended by the Environmental Protection Agency of the United States, we assume that the one tenth of loss contents combusted and exploded completely. Using the TNT explosion and overpressure model mentioned in part 2.2, rough assessment has been drawn according to the distance listed in table 1. The overpressure varies due to distance is shown in figure 1. Equipments damage probability due to various types of equipments is shown in figure 2. From figure 2 we can see that the probability is 1 or near to 1 at distance of 20 meters from the explosion center. In other words, collapsed and smashed, broken, displacement or other structural damage could happen in a range of 20 meters. As the distance increases, different types of equipments have gradually reduced the probability of damage. This analysis shows that if there are equipments in the range which could cause damage, they will get various degrees of damage, then dangerous contents leak from the damaged equipments, and this contents will explode when meet ignition sources, thereby creates a domino chain of accidents. The analysis by our domino safety assessment method complies very well with the actual accidents.

Table 1 overpressure of a tank explosion (containing 25 tons of nitrobenzene)

Actual distance R/m	14	18	22	30	38	46	54	60
overpressureΔP^0/KPa	197	106	70	39	26	20	16	13

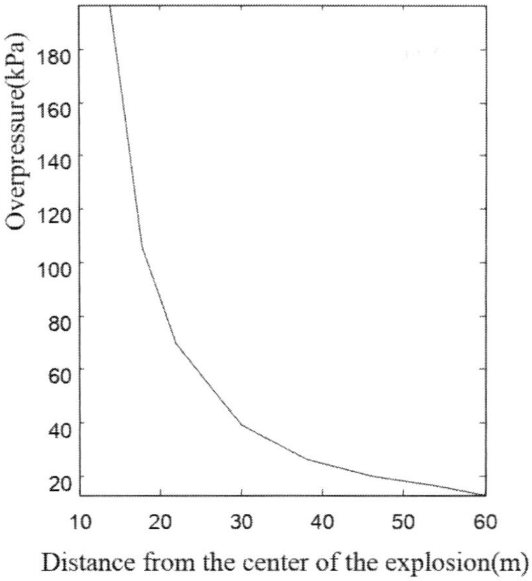

Fig 1 overpressure attenuation due to distance

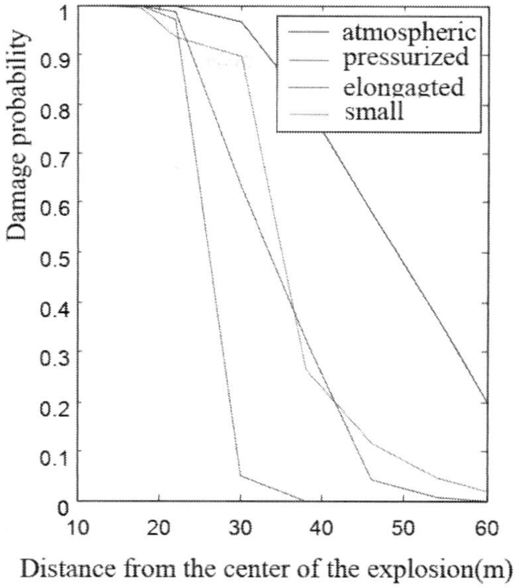

Fig 2 damage probability of different types of equipments

4 .Conclusion

The study presents the concept of domino safety assessment method. Domino safety assessment method could analyze petrochemical accidents more appropriately and it could provide more useful techniques to prevent important domino accidents. Domino evaluation method is a useful supplement for the existing safety assessment methods.

References

1. Khan Faisal I., Abbasi S.A., Estimation of probabilities and likely consequences of a chain of accidents (domino effect) in Manali Industrial Complex. Journal of Cleaner Production, 9 (2001): 493-508
2. Cozzani V., Salzano E., Threshold values for domino effects caused by blast wave interaction with process equipment. Journal of Loss Prevention, 17 (2004): 437-447
3. Cozzani V., Gubinelli G., Salzano E., Escalation thresholds in the assessment of domino accidental events. Journal of Hazardous Materials.A129(2006):1–21.
4. Cozzani V., Gubinelli G., Antonion G.etc, The assessment of risk caused by domino effect in quantitative area risk analysis. Journal of Hazardous Material,A127 (2005): 14–30
5. Cozzani V., Salzano E.,The quantitative assessment of domino effects caused by overpressure Part I Probit Models. Journal of Hazardous Material, A107 (2004): 67-80

6. Cozzani V., Salzano E.,The quantitative assessment of domino effect caused by overpressure Part II Case studies. Journal of Hazardous Material, A107 (2004): 81-94
7. Gubinellia G., Zanellia S., Cozzanib V., A simplified model for the assessment of the impact probability of fragments. Journal of Hazardous. Material, A116 (2004): 175-187

148o GE Gas Turbine Design in O&G Applications According to IEC62061 Requirements Vs CE Marking

Vincenzo Amato and Elda Casamassima. General Electric O&G, Via Felice Matteucci, Florence, 2 / 50100, Italy

Safety Instrumented Systems are used in gas turbine (GT) systems design to mitigate potential Hazardous situations that can arise from machine deviations. International standard IEC62061, (released in 2005) is machinery sector within the framework of IEC61508.

Starting from 31st December 2005 the IEC62061 came harmonized standard for machinery directive 98/37/CE. All members of UE and EFTA use the machinery directive in order to establish the safety requirements (RES) for machine that will be putted in market.

In this way, the IEC62061 are requested to be satisfied in order to give the presumption of conformity for the loop based on Electrical/Electronic/Programmable Electronic components that are called to ensure safety of machinery. This establishing both a probabilistic approach on safety and, the relative safety integrity of a safety instrumented system to perform its prescribed action over the product life cycle (design, installation, and operation of safety instrumented systems).

This paper describes the application of IEC62061 international standard on GE O&G Gas Turbine for the Safety-Related System (SRS) that are separated from the normal electrical control system. This job allows both ensuring a greater safety factor and faulting independence inside E/E/PE system and satisfies the requirements of machinery directives that, together with others applicable European directives, bring to CE marking.

AUTHOR INDEX

AUTHOR INDEX